Introduction to Thermodynamics

Second Edition

Kurt C. Rolle

Charles E. Merrill Publishing Company
A Bell & Howell Company
Columbus Toronto London Sydney

Published by Charles E. Merrill Publishing Co.
A Bell & Howell Company
Columbus, Ohio 43216

This book was set in Times Roman
Cover Design Coordination: Will Chenoweth
Cover photograph: Combustion chamber of the Detroit Diesel Allison Fuel Pincher engine. Courtesy of Detroit Diesel Allison Division, General Motors Corporation.

Library of Congress Catalog Card Number: 79–90390

International Standard Book Number: 0–675–08268-4

Printed in the United States of America

3 4 5 6 7 8 9 10—85 84

To my wife

Joy

and to our children

Kurt, Loreli, Timothy, Heidi, Charity, and Sunshine

Their encouragement, understanding,
patience, and love made this book possible

Preface

The objective of this edition, as in the first edition, has been to present the basic concepts of thermodynamics within a context of mechanical engineering applications. Emphasis is placed on showing the methods of solving a wide range of technical problems using the underlying physical laws and principles. It is assumed that the readers of this book have had introductory courses in physics mechanics, physics heat and light, and college algebra. It is desirable if the reader has had (or is simultaneously taking) a course in calculus, but most of the problems in this book can be solved and all of the concepts understood without it. The chapters devoted primarily to applications are self-contained, and the order in which they are studied could be changed without loss of continuity.

This revised edition uses both SI (Système International) and English units. In spite of the efforts to convert to metric usage in the United States, today's students must be prepared for industry, where a knowledge of English units is also essential. Therefore, this text is split almost evenly between SI and English units, but it can be used with emphasis placed on either system. Practice problems are keyed for both systems, and there are example problems and practice problems that emphasize the conversion of units between the two systems.

The sequence of presentation is definitions, statements of laws or principles, and applications. The importance of understanding the definitions cannot be understated. The vocabulary of thermodynamics contains many words of common usage (such as temperature, heat, and work) which are given precise meaning through definitions. Without this precision, most technical problem solving would be vague or impossible. Laws or principles are stated as truths which have no observed contradictions in nature. Applications of the laws are then presented to give the reader a sampling of the type of problems clarified by the thermodynamic approach.

In proceeding from the basic laws to specific applications, the reader is presented a common methodology for all problems of a thermodynamics nature. From the statements of laws or principles, a few precise equations from which students can proceed are developed. They are shown, for a given problem, how to make statements regarding the physical characteristics of the material involved, and how to make simplifying but realistic assumptions which will reduce general equations to specific. They will then be given the tools for calculating an answer from applicable equations. There are presented many examples of developing a particular from a general equation, using calculus in most of these developments. It should be emphasized that an understanding of the underlying assumptions which allowed the derivation is most important for students; the calculus is merely the way of showing why the use of "any old equation" is a serious error in applying physical laws to problem solving.

A book for engineering technology, treating such a widely written subject as thermodynamics, cannot be expected to present much new and original scientific work. What is presented is well known in the scientific community but, hopefully, is clearer and more accessible for students.

Chapter 3 exposes heat and heat transfer as a single phenomenon and shows the methods used in traditional calculations of heat transfer. This is not a heat transfer text and no expertise is expected of students regarding this process, but some confusion does arise, and an early exposition of terms can help students get a better picture of the field of engineering and technology.

The first law of thermodynamics is presented as a conservation principle applicable to a system. It has been chosen to consider the system as open, closed, or isolated. Some arguments can be presented against this terminology but it is felt that the language is reduced to fewer new words, such as *control volume*—a proper approach for the students' sake.

Entropy is presented only after students have some feeling for energy, work, and heat transfer. It is exposed intuitively and, hopefully, makes sense to a student not desiring mathematical proof. Entropy is used abundantly in the printed literature, and if students are to use thermodynamics in their professional endeavors, they should have some feeling for it.

Availability or available energy is presented in the context of the second law of thermodynamics. A consideration of energy alone, without available energy, can lead to some apparent contradictions regarding the capabilities of energy transfer devices such as heat engines, batteries, and fuel cells.

Chapters 10 through 16 on applications can be approached in any sequence. Each chapter gives a brief exposition of the existing technology and provides students with methods of analyzing the system involved. No attempt is made to produce experts in steam turbine power plants, internal combustion engines, jet engines, rocketry, or meteorology. This expertise can be gained by individual experience and study. However, this book will provide basic tools to start students along any of these specialized paths if they so choose.

In studying this book, the following order of chapters is recommended for a 2- or 3-semester hour course; Chapters 1, 2, 3, 4, 5, 6, 7, 8, and at least Chapters 10, 13, and 14. Additional hours would include Chapter 9 after the first 8 chapters, and then the remaining chapters on applications.

Consistent notation of terms and variables has been used throughout this book, and these notations have been tabulated early in the text. Students generally have more trouble with the myriad of terms than with any other single aspect of thermodynamics, and it is hoped that attention to this item will relieve the difficulty.

The problems at the end of each chapter were selected to demonstrate the wide range of applicability of thermodynamics. Calculus is needed for only a few of the problems in this book, and each is so indicated by an asterisk preceding the problem number. This should ease the task of selecting assignments for the student.

I would like to thank all those who assisted me in many ways during the work of this edition. Particular thanks must go to those reviewers, Professors Stanley Brodsky, N. Y. C. Community College, Fred Emshousen, Purdue University, and Donald L. Miller, Pennsylvania State University, who provided a much needed critique of the theory as well as pointed out mechanical errors. Also, I owe deep gratitude to all the students and professors who uncovered errors in the first edition and took the time to indicate them to me. Finally, my associates at the University of Dayton continue to provide that stimulating atmosphere so necessary for creative work and I thank them for this. Special thanks must go to Professor Jesse H. Wilder, who used the text in the classroom and has contributed much to its development.

Kurt C. Rolle

Contents

1

Introduction

1.1 Energy and Thermodynamics

The most significant contribution to the developing and maintaining of a modern technological society has been human ability to extract large amounts of energy from nature, thereby gaining capacities for work or power. The science which explains and predicts this extraction is called *thermodynamics.* We will be concerned with energy (*therme*, heat) as it changes or changes something else (dynamic). From studies in physics, we know that there are two forms of energy: *kinetic* and *potential.* However, there is at least one other form, *internal energy*, which helps to explain the hotness and coldness of a body. We will investigate these three forms of energy more fully in later chapters, but for now let's define energy.

Energy: *The capacity of a given body to produce physical effects external to that body.*

An explanation is in order regarding the meaning of "effects." In this text physical effects will mean movement or changes in size, color, temperature, or numerous other changes in the physical character of objects.

We define our subject as follows:

Thermodynamics: *That branch of science which treats (1) the conversion of energy from one form to another, and (2) the conveyance of energy from one place to another.*

We will, in fact, be more concerned with converting energy than with conveying it. For example, we will want to know how much of the energy in one pound or kilogram (kg) of coal can be converted into electrical power at the district power plant, how much electrical power is needed to keep a refrigerator at 38°F or 3°C, how much energy is needed to boil one pound or kilogram of water, or how much energy of one gallon or one liter of gasoline we can extract to move an automobile. We can cite other equally valid examples of questions needing answers in both technical and social problems. These questions will be answered with the tools of thermodynamics exposed in the following chapters.

We will be treating ideas and areas such as those just mentioned on a macroscopic scale, and for our purposes, we have arbitrarily divided the physical world into the following three regions:

1. A microscopic world of atoms, protons, electrons, bacteria, viruses, and generally those bodies invisible to the human senses.
2. A macroscopic world, the size of which man can appreciate with his senses. All our directly usable quantities here on earth are in this region.
3. A universal or infinite world: the earth's solar system, the universe, galaxy, and cosmos. This world, due to the efforts of astronomers and astro-physicists, is becoming more comprehensible, but it is infinite in size and numbers.

An understanding of the tremendous effects that these three worlds have on each other is necessary. In particular, we will need some intuitive appreciation of the microscopic (or atomic) structure of matter as it affects the macroscopic world. Thermodynamics is equally at home in all three worlds, but here we will concern ourselves with macroscopic quantities.

Sometimes the precise subject we are studying is called *engineering thermodynamics, general thermodynamics*, or *thermostatics*. There are two other fields which need to be mentioned: *statistical thermodynamics* and *irreversible thermodynamics*. Statistical thermodynamics is the application of mathematical statistics in conjunction with thermodynamics to the microscopic atomic world. The results from this endeavor are enlightening and useful in thermodynamics as well, but we do not have the space or background to present it here. Irreversible thermodynamics concerns itself with systems in states far removed from a macroscopic balance. Examples such as capillary action of atoms creeping up a narrow tube (an apparent contradiction of natural law), diffusion of atoms through a fine plug, severe shock waves induced by supersonic aircraft, and electron flow through a thin wire are a few phenomena which are treated through the concepts of irreversible thermodynamics.

A firm understanding of the tools and concepts of thermodynamics, however, is a prerequisite to proceed intelligently to the seemingly more exotic realms of statistical and irreversible thermodynamics. More importantly, an awareness of the concepts of thermodynamics is important and mandatory for the solution of everyday problems encountered in many technical fields.

1.2 Historical Background of Thermodynamics

The development of thermodynamics can be traced back to the earliest recorded dates in human history. Central to the theme of this development is human desire to ease or replace manual efforts with additional animate or inanimate sources of power. What follows in this section is a brief sketch of the history of the present science of thermodynamics. It is obviously not a complete exposition; it is meant only to give you some historical insight into how the ideas of thermodynamics originated and expanded.

Human use of animate power, such as horses and oxen, began around 4000 B.C. and represented the major source of energy through the nineteenth century A.D. By 3500 B.C., wheeled vehicles were used in Mesopotamia to ease the burden of man and beast. Watermills, steam jets, and various mechanical devices were in use during Christ's lifetime, and around 150 A.D. Hero's turbine was invented. This turbine was a globe containing water from which hot steam could escape through two nozzles, as shown in figure 1-1. A fire placed under the device boiled

Figure 1–1 Hero's turbine or aeolipile. Revised from A. Sinclair, *Development of the Locomotive Engine* (Cambridge, Massachusetts, 1970), p. 2; with permission of MIT Press.

water in the flask, and the steam traveled up the vertical tubes and into the globe. Once in the globe, the steam was expelled through nozzles, thus causing the globe to rotate. It was really nothing but a novelty toy at the time, but it represents a thermodynamic concept of converting inanimate energy from a fuel into an effect (motion).

The science of thermodynamics probably began around 1592 when Galileo used a thermometer to make the first measurement of temperature. The inaccurate and fickle human sense of touch was thus circumvented and replaced by quantitatively describing the hotness or coldness of objects. Meanwhile, the first use of steam for furnishing significant amounts of power for social needs occurred in the late seventeenth century. In 1698, Thomas Savery devised an arrangement of tanks and hand operated valves to utilize steam and its energy to pump water from a well. (See figure 1–2.) In the pump steam was produced in a boiler (*a*) and conducted to the two reservoirs (*b*) through hand operated valves (*c*). The steam was furnished to the reservoirs alternately, in turn pushing water in the reservoir out through pipe (*d*) and to the top. Valve (*c*) would then be closed and a trickle of cold water would condense the

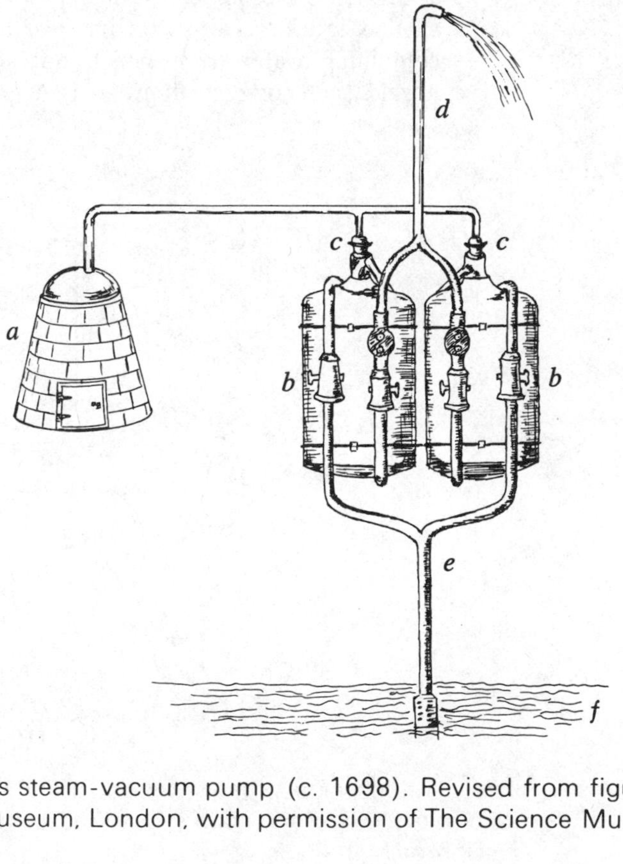

Figure 1–2 Savery's pistonless steam-vacuum pump (c. 1698). Revised from figure of Photo. Science Museum, London, with permission of The Science Museum, London, England.

steam in the reservoir, thus causing a vacuum to be created. This vacuum would allow water to flow up pipe (*e*) from a low water supply (*f*) into the reservoir. Valve (*c*) would be then opened again to repeat the cycle. By alternating between the right and left reservoirs a continuous flow of water was created at the top. While representing a historical first this device was little improvement over animate power. Thomas Newcomen, in 1712, developed a steam-piston engine which was a logical replacement of animate power for pumping water. This arrangement, shown in figure 1–3, provided a cycling motion which we call a *heat engine*. Known as the "atmospheric engine" because a vacuum and atmospheric pressure combined to provide the power stroke, this engine was used for pumping water. Steam produced in the boiler (*a*) was conducted through a hand valve (*b*) to the piston-cylinder (*c*). The steam would push the piston up to the position shown, allowing the pump rod (*d*) to descend into a water supply. The valve (*e*) was then opened to allow a spray of water to condense the steam in the cylinder, causing a vacuum to be created. The piston was then pushed down by atmospheric pressure, the pump rod raised, and water pumped up out of the water supply (*f*).

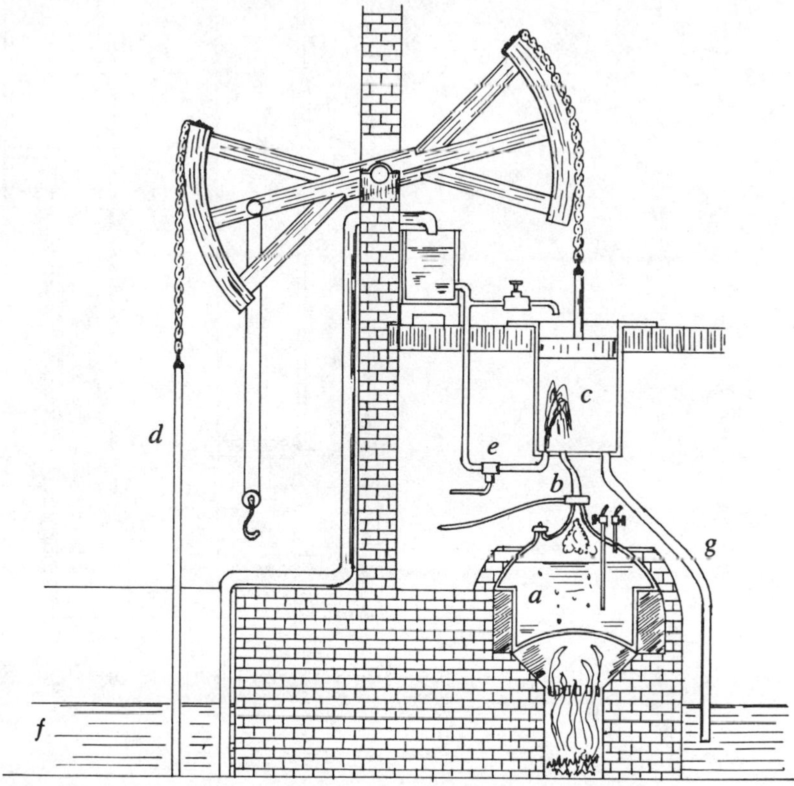

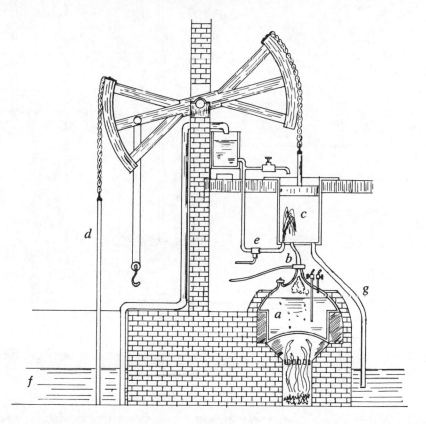

Figure 1–3 Newcomen's steam engine (1712). Revised, with permission, from Sinclair, *Development of the Locomotive*, p. 7.

Valve (*e*) was closed, valve (*b*) opened, and the process repeated. Line (*g*) was intermittently opened to allow the condensed steam to flow out of the cylinder.

With improved machining techniques available for manufacturing parts, James Watt developed a steam-piston engine which represented a significant improvement over Newcomen's. Watt's engine, first operated in 1775 to pump water, represents the forerunner to the steam engines used for railroads, ships, and numerous other applications (figure 1–4).

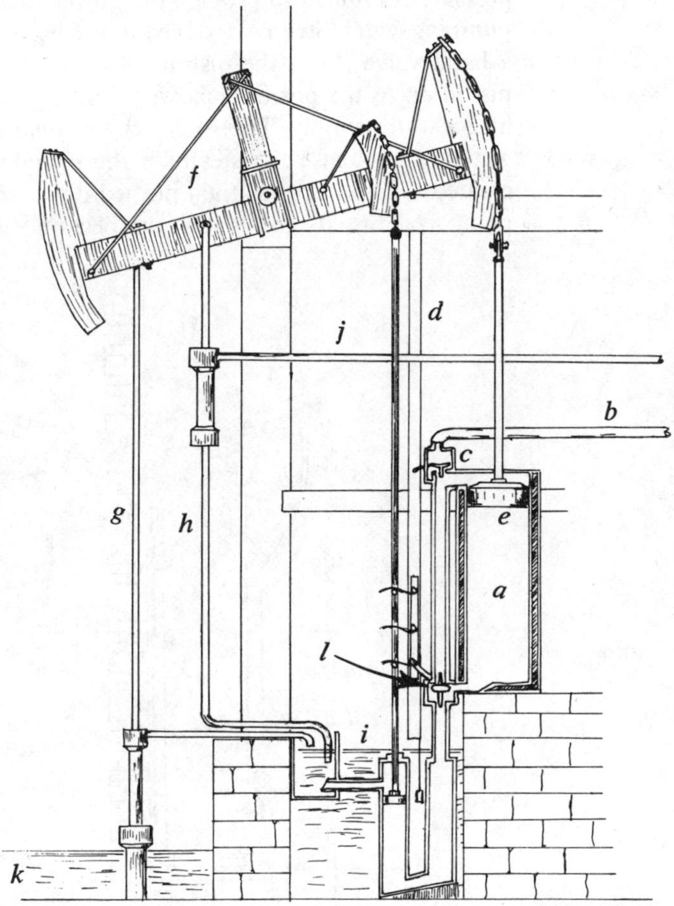

Figure 1–4 Watt's steam engine. Revised, with permission, from Sinclair, *Development of the Locomotive Engine,* p. 10.

Unlike Savery's or Newcomen's devices which condensed steam inside the working cylinder chamber, Watt's steam engine condensed the spent steam external to the cylinder (*a*). Steam was furnished from a boiler through a pipe (*b*). Valve (*c*), controlled from a tappet rod (*d*), allowed steam to enter the top side of the piston (*e*). This pushed the piston down and, through the walking beam (*f*), raised the pump rods (*g*) and (*h*). This motion drew water out of the reservoir (*i*), through the pipe (*j*) and from a reservoir (*k*) to reservoir (*i*). Valve (*l*) was then shifted to allow steam to enter the bottom of the piston; thus equilibrated, the piston moved to the top to begin a new cycle.

Thermodynamic theory began to take form in 1693, when G. W. Liebnitz pronounced the conservation of mechanical energy (kinetic and potential). Sadi Carnot published a treatise in 1824 which described cycling devices or heat engines and which alluded to the first and second laws of thermodynamics. Twenty-six years later, in 1850, Rudolph Clausius formally stated these two laws of thermodynamics, and in 1854 he identified and defined the property now called *entropy*.

From 1840 to 1848, James Joule experimentally proved the equivalence of heat and work, thus making thermodynamics a quantitative science in the best tradition of Galileo. The internal combustion gasoline engine, later used to provide power for automobiles, trucks, and numerous other devices, was developed around 1860 by Lenoir. This type of engine was first used in vehicles around 1876 by Otto and Benz.

Around 1884, Parson introduced a steam turbine capable of developing significant amounts of power. This type of device, utilizing the popular medium steam, has been a most durable power generator and seems more popular today than ever. In the early 1900s, Nernst and Planck separately enunciated the earliest definition of the third law of thermodynamics. These statements have since been refined and revised by various theoreticians.

These advancements in both theory and technology reflect the applicability of thermodynamics to practical efforts; this usefulness has stimulated much of the interest in further expanding that body of knowledge known as the *science of thermodynamics*.

We see that thermodynamics developed through theory, experiments, and practice. Theoretical advancements came from the giants of thought; many men such as Joseph Black, Lord Kelvin, J. W. Gibbs, James Maxwell, L. Boltzmann, H. L. F. Helmholtz, and Albert Einstein contributed to thermodynamics at least as significantly as those mentioned previously. However, without the experimentation, design, creativity, and artisan ability to machine and fabricate parts precisely, useful engines providing significant amounts of power and devices to use this power would be nonexistent.

Our concern in this book is to understand and use the concepts of thermodynamics, clarified by theoreticians both past and present, for the solution of engineering and technological problems.

1.3 Thermodynamic Calculations

This book contains many examples and practice problems that are characteristic of those encountered in engineering applications. The solutions to these examples and practice problems are commonly numerical answers which are calculated from governing mathematical equations. Thus, the reader must be able to perform algebraic and arithmetic operations to gain a clear understanding of the principles of thermodynamics. Further, the reader is urged to include all units when writing out equations for computation. The units will then provide a means of checking the equations' accuracy as well as serving as a check for the proper substitution of terms.

The following examples show the types of problems one will encounter in the following chapters.

Example 1.1 A perfect gas satisfies the relationship $pV = mRT$. If $p = 1.01 \times 10^5$ N/m², $m = 3$ kg, $R = 0.287$ N · m/kg · K, and $T = 300$ K, determine the value of V.

Solution We may algebraically solve for the term V so that

$$V = \frac{mRT}{p}$$

Substituting the values into the proper terms we obtain

$$V = \frac{(3\ \text{kg})(0.287\ \text{N} \cdot \text{m/kg} \cdot \text{K})(300\ \text{K})}{(1.01 \times 10^5\ \text{N/m}^2)}$$

$$= 0.00256\ \text{m}^3$$ *Answer*

Example 1.2 The amount of work achieved or expended during a particular action or process is

$$wk = \frac{1}{1-n}(p_2 V_2 - p_1 V_1)$$

If $n = 1.4$, $p_2 = 220 \times 10^5$ N/m², $V_2 = 0.01$ m³, $p_1 = 16 \times 10^5$ N/m², and $V_1 = 0.09$ m³, determine the amount of work wk.

Solution Here we may readily substitute numerical values into the relationship to obtain

$$wk = \frac{1}{1-1.4}(220 \times 10^5\ \text{N/m}^2 \times 0.01\ \text{m}^3$$

$$- 16 \times 10^5\ \text{N/m}^2 \times 0.09\ \text{m}^3)$$

$$= -190{,}000\ \text{N} \cdot \text{m}$$ *Answer*

Example 1.3 Air is commonly assumed to behave as a perfect gas, describable by the equation $pv = RT$. The gas constant for air R can be taken to be 53.3 ft-lbf/lbm · °R. If air is at a pressure p of 2100 lbf/ft^2 and at a temperature T of 600°R, what is the specific volume v?

Solution We may observe that the specific volume can be solved from the perfect gas relationship. Thus

$$v = \frac{RT}{p}$$

and we can then obtain

$$v = \frac{(53.3 \text{ ft-lbf/lbm} \cdot {}^\circ\text{R})(600^\circ\text{R})}{2100 \text{ lbf/ft}^2}$$

$$= 15.2 \text{ ft}^3/\text{lbm} \qquad \textit{Answer}$$

Example 1.4 During a particular process the relationship between the variables p and V has been found to be $pV^n = \text{constant} = C$, where $n = 1.29$. For two conditions, we know the value of p, say p_1 and p_2. Also we know the value for V_2, and we need to know the value for V_1. The known values are $p_1 = 14$ psi, $p_2 = 280$ psi, and $V_2 = 0.02$ ft^3.

Solution We may write the above relationship as follows:

$$p_1 V_1^n = C = p_2 V_2^n$$

or

$$V_1^n = V_2^n \left(\frac{p_2}{p_1}\right)$$

and

$$V_1 = V_2 \left(\frac{p_2}{p_1}\right)^{1/n}$$

Substituting the values into this equation we obtain

$$V_1 = (0.02 \text{ ft}^3)\left(\frac{280 \text{ psi}}{14 \text{ psi}}\right)^{1/1.29}$$

$$= 0.204 \text{ ft}^3 \qquad \textit{Answer}$$

Readers may find their final answers computed for the above examples slightly different, depending on the method of calculating. For most engineering purposes, answers given to three significant figures are sufficient. For instance, the answer to example 1.4 is given to three significant figures whereas the values for p_1, p_2, and V_2 are all given to two significant figures.

It is recommended that the reader, to more efficiently solve problems and to gain a clearer understanding of the concepts, keep the following items in mind:

1. Use one sheet of paper per problem. Number or otherwise identify each problem and include the date of doing the work on the paper.
2. Keep a margin on top, bottom, and both sides of the paper.
3. State the problem or situation in your own words. Use suitable sketches if needed to help describe the problem.
4. Carefully list the known values and state those unknowns that will solve the problem.
5. List the assumptions you choose to make to allow for a method of solution. These assumptions may be listed with the knowns and unknowns or presented as needed during the solution.
6. Solve the problem using the governing equations. Indicate clearly the answers sought. (See preceding examples.)
7. Be neat and thorough, erasing errors and writing clearly and legibly.

1.4 Summary

Since you now are introduced to what thermodynamics is and how it evolved, let's see how the material that is to follow is presented, and let's give you some idea of what to expect.

Chapter 2 introduces the concept of taking a specific part of nature (called the *thermodynamic system*) and describing it with numbers or properties. Chapter 3 describes the interaction of this system with its surroundings. The first law of thermodynamics and the conservation principles are introduced in chapter 4, followed by some classic uses of the first law in chapters 5 and 6. The important thermodynamic property, entropy, is introduced and explained in chapter 7. The idea of a cycling device, or a motor which runs as long as we wish, is presented in chapter 8, and the second law of thermodynamics is presented in chapter 9. In chapter 9 are also shown some significant implications of the second law which must be considered in any complete analysis of an energy-producing or energy-expending device.

Chapters 10 through 15 are devoted to some of the more traditional applications of thermodynamics. Each chapter could be read and studied separately, and each contains topics of direct application, such as the internal combustion engine, steam and gas turbines, the jet engine, refrigerators, and mixtures of two or more different liquids or gases. Chapter 16 introduces some other applications of thermodynamics to stimulate your imagination.

Practice Problems

1.1. Describe as many energy conversion devices as you can.
1.2. From your previous knowledge of physics, can you define *work?* Refer to chapter 3 and see if you agree.
1.3. From your introduction to physics, define *heat*. Compare with the definition in chapter 3.
1.4. Define *energy* without referring to section 1.1.
1.5. Solve the equation of a perfect gas $pV = mRT$ for T.
1.6. For the equation $xy^{1.6} = 2.3$, solve for x in terms of y then y in terms of x.

Calculate to three significant figures the following quantities.

1.7. [(3.70)(40.1)]/[(136)(270)(3)].
1.8. (1870)(26.0)(9.80).
1.9. $(260)^2$.
1.10. $(260)^{1/4}$.
1.11. $(62.1)(35.1/26.1)^{1.6}$.
1.12. (333)[1/(1 − 1.2)].

2

The Thermodynamic System

In any scientific or technical analysis the first steps should be to focus attention on an object and identify it quantitatively. The science of thermodynamics is no exception to this rule. In this chapter, we will focus mainly on the concept of a *system* and will discuss how to establish the parameters necessary to measure a system. We will be defining terms extensively, including the necessary terminology for treating systems as they change in time or space, that is, as systems interact with their surroundings.

2.1 System

The thermodynamic system: *Any region which occupies a volume and has a boundary.*

In solving a technical problem through thermodynamics, you must identify the system and its boundaries. For instance, suppose you wish to know the power required to operate a refrigerator. In this circumstance the system boundary would be the outside surface of the refrigerator; everything included inside this surface would be the system. On the other hand, if you are concerned only with the operation of a compressor within the refrigerator, then the compressor itself is the system.

As another example, let's consider the internal combustion reciprocating engine of the present day automobile. If you are interested in the total operation of the automobile, your system might contain the whole vehicle, including the engine, fuel tank, battery, controls, and maybe even the passengers. However, if you wish to study the detailed manner in which power is extracted from the fuel and converted into mechanical energy, the system might be only one cylinder of the engine itself and not even the actual surfaces of the cylinder. This one-cylinder system or *piston-cylinder* for our purposes, is shown in figure 2–1, with the boundary indicated by a dashed line. This figure illustrates some important characteristics that a boundary has, and as you look at it, you should recognize that it represents a dynamic system where the piston is in motion at all times. In addition, the valves open and close at opportune instances, either allowing fuel and air to enter the system or exhausting the burned gases. Now, obviously the boundary can move (since the piston and valves move), and we can even shuffle fuel, air, and exhaust gases across the boundary; but not all system boundaries have this capability, and we will see in section 3.8 that the difference between two types of systems — open and closed — will be determined by whether matter crosses the boundary or not. Open and closed systems will then be an important part of the thermodynamic analysis, each handled in a slightly different manner.

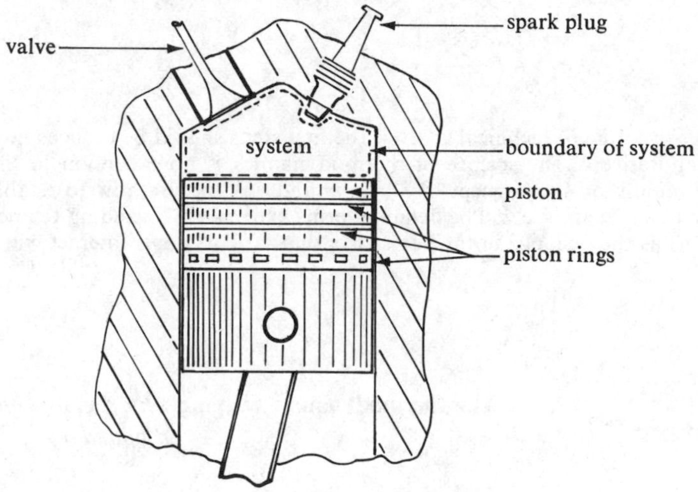

Figure 2–1 Piston-cylinder system.

Another system is identified in figure 2–2. This is a balloon made of thin rubber which can stretch and contract. The system boundary of the right side configuration is drawn outside the balloon surface and envelopes two separate materials: air in the balloon and the rubber balloon itself. In analyzing this system, you must be careful. In figure 2–1 the system was homogeneous at any given time, but the balloon in figure 2–2 is not. By *homogeneous*, we mean that there is one and only one distinct uniform material throughout the system at any instant in time. In figure

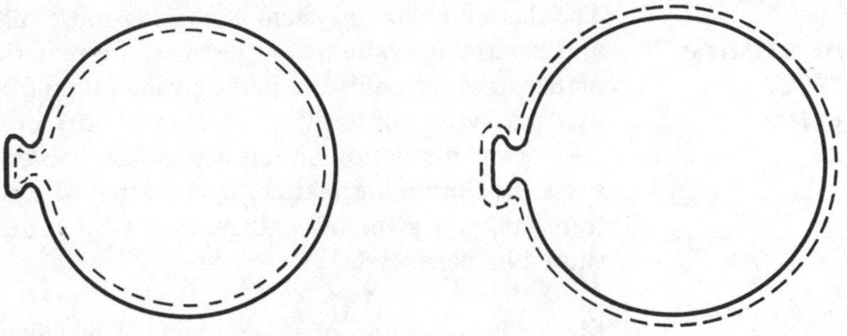

Figure 2–2 Inflated balloon with system defined by boundary inside or outside balloon material.

2–2 we have two different materials, the balloon and the air, each of which has distinct ways of handling energy, thereby affecting the analysis differently.

In defining or establishing a system, whenever possible, select a boundary which will enclose homogeneous materials. In the balloon problem, we were concerned with rubber as well as the air, so that ignoring all but the air would be undesirable. If we wished, however, to specify a homogeneous system in the balloon we could do so by denoting a boundary inside the balloon, thus defining a system composed of air alone, as indicated in the left side configuration of figure 2–2.

As we have seen, the boundary and the enclosed volume of a system can be arbitrary, and the possible variations are infinite. Let's allude to the physical world. The volume outside the boundary we will call the *surroundings* of the system, and the sum of the system and its surroundings we will call the *universe*. The universe, if we are completely rigorous and correct, is infinite, but we need not in this book concern ourselves with this aspect. We will consider the surroundings as only that physical part of the universe which is capable of affecting our system during the time period in which we are concerned. The point, though, is that we will directly concern ourselves only with the system and its boundary, thereby circumventing the problem of infinite universes.

In addition, once we have assured ourselves of either a homogeneous system or a simple composition of homogeneous systems (like the balloon problem), then the internal details of the dynamics or design of the system are not needed. That is, for the piston-cylinder system of figure 2–1, the materials, dimensional sizes, or velocities and accelerations of the valves, pistons, spark plugs, or cylinder wall are not necessarily important in the thermodynamic analysis, provided we know that the system is some homogeneous gas inside the cylinder.

2.2 Elementary Matter Theory

The thermodynamic system is nothing more than some volume of space and invariably contains matter or some material. If the volume has no matter, it is, of course, a perfect vacuum, and a very attractive thermodynamic system in terms of producing work or power. However, we will reserve any discussion on this aspect for chapter 9 and here consider only a system containing matter. It is matter which contains, collects, sorts, transfers, and gives off energy, and a brief description of its structure should be beneficial.

Matter is composed of *atoms*, each of which in turn is composed of a cluster (one or more) of *protons* and *neutrons* very tightly compressed into a *nucleus*. The nucleus is surrounded by a cloud of one or more very small particles, called *electrons*, found in different concentric levels, or shells, encircling the nucleus. Protons have a positive charge; electrons, a negative charge; and neutrons, as the name implies, have no charge, that is, they are neutral. The electrons, protons, and neutrons of different atoms vary in number and arrangement, and each unique arrangement is called an *element*. There are naturally occurring and fabricated elements. (See table B.18 of appendix B.) A simplistic diagram of an atom is shown in figure 2–3, a sketch of a hydrogen atom.

Each shell of an atom can hold a certain maximum number of electrons, for example, the first shell can hold two electrons, and the second, eight.

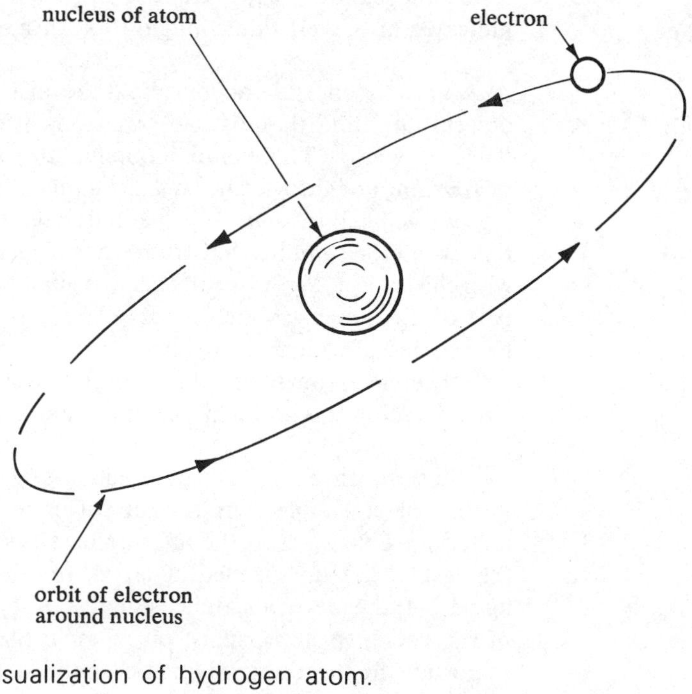

Figure 2–3 Simplified visualization of hydrogen atom.

When the outer shell, that is, the last shell that contains electrons, contains the maximum number it can hold, that atom is in a stable state and does not readily combine with other atoms. For example, neon atoms have two electrons in their first shell and eight in their second and are thus stable. However, if an atom has an incomplete outer shell and does not contain the maximum of electrons it can hold, it is unstable and will readily combine with other atoms to make its outer shell complete. When two atoms combine, one gains the electrons which the other loses; in combining, atoms form different types of chemical bonds, such as ionic or metallic (figure 2–4). When atoms combine, they form what are called *molecules*.

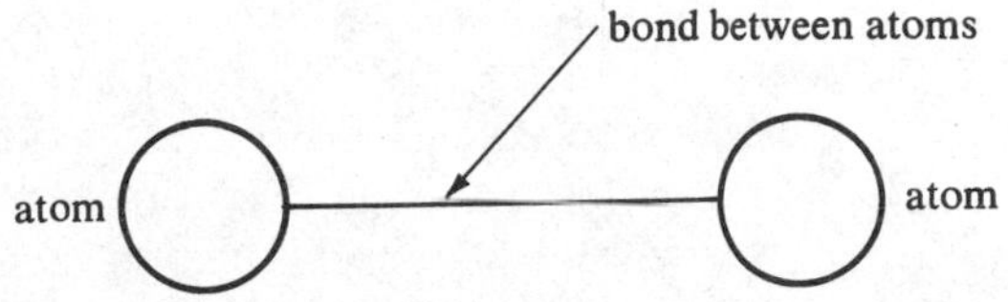

Figure 2–4 Simplified view of metallic or ionic bond in molecule composed of two atoms.

In addition to gaining or losing electrons, atoms can form bonds by "sharing" one or more pairs of electrons; this type of bond is called *covalent*. In this sharing of electron pairs, each electron in a pair comes from a different atom. A hydrogen molecule is a good example of covalent bonding, or electron sharing (figure 2–5). Each hydrogen molecule contains two atoms, each of which when it exists singly without the other, has only one electron in its outer shell and needs one more to make it complete. However, when the atoms combine to form a molecule, each shares the other's electron so that each has two electrons and thus a completed outer shell.

Atoms and molecules have mass, which is measured in *atomic weight*. Notice in table B.18 that each element has an atomic weight. This number represents the mass of one gram-mole of the element. The gram-mole is a mass which contains 6×10^{23} atoms of a given element under standard atmospheric conditions and based on the mass units of grams. This number is called Avogadro's number and is an important chemical quantity. For instance, 6×10^{23} molecules of hydrogen make up 1 gram-mole of hydrogen and have a mass of 1 gram. Similarly, 6×10^{23} atoms of carbon make up 1 gram-mole of carbon, and they have a mass of 14 grams. We can also speak of pound-moles of an element, which is the mass of an element based on the pound-mass unit of mass. (Refer to section 2.7 for a discussion of the pound-mass unit.)

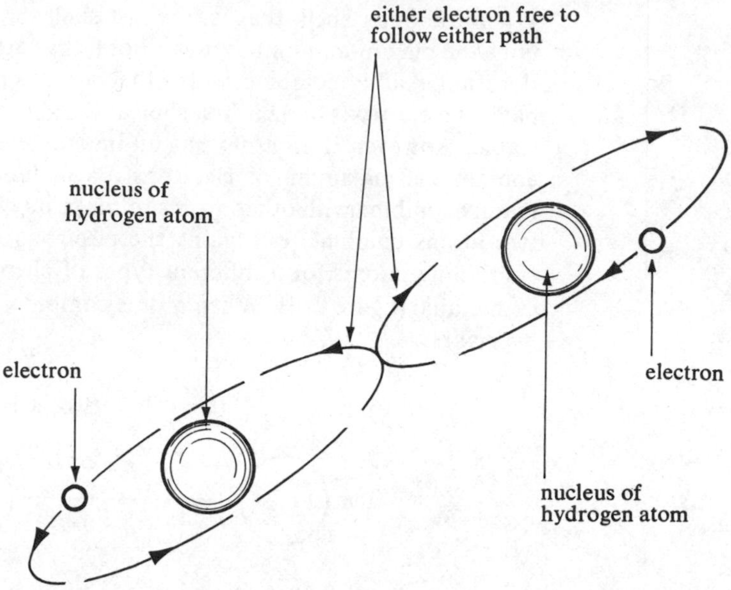

Figure 2–5 Hydrogen molecule.

In our approach to thermodynamics we will not be concerned with counting or otherwise describing individual atoms or molecules, but we will consider very large amounts of atoms or molecules mixed together, called a *substance*. If the substance is composed of only one type of molecule, we will call this a *pure substance*. Water, composed of molecules made up of two hydrogen (H) atoms and one oxygen (O) atom and written H_2O, is an example of a pure substance.

The atoms and molecules of substances behave differently, depending on the conditions of the surroundings in which the substance is immersed.

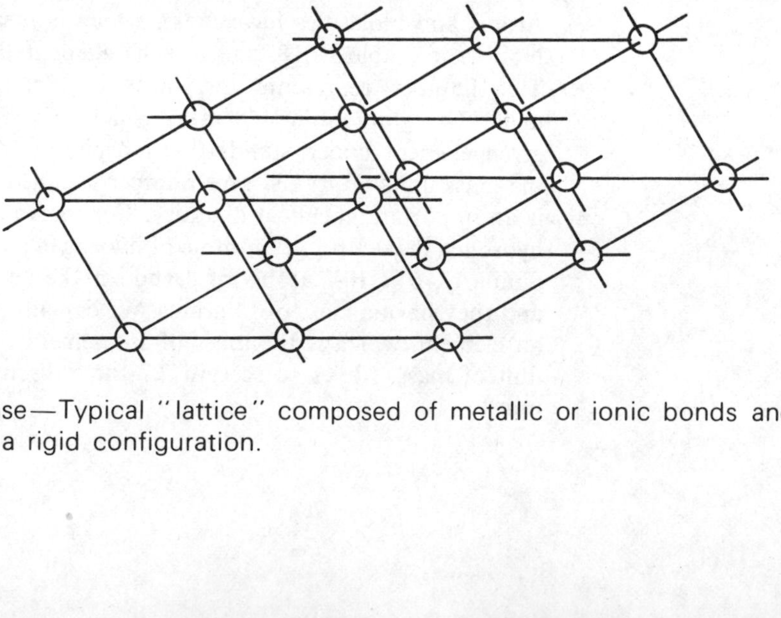

Figure 2–6 Solid phase—Typical "lattice" composed of metallic or ionic bonds and atoms in a rigid configuration.

Consequently, we will want to know the arrangement and interaction of these particles, the *phase* of the element or atoms. We will concern ourselves with three phases of a substance: solid, liquid, and gaseous. The solid phase of a substance is characterized by rigid bonds between atoms or molecules, bonds which can generally stretch, bend, and break (figure 2–6). Also, the solid phase will, under many conditions, be characterized by a crystal structure.

The gaseous phase has free-floating molecules with some, although rather insignificant force between them (figure 2–7). We will be discussing a substance called a *perfect gas*, which is a gas that has no forces, other than collision forces, between the individual atoms or molecules. The collision for a perfect gas is assumed to be "perfect" in the sense that no permanent distortion occurs among the particles after colliding.

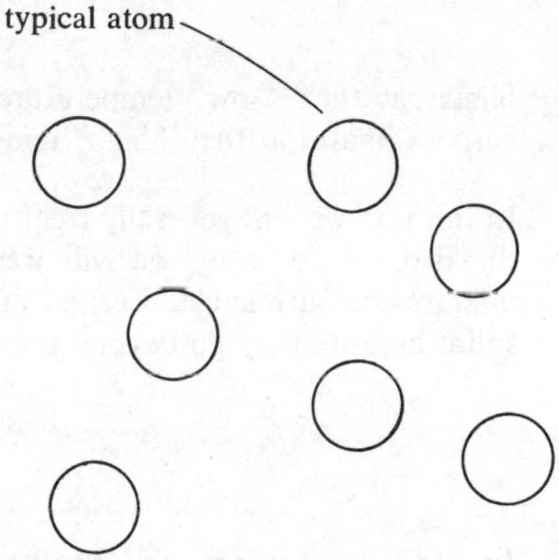

Figure 2–7 Gaseous phase — Little or no evidence of bonds between atoms: atoms free to move independently in any direction.

The liquid phase is behaviorally like a mixture of the solid and gaseous phases. The molecules are more closely packed than in a gas, but the bonds between the molecules are passive. That is, the molecules will adhere to one another under very light forces but can easily slide apart and shift arrangements without great forces, as shown in figure 2–8. We could characterize liquids as being able to retain some volume (one gallon for instance), but not shape. If left alone in space, gaseous substances cannot even retain their volume, let alone their shape, whereas solids retain volume and shape.

At low temperatures, materials generally are in a solid or a liquid phase; as the temperature rises, they will pass to a gaseous phase. We will see that "high" or "low" temperature has no absolute value thermodynamically, so some materials, such as air, do indeed have liquid and solid

typical atom

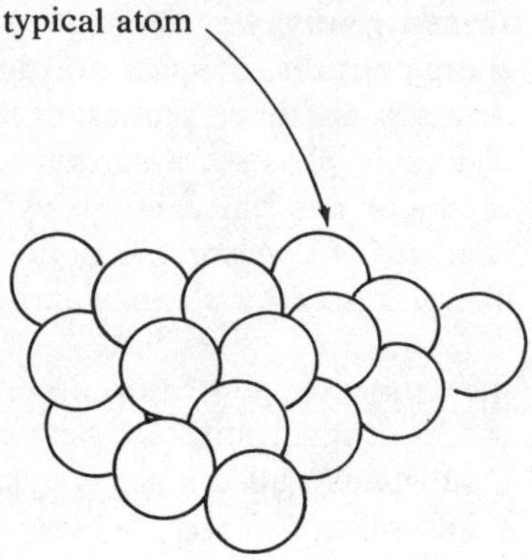

Figure 2–8 Liquid Phase — Atoms attracted by bonds which allow sliding between atoms but prevent separation of individual atoms.

phases at their "low" temperatures, and some solids have liquid and gaseous phases at their "high" temperatures.

In this text we will generally treat gas and liquid phases and mixtures of the two. In addition, we will treat pure substances or homogeneous mixtures of pure substances, as mentioned earlier. We will not discuss solids because they go beyond the scope and depth of this text.

2.3 ✓
Property

In order to describe and analyze a system, we must know some of the quantities that are characteristic of it. These quantities are called *properties*, and include volume, mass, weight, pressure, temperature, density, shape, position in space, velocity, energy, specific heat, color, taste, and odor. The list could go on and on, and the longer the list, the better description we could give our system.

We shall separate properties into two general classifications: *intensive properties* and *extensive properties*. An intensive property is independent of the mass or total amount of the system, for example, color, taste, odor, velocity, density, temperature, and pressure. Extensive properties are all those properties which are dependent on the total amount of the system, for example, mass, weight, energy, and volume. It is worth noting that any extensive property can be made intensive by dividing by the mass. In this book we shall generally denote extensive properties by capital letters and intensive by lowercase letters. Exceptions to this notation will be the mass m, which is an extensive property, and velocity $\bar{V}$, which is intensive. Also, we shall frequently use the term *specific* to denote an intensive property. Thus, we use the term *specific energy* to describe the energy per unit of mass of some material. The term *total* will be used to describe extensive properties, such as *total energy* to denote the energy in a given amount of mass.

2.4 State of a System

A complete list of the properties of a system describes its state. In order that a list does not get too long and confusing, we will generally assume that the system is composed of pure substances of one phase, or of simple inert mixtures of the three phases, gaseous, liquid, and solid. We will see that only a few properties need to be known under these conditions, and these are generally as follows:

Type of substance (element)
Volume
Weight or mass
Pressure
Density or specific volume
Temperature
Energy

2.5 Process

The primary reason for describing the state of a system is to analyze the system as a power-producing or power-consuming quantity. *Process* is a change of state, which can occur in a number of ways. For example, if we change one or more of the properties such as energy, pressure, temperature, or volume, we will have gone through a process. Our concern will be evaluating the state immediately before and after this change. An important point to remember is that work, power, and heat can occur only during processes and only across the boundary of the system. We will return to this later.

2.6 Cycle

Having a system which changes its state, thereby producing or using work and/or heat, is all well and good, but if we want an engine that runs continuously without seemingly changing its state further and further, then we must occasionally return to our stabilizing point or initial state.

Cycle: *A combination of two or more processes which, when completed, return the system to its initial state; a system operating on a cycle is called a* **cyclic device**.

Notice that there is no change in energy or property over a complete cycle. There can be, however, work and heat added or extracted, which is the essence of the power-producing and power-consuming devices of our society—we return the engine (or cyclic device) to its initial condition periodically (for example, 300 times per minute if we have an engine running at 300 revolutions/minute), and it still continues to produce work.

The special form of the cyclic device which transfers heat into work is called a *heat engine* and is of major concern to us in the study of thermodynamics. Examples of heat engines are the internal combustion engine of an automobile, the jet engine used to power aircraft, and the steam turbine electric power generating system.

2.7 Weight and Mass

In describing a system, one of the first properties which we think of is weight. We will denote weight by the symbol W and express it in the units of newtons, N. In the English system of units, the force or weight is called the pound-force, lbf.

When we measure the weight of a system, we use a weight scale and determine the deflection, or "stretch," of a spring. This stretch is directly proportional to a force applied to the spring. Figure 2–9(a) shows a

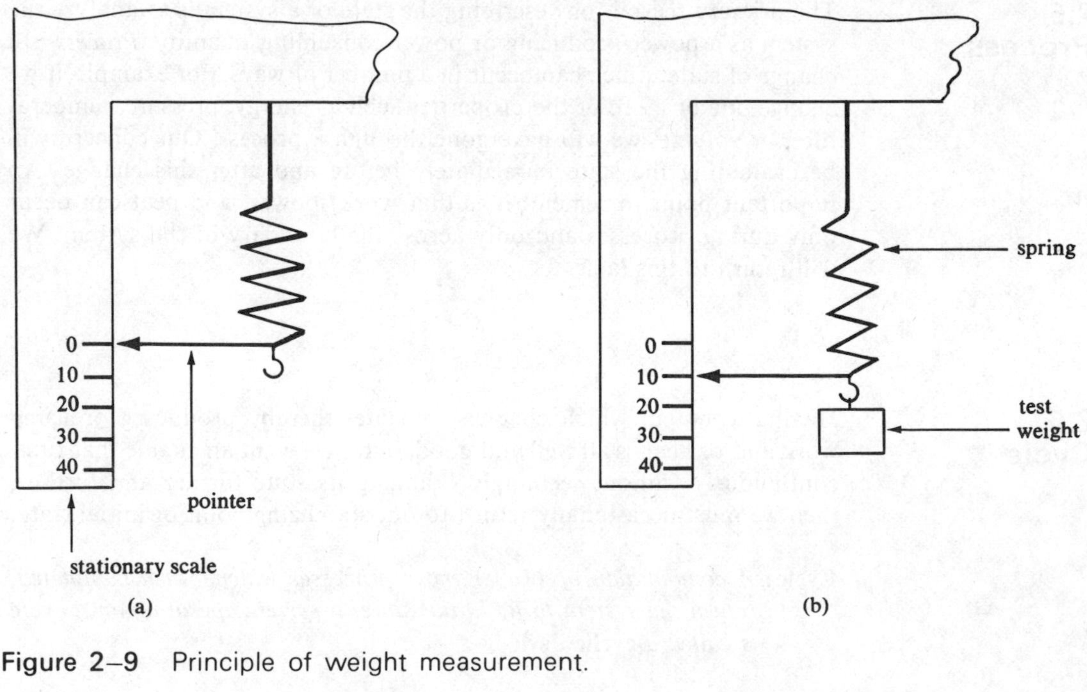

Figure 2–9 Principle of weight measurement.

typical weight scale which is composed of a spring, a pointer attached to the spring, and a stationary scale. Figure 2–9(b) shows this same device after a test weight (our system) has been attached. In order to determine the system weight, note that the spring has extended in length (a distance proportional to the system weight W), and the pointer has moved from a position indicating no weight (0 on the stationary scale) to a new position where we say the weight is 10 newtons. If the system had weighed 20 newtons, then, of course, the pointer would have indicated 20 newtons on the stationary scale, and any other weight would as well register that particular number on the stationary scale.

Note in figure 2–9 that we are not directly measuring the weight of a system. First, we are measuring spring deflection, which is proportional to the force on the spring. This force is equal to the weight of the system since there is no motion after the spring has deflected. Figure 2–10 shows the forces acting of the system for the situation of figure 2–9. From physics mechanics we know that under static (motionless) conditions the sum of all forces equals zero.

$$\Sigma \text{ Forces} = 0$$

or

$$F - W = 0 \tag{2–1}$$

and

$$F = W$$

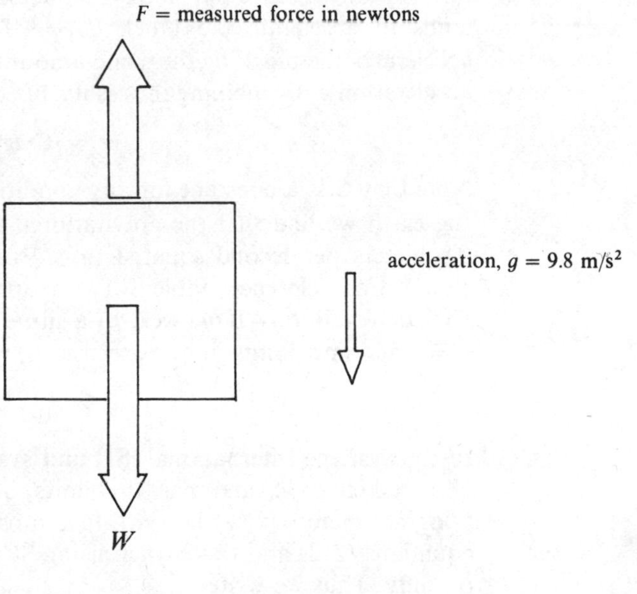

Figure 2–10 Forces acting on system when measuring weight.

This result was stated before. In addition, since any force is a vector (a vector has magnitude and direction), we see that weight is a vector.

Mass, *m*, is a scalar (magnitude only) quantity and is, along with the volume of a system, the primary measure of the amount of matter in the system. (See section 2.2.) The units of mass are kilograms (kg) or in the English system, pound-mass (lbm) or slugs. Mass can be related to the weight by Newton's second law of motion which says that an external force applied to a body is proportional to the mass of that body times the acceleration, and that the force will have the same direction as the acceleration. Algebraically this is

$$F \backsim ma \tag{2–2}$$

where

m = mass of the block
a = acceleration of the block

and

F = external force applied to the block

We can easily make equation (2–2) an equality by inserting a constant C so that

$$F = Cma \tag{2–3}$$

In figure 2–10 we have a static condition which means that no acceleration is present. Since the block is immersed in a gravitational field which tends to accelerate the block *down*, the external measured force F accelerates the block *up* the same amount; we will call this gravitational acceleration g. Combining the results of equations (2–1) and (2–3), we get

$$W = Cmg \tag{2–4}$$

Note that C is a constant for any condition whereas g is a variable. On the earth we find that the gravitational acceleration g is approximately 9.8 meters per second squared (m/s^2) or 32.2 feet per second squared (ft/s^2). For reference, table B.17 in appendix B lists values of g for various conditions. If we were in a situation (like space travelers) where $g = 0$ at some points in our excursions, we would have no weight since

$$W = Cm(0) = 0$$

In the Systeme International (SI) unit system, the value for the constant C is taken as 1 and has the units of newton-second squared per kilogram-meter ($N \cdot s^2/kg \cdot m$). It is most convenient to delete C from equations (2–3) and (2–4) when using SI units since C does have a value of unity. Thus we write

$$W = mg \tag{2-5}$$

In the English system, the pound-force (lbf) and pound-mass (lbm) are defined as having equal values (but not units) on the earth's sea level where $g = 32.17$ ft/s^2. Based on this convention the value of C must be 1 lbf · s^2/32.17 lbm-ft which can be seen by substituting into equation (2–4). Thus, if mass is, say, 1 lbm by the convention, then weight must be 1 lbf when $g = 32.17$ ft/s^2. From equation (2–4), then

$$C = \frac{W}{mg} = \frac{(1 \text{ lbf})}{(1 \text{ lbm})(32.17 \text{ ft/s}^2)}$$

$$= \frac{1}{32.17} \frac{\text{lbf} \cdot \text{s}^2}{\text{lbm-ft}}$$

For convenience we will say

$$C = \frac{1}{g_c} \tag{2-6}$$

and call g_c the universal gravitational constant equal to 32.17 ft/s^2 lbm/lbf.

Then, if we use pound-mass and pound-force units, equation (2–4) is

$$W = \frac{1}{g_c} mg \tag{2-7}$$

and Newton's law, equation (2–3), becomes

$$F = \frac{1}{g_c} ma \tag{2-8}$$

or, from equation (2–7),

$$F = \frac{W}{g} a \tag{2-9}$$

Another unit of mass used in the English system is the slug. The value of the slug is defined as

$$1 \text{ slug} = 32.17 \text{ lbm} \tag{2-10}$$

and the value of C must be recalculated. From equation (2–4) again we have

$$C = \frac{W}{mg} \tag{2-11}$$

When W has a value of 1 lbf we know that m must equal 32.17 lbm where $g = 32.17$ ft/s^2. But, from equation (2–10) we have that 1 lbm equals 1/32.17 slug. Thus, if these values are substituted into equation (2–11), we obtain

$$C = \frac{(1\ \text{lbf})}{(1/32.17\ \text{slug})(32.17\ \text{ft/s}^2)}$$

$$= \frac{1\ \text{lbf} \cdot \text{s}^2}{\text{slug} \cdot \text{ft}}$$

Thus, we see that equations (2–3) and (2–4) take the form

$$F = ma \quad \text{and} \quad W = mg \tag{2–12}$$

since C has the value of unity.

Example 2.1 A man has a mass of 100 kilograms. Determine his weight at 40° latitude and 1500 meters above sea level.

Solution From Table B.17 we find that g has a value of 9.7976 m/s² at 40° latitude and 1500 meters elevation. Thus, from equation (2–5) we have

$$W = mg$$

$$= (100\ \text{kg})(9.7976\ \text{m/s}^2)$$

$$= 979.76\ \text{kg} \cdot \text{m/s}^2 = 979.76\ \text{N}$$ *Answer*

or

$$= 980\ \text{N}$$ *Answer*

Example 2.2 An automobile engine weighs 300 lbf at 1000 feet elevation and 20° latitude. What is the mass of the engine and what would it weigh if it were located at 20° latitude and 4000 feet?

Solution From Table B.17 we find the value of g to be 32.105 ft/s². Then, since the weight was expressed in pound-mass units, we use equation (2–7) and obtain

$$m = \frac{W}{g} g_c$$

$$= \frac{(300\ \text{lbf})(32.17\ \text{ft-lbm/s}^2 \cdot \text{lbf})}{(32.105\ \text{ft/s}^2)}$$

$$= 300.6\ \text{lbm}$$ *Answer*

If the engine were located at 4000 feet elevation, the value for g would be 32.096 ft/s², and the weight would be found from equation (2–7) to be

$$W = \frac{(300.6\ \text{lbm})(32.096\ \text{ft/s}^2)}{(32.17\ \text{ft-lbm/lbf} \cdot \text{s}^2)}$$

$$= 299.9\ \text{lbf}$$ *Answer*

Example 2.3 What is the weight W of a suitcase having a mass of 1 slug when the system is:

(a) At sea level and 40° latitude.
(b) At 1000 feet above sea level and 40° latitude.

Solution (a) The value of g is 32.158 ft/s² at sea level and 40° latitude. (See table B.17.) Then from equation (2–12)

$$W = mg$$
$$= 1 \text{ slug} \times 32.158 \text{ ft/s}^2$$
$$= 32.158 \text{ lbf} \qquad \textit{Answer}$$

We see that

$$1 \text{ slug} = 1 \text{ lbf} \cdot \text{s}^2/\text{ft}$$

(b) At 1000 ft we see from table B.17 that g is 32.155 ft/s², so

$$W = 1 \text{ slug} \times 32.155 \text{ ft/s}^2$$
$$= 32.155 \text{ lbf} \qquad \textit{Answer}$$

Alternatively, the mass of the system in lbm units is

$$m = 1 \text{ slug} \times 32.17 \text{ lbm/slug}$$

from equation (2–10), and from equation (2–7)

$$W = \frac{1}{32.17 \text{ ft/s}^2 \times \text{lbm/lbf}} \times 1 \text{ slug} \times 32.17 \text{ lbm/slug} \times 32.155 \text{ ft/s}^2$$
$$= 32.155 \text{ lbf}$$

While the mass unit of slugs provides somewhat simplified equations by eliminating a factor of $1/g_c$ from the force-mass relationships as shown by equations (2–11) and (2–12) as compared to (2–9) and (2–7), we will generally use pound-mass (lbm) units unless otherwise specified. The slug unit is used in some reference literature, but the pound-mass unit is generally used for reporting thermodynamic data. This is our primary reason for using the pound-force and pound-mass units.

2.8 Volume, Density, and Pressure

Volume We have seen that volume and mass are the primary measures of the quantity of matter in a system.

Volume: *An extensive and geometric property having a value characterized by a length times a height times a width, simply "length cubed."*

Volume will be denoted by the letter V and expressed in the units of cubic meters (m^3). Other units which may be used to express volume are liters (l) and cubic centimeters (cm^3 or cc). In the English system the common units for volume are cubic feet (ft^3), gallons (gal), and cubic inches (in^3).

Specific Volume

Many times we will want to know the volume occupied by a unit mass of the system constituent. This is called the *specific volume* and is an intensive property denoted by the letter v. If the matter is homogeneous then

$$v = \frac{V}{m} \tag{2–13}$$

Specific volume will be described with the units of cubic meters per kilogram (m^3/kg) or liters per kilogram (l/kg). In English units specific volume will be described by cubic feet per pounds-mass (ft^3/lbm), cubic feet per slug (ft^3/slug) or gallons per pounds-mass (gal/lbm).

Density

Density is a property used frequently in describing the mass of a system. It is denoted in this book by the Greek letter ρ, and for homogeneous materials we have

$$\rho = \frac{m}{V} \tag{2–14}$$

Note that density is an intensive property and is the inverse of specific volume, that is

$$\rho = \frac{1}{v} \tag{2–15}$$

and will be expressed in the units of kilograms per cubic meter (kg/m^3), kilograms per liter (kg/l), and grams per cubic centimeter (g/cm^3). The English units used to describe density are pounds-mass per cubic feet (lbm/ft^3), slugs per cubic feet (slugs/ft^3), and pounds-mass per cubic inches (lbm/in^3).

Specific Weight

Occasionally a property given by the weight per unit volume is desired. This we call the *specific weight*, denoted by the Greek letter γ. Specific weight for homogeneous materials is calculated from

$$\gamma = \frac{W}{V} \tag{2–16}$$

The units of specific weight are newtons per cubic meter (N/m^3), newtons per liter (N/l), and newtons per cubic centimeter (N/cm^3). In the English unit system the common units for specific weight are pounds-force per cubic foot (lbf/ft^3), pounds-force per cubic inch (lbf/in^3), and pounds-force per gallon (lbf/gal). The numerical value of the specific weight is very nearly equal to the density when using the pound-mass and pound-force units. In this case the specific weight may be accurately computed from the equation

$$\gamma = \frac{g}{g_c}\rho \qquad \textbf{(2–17)}$$

In SI units the specific weight of a material may be computed from the equation

$$\gamma = \rho g \qquad \textbf{(2–18)}$$

Specific Gravity

Liquids are sometimes described by their specific gravity. This quantity, denoted by the symbol sg, is the ratio of the density of the described fluid to the density of water where the water is at 4°C (39.2°F). The density of water at this temperature has been found to be 1000 kg/m^3 (62.43 lbm/ft^3), so the specific gravity may be computed from the equations

$$sg = \frac{\rho}{1000} \qquad (\rho \text{ in kg/m}^3) \qquad \textbf{(2–19)}$$

$$sg = \frac{\rho}{62.43} \qquad (\rho \text{ in lbm/ft}^3) \qquad \textbf{(2–20)}$$

Note that specific gravity is unitless and should have the same value in all unit systems. Table 2–1 lists values of specific gravity for some common liquids.

Table 2–1

Density and Specific Gravity of Some Liquids at 20°C (68°F) and 1.01 bars (14.7 psi)

Liquid	*Density* (ρ, *rho*) kg/m³	lbm/ft³	*Specific Gravity* (sg)
Benzene	879	54.9	0.8790
Carbon tetra-chloride	1,587	99.1	1.5950
Ethyl alcohol	788	49.2	0.7893
Glycerine	1,259	78.6	1.2600
Mercury	13,536	845.0	13.5460
Methyl alcohol	793	49.5	0.7928
Water	998	62.3	0.9982

Pressure

One of the properties which we will use extensively in this text is pressure. For our purposes we will consider pressure to be a force per unit area. Specifically, we will define it as

$$p = \frac{F}{A} \qquad \textbf{(2–21)}$$

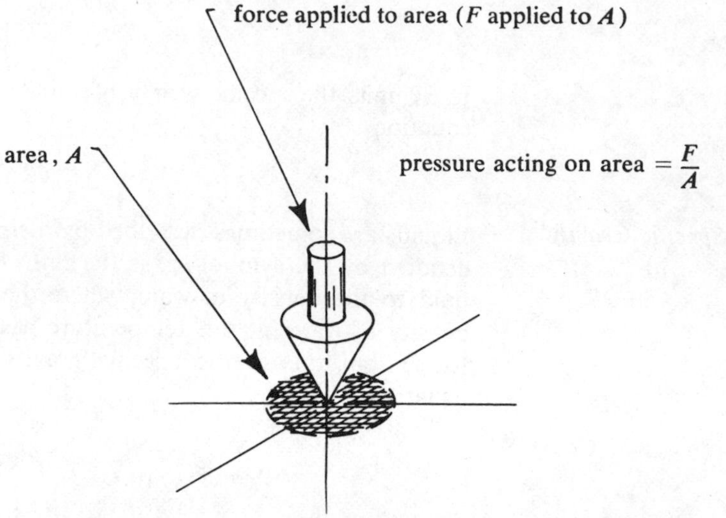

Figure 2–11 Pictorial representation of pressure.

where p denotes the pressure and F is the perpendicular or normal force to the area A as indicated in figure 2–11.

From this equation we can see that the units of pressure must be force per unit of area or newtons per square meter (N/m^2). This unit is called the pascal (Pa), so

$$1 \text{ Pa} = 1 \text{ N/m}^2$$

In much of the published thermodynamic work the unit bar is used, and this is defined as

$$1 \text{ bar} = 1 \times 10^5 \text{ Pa} \tag{2–22}$$

A convenient memory device is to remember that 1 bar is approximately equal to the atmospheric pressure on the earth. Actually, 1.01 bars equal 1 standard atmosphere, or 1 atm. In English units pressure is described by lbf/in^2 (psi), lbf/ft^2, inches of mercury (in Hg), and feet of water (ft H_2O).

From an important physical principle known as Pascal's law it is known that in gases or liquids a pressure at one point in one direction (such as pressure p_1 in figure 2–12) induces a pressure in all directions at that same point as indicated in figure 2–12. We can also note that the above definition for pressure corresponds to a definition for stress in solid materials. There are, however, conditions which make Pascal's law for solids much more complicated than for liquids and gases.

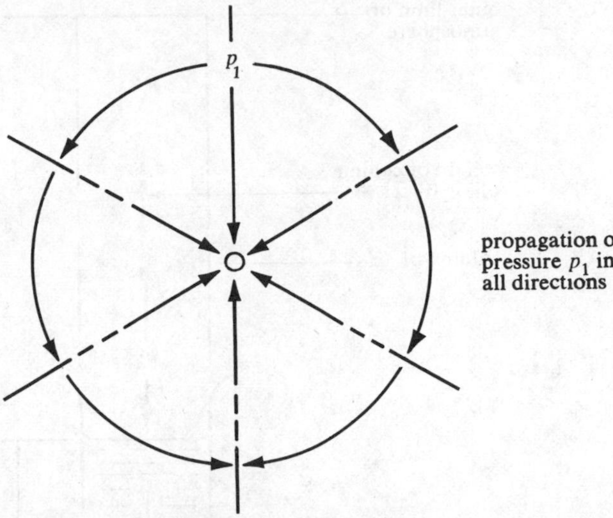

Figure 2–12 Pascal's law for liquids and gases.

To understand how pressure can differ within a system, let's look at a vertical tube of cross-sectional area A, and height (*l*), shown in figure 2–13. The tube is filled with water at 4°C (39.2°F) and the pressure at the top of the tube is observed to be the weight of a volume of air (the atmosphere) divided by A. This pressure we call atmospheric pressure p_a, and its value is near 1.01 bars or 14.7 lbf/in^2 on the surface of the earth. Table B.16 gives the U.S. standard day values of atmospheric pressure for various locations on or near the earth. Standard day is defined as annual averages at 45° latitude within the U.S. (probably never realized exactly).

At the bottom of the water column now, we see that the pressure p_b is the pressure due to the weight of the water column W_w plus p_a.

$$p_b = \frac{W_w}{A} + p_a$$

We could also substitute for W_w

$$W_w = \gamma_w \times l \times A$$

where γ_w = specific weight of water, yielding

$$p_b = \gamma_w l + p_a$$

Additionally, if we look at some other point in the water level, say point C at x in figure 2–13, the pressure will be

$$p_C = \gamma_w x + p_a$$

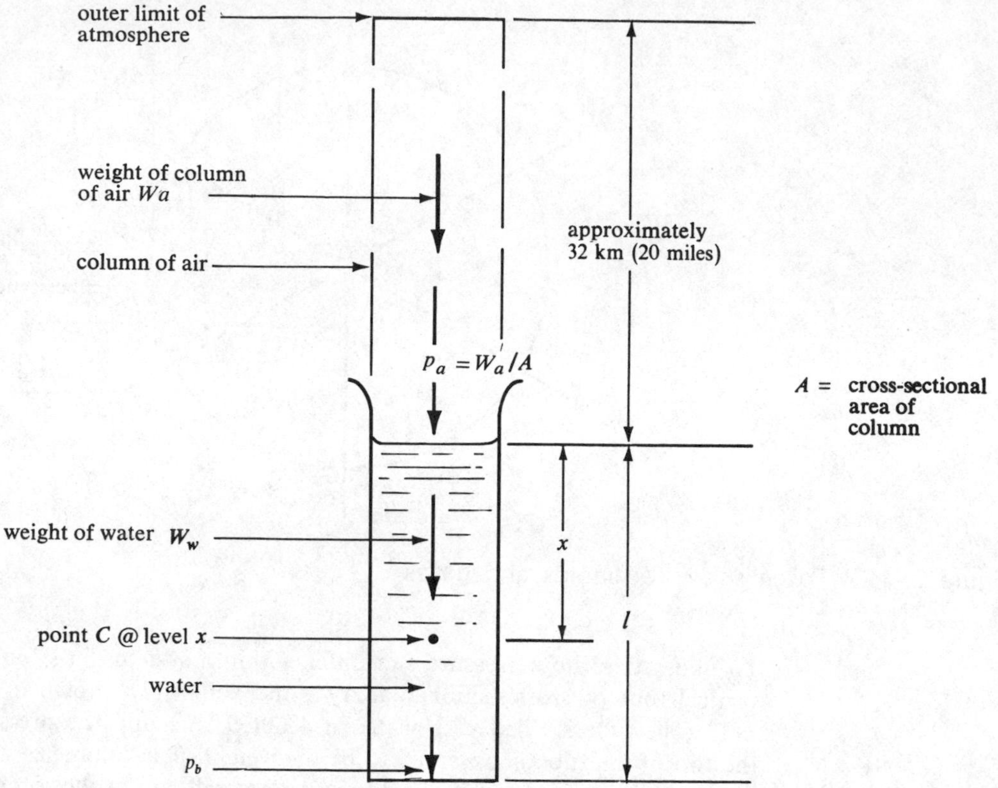

Figure 2–13 Pressures acting on water in tube.

Interestingly, a consequence of Pascal's law is that the pressure in a gas or liquid is dependent only on the vertical depth (x in figure 2–13) called the *pressure head* and the *atmospheric pressure*. In figure 2–14, the pressure at level (1) is the same in all the columns—A, B, C, and D.

$$p_1 = \gamma_w x_1 + p_a$$

Similarly, at level (2)

$$p_2 = \gamma_w x_2 + p_a \text{ for all four columns.}$$

For systems which contain gases, we will generally neglect the pressure changes due to elevation differences because the specific weight times the elevation change ($\gamma \times h$ term) will be several orders of magnitude less than the system pressure at the top; that is, we will say

$$p_1 = p_2 = \text{constant as identified in the system of figure 2–15}$$

There are various schemes for measuring pressure of a system, but we will look only at the U-tube manometer since it is simple and contains the

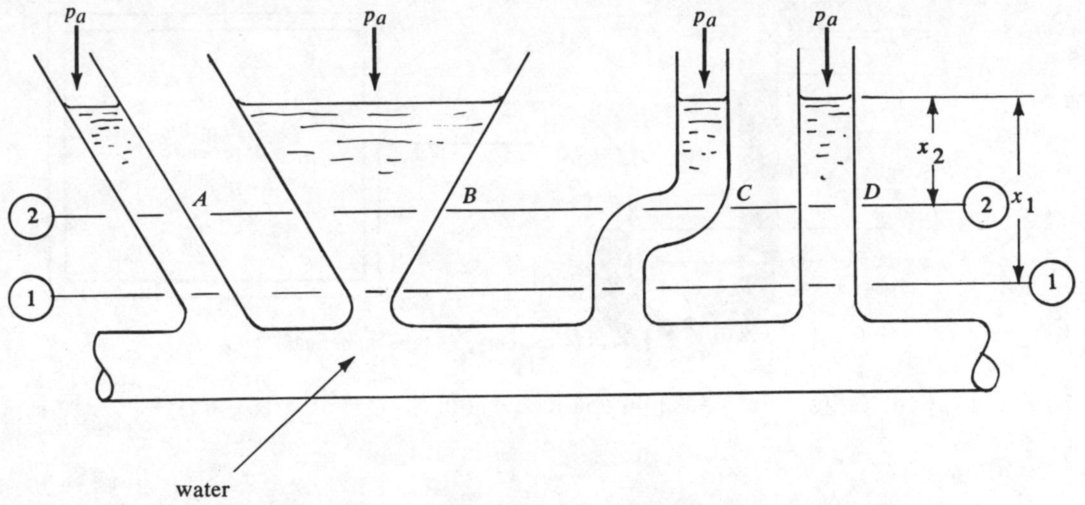

Figure 2–14 Illustration of complete dependence of pressure to elevation.

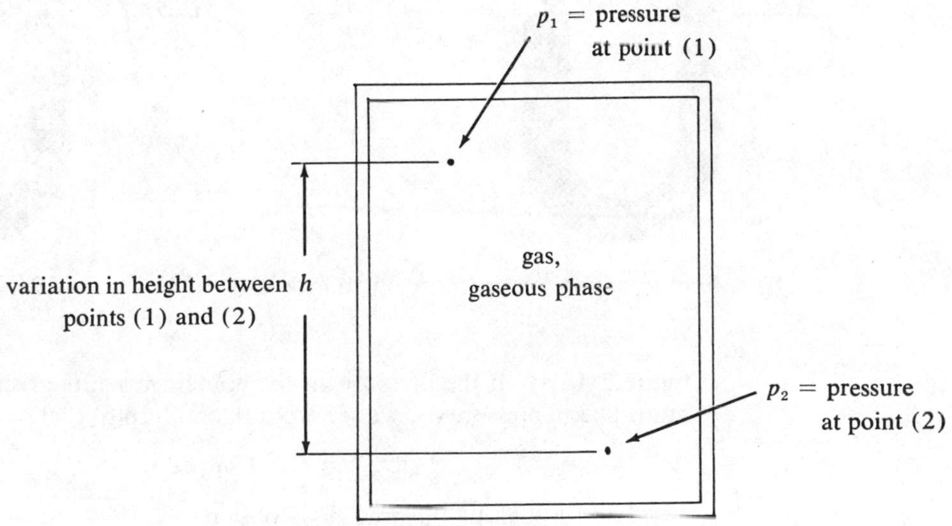

Figure 2–15 Illustration of independence of pressure with elevation in gaseous systems.

essence of all pressure-measuring devices. It is merely a glass or transparent tube having a constant cross section area A and formed into a "U" as shown in figure 2–16(a). This tube can contain water or any other liquid, but more commonly mercury (Hg) is used. Now, let's use this gage to measure the pressure inside a closed container (system B in

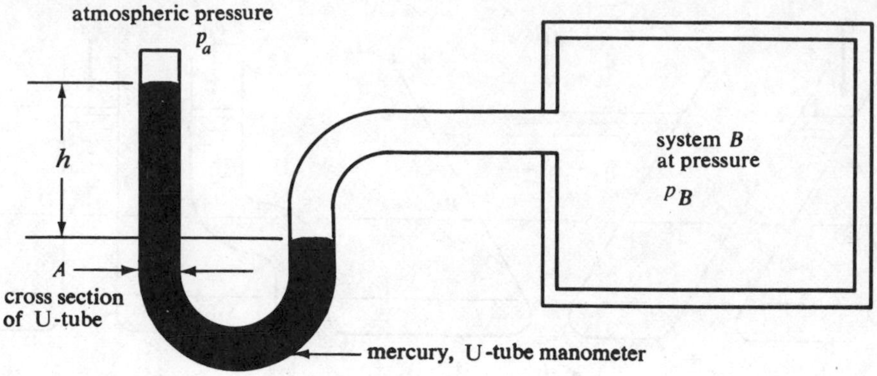

Figure 2–16(a) Pressure measurement of system.

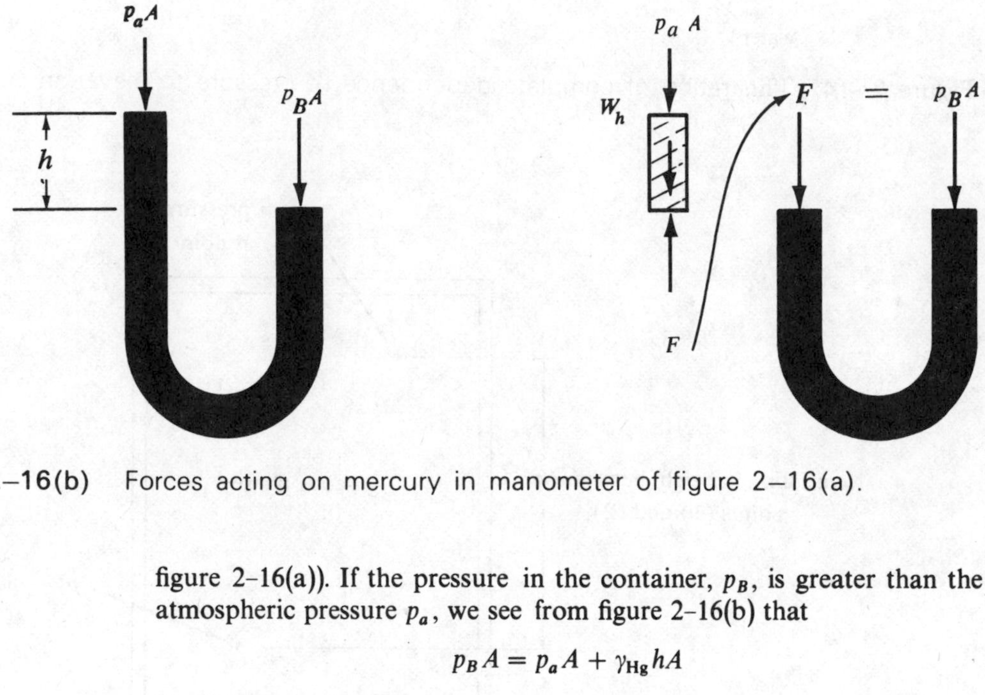

Figure 2–16(b) Forces acting on mercury in manometer of figure 2–16(a).

figure 2–16(a)). If the pressure in the container, p_B, is greater than the atmospheric pressure p_a, we see from figure 2–16(b) that

$$p_B A = p_a A + \gamma_{Hg} h A$$

Dividing through by the area A we obtain

$$p_B = p_a + \gamma_{Hg} h$$

Notice that we are physically determining the term h in this measurement, but we will call the term γh the gage pressure, p_g, so that

$$p = \text{pressure of a system}$$

or

$$p = p_g + p_a \tag{2–23}$$

where

$$p_g = \gamma h \tag{2–24}$$

The term h is described in millimeters (mm) of mercury (Hg). The specific weight of mercury can be computed from equation (2–18) and by using the density of mercury from table 2-1. Thus,

$$\gamma = \rho g$$
$$= (13{,}536 \text{ kg/m}^3)(9.8 \text{ m/s}^2)$$
$$= 132{,}650 \text{ N/m}^3$$

If h is arbitrarily set at 1 mm Hg, then from equation (2–24) we obtain, where 1 mm = 1×10^{-3} m

$$p = (132{,}650 \text{ N/m}^3)(1 \times 10^{-3} \text{ m})$$
$$= 132.65 \text{ N/m}^2$$
$$= 0.00132 \text{ bar}$$

Clearly, we have a conversion factor for pressure given by

$$1 \text{ mm Hg} = 0.00132 \text{ bar}$$
$$= 1.32 \times 10^{-3} \text{ bar} \tag{2–25}$$

Similarly, in the English system the gage pressure is commonly measured or read in inches of mercury (in Hg). It can be shown that this is equivalent to 0.49 psi. Thus

$$1 \text{ in Hg} = 0.49 \text{ psi} \tag{2–26}$$

Example 2.4 The water tank shown in figure 2–17 has a pressure gage which reads 10 mm Hg. What is the pressure in the tank in bars if the atmospheric pressure is 1.01 bars?

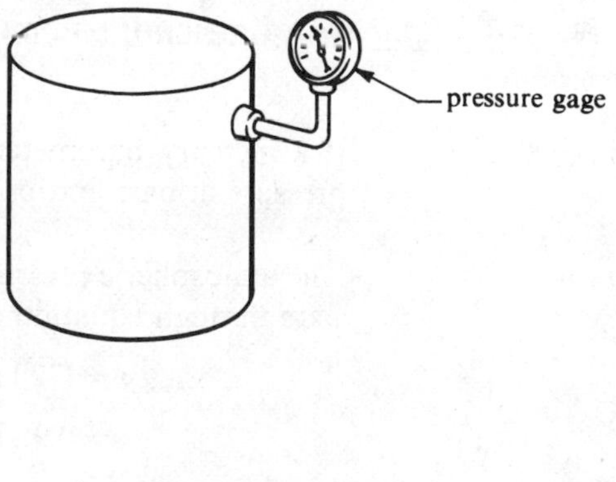

Figure 2–17 Water pressure tank.

Solution The pressure indicated by the gage is the gage pressure, 10 mm Hg. We may readily convert to the units of bars by using equation (2–25). Thus

$$p_g = (10 \text{ mm Hg})(1.32 \times 10^{-3} \text{ bar/mm Hg})$$
$$= 0.0132 \text{ bar}$$

Using equation (2–23) we then obtain

$$p = 0.0132 \text{ bar} + 1.01 \text{ bars}$$
$$= 1.0232 \text{ bars}$$ *Answer*

We see that the pressure of a system may be found by adding the atmospheric pressure to the gage pressure as indicated by equation (2–23). The atmospheric pressure is measured using a barometer, a device specifically designed for this measurement. There are various types of barometers, but the mercury barometer, shown schematically in figure 2–18, demonstrates most clearly the principle of operation of all barometers. A glass column, which is as complete a vacuum as possible, is immersed in a shallow pan of mercury. The mercury will then rise in the glass column due to the atmospheric pressure acting on the pan, and this height will indicate the pressure of the atmosphere. The height of the mercury column is commonly around 760 mm Hg or 29.8 in Hg on the surface of the earth. Table B.16 lists values for the earth's atmosphere at various elevations or altitudes.

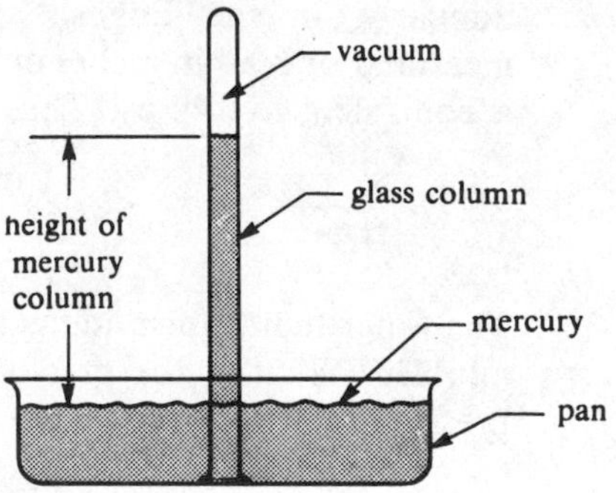

Figure 2–18 Principle of mercury barometer.

Example 2.5 If a mercury barometer reads 720 mm Hg, determine the atmospheric pressure in bars and psi.

Solution The atmospheric pressure is given as 720 mm Hg, so we may convert to bars through equation (2–25).

$$p = (720 \text{ mm Hg})(1.32 \times 10^{-3} \text{ bar/mm Hg})$$
$$= 0.95 \text{ bar}$$ *Answer*

In pressure units of psi we may use the conversion factor listed in Table B.15, 1 psi = 0.06895 bar. Thus

$$p = (0.95\ \text{bar})\left(\frac{1\ \text{psi}}{0.06895\ \text{bar}}\right)$$

$$= 13.8\ \text{psi} \qquad\qquad \textit{Answer}$$

Example 2.6 Two pipes *A* and *B* are used to convey water and air respectively as shown in figure 2–19. Pressure gage *A* indicates a pressure of 1.01 psig (pounds per square inch gage) in the water-filled pipe. If the atmospheric pressure is 14.69 psi, find

(a) Absolute pressure at center of pipe *A*.

(b) Absolute pressure in air-filled pipe *B*.

(c) Gage pressure measured by pressure gage *B*.

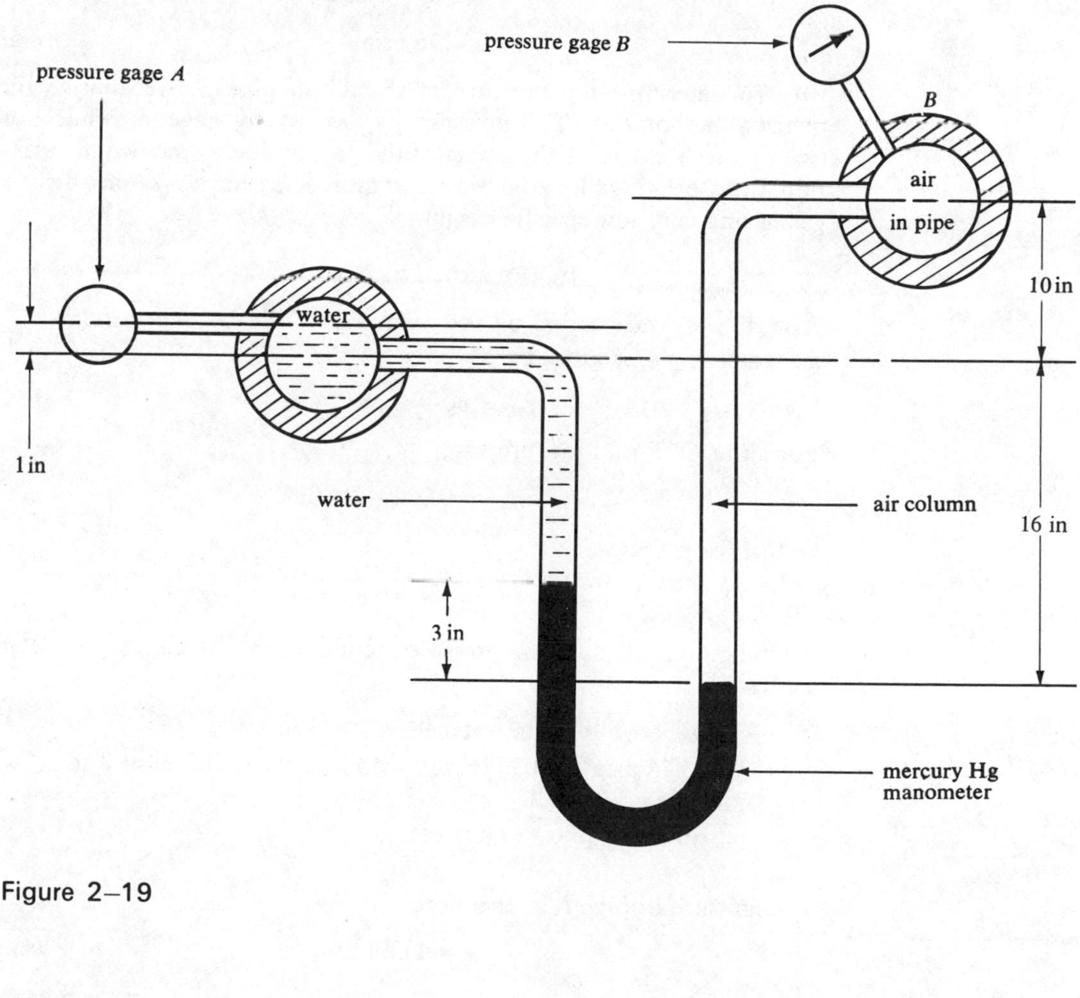

Figure 2–19

Solution (a) Pressure gage A (which may be a mercury U-tube manometer) is located 1 inch above the center of the pipe. The absolute pressure at the gage location is, from equation (2–23)

$$p = p_g + 14.69 \text{ psi} = 1.01 \text{ psig} + 14.69 \text{ psi}$$

or

$$p = 15.70 \text{ psia}$$

Corrected to the center of the pipe A, the pressure is, from equation (2–22)

$$p_A = 15.70 \text{ psia} + \gamma_w \times 1 \text{ in}$$

and where γ_w is the specific weight of the water in pipe A. We have

$$\gamma_w = 62.4 \text{ lbf/ft}^3 \cong 0.036 \text{ lbf/in}^3$$

so

$$p_A = 15.70 + 0.036 \times 1$$
$$= 15.736 \text{ psia}$$ *Answer*

(b) To determine the pressure of the air in pipe B, we analyze the mercury manometer. The pressure p_B, sensed by gage B, while not exactly the pressure at the center of the pipe B due to the height variation, is close enough for engineering or most scientific work since air is a gas having very low specific weight:

$$p_B = \text{constant in pipe } B$$

Also, the pressure acting on the right side of the U-tube mercury manometer, p'_B, is given by

$$p'_B = p_B + \gamma_{\text{air}} \times 26 \text{ in}$$

For air at 68°F and low pressure

$$\gamma_{\text{air}} \approx 0.00004 \text{ lbf/in}^3$$

so that for our case

$$p'_B \approx p_B$$

From figure 2–20, showing pressures acting on the mercury manometer, we have

$$\begin{aligned} p'_B &= p_A + \gamma_w \times 13 \text{ in water} + \gamma_{\text{Hg}} \times 3 \text{ in Hg} \\ &= 15.74 \text{ psia} + 0.036 \text{ lbf/in}^3 \times 13 \text{ in} + 0.491 \text{ lbf/in}^3 \times 3 \text{ in} \\ &= 15.74 + 0.47 + 1.47 \\ &= 17.68 \text{ psia} \end{aligned}$$

The pressure in pipe B is, therefore

$$p_B = 17.68 \text{ psia}$$ *Answer*

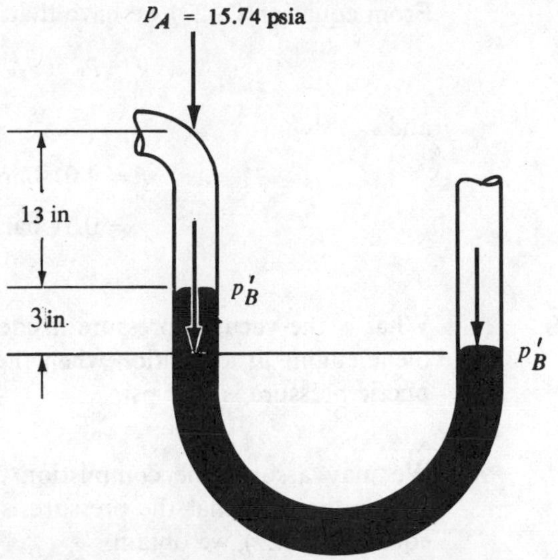

Figure 2–20 Pressures acting on mercury manometer in example 2.6.

(c) The gage pressure in pipe B must be, from equation (2–23)

$$p = p_B - 14.69 = 2.99 \text{ psig}$$ *Answer*

If the system pressure, p, is less than p_a, then we have a vacuum. Referring to figure 2–16, we would then physically have a case of p_B and p_A interchanged so that

$$p_a A = p_B A + \gamma_{Hg} h A$$

or

$$p_a = p + \gamma_{Hg} h = p + p_{gv}$$

The pressure in a vacuum chamber is then written

$$p = p_a - p_{gv} \tag{2–27}$$

where p_{gv} is the vacuum gage pressure. Notice that this term, p_{gv}, can be no more than p_a; if $p = 0$, then $p_a = p_{gv}$ atmospheric pressure and p cannot be less than zero.

Remember that the gage or vacuum gage pressure is the quantity measured in all pressure gages. The atmospheric pressure must be included as shown by equations (2–23) or (2–27) to determine the absolute or true pressure of the system. In these equations, the system pressure is calculated at the interface with the mercury or gage fluid. If the system contains a liquid then the pressure, of course, varies inside the system as predicted by its height (2–22).

Example 2.7 The steam leaving a turbine is at a pressure of 0.7 bar vacuum. Determine the absolute pressure if the atmospheric pressure is 1.01 bars.

Solution

From equation (2–27) we have that

$$p = p_a - p_{gv}$$

and

$$p = 1.01 \text{ bars} - 0.7 \text{ bar}$$
$$= 0.31 \text{ bar}$$ *Answer*

Example 2.8

What is the vacuum pressure inside the combustion chamber of a gasolene engine at a position when the pressure is 14.2 psia? The atmospheric pressure is 14.7 psi.

Solution

We may assume the combustion chamber to be a vacuum chamber during the time that the pressure is less than atmospheric. Then, using equation (2–27), we obtain

$$p_{gv} = p_a - p$$
$$= 14.7 \text{ psi} - 14.2 \text{ psia}$$
$$= 0.4 \text{ psi vacuum}$$ *Answer*

2.9 Equilibrium and the Zeroth Law of Thermodynamics

In mechanics we utilize equilibrium to determine forces acting on bodies. This type of equilibrium we call *mechanical equilibrium*, and it is indeed an important part of our scientific efforts. In thermodynamics we will be concerned with *thermal equilibrium*. The property which determines this type of equilibrium we call *temperature* and we postulate the following:

Zeroth law of thermodynamics: *Two separate bodies which are in thermal equilibrium with a third body are also in thermal equilibrium with each other.*

This law tells us that we can measure temperature through thermal equilibrium of bodies and be assured that it is independent of the materials involved.

Remember that if two separate bodies at different temperatures are brought into contact with each other, thermal equilibrium will be reached and retained when the temperature is the same in both bodies. Interestingly, temperature equilizes in this condition, but the energies of the two bodies do not necessarily equalize.

2.10 Temperature and Thermometers

When the temperature is determined for a system (such as the air temperature in a room), the characteristic method involves allowing a temperature meter (thermometer) to come to thermal equilibrium with the observed system and subsequently measuring the change in some property of the thermometer which we will call the thermometric property. Table 2–2 tabulates a few of the more prevalent types of thermometers with their associated thermometric properties.

Table 2–2

Thermometric Properties for Temperature Measurement

Type of Thermometer	*Thermometric Property*
Mercury-in-glass	Volume
Thermistor	Electrical resistance
Thermocouple	Voltage
Constant volume gas	Pressure of gas
Constant pressure gas	Volume of gas

The mercury-in-glass is the most common, as it is used to measure human body temperatures, normal atmospheric temperatures, and temperatures of objects used in everyday activities. The temperature is recorded in degrees Fahrenheit (°F), which is the English unit, or degrees Celsius (°C), the SI unit. These scales are arbitrary, but are both characterized by being defined at the boiling and freezing (or melting) points of pure water at a pressure of 1.01 bars. These are as follows: Boiling point is 100°C which equals 212°F, and the freezing point is 0°C which equals 32°F. The setting t_F is the temperature in degrees Fahrenheit, and t_C is the temperature in degrees Celsius.

We see by plotting t_F vs. t_C as shown in figure 2–21 that

$$t_F = \frac{180°F}{100°F} t_C + 32°$$

$$= \frac{9}{5} t_C + 32°$$

or

$$t_C = \frac{5}{9}(t_F - 32°) \tag{2–28}$$

If we were to correlate the two mechanical properties, pressure and volume, of a material with the thermal property of temperature, we would find that for many vapors or gases, the equation

$$pV = mRT \tag{2–29}$$

is very descriptive of the material. This equation is called the *perfect gas equation* and is the simplest example of an *equation of state.* Other

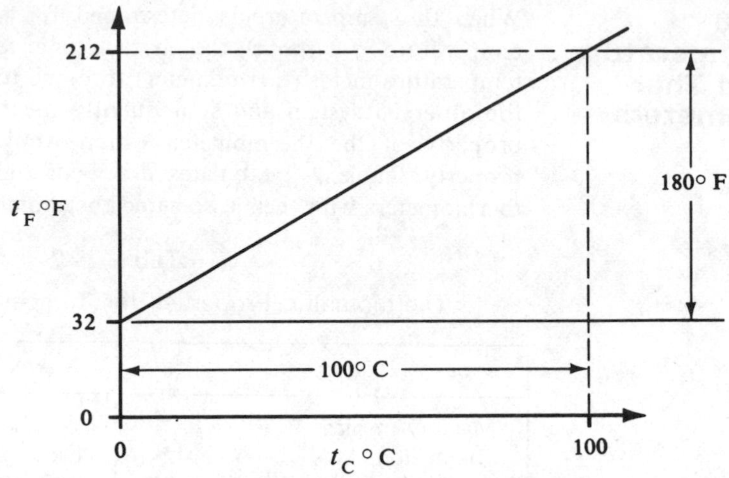

Figure 2–21 Relation of Celsius and Fahrenheit temperature scales.

examples of equations of state and a further discussion of the perfect gas will be given in chapter 5. At this point, however, note that if we fill a rigid chamber (constant volume container) with a perfect gas, following equation (2–29), and if we seal the chamber except for a manometer tube as shown in figure 2–22, we have what is called a gas thermometer. For this apparatus we then apply equation (2–29)

$$pV = mRT$$

Solving for the pressure we obtain

$$p = \frac{m}{V} RT = \rho RT \tag{2–30}$$

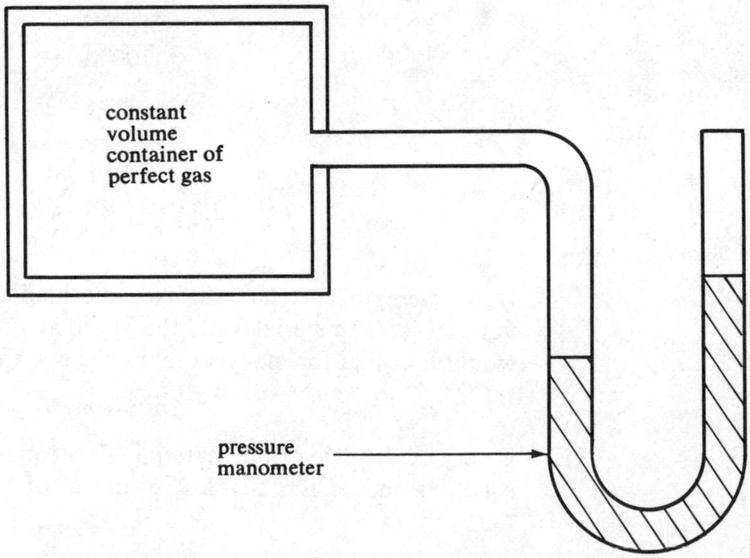

Figure 2–22 Gas thermometer.

where ρ is the density defined in equation (2–14). But density and the gas constant R are constant so that we may write

$$p = (\text{constant})T \qquad \textbf{(2–31)}$$

Suppose now we use the apparatus of figure 2–22 to measure temperature-pressure relationships of the three gases, A, B, and C. We could easily generate pressure-temperature data points shown in figure 2–23. Some of the data points fit straight lines, and the point where these lines intersect with the vertical axis (point Z in figure 2–23) we define as the *absolute zero point* of temperature. If the temperature scale in figure 2–23 had been in degrees Fahrenheit, we would have found

$$Z = -459.4°\text{F}$$

Similarly, if we had used degrees Celsius, then

$$Z = -273°\text{C}$$

From these values of absolute zero, we then say the temperature of a body, T, is

$$T = (t_F + 459.4) \text{ degrees Rankine (°R)} \qquad \textbf{(2–32)}$$

or

$$T = (t_C + 273) \text{ kelvin} \qquad \textbf{(2–33)}$$

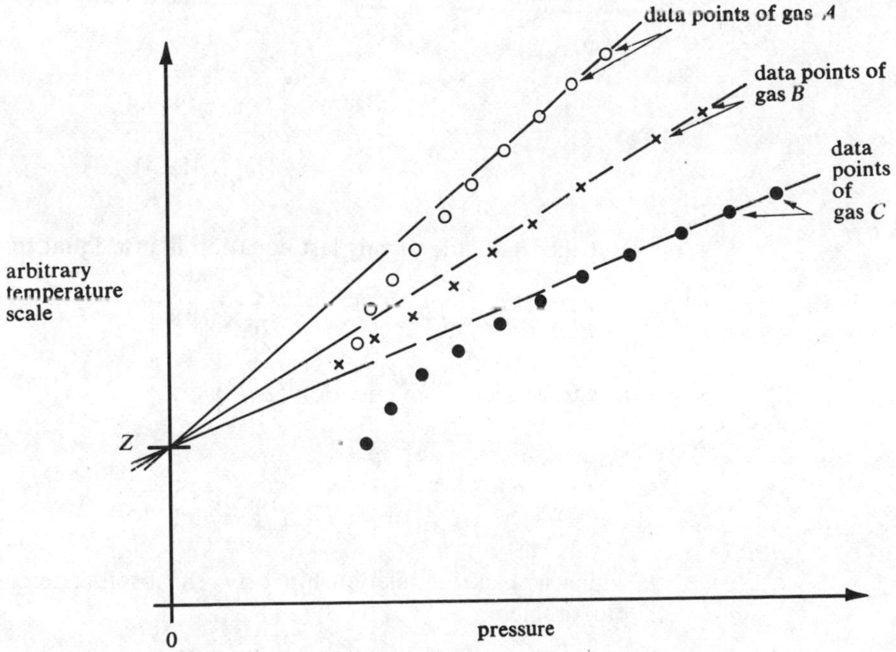

Figure 2–23 A method for determining absolute zero temperature.

The reason some of the data points at the lower temperatures do not align with the appropriate straight lines is that all gases depart from a perfect gas at the lower temperatures. Thus equation (2–30) is not a correct description of the physical situation at a low temperature but is correct at higher temperatures.

Example 2.9 A thermodynamic system composed of liquid nitrogen is at 72 kelvin. Convert this to degrees Celsius, Fahrenheit, and Rankine.

Solution From equation (2–33), we have

$$t_C = T - 273 \text{ K}$$
$$= 72 - 273$$

So

$$t_C = -201°\text{C} \qquad \textit{Answer}$$

Now t_C can be easily converted to degrees Fahrenheit by using equation (2–28). Let us, however, approach this problem somewhat differently and, consequently, observe an interesting relationship. Substituting equation (2–28) into equation (2–33) yields

$$T \text{ K} = \frac{5}{9}(t_F - 32) + 273$$

or

$$T \text{ K} = \frac{5}{9} t_F + \frac{5}{9}\left(\frac{9}{5} 273 - 32\right)$$

then,

$$T \text{ K} = \frac{5}{9} t_F + \frac{5}{9}(459.4)$$
$$= \frac{5}{9}(t_F + 459.4)$$

But the right side of this last equation is just equal to

$$\frac{5}{9} \times T°\text{R}$$

as can be seen from equation (2–32).

So, we have

$$T \text{ K} = \frac{5}{9} T°\text{R} \qquad \textbf{(2–34)}$$

which is a useful relationship between absolute temperature scales. In our problem

$$T \text{ K} = 72 \text{ K}$$

so, from equation (2–34)

$$72\ \text{K} = \frac{5}{9} T°\text{R}$$

which gives us

$$T°\text{R} = \frac{9}{5} \times 72\ \text{K} = 129.6°\text{R} \qquad \textit{Answer}$$

To find the Fahrenheit temperature, we use equation (2–32) and solve algebraically for t_F:

$$t_F = T°\text{R} - 459.4$$

Substituting for T we have

$$t_F = 129.6 - 459.4$$

or

$$t_F = -329.8°\text{F} \qquad \textit{Answer}$$

2.11 Energy

In chapter 1 we defined energy as a certain capacity of a body; that is, energy is a property of the system or body in which it resides. At this point, let us seek a better grasp of energy in quantitive terms. There are many forms of energy, but we will be primarily concerned with *kinetic*, *potential*, and *thermal*, or *internal energy*.

Kinetic Energy

A body or particle has energy by virtue of its movement. As an example, a stone lying on the ground is motionless and incapable of effecting change. Pick the stone up, however, and throw it against a window. This will produce a change directly attributed to the motion of the stone. This energy is called *kinetic energy* (KE) and is given by

$$\text{KE} = \frac{1}{2} m\bar{V}^2 \qquad \textbf{(2–35)}$$

where $\bar{V}$ is the velocity of the stone. If we substitute equation (2–5) into this result, we obtain

$$\text{KE} = \frac{1}{2} \frac{W}{g} \bar{V}^2 \qquad \textbf{(2–36)}$$

from which we can see that the units are newton-meters (N · m). A newton-meter is defined as the joule (J). In the English system the units for energy are foot-pounds (ft-lbf) or British thermal units (Btu).

The specific kinetic energy or kinetic energy per unit mass is given by

$$\text{ke} = \frac{\text{KE}}{m} \qquad \textbf{(2–37)}$$

and has the units of N · m/kg or J/kg. Equation (2–37) may be combined with equation (2–35) to give

$$ke = \frac{1}{2}\bar{V}^2 \text{ (J/kg)} \tag{2–38}$$

In the English system one must use equation (2–7) in equation (2–36) to obtain

$$KE = \frac{m}{2g_c}\bar{V}^2 \text{ (ft-lbf)} \tag{2–39}$$

Further, the specific kinetic energy is then given by the relationship

$$ke = \frac{1}{2g_c}\bar{V}^2 \text{ (ft-lbf/lbm)} \tag{2–40}$$

Example 2.10

A stream of water coming from a hose is found to have a velocity of 30 m/s. Determine the kinetic energy per kilogram of this water.

Solution

The kinetic energy per kilogram is computed from equation (2–38) as

$$ke = \frac{1}{2}\bar{V}^2$$

where the velocity $\bar{V}$ is 30 m/s for the water. Then

$$ke = \frac{1}{2}(30 \text{ m/s})^2$$

$$= 450 \text{ J/kg}$$ *Answer*

Example 2.11

A piston of an internal combustion engine is found to have a velocity of 180 ft/s at a particular instant in its travel. If the piston weighs 2 lbf, determine its KE and ke. Use the value of 32.2 ft-lbm/lbf · s² for g_c and 32.2 ft/s² for g.

Solution

From equation (2–40) the kinetic energy per pound-mass is readily computed.

$$ke = \frac{(180 \text{ ft/s})^2}{(2)(32.2 \text{ ft-lbm/lbf} \cdot \text{s}^2)}$$

$$= 503.1 \text{ ft-lbf/lbm}$$ *Answer*

The kinetic energy is then easily found from equation (2–36) as

$$KE = \frac{(2 \text{ lbf})(180 \text{ ft/s})^2}{(2)(32.2 \text{ ft/s}^2)}$$

$$= 1006 \text{ ft-lbf}$$ *Answer*

Potential Energy

Objects tend to attract each other because they all have mass. This attracting force due to mass is the essence of Newton's law of gravitation.

Here on the earth the attraction between the earth and a system or an object represents a "potential" of motion. That is, the earth and the object are tending to move closer together; we call this capability *total potential energy*, denote it by PE, and calculate it from

$$\text{PE} = Wz \tag{2-41}$$

where z is a vertical reference distance between the center of the earth and the center of the object which we say has the potential energy. The units of potential energy are easily seen to be foot-pounds and joules. The potential energy per unit of mass, or the specific potential energy, is given by the relationship

$$\text{pe} = \frac{\text{PE}}{m} = \frac{Wz}{m} \tag{2-42}$$

The units of the specific potential energy are joules per kilogram (J/kg) or foot-pounds per pounds-mass (ft-lbf/lbm). Additionally, equation (2–40) can be simplified by substituting equation (2–5). Thus

$$\text{pe} = gz \text{ (J/kg)} \tag{2-43}$$

Similarly, for English units we can substitute equation (2–7) into (2–40) to obtain

$$\text{pe} = gz/g_c \text{ ft-lbf/lbm} \tag{2-44}$$

Example 2.12

Water dropping over a 300-m fall loses potential energy. If we say that the potential energy is zero at the bottom of the fall, what is the water's potential energy per kilogram at the top?

Solution

From equation (2–42) we obtain

$$\text{pe} = (9.8 \text{ m/s}^2)(300 \text{ m})$$

$$= 2940 \text{ J/kg}$$ *Answer*

Example 2.13

A 5-ton elevator rises 10 stories, where each story is 10 feet high. What increase in potential energy does the elevator have in this motion upward?

Solution

A 5-ton elevator weighs 5 tons × 2000 lbf/ton, or 10,000 lbf. The increase in potential energy is found from equation (2–40) to be

$$\text{PE} = (10{,}000 \text{ lbf})(10 \text{ stories} \times 10 \text{ ft})$$

$$= 1 \times 10^6 \text{ ft-lbf}$$ *Answer*

Internal Energy

The hotness or coldness of a body can physically affect the surroundings of that body. This capacity, sometimes called *heat* or *thermal energy*, is quite naturally indicated by the temperature of the body. In this text we will call this property *total internal energy* and designate it by U. Internal energy will thus be the measure of the "hotness" of a body; "coldness" will represent the absence of internal energy. We will not, under any circumstance, use the word *heat* to describe this property but may on occasion use the term *thermal energy*. *Heat* will be assigned a much different meaning in the next chapter.

Internal energy, being a form of energy, has the same units associated with it as the other forms of energy, kinetic and potential. Thus we will use joules, foot pounds, and British thermal units, to describe internal energy. The specific internal energy u is given by the equation

$$u = \frac{U}{m} \tag{2-45}$$

and has the units of joules per kilograms, foot-pounds per pounds-mass, and British thermal units per pounds-mass.

Internal energy is frequently described as the property which reflects mechanical energy of the molecules and atoms of the material. Generally the contributions to internal energy are as follows:

1. Translational or kinetic energy of the atoms or molecules as indicated in figure 2–24(a).
2. Vibrational energy of the individual molecules due to straining of the atomic bonds at increasing temperatures (figure 2–24(b)).
3. Rotational energy of those molecules which spin about an axis as shown in figure 2–24(c).

There are other forms of energy, such as electromagnetic energy, chemical energy, and strain energy (caused by the stretching of solid materials), which must be included in any complete analysis of a thermodynamic problem; but in this text we will normally consider the total energy E of a system to be given by

$$E = \text{KE} + \text{PE} + U \tag{2-46}$$

and the energy e,

$$e = \text{ke} + \text{pe} + u \tag{2-47}$$

If the above equations are substituted into the appropriate terms of equation (2–47) we have

$$e = \frac{1}{2}\bar{V}^2 + gz + u \ \text{(J/kg)} \tag{2-48}$$

$$= \frac{\bar{V}^2}{2g_c} + \frac{g}{g_c}z + u \ \text{(ft-lbf/lbm)} \tag{2-49}$$

Example 2.14

A 2-ton automobile travels at 50 mph on a highway 1000 feet above sea level. Determine the total mechanical energy of the auto with respect to sea level and the energy per unit mass of the auto.

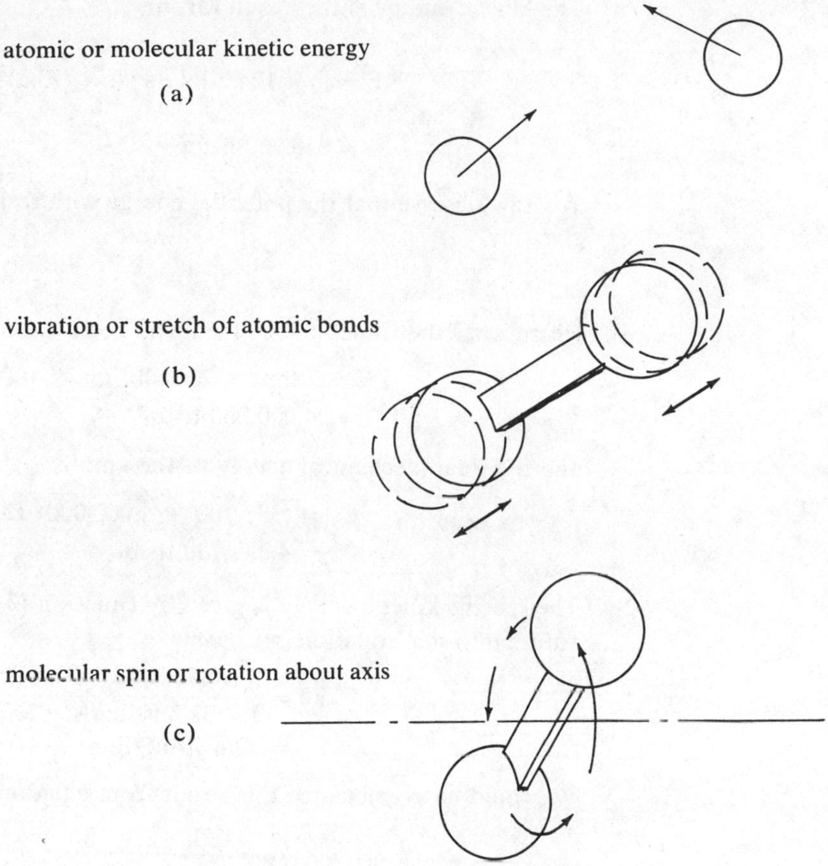

Figure 2–24 Molecular motions contributing to internal energy.

Solution The total mechanical energy is the total energy minus the internal energy, denoted as E_m. It is found from the equation

$$E_m = \text{KE} + \text{PE}$$

For the kinetic energy we have

$$\text{KE} = \frac{1}{2} m\overline{V}^2$$

and the mass is found from

$$m = \frac{2 \text{ tons} \times 2000 \text{ lbf/ton}}{32.2 \text{ ft/s}^2}$$

$$= 124.2 \text{ lbf} \cdot \text{s}^2/\text{ft}$$

$$= 124.2 \text{ slugs}$$

The velocity is

$$\overline{V} = 50 \text{ mph} \times 5280 \text{ ft/mile} \times 1 \text{ hr/3600 s}$$
$$= 73.3 \text{ ft/s}$$

The kinetic energy is then found from

$$KE = \frac{1}{2} \times 124.2 \text{ lbf} \cdot \text{s}^2/\text{ft} \times (73.3)^2 \text{ ft}^2/\text{s}^2$$
$$= 333{,}700 \text{ ft-lbf}$$

We now observe that the potential energy with respect to the sea level is given by

$$PE = Wz$$

where z has the value of 1000 feet. The potential energy is then equal to

$$PE = 2 \text{ tons} \times 2000 \text{ lbf/ton} \times 1000 \text{ ft}$$
$$= 4{,}000{,}000 \text{ ft-lbf}$$

and the total mechanical energy of the auto is

$$E_m = (333{,}700 + 4{,}000{,}000) \text{ ft-lbf}$$
$$= 4{,}333{,}700 \text{ ft-lbf} \qquad \textit{Answer}$$

The specific kinetic energy is given by equation (2–40), and substituting values into this equation, we have

$$ke = \frac{(73.3)^2 \text{ ft}^2/\text{s}^2}{2 \times 32.2 \text{ ft-lbm/s}^2 \cdot \text{lbf}}$$
$$= 83.4 \text{ ft-lbf/lbm}$$

We could have calculated this result from equation (2–37), which is

$$ke = \frac{KE}{m}$$
$$= \frac{333{,}700 \text{ ft-lbf}}{124.2 \text{ slugs}}$$
$$= 2687 \text{ ft-lbf/slug}$$
$$= \frac{2687 \text{ ft-lbf/slug}}{32.2 \text{ lbm/slug}}$$
$$= 83.4 \text{ ft-lbf/lbm}$$

which agrees with our previous result. Similarly, for potential energy we have

$$pe = \frac{g}{g_0} z$$
$$= \frac{32.2 \text{ ft/s}^2}{32.2 \text{ ft-lbm/s}^2 \cdot \text{lbf}} \times 1000 \text{ ft}$$
$$= 1000 \text{ ft-lbf/lbm}$$

and, again, an alternate method of solution is

$$pe = \frac{PE}{m}$$
$$= \frac{4{,}000{,}000 \text{ ft-lbf}}{124.2 \text{ slugs} \times 32.2 \text{ lbm/slug}}$$
$$= 1000 \text{ ft-lbf/lbm}$$

Finally, the specific mechanical energy for the automobile is

$$\frac{E_m}{m} = 83.4 \text{ ft-lbf} + 1000 \text{ ft-lbf/lbm}$$

$$= 1083.4 \text{ ft-lbf/lbm} \qquad \textit{Answer}$$

In this example problem let us observe an important principle. Suppose the highway on which the auto was traveling sloped down to sea level and the brakes were not applied in the auto as it moved down the highway. What would happen in this condition? The auto would lose its potential energy as it rolled down the highway, but it would increase its velocity and thereby increase its kinetic energy. The point is that the total mechanical energy of the auto remains constant, but the potential energy is converted into kinetic energy. We will see in chapter 4 that energy is conserved in any process; this principle is the first law of thermodynamics.

2.12 Units

In sections 2.7, 2.8, 2.10, and 2.11, important thermodynamic properties were introduced, and invariably the unit used to describe the given property was stated. The two systems of units used in these discussions were the Système International (SI) and the English. Both of these unit systems are composed of four fundamental dimensions: length (L), mass (m), time (τ), and temperature (t and T). All other units are then combinations of these four basic units. In table 2–3 are listed the common units for each of the fundamental dimensions and force and energy.

Table 2–3

Units and Symbols

Quantity	*Système International*	*English*
Length	meter (m)	foot (ft)
Mass	kilogram (kg)	pound-mass (lbm)
Force	newton (N)	pound-force (lbf)
Time	second (s)	second (s)
Temperature	K or °C	°R or °F
Energy	joule (J)	foot-pound (ft-lbf)

A number of the important derived quantities used in thermodynamics are listed in table 2–4. It is strongly suggested that problems be solved by including all units and replacing the special units by their derivation. As an example, the force unit of newtons is equivalent to the units of kilogram-meters per second squared, and in many instances it will be most helpful to replace newtons by this derivation. A familiarization of the proper derived quantities thus provides a future check on the accuracy of the answer both conceptually and numerically.

This book will use the pressure unit of bar. Notice that in table 2–4 the pascal (Pa) represents the natural derived unit of pressure (N/m^2), but the magnitudes of the unit are so small as to make it inconvenient. Thus, the bar represents a more practical increment to use. We have seen that the atmospheric pressure is of the order of 1 bar (commonly 1.01 bars), while this same pressure would be 1×10^5 Pa. It is just more convenient to think of "1.01" rather than "1.01×10^5".

Table 2–4

Derived Thermodynamic Units

	Système International		*English System*	
Quantity	Unit	Derivation	Unit	Derivation
Area	m^2	m^2	ft^2	ft^2
Volume	m^3	m^3	ft^3	ft^3
Density	kg/m^3	kg/m^3	lbm/ft^3	lbm/ft^3
Force	N	$kg \cdot m/s^2$	lbf	$\frac{1}{32.17}$ lbm-ft/s^2
Pressure	Pa*	$kg \cdot m/m^2 \cdot s^2$	psi	$\frac{1}{144}$ lbf/ft^2
Energy	J	$kg \cdot m^2/s^2$	ft-lbf	$\frac{1}{32.17}$ lbm-ft^2/s^2
Specific energy	J/kg	$kg \cdot m^2/s^2 \cdot kg$	ft-lbf/lbm	
Entropy	J/K	$kg \cdot m^2/s^2 \cdot K$	ft-lbf/°R	$\frac{1}{32.17}$ lbm-$ft^2/s^2 \cdot$ °R
Specific entropy	$J/kg \cdot K$	$kg \cdot m^2/s^2 \cdot kg \cdot K$	ft-lbf/lbm · °R	
Power	W	$kg \cdot m^2/s^3$	horsepower (hp)	550 ft-lbf/s

* 1×10^5 Pa = 1 bar.

Many times a conversion from one system of units to the other is required. To facilitate this, the following equivalents are useful:

Length: 1 inch = 2.54 centimeters (cm)
1 foot = 30.48 centimeters (cm) = 0.3048 meter (m)

Force: 1 pound-force (lbf) = 4.448 newtons (N)

Mass: 1 pound-mass (lbm) = 0.454 kilogram (kg)

A more complete listing of conversion factors is given in table B.15.

Example 2.15 Determine the specific kinetic energy of air moving at 30 m/s. Then convert the answer to English units.

Solution The specific kinetic energy can be computed from equation (2–38).

$$\text{ke} = \frac{1}{2}V^2$$

$$= \left(\frac{1}{2}\right)(30\ \text{m/s})^2$$

$$= 450\ \text{m}^2/\text{s}^2 = 450\ \text{kg} \cdot \text{m}^2/\text{s}^2 \cdot \text{kg}$$

$$= 450\ \text{N} \cdot \text{m/kg} = 450\ \text{J/kg} \qquad \textit{Answer}$$

The conversion of units to the English system can be made with the factors listed in table B.15. Thus,

$$334.5\ \text{ft-lbf/lbm} = 1\ \text{kJ/kg}$$

and we obtain

$$\text{ke} = 450\ \text{J/kg} \times \frac{334.5\ \text{ft-lbf/lbm}}{1\ \text{kJ/kg}}$$

$$= 150.5\ \text{ft-lbf/lbm} \qquad \textit{Answer}$$

Example 2.16 Convert 200 psia pressure to SI units.

Solution From table B.15 we find

$$1\ \text{psi} = 0.06895\ \text{bar}$$

so

$$200\ \text{psia} = (200\ \text{psia})(0.06895\ \text{bar/psi})$$

$$= 13.9\ \text{bar} \qquad \textit{Answer}$$

$$= 13.9 \times 10^5\ \text{N/m}^2 \qquad \textit{Answer}$$

The SI system uses prefixes to describe quantities that are inconveniently large or small in the base units. These prefixes are listed in table 2–5. Thus, as we have seen in the discussion of weight and mass, 1 kg = 1000 g. Similarly, 1 km = 10000 m and 1000 mm = 1 m. Caution in the use of prefixes must be used since the prefix letters frequently can be confused with base unit descriptions. The reader is encouraged again to use the following rule:

Always include units when writing equations and solutions.

Table 2–5
SI Unit Prefixes

Amount	*Multiple*	*Prefix*	*Symbol*
1 000 000 000	10^9	giga	G
1 000 000	10^6	mega	M
1 000	10^3	kilo	k
100	10^2	hecto	h
10	10	deka	da
0.1	10^{-1}	deci	d
0.01	10^{-2}	centi	c
0.001	10^{-3}	milli	m
0.000 001	10^{-6}	micro	μ
0.000 000 001	10^{-9}	nano	n

2.13 Summary

In this chapter, the concept of the system has been introduced and identifying the boundaries of this system has been stressed. Once the system has been identified (not too easy a task in many cases), its quantitative description is determined through the system properties, the total list of which describes the state of the system. The important properties for a thermodynamic analysis are as follows: mass, weight, density, volume, specific volume, pressure, temperature, and energy. Energy has been typified as potential, kinetic, or internal.

The following list of formulas describes the important points of this chapter. The symbols correspond to those in table 2–6.

mass/weight:

$$F \text{ (newtons)} = ma \qquad m \text{ in kg}$$
$$W \text{ (newtons)} = mg \qquad m \text{ in kg}$$
$$F \text{ (lbf)} = \frac{ma}{g_c} \qquad m \text{ in lbm}$$
$$W \text{ (lbf)} = \frac{mg}{g_c} \qquad m \text{ in lbm}$$

density, volume/pressure:

$$v = \frac{1}{\rho} = \frac{V}{m}$$
$$p = p_g + p_a$$
$$p = p_a - p_{gv}$$

temperature/energy:

$$T^\circ\text{R} = t^\circ\text{F} + 460^\circ$$
$$T^\circ\text{K} = t^\circ\text{C} + 273^\circ$$
$$T^\circ\text{R} = \frac{9}{5} T^\circ\text{K}$$

$$\text{KE} \quad (\text{J}) = \frac{1}{2} m\bar{V}^2 \qquad m \text{ in kg}$$

$$\text{ke} \quad (\text{J/kg}) = \frac{1}{2} \bar{V}^2$$

$$\text{KE} \quad (\text{ft-lbf}) = \frac{m\bar{V}^2}{2g_c} \qquad m \text{ in lbm}$$

$$\text{ke} \quad (\text{ft-lbf/lbm}) = \frac{\bar{V}^2}{2g_c}$$

$$\text{PE} \quad (\text{J}) = Wz = mgz \qquad m \text{ in kg}$$

$$\text{pe} \quad (\text{J/kg}) = gz$$

$$\text{PE} \quad (\text{ft-lbf}) = \frac{mgz}{g_c} \qquad m \text{ in lbm}$$

$$\text{pe} \quad (\text{ft-lbf/lbm}) = \frac{gz}{g_c}$$

$$u = \frac{U}{m}$$

Table 2–6 should be helpful in clarifying notations, units and concepts of this book. On the far right of the table, notice that the chapter in which the given symbol or concept is first introduced is listed.

Table 2–6

Thermodynamic Notation

Symbol	*Variable*	*Units*		*Chapter Introduced*
		English	SI	
a	Acceleration	ft/s²	m/s²	2
A	Area	ft²	m²	2
b	Spring modulus	lbf/in	N/cm	3
c	Concentration ratio	lbm/lbm	kg/kg	15
C, c	Constants	varied	varied	
C_v	Total specific heat @ constant volume	Btu/°R	kJ/K	5
c_v	Specific heat @ constant volume	Btu/lbm · °R	kJ/kg · K	5
C_p	Total specific heat @ constant pressure	Btu/°R	kJ/K	5
c_p	Specific heat @ constant pressure	Btu/lbm · °R	kJ/kg · K	5
D	Diameter	ft	m	
E	Energy	ft-lbf	kJ	2
e	Specific energy	ft-lbf/lbm	kJ/kg	2
F	Force	lbf	N	2
g	Local gravitational acceleration	ft/s²	m/s²	2
g_c	Gravitational constant	32.17 ft-lbm/lbf · s²		2
H	Enthalpy	Btu or ft-lbf	kJ	4
h	Specific enthalpy	Btu/lbm	kJ/kg	4

Table 2–6 (continued)

Symbol	*Variable*	*Units*		*Chapter Introduced*
		English	SI	
J	Mechanical–thermal conversion factor	778 ft-lbf/Btu		3
k	c_p/c_v	unitless		5
K_h	Coefficient of heat transfer	Btu/hr · ft² · °F	W/m² · K	3
KE	Kinetic energy	ft-lbf	kJ	2
ke	Specific kinetic energy	ft-lbf/lbm	kJ/kg	2
L, l	Length	ft	m	2
m	Mass	lbm	kg	2
$\dot{m}$	Mass flow rate	lbm/s	kg/s	4
n	Polytropic exponent			6
p_a	Atmospheric pressure	psia	Pa or bar	2
p_g	Gage pressure	psig	bar	2
p_{gv}	Vacuum gage pressure	psiv	bar	2
p	Pressure	psia	bar	2
PE	Potential energy	ft-lbf	kJ	2
pe	Specific potential energy	ft-lbf/lbm	kJ/kg	2
Q	Heat or heat transfer	Btu	kJ	3
q	Specific heat transferred	Btu/lbm	kJ/kg	3
$\dot{Q}$	Rate of heat transfer	Btu/s	kJ/s	3
R	Gas constant	ft-lbf/lbm · °R	N · m/kg · K	2
R_u	Universal gas constant	ft-lbf/lbm · mole · °R	N · m/kg · mole · K	5
r	Radius	ft	m	3
r_p	Pressure ratio	unitless		10
r_v	Compression ratio	unitless		10
S	Entropy	Btu/°R	kJ/K	7
s	Specific entropy	Btu/lbm · °R	kJ/kg · K	7
SE	Strain energy	ft-lbf	N · m	6
sg	Specific gravity	unitless		2
T	Absolute temperature	°R	K	2
t	Thermometric temperature	°F	°C	2
U	Internal energy	ft-lbf	kJ	2
u	Specific internal energy	ft-lbf/lbm	kJ/kg	2
V	Volume	ft³	m³	2
v	Specific volume	ft³/lbm	m³/kg	2
$\dot{V}$	Volume flow rate	ft³/s	m³/s	4
$\bar{V}$	Velocity	ft/s	m/s	2
W	Weight	lbf	N	2
Wk	Work	ft-lbf	kJ	3
Wk_{cs}	Closed system work	ft-lbf	kJ	4
Wk_{os}	Open system work	ft-lbf	kJ	4
wk	Work per unit mass	ft-lbf/lbm	kJ/kg	3
$\dot{Wk}$	Power	ft-lbf/s	kJ/s or kW	3
x, y	Length	ft	m	6
Y	Young's modulus	psi	MPa	6
z	Reference elevation above plane of zero potential energy	ft	m	2

Table 2–6 (continued)

Symbol	*Variable*	*Units*		*Chapter Introduced*
		English	SI	
alpha, α	Seebeck coefficient	joules/coulomb		16
beta, β	Relative humidity	percent		15
gamma, γ	Specific weight	lbf/ft^3	N/m^3	2
epsilon, ε	Emissivity	unitless		3
eta, η	Efficiency	percent		8
delta, Δ	Change in variable			2
kappa, κ	Thermal conductivity	Btu/hr · ft · °F	kW/m · K	3
lambda, Λ	Function of availability			9
mu, μ	Chemical potential	Btu/lbm	kJ/kg	15
rho, ρ	Density	lbm/ft^3	kg/m^3	2
sigma, σ	Stefan-Boltzmann constant	$Btu/hr \cdot ft^2 \cdot {}^\circ R^4$	$W/m^2 \cdot K^4$	3
tau, τ	Time	seconds		3
phi, Φ	Total availability	Btu or ft-lbf	kJ	9
phi, ϕ	Availability	Btu/lbm or ft-lbf/lbm	kJ/kg	9
chi, χ	Quality	percent		8
omega, ω	Specific humidity	$\frac{\text{lbm vapor}}{\text{lbm dry air}}$	$\frac{\text{kg vapor}}{\text{kg dry air}}$	15
xi, ξ	Strain			6
$\mathfrak{D}$	Diffusivity	ft^2/s	m^2/s	15
$\mathcal{E}$	Electric potential	volts		16
$\mathcal{F}$	Faraday's constant	coulombs/g · mole		16
G'	Total Gibbs free energy	Btu or ft-lbf	kJ	9
g'	Gibbs free energy	Btu/lbm or ft-lbf/lbm	kJ/kg	9
H'	Total Helmholtz free energy	Btu or ft-lbf	kJ	9
h'	Helmholtz free energy	Btu/lbm or ft-lbf/lbm	kJ/kg	9
$\mathcal{I}$	Electric current	amperes		16
$\mathcal{Q}$	Electric charge	coulombs		16
$\mathcal{R}$	Resistance	ohms		16
$\mathcal{T}$	Thomson coefficient	joules/coulombs · K		16
$\mathcal{V}$	Voltage	volts		10

Practice Problems

Système International (SI) units are used in those problems with the notation (M) for metric under the number of the problem. These problems that use the English units are indicated with (E), and mixed unit problems are listed with (C) for combined under the problem number.

Section 2.7

2.1. **(M)** A 2-kg cube of gold is weighed on a spring scale in two locations. The first location has a local gravitational acceleration of 9.80 m/s^2 and the second has a gravitational acceleration of 9.78 m/s^2. At which location will the gold's weight W in newtons be greatest?

2.2. (M) Three kilograms of water are vaporized in a boiler. What is the weight of this water in newtons, if the local acceleration of gravity is 9.79 m/s^2?

2.3. (E) A weight scale located on the earth's sea level indicates a gallon of water weighs 8.333 lbf. Determine the mass of 1 gallon in units of slugs and pounds-mass.

2.4. (E) A battery weighing 32 lbf on earth's sea level is transported to the moon. Determine the weight of the battery on the moon where $g = 5.47$ ft/s^2.

2.5. (C) Referring to necessary tables, convert the following masses:
(a) 1 lbm to grams.
(b) 2 lbm to kilograms.
(c) 20 slugs to pounds-force on earth's sea level.
(d) 100 grams to dynes on earth's sea level.
(e) 200 kilograms to pounds-force on earth's sea level.

2.6. (M) A cylindrical tank filled with a fluid has a diameter of 1 meter and a length of 1.5 meters. It weighs 6000 newtons (N) where the gravitational acceleration is 9.82 m/s^2. Determine:
(a) Volume occupied by the fluid.
(b) Specific weight of the fluid.
(c) Density of the fluid.
(d) Specific gravity of the fluid.

2.7. (M) A balloon is filled with a gas having a specific volume of 0.9 m^3/kg. If the inside volume of the balloon is 4800 cm^3, what is the weight of the gas-filled balloon? Ignore the weight of the balloon itself and assume $g = 9.78$ m/s^2.

2.8. (M) A tank is filled with hydrogen gas and a gage indicates that the pressure inside the tank is 0.01 bar.
(a) If the atmospheric pressure is 1.01 bars, what is the pressure inside the tank?
(b) What is the pressure in the tank if the atmospheric pressure is 768 mm Hg?

2.9. (M) A tank 5 meters high is half full of water, and air at 0.13 bar gage pressure is occupying the remaining volume. The tank is 1.5 meters in diameter and the contents are at 20°C.
(a) What is the gage pressure at the top of the water?
(b) What is the gage pressure at the bottom of the tank?
(c) If the atmospheric pressure is 1.01 bars, find the absolute pressure of (a) and (b).

2.10. (E) Steam has a specific volume of 10.07 ft^3/lbm at a pressure of 80 psia and 900°F. At this state, assuming $g = 32.1$ ft/s^2, determine the following:
(a) Density.
(b) Specific weight.
(c) Specific gravity.

2.11. (E) Two tanks, A and B, contain air (figure 2–25). Tank A is at a pressure of 20 psig and tank B is at a pressure of 18 psig. If the two tanks are connected by a U-tube manometer, as shown, what is the difference in height of the mercury column, h? Note: Specific weight of mercury is 845 lbf/ft^3 and specific weight of air is 0.076 lbf/ft^3.

2.12. (E) A hydraulic pump produces a pressure of 250 psig in an oil line as shown (figure 2–26). What force in pounds-force will be applied to the 1-ft diameter piston?

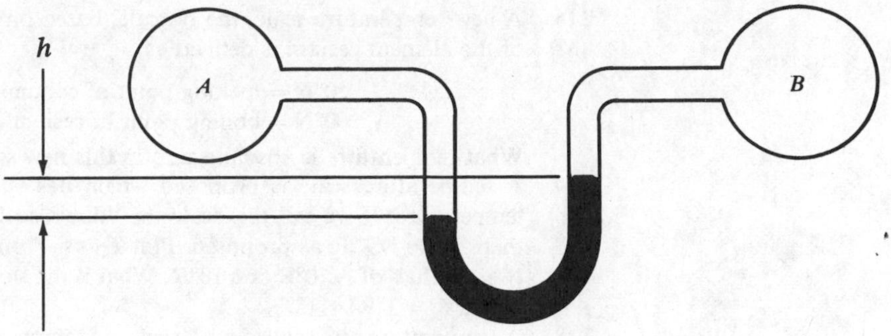

Figure 2–25

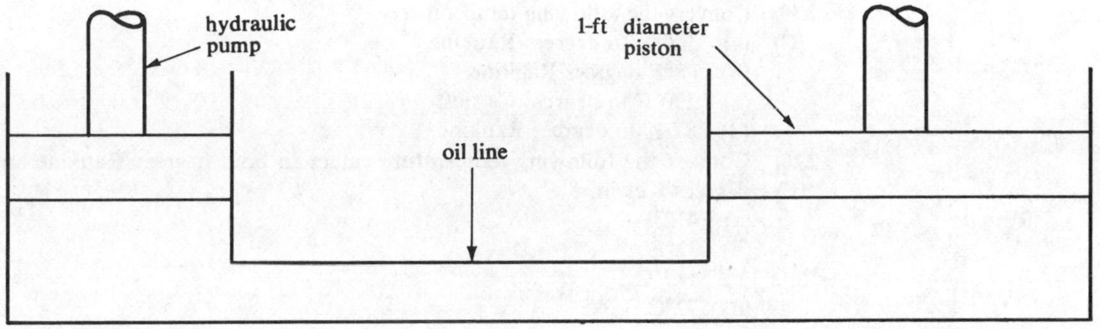

Figure 2–26

2.13. **(E)** Determine the maximum vacuum pressure possible inside a tank with the following surrounding conditions:

(a) An atmospheric pressure of 14.8 psia.

(b) An atmosphere at a pressure of 14 in Hg.

2.14. **(C)** Convert the following pressures:

(a) 14.7 psi to inches of mercury.

(b) 460 mm Hg to bars.

(c) 300 in Hg to pounds per square inch.

(d) +50 psi to bars.

(e) 20 bars to pounds per square inch.

Sections 2.9 and 2.10

2.15. **(M)** A thermometer indicates 70°C when it is in thermal equilibrium with block *A* and 100°C when in thermal equilibrium with block *B*. Are blocks *A* and *B* in thermal equilibrium? State a condition when blocks *A* and *B* are in thermal equilibrium (figure 2–27).

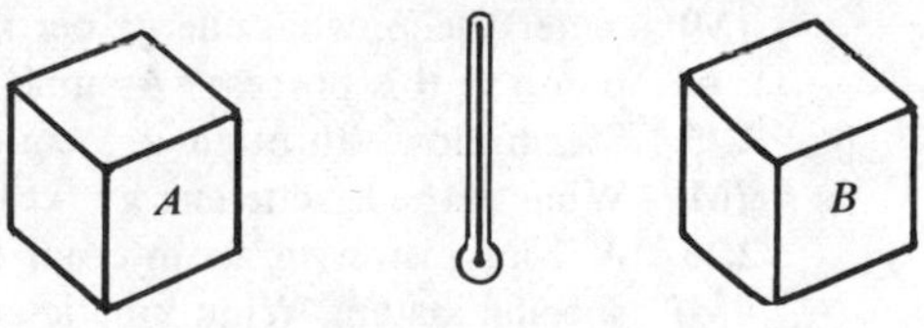

Figure 2–27

2.16. **(M)** A new temperature scale, the N scale, based on the melting and freezing of the element cesium is defined as:

$$0°N = \text{melting point of cesium} = 28.5°C$$
$$100°N = \text{boiling point of cesium} = 690°C$$

What temperature is absolute zero in this new scale?

2.17. **(E)** A temperature scale is proposed which has values of inverse Rankine temperature. If we call this scale the "down scale" and identify it as T_D, then $T_D = 1/T°R$ as proposed. Plot T_D vs. T on regular graph paper between values of $1/10°R$ and $10°R$. What is the slope of the curve at a point where $T = 1°R = 1°T_D$?

2.18. **(C)** A temperature scale called an L scale is suggested, which has the values of $T_L = \log T°R$. Determine the value of T_L in terms of the Kelvin scale (K).

2.19. **(C)** Convert the following temperatures:
(a) 140°F to degrees Rankine.
(b) 88°F degrees Rankine.
(c) 230°F to degrees Celsius.
(d) 87 K to degrees Rankine.

2.20. **(C)** Convert the following temperature values to both degrees Rankine and degrees Kelvin.
(a) 412°F.
(b) 32°F.
(c) 117°C.
(d) 72°C.

Section 2.11

2.21. **(M)** A 45,000-kg aircraft travels at 1000 km/h at an altitude of 3000 meters. The local gravitational acceleration at this altitude is 9.81 m/s². Determine the following:
(a) Kinetic energy of the aircraft.
(b) Potential energy of the aircraft if the sea level is considered as the plane of zero potential energy.

2.22. **(M)** Assume that you are taking a trip on a train. You carry a filled suitcase which weighs 170 newtons and during the course of your trip you are told that the train is moving at 140 km/hr.
(a) What do you observe the kinetic energy of the suitcase to be, before you are told the velocity of the train?
(b) What do you interpret the kinetic energy of the suitcase to be after knowing the train's velocity?

2.23. **(M)** A 1-kg piece of wood and a 1-kg piece of steel are dropped simultaneously from a bridge which is 40 meters high into 20-m-deep water. What is the change in potential energy or the available potential energy of
(a) The wood?
(b) The steel?

2.24. **(M)** A pump is used to remove water from a well into a tank. If the well is 75 meters deep, what energy per kilogram of water must be supplied by the pump in this process? Assume $g = 9.8$ m/s².

2.25. **(M)** Steam flows through a 5-cm-diameter pipe with a velocity of 24 m/s. What is the kinetic energy (ke) of the steam per unit mass?

2.26. **(M)** A 1-kg piston in an internal combustion engine travels at 60 m/s at a specific instant. What kinetic energy does the piston possess?

2.27. (M) Angular kinetic energy (AKE) is a form of mechanical energy which is usually not accounted for in the kinetic energy. A system has the following amounts of energy:

$$AKE = 150 \text{ kJ}$$
$$KE = 100 \text{ kJ}$$
$$PE = 20 \text{ kJ}$$
$$U = 35 \text{ kJ}$$

(a) Determine the total energy of the system.
(b) Determine the total mechanical energy of the system.

2.28. (E) A balloon weighing 10 ounces at a location where $g = 31.7 \text{ ft/s}^2$ is released at 20° latitude and floats to an altitude of 6000 ft above sea level. If we released the balloon from 1000 feet above sea level and defined this elevation as where potential energy is zero, determine the following:
(a) Total potential energy of balloon when floating.
(b) Total potential energy of balloon if it had been at sea level.
(c) Total potential energy of balloon at release.
Note: 16 ounces = 1 lbf

2.29. (E) One pound-mass of mercury at 426°F has 150 Btu of internal energy under a certain condition, while it also has 28 Btu of KE and 2 Btu of PE. Determine the total energy of the mercury.

2.30. (E) During a windstorm, the velocity of the wind is measured at 70 mph. What is the specific kinetic energy of the air under this condition?

2.31. (E) Ten pounds-mass of steam flowing through a pipe are found to have 15,000 Btu of internal energy U. If the kinetic energy ke is 500 Btu/lbm, and the potential energy pe is 100 Btu/lbm, determine the total energy of the 10-lbm steam, E, and the energy e.

2.32. (E) Refrigerant R-22 flows through a circuit of pipes and components shown in the figure 2–28. If the refrigerant R-22 has a velocity of 2 ft/s at both points (1) and (2) in the circuit, what is the difference in energy per pound-mass of the refrigerant R-22 between (1) and (2), if internal or thermal energy is neglected? Assume $g = g_c$.

2.33. (E) For problem 2.32, determine the kinetic energy at points (1) and (2) (figure 2–28).

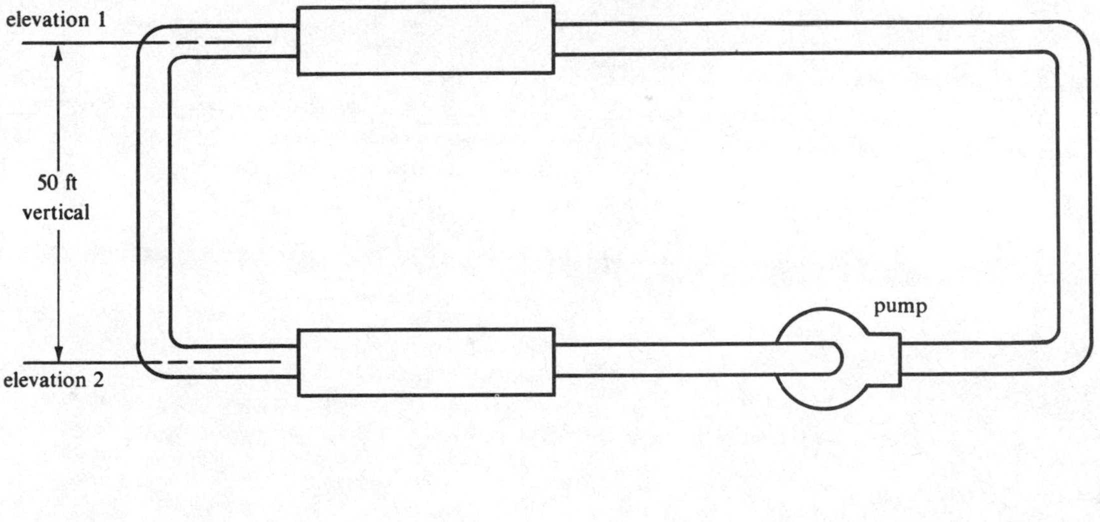

2.34. For problem 2.32, determine the specific potential energy at points (1)
(E) and (2), if the plane of zero potential energy is assumed to be 100 feet below point (2).

Section 2.12

2.35. The Reynold's number, R_e, is a dimensionless (unitless) number frequently
(C) used in fluid mechanics, thermodynamics, and physics. This number is found by the defining equation

$$R_e = \frac{\rho \bar{V} D}{\mu}$$

where D is the characteristic length having units of meters or feet and μ is the viscosity. In SI and English units, what dimensions does viscosity have?

2.36. In terms of SI and English units, determine the dimension x in the follow-
(C) ing equations, using tables 2–3, 2–4, or 2–5 when necessary:
(a) $x = T\Delta s/\Delta\tau$.
(b) $x = \Delta h/\Delta T$.
(c) $x = T\Delta S$.
(d) $x = RT/(v - c)^2$.
(e) $x = pv/T$.

2.37. Show that both sides of the following equation agree dimensionally:
(C)

$$\frac{\Delta p}{L} = \frac{\rho \bar{V}^2}{D g_c} \left[C \frac{\mu}{D \bar{V} p} \right]$$

where C = dimensionless constant and L has units of length. See problem 2.35 for definitions of μ and D.

2.38. What are the dimensions of C in the following equations?
(C) (a) $C = pv^{1.7}$.
(b) $C = pv^{1.3}$.
(c) $C = pv/v^{2.3}$.
(d) $C = p$.
(e) $C = T$.

3

Work, Heat, and Reversibility

In this chapter we will define *work* and develop some useful relationships for calculating its magnitude. We will make some calculations that require the use of calculus and some that do not; those that do require calculus will be clearly indicated in the text. The mechanisms of heat transfer — *conduction, convection*, and *radiation* — will then be presented. The concept of *reversibility* will be introduced, along with the items or causes for irreversibilities in processes. Following this, the concepts of the equivalence and differences of work and heat will be discussed. The thermodynamic system, which was introduced in chapter 2, will here be classified into three general types: *open, closed*, and *isolated*. This classification is introduced now to allow for better understanding of the motivation behind typifying systems. The chapter will end with a tabulation of the forms of energy, providing the foundation for the *conservation of energy* principle presented in subsequent chapters.

3.1 Work

Work connotes an active or dynamic state, during which some mechanical effort has been exerted. This visualization is embodied in our definition of work.

Work: *Force times distance through which the force acts constantly.*

$$Wk = F\Delta x \tag{3–1}$$

In order to provide for cases where the force acting through a distance is not constant, we will prefer to consider a small amount of work, dWk, resulting from a force which, while varying over a finite distance Δx, is considered to be constant over a small distance dx and acting in the same direction as dx. Then

$$dWk = Fdx \tag{3-2}$$

and using integral calculus we say that the work done over that finite distance x_1 to x_2 is

$$Wk = \int dWk = \int_{x_1}^{x_2} Fdx \tag{3-3}$$

If we plot force F vs. distance or displacement x we might have a graph which looks like that shown in figure 3–1. This graph (or curve) describes a force which varies with the distance. Using the concepts of calculus then, we can show that the area under the curve in a force-displacement graph is equal to the work Wk since the area under the curve is also equal to $\int Fdx$ between the two values x_1 and x_2.

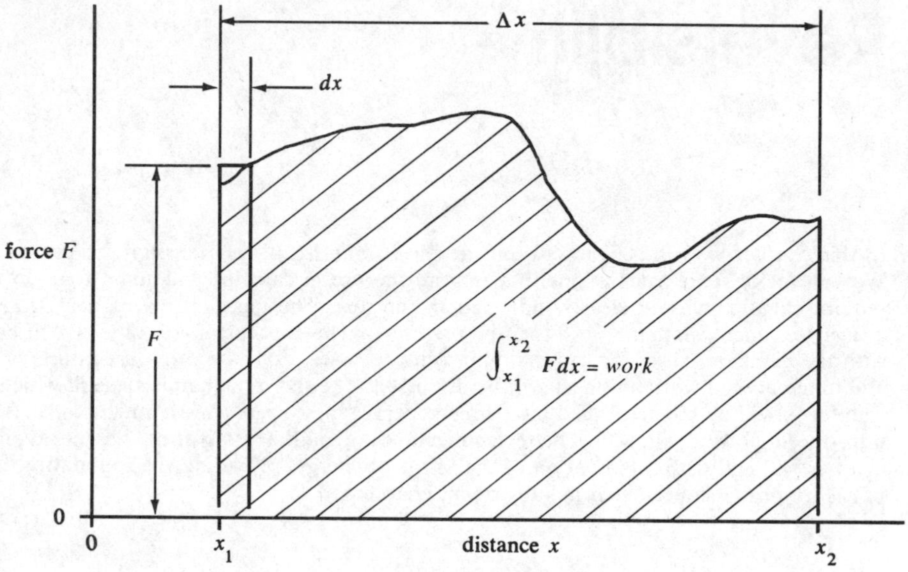

Figure 3–1 Typical force-distance relationship for a process involving work.

Example 3.1 Determine the work done in lifting a 60-kg block vertically 3 meters.

Solution The force required to lift the block is just equal to the weight W given by equation (2–5).

$$\begin{aligned} W &= mg \\ &= (60 \text{ kg})(9.8 \text{ m/s}^2) \\ &= 588 \text{ N} \end{aligned}$$

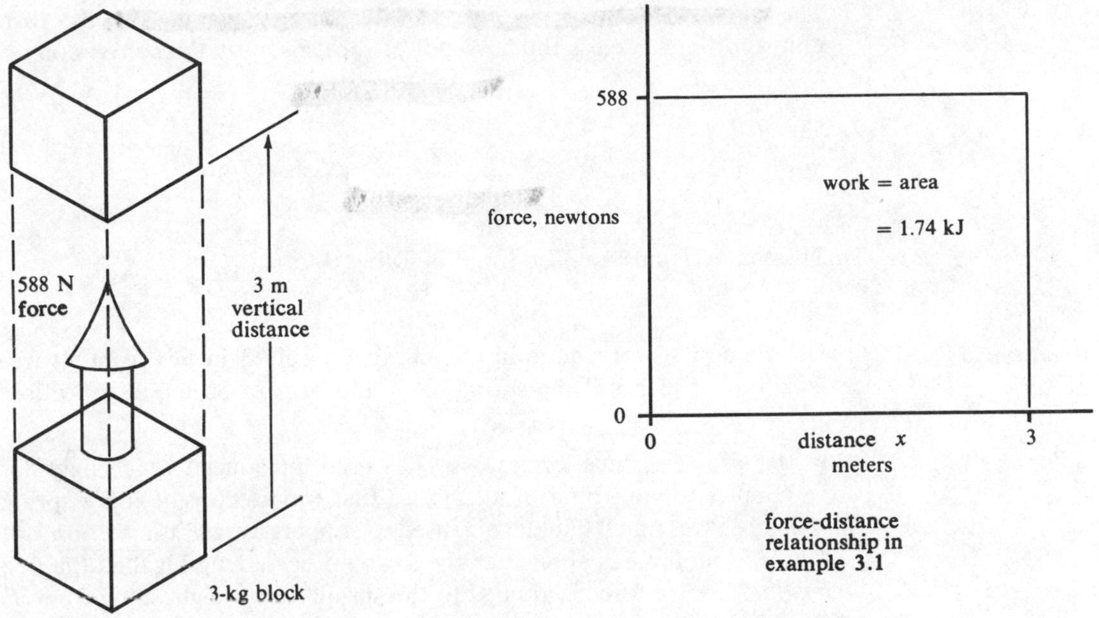

Figure 3–2 Example 3.1.

The force F is then 588 newtons, and the distance x is 3 meters. Then, using equation (3–1) we obtain

$$Wk = (588 \text{ N})(3 \text{ m})$$

$$= 1764 \text{ N} \cdot \text{m} = 1764 \text{ J} = 1.76 \text{ kJ} \qquad \textit{Answer}$$

Notice that the units on work are the same as those assigned to energy and that the potential energy of the weight has increased by an amount equal to the work done, 1.76 kJ. The system in this problem is the 60-kg block, and we can define work in a slightly more general manner.

Work: *Energy in transition across the boundary of a system, which can always be identified with a mechanical force acting through a distance.*

We see then that work is the boundary effect on a system as opposed to the internal property of energy.

The rate of doing work is called *power* and is denoted by the symbol $\dot{W}k$. The "dot" represents "per unit of time" and is equivalent to the derivative dWk/dt. Thus

$$\dot{W}k = \frac{dWk}{dt} \tag{3–4}$$

The common unit of power is the watt (W) or kilowatt (kW). Another unit sometimes used is the horsepower (hp), given by the conversion

$$1 \text{ hp} = 0.746 \text{ kW}$$

or

$$1.34 \text{ hp} = 1 \text{ kW}$$

These conversions are *also* listed in Table B.15.

Example 3.2 Determine the work done in stretching a spring 3 inches from its free length of 12 inches if the spring has a modulus of 30 lbf/in deflection.

Solution In this problem some terminology has been introduced which may not be familiar to you. First of all, notice that we are considering a spring which behaves much like the one used to support a weight in section 2.7; that is, if no force is applied to the spring it has a length called the *free length.* As some force is applied to the spring (or a weight hung from it) the spring will stretch some distance. As the force is increased, the spring continues to stretch further and the force per unit amount of stretch or deflection is called the *modulus* of the spring. If we plot on a sheet of graph paper what has just been stated we have a relation like that shown in figure 3–3. The slope of the line is the change in force divided by the corresponding change in deflection, Δx. Obviously, this problem repre-

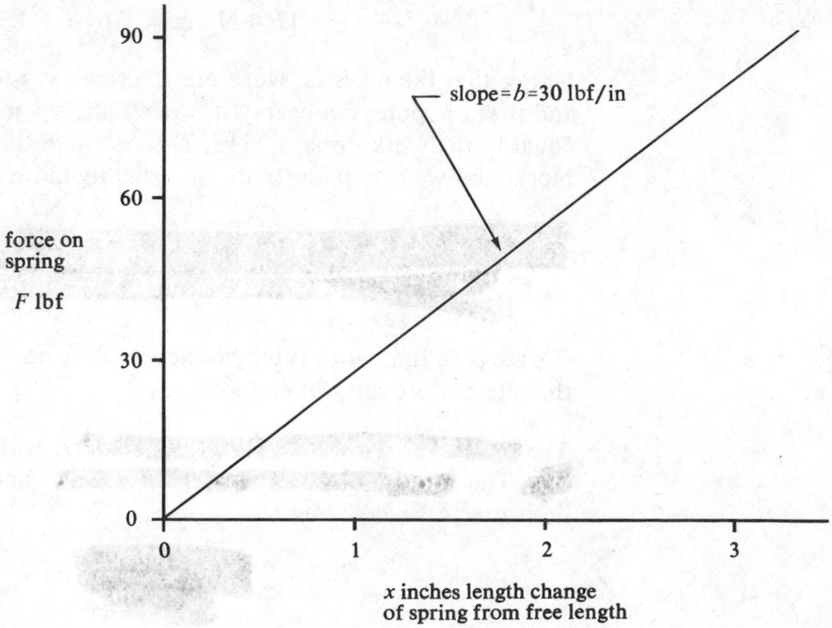

Figure 3–3 Example 3.2.

sents a case of varying force with distance, and we therefore find the work done by referring to equation (3–3):

$$Wk = \int_{x_1}^{x_2} F dx$$

In this problem, the initial distance x_1 is zero and the final distance x_2 is 3 inches. From figure 3–3 we see

$$F = bx \quad \text{where} \quad b = 30 \text{ lbf/in}$$

and we substitute this relation into our work equation to obtain

$$\begin{aligned} Wk &= b\int_{x_1}^{x_2} x\,dx \\ &= \frac{1}{2}bx_2^2 - \frac{1}{2}bx_1^2 \\ &= \frac{1}{2}(30 \text{ lbf/in})(3 \text{ in})^2 \\ &= 135 \text{ in-lbf} \\ &= 11.25 \text{ ft-lbf} \end{aligned}$$

Answer

This then is the work required to stretch the spring 3 inches. It corresponds to the area under the curve in figure 3–3 between $x = 0$ and $x = 3$ inches. We can say the work done to change the length of a spring which has a force-length relationship pictured in figure 3–3 is

$$Wk_{\text{spring}} = \frac{1}{2}bx^2 \tag{3–5}$$

where x is the change in spring length from the free length of the spring.

Example 3.3 Determine the average power required in kilowatts and horsepower to move an elevator weighing 2000 lbf vertically through 40 feet in 10 seconds.

Solution From equation (3–4) we see that power is the time rate of doing work. The average power required can be computed from the relationship

$$\dot{Wk}_{\text{av}} = \frac{Wk}{\tau}$$

The amount of work done is

$$\begin{aligned} Wk &= 2000 \text{ lbf} \times 40 \text{ ft} \\ &= 80{,}000 \text{ ft-lbf} \end{aligned}$$

and this work is done during an interval of 10 seconds. Thus,

$$\begin{aligned} \dot{Wk}_{\text{av}} &= \frac{80{,}000 \text{ ft-lbf}}{10 \text{ s}} \\ &= 8000 \text{ ft-lbf/s} \end{aligned}$$

Using the conversion listed in table B.15 that 550 ft-lbf/s = 1 hp, we obtain

$$\dot{W}k_{\text{av}} = 14.5 \text{ hp} \qquad \textit{Answer}$$

Further, from table B.15 we have

$$\dot{W}k_{\text{av}} = \frac{14.5 \text{ hp}}{1.34 \text{ hp/kW}}$$

$$= 10.8 \text{ kW} \qquad \textit{Answer}$$

Let us now look at a type of apparatus which we will see often in the remainder of this book. This apparatus includes a freely sliding frictionless piston of radius r, and a gas which pushes the piston out of the cylinder or which is compressed when the piston is retracted into the cylinder. This arrangement is shown in figure 3–4 and since the gas is not free to escape from the container, we will call this a *closed system.* (See section 3.8.) The apparatus we call a *piston-cylinder.* In this case let us assume that the gas contained in the cylinder is at some high pressure and subsequently moves the piston out of the cylinder. If the gas itself is to be considered as the system in this process, we see that work is being done by the system. In examples 3.1 and 3.2, work was done on the system. Obviously then, work is a directional phenomenon in the sense that it is energy crossing the boundary of the system into or out of the system. We can account for this direction by stating that *positive work is work done by the system and negative work is work done on the system.* We should note that work is not a vector. Although it is a product of two separate vectors (*force* and *distance*, or *displacement*), it does not have a vector direction. That is, the precise path followed by the work is not known to us, so we must conclude that work is a scalar product.

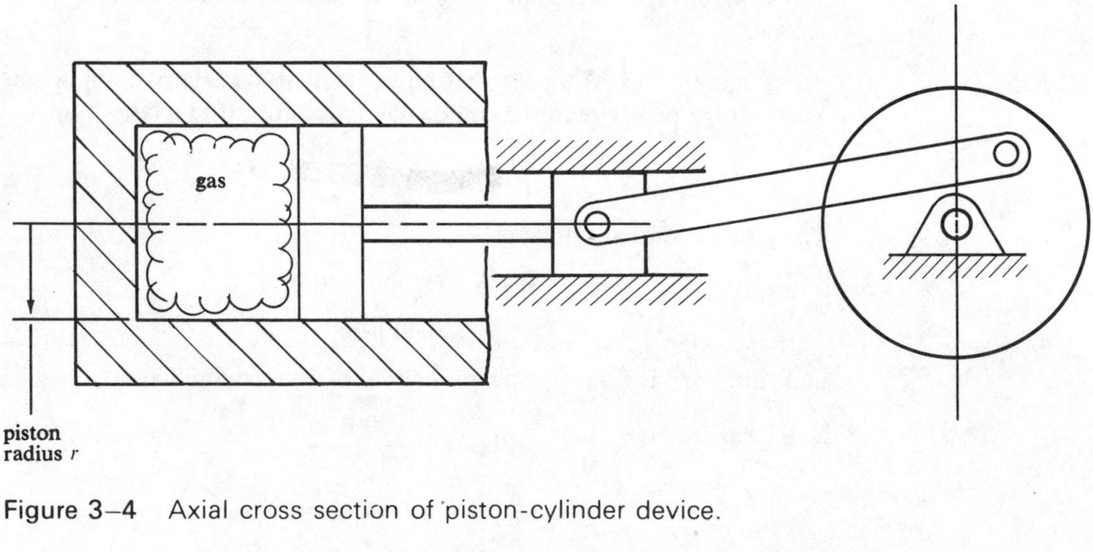

Figure 3–4 Axial cross section of piston-cylinder device.

Now, we notice that as the gas in the piston-cylinder expands and pushes the piston out, the pressure will likely drop and since the force applied to the piston, F, is

$$F = p \text{ (circular cross-sectional area of piston)} = p \times A \qquad (3\text{–}6)$$

the force will also drop. We, therefore, will have a case of varying force, so the work of the closed system piston-cylinder can be described with equation (3–2).

$$dWk_{cs} = Fdx$$

The subscript *cs* denotes *closed system* here. We know that the force, in terms of pressure, is given by equation (3–6), so

$$dWk_{cs} = pAdx$$

Now, A is the circular area of the face of the piston so that Adx is a small change in the volume dV of the gas itself. We therefore write

$$dWk_{cs} = pdV \qquad (3\text{–}7)$$

which will be used many times. In order to find the work done during some finite change in volume of the gas system, let us plot pressure vs. volume for the case where the piston is extended out of a cylinder, thereby increasing the volume of gas from V_1 to V_2 as indicated in figure 3–5. From equation (3–3) we see that the work will be

$$Wk_{cs} = \int dWk_{cs} = \int_{V_1}^{V_2} pdV \qquad (3\text{–}8)$$

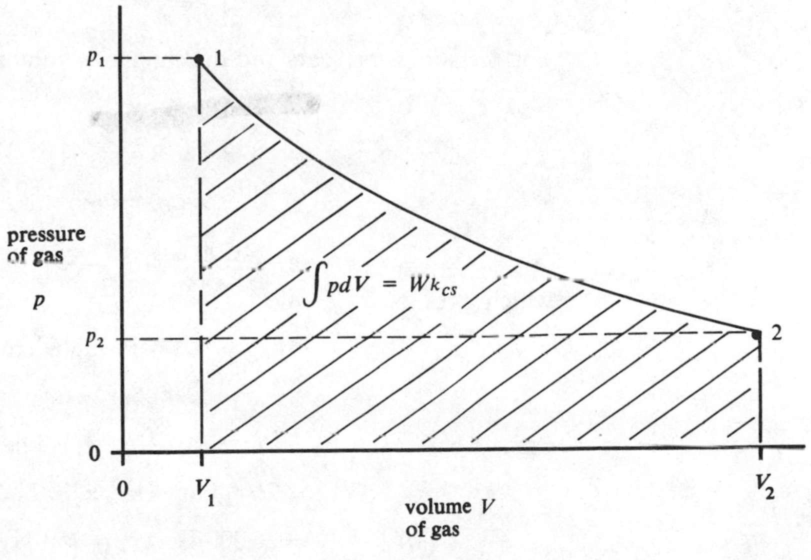

Figure 3–5 Pressure-volume, p-V, diagram of closed thermodynamic system involved in a work process.

Note that the area under the curve in a p-V diagram is the work of a frictionless closed system Wk_{cs}. We will frequently say that the work of a closed system is $\int pdV$—implying the limits V_1 and V_2.

A relation which will be used frequently is the work per unit of system mass. This is, for the closed system

$$wk_{cs} = \frac{Wk_{cs}}{m} = \int \frac{pdV}{m}$$

or

$$wk_{cs} = \int pdv \tag{3–9}$$

We will see that many specialized relationships for work come from either equation (3–8) or (3–9), among which is the following example of a process which has great utility.

Example 3.4 A 9-cm diameter piston-cylinder contains a gas which, under constant pressure, extends the piston 8 centimeters. Determine the work of this process if the gas pressure is 5.1 bars.

Solution In this case the pressure is constant, so equation (3–8) can be written

$$Wk_{cs} = p \int_{V_1}^{V_2} dV$$
$$= p(V_2 - V_1)$$
$$= p\Delta V$$

The pressure is 5.1 bars and the change in volume ΔV is calculated from

$$\Delta V = A \times 8 \text{ cm}$$
$$= \pi r^2 \times 8 \text{ cm}$$
$$= \pi \times (4.5)^2 \text{ cm}^2 \times 8 \text{ cm}$$
$$= 509 \text{ cm}^3$$

which gives us

$$Wk_{cs} = 5.1 \text{ bars} \times 509 \text{ cm}^3$$
$$= 2600 \text{ bar} \cdot \text{cm}^3$$

Since 1 bar $= 1 \times 10^5$ N/m^2 $= 10$ N/cm^2, we have

$$Wk_{cs} = 2600 \text{ bar} \cdot \text{cm}^3 \times 10 \text{ N/cm}^2\text{/bar}$$
$$= 26{,}000 \text{ N} \cdot \text{cm} = 260 \text{ N} \cdot \text{m}$$
$$= 260 \text{ J} = 0.26 \text{ kJ}$$

Answer

3.2 Heat

Heat is a word which probably has been more mistreated in technological language than any other single word. Following the manner of defining work in section 3.1, we will define heat.

Heat: *Energy in transition across the boundary of a system, which cannot be identified with a mechanical force acting through a distance.*

Heat occurs in a process when there is some temperature difference between the system and its surroundings. The direction of energy transition is always toward the area of lesser temperature. Heat will leave a system if it is hotter than its surroundings; if it is cooler, heat will enter the system. This energy transition will continue in the same direction until the system and its surroundings are separated or until thermal equilibrium is reached.

Heat will be identified by the symbol Q and the heat per unit of mass by q. In the old metric system of units, the term *calorie* was used to describe heat. This unit is defined as follows:

1 Calorie: *The amount of heat required to raise the temperature of 1 gram of water 1°C at 4°C.*

Frequently the kilocalorie, equal to 1000 calories, is used and called the "large calorie." The kilocalorie is often used to describe the energy associated with food and thus one speaks of consuming "so many" calories, that is, kilocalories.

The calorie is related to the customary unit of energy, the joule, by the conversions listed in Table B.15: 4.1868 J = 1 calorie. Since the SI system does not use the calorie as a proper unit, this text will use only joules (J) or kJ to describe heat.

In the English system the common unit used to describe heat is the British thermal unit (Btu), given by the following definition:

1 Btu: *The amount of heat required to raise the temperature of 1 lbm of water 1°F when the water is at 39°F.*

The manner in which heat occurs in a process is called *heat transfer* and is synonymous with *heat*. That is, we will use the two terms to mean the same thing. Heat transfer is a branch of science which is well documented and has found many applications in technology. There are three forms of heat transfer: *conduction*, *convection*, and *radiation*. (They will be discussed individually in the following three sections.) Each of these represents a method of calculating Q, depending on the mode of heating, much like we calculated work in the previous section. There is one distinction, however, between the approach of the following three sections and the rest of this text where heat is calculated; here we must know the temperature of both the surroundings *and* the system, whereas in chapter 7 the property referred to as *entropy* will be introduced which will allow heat or heat transfer to be calculated from a knowledge of either the system *or* the surrounding temperature, but not necessarily both.

3.3 Conduction Heat Transfer

Suppose a pot of water is placed on a stove burner which has a temperature of between 400 K and 1600 K depending on the type of burner and the operating conditions. Obviously, the water will begin to warm up and, if left on the stove, will vaporize (boil).

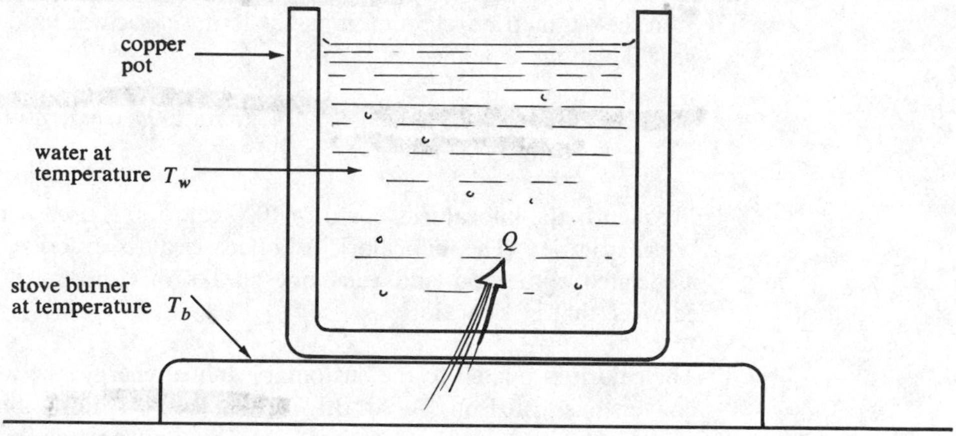

Figure 3–6 Physical arrangement for example of heat transfer by conduction.

In the arrangement shown in figure 3–6, a change is certainly being effected on the water in the pot. We can easily identify this change as heat or a heat transfer process in which internal energy of the water is increased. Energy is being lost from the surface of the burner and, in the form of heat, is transferred or conducted through the copper pot into the water itself. The heat transfer through the walls of the copper pot is called *conduction*, and the rate of conduction is given by Fourier's law of conduction:

$$\dot{Q} = \left(\frac{\kappa A \Delta T}{\Delta x}\right) \tag{3–10}$$

where $\dot{Q}$ is a rate of heat transfer with respect to time, given by an amount of heat Q, divided by time τ, or more precisely

$$\dot{Q} = \frac{dQ}{d\tau} \tag{3–11}$$

Table 3–1

Thermal Conductivity of Common Materials at Characteristic Temperatures

			Thermal Conductivity (κ, kappa) (Btu/ft·°F·hr*)						
Material	*Phase*	*Type of Thermal Conductor*	20°F	32°F	200°F	400°F	1000°F	1500°F	2000°F
Air	Gas	Insulation	0.0137	0.0139	0.0178	0.0219	....	....	....
Glass wool	Solid	Insulation	0.0217	0.0231	0.0435	....	....	....	....
Rock wool	Solid	Insulation	0.0321	0.0355	0.0373	....	....	....	....
Hydrogen	Gas	Insulation	....	0.0990	0.1240	....	....	....	....
Helium	Gas	Insulation	....	0.0820	0.0970	....	....	....	....
Pennsylvania fire brick	Solid	Refractory	....	....	....	....	0.71	0.76	0.84
Magnesite brick	Solid	Refractory	....	....	....	....	2.90	2.32	2.14
Water	Liquid	Conducting	....	0.320	0.392	....	....	....	....
Aluminum	Solid	Conducting	....	132.0	132.0	131.0	130.0	....	....
Iron	Solid	Conducting	....	43.0	40.0	36.0	25.0	....	....
Silver	Solid	Conducting	....	233.0	226.0	217.0	218.0	....	....
Copper	Solid	Conducting	....	220.0	218.0	215.0	214.0	....	....

* 1 Btu/ft · °F · hr = 1.73 W/m · K.

where dQ is a small amount of heat transferred during a small time period, $d\tau$. In equation (3–10), A is the surface area of the burner in contact with the corresponding area of the copper pot; ΔT is the temperature drop across the bottom of the pot which is $T_b - T_w$; Δx is the pot thickness; and κ is the thermal conductivity of the pot. In table 3–1 the thermal conductivity is listed for various types of material and temperatures. A more complete listing of the conductivity of materials can be found in various technical references, but note from table 3–1 that κ varies with temperature for all materials. Those materials listed as refractory type are characterized by their minimum dependence of κ on temperature, and are used in very high temperature applications such as furnaces and incinerators. Insulating materials have low thermal conductivities and are thus capable of retarding the heat transfer, whereas conducting materials have relatively high values of thermal conductivity for high heat transfer rates. Note from equations (3–10) and (3–11) that the material's conductivity determines the rate of heat transfer, but it does not determine the amount of heat transfer if the process can continue indefinitely.

Let us now look at the water that was placed on the stove burner.

Example 3.5

The thickness of the pot is 1.5 millimeters and the water in the pot is at 20°C initially and 60°C after being on the stove 8 minutes. Determine the rate of heat transfer when the pot is put on the stove and the rate after 8 minutes. Assume the pot diameter is 20 centimeters and the surface temperature of the stove burner is 150°C.

Solution

The heat transfer rates can be found by using equation (3–10). The terms in this equation have the following values initially:

$$A = \frac{\pi D^2}{4} = \frac{(\pi)(20\text{ cm})^2}{4} = 314\text{ cm}^2$$

$$= 0.0314\text{ m}^2$$

$$\Delta T = (150 + 273)\text{ K} - (20 + 273)\text{ K}$$

$$x = 1.5\text{ mm} = 0.0015\text{ m}$$

and the thermal conductivity is found from Table 3–1. By interpolating between 32°F (0°C) and 200°F (93°C) we obtain

$$\kappa = 219\text{ Btu/ft}\cdot{}^\circ\text{F}\cdot\text{hr}$$

$$= 219 \times 1.73\text{ W/m}\cdot\text{K}$$

$$= 379\text{ W/m}\cdot\text{K}$$

Using equation (3–10) we find

$$\dot{Q} = (379\text{ W/m}\cdot\text{K})(0.0314\text{ m}^2)\frac{(130\text{ K})}{(0.0015\text{ m})}$$

$$= 1030\text{ kW}$$ *Answer*

After 8 minutes we have

$$A = 0.0314\text{ m}^2$$

$$T = (150 + 273)\text{ K} - (60 + 273)\text{ K}$$

$$= 90\text{ K}$$

$$\kappa = 379\text{ W/m}\cdot\text{K}$$

and

$$\dot{Q} = (379\text{ W/m}\cdot\text{K})(0.0314\text{ m}^2)\frac{(90\text{ K})}{(0.0015\text{ m})}$$

$$= 714\text{ kW}$$ *Answer*

Many heat transfer problems involve conduction through more than one type of material. For example, the wall of a heated building is normally composed of an outside wall, an inside wall, and insulation and/or an air space between these surfaces. For flat surfaces it can be shown that the heat transfer through such a composite of materials is given by

$$\dot{Q} = \frac{A\,\Delta T}{\Delta x_{12}/k_a + \Delta x_{23}/k_b + \Delta x_{34}/k_c} \qquad \textbf{(3–12)}$$

where the subscripts refer to those indicated in figure 3–7 and ΔT equals $T_1 - T$. Obviously, for a composite structure of more or less than three materials equation (3–12) would need to be altered by adding or deleting terms $(\Delta x)/k$.

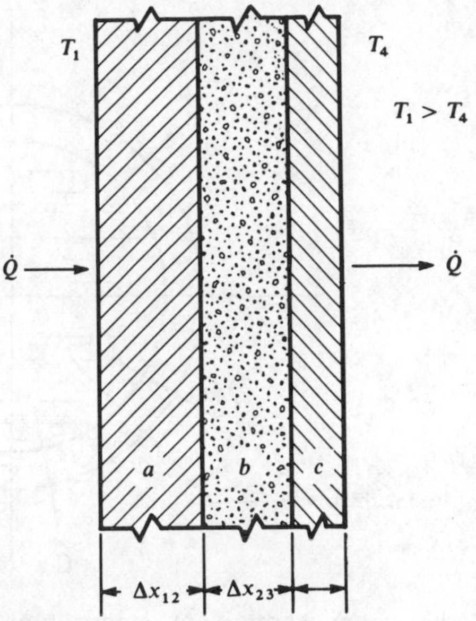

Figure 3–7 Composite wall heat transfer.

Example 3.6 A wall of a house is composed of 4-in thick brick on the outside, $\frac{3}{4}$-in thick plaster board on the inside, and 3 inches of rock wool insulation between these two surfaces. Estimate the heat loss per square foot of surface area exposed to the outside weather at 0°F if the inside wall surface is to be at 55°F. Assume that the plaster board has a thermal conductivity of 0.2 Btu/ft · °F · hr.

Solution Using equation (3–12) we obtain

$$\frac{\dot{Q}}{A} = \frac{\Delta T}{\Delta x_{12}/k_a + \Delta x_{23}/k_b + \Delta x_{34}/k_c}$$

where

$$\Delta T = 55°\text{F} - 0°\text{F} = 55°\text{F}$$
$$\Delta x_{12} = \tfrac{3}{4}\text{ in} = \tfrac{3}{48}\text{ ft}$$
$$\Delta x_{23} = 3\text{ in} = 0.25\text{ ft}$$
$$\Delta x_{34} = 4\text{ in} = \tfrac{4}{12}\text{ ft}$$
$$k_a = 0.2\text{ Btu/ft} \cdot °\text{F} \cdot \text{hr}$$

From Table 3–1 we estimate

$$k_b = 0.035\text{ Btu/ft} \cdot °\text{F} \cdot \text{hr}$$

$$k_c = 0.32\text{ Btu/ft} \cdot °\text{F} \cdot \text{hr}$$

and we then obtain

$$\frac{\dot{Q}}{A} = \frac{55°\text{F}}{(3/48\text{ ft})/(0.2\text{ Btu/ft} \cdot °\text{F} \cdot \text{hr}) + (0.25)/(0.035) + (4/12)/(0.32)}$$
$$= 6.47\text{ Btu/hr} \cdot \text{ft}^2$$

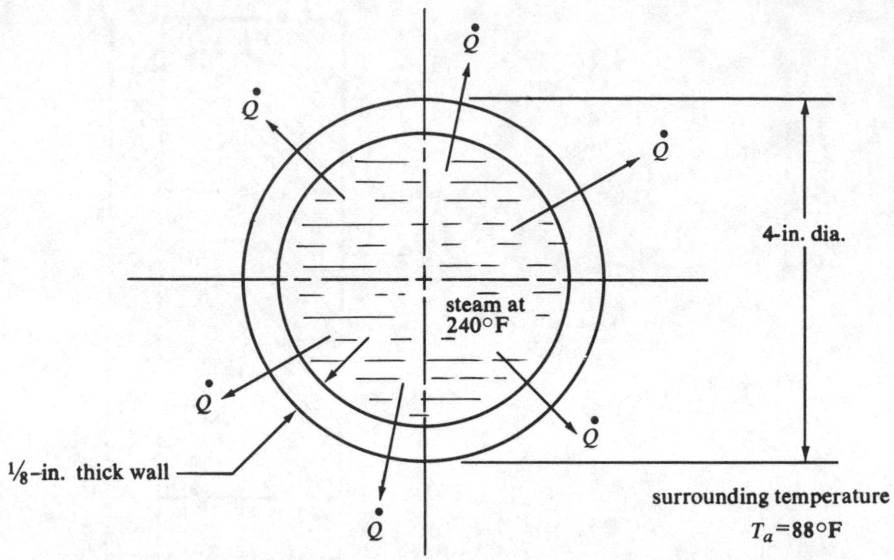

Figure 3–8 Radial cross section of steam pipe with conduction heat transfer to surroundings.

Frequently heat transfers occur through curved surfaces and, in particular, through circular or cylindrical surfaces such as pipes and tubes. For the case of conduction heat transfer through cylindrical sections, as shown in figure 3–8, the heat transfer rate becomes (see problem 3.20)

$$\dot{Q} = \frac{2\pi l k(\Delta T)}{\ln(d_o/d_i)} \tag{3–13}$$

where l is the axial length of the cylindrical section, d_o is the outside diameter, and d_i is the inside diameter. The heat transfer rate per unit of axial length is then given by

$$\frac{\dot{Q}}{l} = \frac{2\pi k(\Delta T)}{\ln(d_o/d_i)} \tag{3–14}$$

If the cylindrical section is composed of, say, two concentric materials as shown in figure 3–9 one can obtain for the heat transfer rate per unit of axial length

$$\frac{\dot{Q}}{l} = \frac{\pi(\Delta T)}{(1/2k_a)\ln(d_2/d_1) + (1/2k_b)\ln(d_2/d_1)} \tag{3–15}$$

Example 3.7

Iron pipe is used to convey steam from a steam turbine to a condenser where the steam will change from a vapor to a liquid. During the flow the steam is found to be 240°F at a certain point in the length of the pipe. Assuming that the atmosphere surrounding the pipe is 88°F, determine the rate of heat loss per foot length of pipe. The pipe has an outside diameter of 4 inches and a wall thickness of $\frac{1}{8}$ inch, as shown in figure 3–8.

Solution The solution to this problem can be obtained by using equation (3–14). The thermal conductivity of the iron pipe, estimated from table 3–1, is

$$\kappa = 40 \text{ Btu/hr} \cdot \text{ft} \cdot {}^\circ\text{F}$$

and the temperature difference ΔT across the pipe wall is 240°F − 88°F = 152°F. The outside diameter d_o is 4 inches, or $\frac{4}{12}$ foot, and the inside is 3.75 inches, or 3.75/12 foot.

Then we obtain

$$\frac{\dot{Q}}{l} = \frac{(40 \text{ Btu/hr} \cdot \text{ft} \cdot {}^\circ\text{F})(\pi)(152^\circ\text{F})}{\ln(4/3.75)}$$

$$= 296{,}000 \text{ Btu/ft} \cdot \text{hr} \qquad \textit{Answer}$$

This example is one of steady state, that is, at a point in space the properties remain the same with time, even though the material passing through (steam) will likely have its properties changed at some other point in space. A state where properties such as temperature change at a given point with time elapsing, is called a *transient condition*, which we have here avoided. Other examples of conduction heat transfer and a more complete description of this phenomenon can be found in the references listed in appendix C.

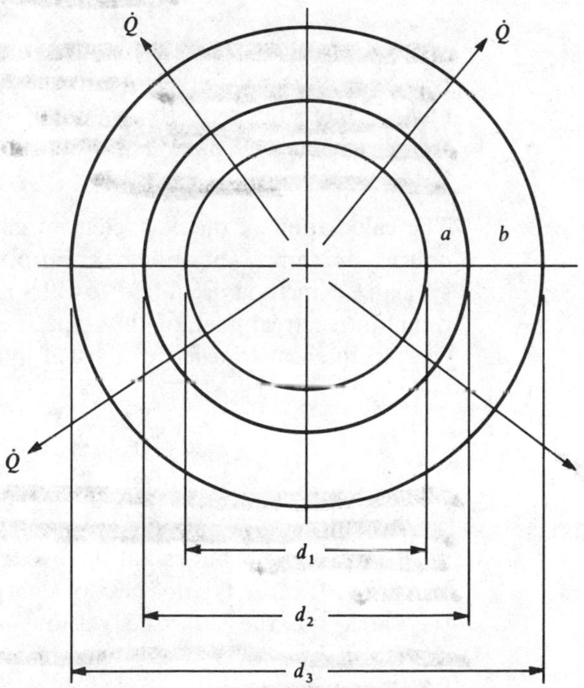

Figure 3–9 Heat flow through concentric cylindrical walls.

3.4 Convection Heat Transfer

In the previous section we considered two examples of conduction heat transfer through a container or a wall. In the first example, water was boiling in a pot, but a closer examination of this problem would reveal that the water at the bottom (closest to the source of energy or the stove burner) would boil first. In fact, although the water will not all be at the same temperature simultaneously, the temperature can be made more uniform by stirring the water. This type of heat transfer then is an example of *forced convection* and is the phenomenon involving conduction and flow of matter due to some outside agent (we stirred the water). Similarly, in example 3.7 the transfer of heat from the 240°F steam through the wall of the pipe is another example of forced convection, since the steam was moving (mass transfer) and there was obviously a conduction of heat at the inside surface of the pipe. There is one other form of convection, called *free convection*, an example of which is the transfer of heat from the pipe to the surrounding air in example 3.7 The natural or free currents of air passing around the outside of the pipe cause a mass transfer and again, conduction at the outer surface of the pipe. If, however, someone were to operate a fan and force air around the pipe, rather than rely on nature, the phenomenon would be considered forced convection.

The rate of heat transfer, $\dot{Q}$, for convection heat transfer is found from the equation known as Newton's law of heat transfer:

$$\dot{Q} = K_h A \Delta T \text{ Btu/hr} \tag{3-16}$$

where K_h is the coefficient of convection heat transfer, a function of the type of materials involved, the magnitude of the motion of the material, and the temperature of the material; and where ΔT is the temperature difference between the temperatures of the two materials convecting heat across a surface area A.

The calculation of the coefficient of convection heat transfer K_h, is, of course, the important and critical problem in convection, and there are specialized methods required to calculate it. For a more complete and quantitative treatment of this type of heat transfer, you may want to refer to the heat transfer texts listed in appendix C.

3.5 Radiation Heat Transfer

Conduction and convection are two types of heat transfer which occur due to intimate contact between materials. Neither type can occur through a vacuum, but radiation, or radiant heat transfer, occurs in spite of matter. It is the type of heat or energy transfer which occurs between the earth and the sun, and without which people could not survive. Radiation is emitted from all matter, and the rate can be determined from the equation

$$\dot{Q} = \epsilon \sigma A T^4 \tag{3-17}$$

where

T = the temperature of the body emitting radiant heat
A = the surface area normal to the radiation
$\sigma = 5.67 \times 10^{-8}\ \mathrm{W/m^2 \cdot K^4}$ | Stefan-Boltzmann
$\sigma = 0.174 \times 10^{-8}\ \mathrm{Btu/hr \cdot ft \cdot {}^\circ R^4}$ | constant
ϵ = emissivity, or coefficient of radiant heat transfer

For an object which has perfect emission and absorption, ϵ has a value of 1. A type of object or material having these characteristics is called a *black body* and can be approximated very nearly in lampblack, water, and and in hollow containers having only one opening through which radiation can be emitted or absorbed. An example of this last type is seen in figure 3–10, where we are considering a beam of light (one of the more common forms of radiation). Notice in the figure that the light beam enters the only opening to the chamber and subsequently reflects and re-reflects continuously. Conversely, this process could be considered in reverse and then all emission from the inner walls of the chamber would exit through the single opening, thus acting as a perfectly emitting body, or a black body.

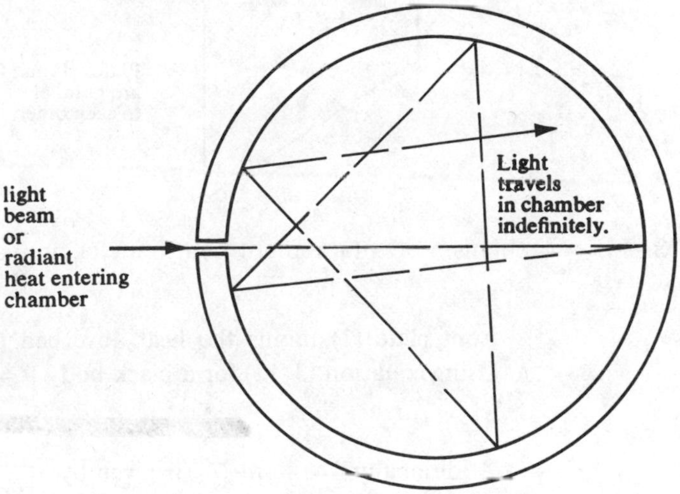

Figure 3–10 Cross section of black body and mechanism for complete absorption of radiant heat, $\epsilon = 1$.

Now let us consider the radiant interaction between two black bodies. Figure 3–11 shows two flat plates which are parallel and which have radiant heat transfer between them. The radiant heat transfer from plate (1) to plate (2) is represented by the term $\dot{Q}_{12}$ and, conversely, $\dot{Q}_{21}$ represents the heat transfer from plate (2) to plate (1). If we assume both plates are black body radiators, then we have $\dot{Q}_{12}$ = the heat emitted

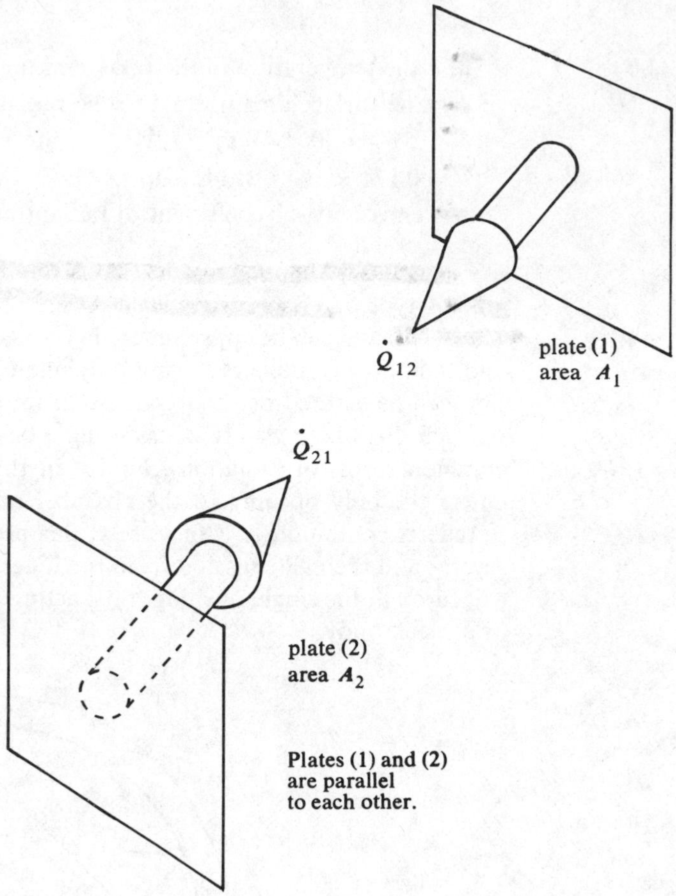

Figure 3–11 Mechanism of total radiant heat transfer between two plates.

from plate (1) minus the heat absorbed by plate (1) from plate (2). Using equation (3–17) for a black body ($\epsilon = 1$), we obtain

$$\dot{Q}_{12} = \sigma(A_1T_1^4 - A_2T_2^4) \tag{3–18}$$

Additionally, the term $\dot{Q}_{21}$ is given by

$$\dot{Q}_{21} = \sigma(A_2T_2^4 - A_1T_1^4) \tag{3–19}$$

A comparison of equations (3–18) and (3–19) indicates that

$$\dot{Q}_{12} = -\dot{Q}_{21}$$

which seems reasonable for a condition of equilibrium.

Example 3.8 Two black body plates are shown in figure 3–11. The plates are parallel and have areas of 3 ft². The temperature of plate (1) is 300°F and 2000°F for plate (2). Find the net heat transfer and state whether it is directed toward plate (1) or plate (2).

Solution The surface temperatures are

$$T_1 = 300°F + 460 = 760°R$$
$$T_2 = 2000°F + 460 = 2460°R$$

and from equation (3–14) we have

$$\dot{Q}_{12} = 0.174 \times 10^{-8} \text{ Btu/hr} \cdot \text{ft}^2 \cdot °\text{R}^4(3 \text{ ft}^2 \times 760^4 - 3 \text{ ft}^2 \times 2460^4)$$
$$= -1.894 \times 10^5 \text{ Btu/hr} \qquad \textit{Answer}$$

In addition,

$$\dot{Q}_{21} = 1.894 \times 10^5 \text{ Btu/hr} \qquad \textit{Answer}$$

We see that the net heat transfer is into plate (1) which is at the lower temperature; that is, there is more radiant energy transferred into plate (1) than is emitted by plate (1).

In this example problem many simplifying assumptions were made. Among these is the condition that heat transfer is only radiated between the two plates. In reality, however, each of the plates would probably radiate energy to other bodies or area and thereby alter the rates of heat transfer between the two. Also, it should be stressed that though heat transfer is conveniently categorized as *conduction, convection,* and *radiation,* any real heat transfer process is composed of all three of these types in varying amounts. A lengthier and more detailed treatment of radiant heat transfer is beyond the purposes of this book and the reader interested in further discussions of heat transfer is here referred to the references in appendix C.

3.6 Reversibility

In section 3.1, the work done by the gas contained in a piston-cylinder device was found to be given by $\int pdV$ or the area under a curve in a p-V diagram. What we want to know, however, is the amount of work available external to the piston-cylinder. This work will be equivalent to an external force F_x times a distance (figure 3–12). But the force F_x is given by

$$F_x = F_g - F_f - F_i$$

in figure 3–12. In all of this discussion, the piston is assumed to have no mass and therefore no inertia. The work transmitted outside the piston cylinder is then

$$Wk_x = \int pdV - \text{friction work} - \text{inertia work} \qquad (3\text{–}20)$$

where *friction work* and *inertia work* are the efforts required to overcome friction of the piston and inertia of the gas and piston, respectively. In reality, inertia is not overcome, but rather detracts from the gas pressure inside the piston. This is equivalent to the result of equation (3–20).

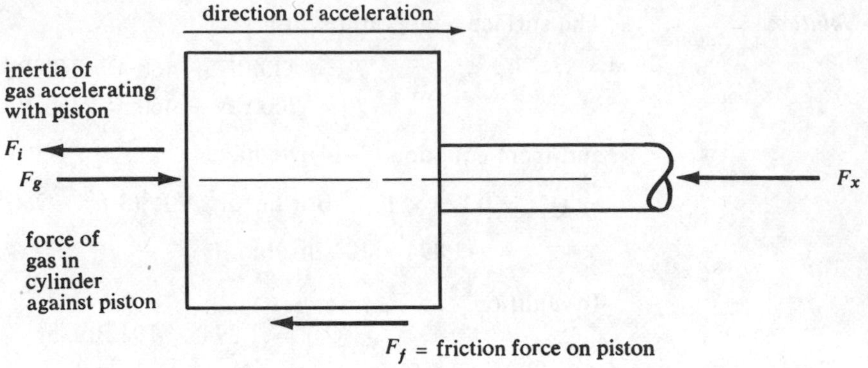

Figure 3–12 Forces acting on massless piston in actual piston-cylinder work process with friction and acceleration.

Recall from physics mechanics that inertia (or inertia force) is given by the mass times the acceleration of the mass; and acceleration is the change of velocity per unit time. Inertia must then be overcome if the piston is to move, starting from rest, in a limited amount of time since there will always be some acceleration and consequently an inertia force of the gas which reduces the pressure acting on the piston.* With the intent of not treating all that is pertinent to the subject, friction and inertia will generally be ignored in this book as they are the two most mentioned items causing irreversibilities in processes. Now we can write equation (3–20) as

$$Wk_x = \int pdV - \text{irreversible work} \tag{3–21}$$

If the piston or mechanism is frictionless and a process is carried out very, very slowly so that the piston is always in a static or quasi-static condition then we can write

$$Wk_x = \int pdV \tag{3–22}$$

since

$$\text{Irreversible work} = 0$$

The process is then said to be *reversible* which means that work can be extracted from or put into the system, depending on the requirement. Of course, there are no actual reversible processes but many are close to this idealization and equation (3–22) is a good description for use in engineering analysis. Keep in mind that the term $\int pdV$ is an area under a

*The student should observe that during deceleration of the piston, the inertia force will be opposite from that shown in figure 3–12, but the gas in the cylinder will not reflect this on the face of the piston. Consequently, inertia will detract from the effective gas pressure but will not, under reversed conditions, add to the pressure.

curve in a *p-V* diagram and it can be found by calculus or simple geometry.

Now that we have considered *reversible* and *irreversible* as adjectives describing energy transfer processes (work) we should consider these words as they describe heat. Reversible work can be determined by neglecting irreversible work, but heat, by its very nature, is one-directional; that is, it always moves from a region of high temperature to a region of lower temperature. Reversible heat transfer is completely unreal at the macroscopic level, and a process which involves heat is irreversible unless the heat transfer is progressing at an infinitesimal rate or unless there is an infinitesimal temperature difference between the bodies involved in the heat transfer. Under these conditions we can have reversible heat transfer but, of course, it will take forever to observe the effects. We will, however, use the concept of reversible heat, and it will have the same connotation as reversible work to allow for simplified analysis of thermodynamic problems.

3.7 The Equivalence of Work and Heat

Work and heat have been defined as mutually exclusive phenomena; that is, if the energy transfer is heat, then it cannot be work, and vice versa. But heat and work are both energies in transition (or energy being transferred) so that work could ultimately affect a system exactly as if the process had involved heat instead of work. The reverse of this statement is not always true, as the second law of thermodynamics will later demonstrate; heat cannot always affect a system exactly like work. There is, however, an equivalence between the common unit of heat, the joule (J), and work, the newton-meter (N · m), or

$$1\ \text{J} = 1\ \text{N} \cdot \text{m}$$

In the English system the common unit for describing heat, the British thermal unit (Btu), and the unit for work, the foot-pound (ft-lbf), are related by the conversion

$$778.16\ \text{ft-lbf} = 1\ \text{Btu}$$

or

$$778.16\ \text{ft-lbf/Btu} = 1$$

Some authors treat the conversion factor 778.16 ft-lbf/Btu as an algebraic term and assign to it the letter *J*. In this text we will refrain from this practice and will consider the conversion from common heat units to common work units as equivalent to converting from feet to inches or any other unit conversion.

Note that when the above conversion factor is recalled we can place heat, work, and energy all in the same units of British thermal units or foot-pounds, as we choose. Let us look at an example of this conversion.

Example 3.9 The internal energy of 3 lbm of air is 60 Btu. How much energy is this in ft-lbf?

Solution This is strictly a unit conversion, so

$$U = 60 \text{ Btu} \times 778 \text{ ft-lbf/Btu}$$
$$= 46{,}680 \text{ ft-lbf} \qquad \textit{Answer}$$

Obviously 1 Btu represents a much greater *amount* of energy than 1 ft-lbf and it is for this reason that the equivalence of the units of heat and work was not readily accepted by the scientific community when first proposed by James Joule in 1842.

3.8 Types of Systems

When the concept of the system was introduced in chapter 2, it was emphasized that the identification of the system was a first step in the thermodynamic method of solving real problems. Here we will classify the system into one of three types, *open, closed,* or *isolated,* indicating that the *second step* in solving a problem is determining what type the identified system is. So that the learner can make the distinction we identify the three types of system:

Open System: *A system whose boundaries allow for mass transfer, heat transfer, and work. That is, the amount of mass and energy in an open system can change.*

Closed System: *A system whose boundaries allow for heat transfer and work, but not mass transfer. That is, the amount of mass of a closed system always remains the same, but the amount of energy can change.*

Isolated System: *A system whose boundaries prevent mass transfer, heat transfer, and work. That is, the amount of mass and energy of an isolated system remains the same.*

Certain peculiarities exist for each of the three systems and will be notated with appropriate subscripts (such as Wk_{cs} for work of a closed system, *cs*) when needed.

3.9 The Forms of Energy

During these past two chapters, *energy* has been a term which entered the discussion frequently. We will see in chapter 4 that conserving energy is the major task of thermodynamics, called the *conservation of energy* or the *first law of thermodynamics.* It involves, naturally, that ubiquitous property of the system, energy; in fact, other properties of the system

Table 3–2
The Forms of Energy

Form	*Type*	*Condition*
Static energy	Potential Kinetic Internal (thermal) Electromagnetic Strain Chemical	System property
Dynamic energy, i.e., energy in transition	Work Heat Heat transfer	Not a system property. Dependent on process. Occurs only during a process.

(such as mass, pressure, volume, and temperature) will be measured so that the amount of energy can be subsequently determined. We know that in open or closed systems, the amount of energy can be increased or decreased; thus energy can be *static* (stationary) or *dynamic* (moving from one place to another).

Table 3–2 concisely lists the forms in which energy exists and makes clear the distinction between static and dynamic energy. In studying the table, note that work and heat exist only when energy is dynamic, that is, only when energy is in a state of motion or transition. (See section 2.5.) As soon as energy becomes static, it changes form, and the forms of heat and work cease to exist. We will see the importance of this change in the next chapter.

Practice Problems

Problems marked with an asterisk (*) are generally more difficult and require the use of differential or integral calculus.

Problems using SI units are indicated with (M) under the problem number, and those using the English units are indicated with (E). Mixed unit problems are listed with (C) under the problem number.

Section 3.1

3.1. **(M)** A force of 20 newtons is required to slide a 30-kg box horizontally across a platform 20 meters long. What work is required?

3.2. **(M)** In problem 3.1, what work is required to lift the 30-kg boxes vertically 20 meters?

3.3. **(M)** A 15-cm-long spring having a modulus of 180 N/cm is deflected an amount which requires 1.8 joules of work. Determine the deflection of the spring.

3.4. (M) For certain reversible processes of a closed system, the pressure-volume relationships are given by the solid lines in figure 3–13. Find the work for these processes.

3.5. (M) In figure 3–13 the dashed lines represent a p-V relationship for a process which involves work. Find the work done.

***3.6.** (M) For a process it is found that $pV^{1.5} = 0.22$ where p is in bars and V is in cubic meters. Determine the work done if the volume increases from 2 to 3 m^3.

3.7. (E) A spring, deflected 1 inch from its free length, is deflected 1 inch further. If the modulus of the spring is 140 lbf/in, what work is required to deflect the spring the second inch?

3.8. (E) A 3000-lbm automobile accelerated at a constant rate from rest to 60 mph in 10 seconds on a flat, straight highway.

(a) How much work was done by the auto's engine in this process? (Hint: force = mass × acceleration; distance = $\frac{1}{2}a\tau^2$.)

(b) If the acceleration to 60 mph required 15 seconds, what was the engine's work output?

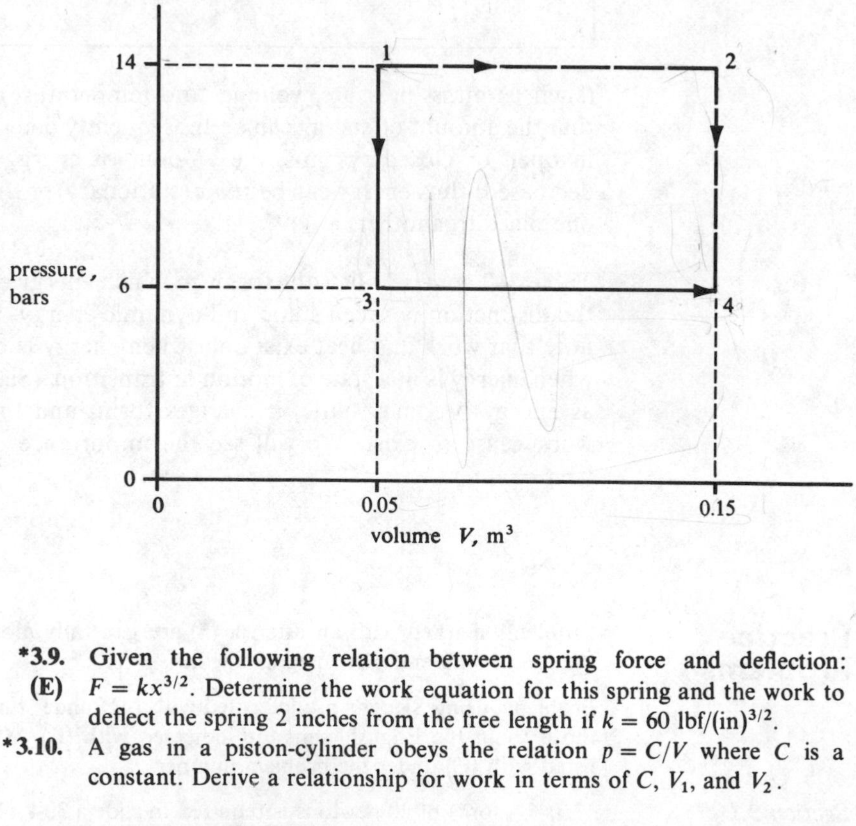

Figure 3–13

***3.9.** (E) Given the following relation between spring force and deflection: $F = kx^{3/2}$. Determine the work equation for this spring and the work to deflect the spring 2 inches from the free length if $k = 60$ lbf/(in)$^{3/2}$.

***3.10.** A gas in a piston-cylinder obeys the relation $p = C/V$ where C is a constant. Derive a relationship for work in terms of C, V_1, and V_2.

Sections 3.2 and 3.3

3.11. Indicate which mode of heat transfer is primarily involved in the following situations:

(a) Frying eggs on a grill.

(b) Heating a room with a fireplace.

(c) Heating a room with an electric space heater.

(d) Broiling a steak.

(e) Welding with an acetylene torch.

3.12. **(M)** A cryogenic sphere of 0.25 meter outside radius holds liquid oxygen at a temperature of $-200°C$. If the sphere is constructed of a 1.5-cm thick steel plate and the atmospheric temperature is 25°C, estimate the heat transfer rate into the sphere. Assume steel and iron have the same thermal conductivities and assume flat surface conduction. For spheres the surface area is $4\pi r^2$.

3.13. **(M)** The rate of heat loss from a 0.1-m^2 area of a furnace wall 40 centimeters thick is 150 watts. If the inside of the furnace is at 1000°C and the wall is composed of fire brick, estimate the outside wall temperature.

3.14. **(M)** The combustion chamber of an internal combustion engine is at a temperature of 1100°C when fuel is burned in the chamber. Assuming that the engine is made of iron and has an average thickness of 6.5 centimeters between the combustion chamber and the outside surface, determine the heat transfer rate per unit of wall area when combustion of fuel occurs. The temperature around the engine is 26°C, but the outside surface of the engine is 38°C.

3.15. **(M)** An insulated steam line has 2 centimeters of rock wool insulation surrounding a 5-cm-outside-diameter iron pipe with a 0.5-cm wall. Determine the heat loss through the line if steam at 110°C is in the line and the surrounding temperature is 10°C.

3.16. **(E)** For figure 3–14, the outside brick wall is at a temperature of 15°F at a certain time in severe winter weather. The inside wallboard, which is separated from the brick wall by a 6-in air gap, is at 78°F. Determine, approximately, the heat rate per square foot area of wall for this condition and indicate the direction, whether from brick to wallboard or vice versa.

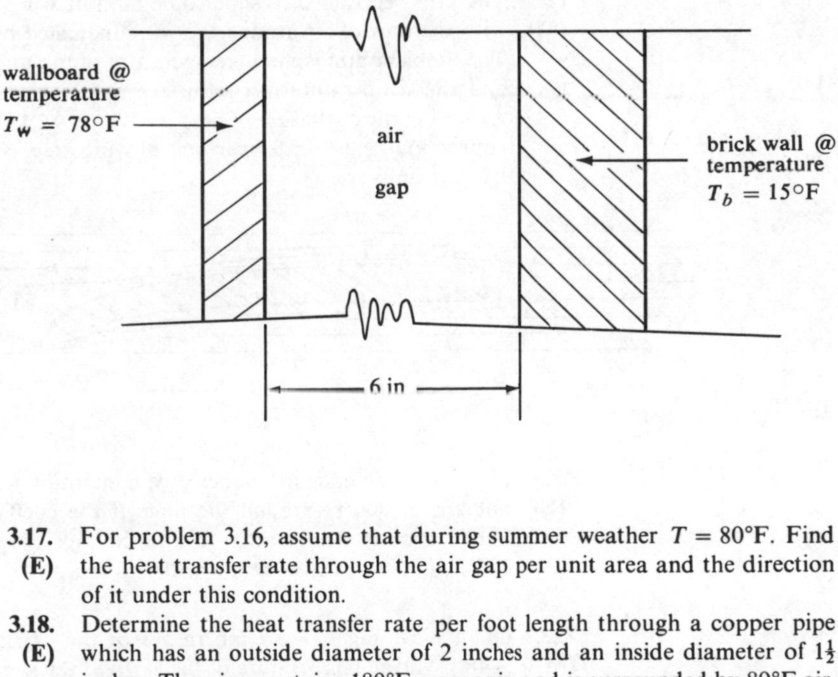

Figure 3–14

3.17. **(E)** For problem 3.16, assume that during summer weather $T = 80°F$. Find the heat transfer rate through the air gap per unit area and the direction of it under this condition.

3.18. **(E)** Determine the heat transfer rate per foot length through a copper pipe which has an outside diameter of 2 inches and an inside diameter of $1\frac{1}{2}$ inches. The pipe contains 180°F ammonia and is surrounded by 80°F air.

3.19. **(E)** Boiler tubes in the superheater section of a steam generating unit are surrounded by combustion gases at 1100°F and the steam flowing through the tubes is at 900°F. If the tubes are made of iron, have an inside diameter of 8 inches, and have $\frac{1}{2}$-inch-thick walls, determine the heat transfer rate to the steam.

3.20. **(C)** Equation (3–10) can be written in differential form as $\dot{Q} = \kappa A\ dt/dx$. Noting that $A = 2\pi rL$ and $dx = dr$, in figure 3–15 derive equation (3–13) for calculating heat transfer rate per unit of length, $\dot{Q}/L$, through the tube if T_o = outside wall temperature and T_i = inside wall temperature.

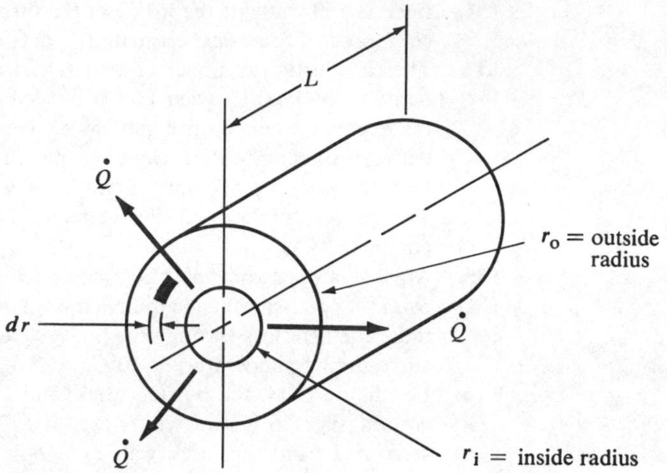

Figure 3–15

Section 3.4

3.21. **(M)** The cross section of a supersonic aircraft wing is shown in figure 3–16, with the flow of air around the wing indicated by lines called *streamlines*. The ambient atmospheric temperature of the undisturbed air t_∞ is -5°C and the temperature of the surface of the wing t_s is found to be 480°C. Assuming the coefficient of heat transfer is 4.2 kW/m² · K, find the rate of convection heat transfer per unit of wing area. Note: answer should be in kW/m² units.

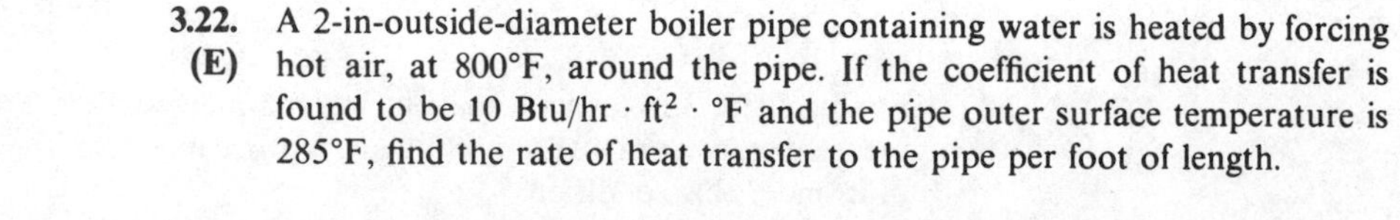

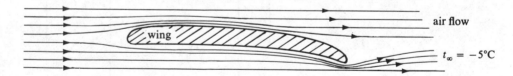

Figure 3–16

3.22. **(E)** A 2-in-outside-diameter boiler pipe containing water is heated by forcing hot air, at 800°F, around the pipe. If the coefficient of heat transfer is found to be 10 Btu/hr · ft² · °F and the pipe outer surface temperature is 285°F, find the rate of heat transfer to the pipe per foot of length.

Section 3.5

3.23. **(M)** On a clear night, the temperature of the sky as seen on the earth is -48°C. If the temperature of the earth at the same time is 15°C and both earth and sky are assumed to be black body radiators, find the net heat transfer rate to the sky per square meter of area.

3.24. **(M)** A radiation pyrometer is a device utilizing radiant heat to measure the temperature of a surface. Assume that a pyrometer has an area of 5 cm^2 and has a temperature of 20°C when it is directed toward a fire having a temperature of 1100°C. Determine the net rate of heat transfer toward the pyrometer, if black body radiation is assumed.

3.25. **(E)** The surface of the sun seems to have a temperature of around 10,000°F. What would you guess the rate of heat emission from the sun per square foot area to be?

3.26. **(E)** Assuming the surface of an electric stove heating element to be 540°F when operating, what would the rate of radiation heat transfer be from the element if it has an emissivity of 0.89? The element has an area of 28 in^2.

Section 3.7

3.27. **(C)** Convert the following:

(a) 17 Btu/lbm to ft-lbf/lbm.
(b) 3350 ft-lbf to British thermal units.
(c) 2,000,000 in · ounces to British thermal units.
(d) 27.8 kilojoules to newton-meters.
(e) 3000 megawatts to basic SI units.

4

The First Law of Thermodynamics

In this chapter, the conservation principles of mass and energy will be introduced as the general vehicles for solving thermodynamic problems. Applying these concepts thus represents the third step in the thermodynamic method, after (1) identifying the system and its boundaries and (2) classifying the system as *open*, *closed*, or *isolated*. The *conservation of mass* will be presented for the general system, and for the *steady state* or *steady flow* condition.

The *conservation of energy*, or the first law of thermodynamics, will be stated and the equations representing this concept will be formulated for the system, with particular emphasis placed on the closed system. The isolated system will be discussed to illustrate the conversion of energy from one form to another when no work or heat is present.

Flow work and *enthalpy* will be introduced and will be used to formulate the equation of the first law of thermodynamics applied to the open system. Particular attention will be given to the steady flow energy equation, so common in engineering problems.

4.1 Conservation of Mass

One of the most fundamental concepts of science is that mass is indestructible; that is, mass can neither be created nor destroyed.* This is the principle known as the *conservation of mass* and for a closed or an isolated system we write that

$$\text{Mass} = \text{constant} \tag{4-1}$$

* An exception to this principle is the theory of relativity which relates, mass and energy, or rest energy of the mass by the equation $E = mc^2$ where c is the velocity of light. Thus, for certain processes such as nuclear reactions, mass and energy are conserved together, but not individually.

If the system is open so that mass can be transferred into or out of it, then the statement of the conservation of mass is written

$$m_{in} - m_{out} = \Delta m_{system} \tag{4–2}$$

where m_{in} is the mass entering the system, m_{out} is the mass leaving the system, and Δm_{system} is the change in mass of the system. (See figure 4–1.)

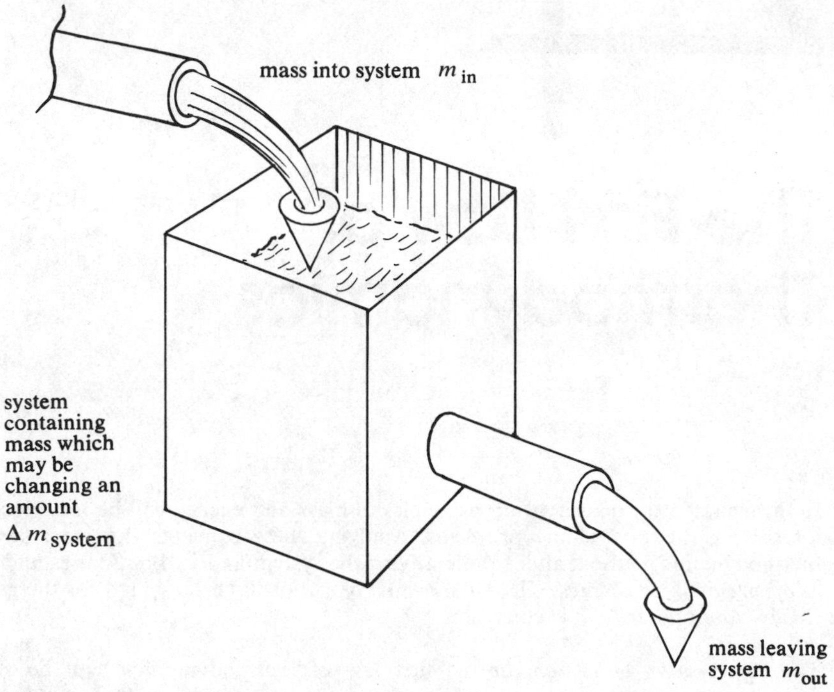

Figure 4–1 Conservation of mass.

The term Δm_{system} is positive if the system is gaining mass, and negative if it is losing mass. Equation (4–2) implies that all terms are related to some common time period, $\Delta\tau$, which can be divided into each of the terms in the equation, giving a new form of the mass balance written

$$\frac{m_{in}}{\Delta\tau} - \frac{m_{out}}{\Delta\tau} = \frac{\Delta m_{system}}{\Delta\tau} \tag{4–3}$$

Now, in mathematical language, as $\Delta\tau$ is made increasingly small, consequently making m_{in}, m_{out}, and Δm_{system} smaller since the time period is decreased, the terms each approach a differential so that we then write

$$\frac{dm_{in}}{d\tau} - \frac{dm_{out}}{d\tau} = \frac{dm_{system}}{d\tau} \tag{4–4}$$

In this text, the differential with respect to time, $d/d\tau$, will be denoted by a "dot" so that equation (4–4) is then written

$$\dot{m}_{in} - \dot{m}_{out} = \dot{m}_{system} \quad \textbf{(4–5)}$$

The two terms $\dot{m}_{in}$ and $\dot{m}_{out}$ are called the *mass flow rate into the system* and the *mass flow rate out of the system*, respectively. The term $\dot{m}_{system}$ is the rate at which the system mass changes with respect to time. The mass flow rate is commonly found from the equation

$$\dot{m} = \rho A \bar{V} \quad \textbf{(4–6)}$$

where A is the cross-sectional area across which mass is moving with average velocity $\bar{V}$ and with density ρ.

To derive equation (4–6), we note that the amount of mass flowing across area A is equal to the volume of the mass times its density. We write this as

$$m = \rho V \quad \textbf{(4–7)}$$

but the volume, V, is also equal to A times the distance through which the leading surface of the mass has traveled during some time period, $\Delta\tau$, as indicated in figure 4–2. If we denote this distance as x then

$$m = \rho A x \quad \textbf{(4–8)}$$

and if this equation is divided by the time period, $\Delta\tau$, on both sides we have

$$\frac{m}{\Delta\tau} = \frac{\rho A x}{\Delta\tau} \quad \textbf{(4–9)}$$

In this last equation $x/\Delta\tau$ is just the velocity of the flowing mass, $\bar{V}$, which then gives us

$$\frac{m}{\Delta\tau} = \rho A \bar{V}$$

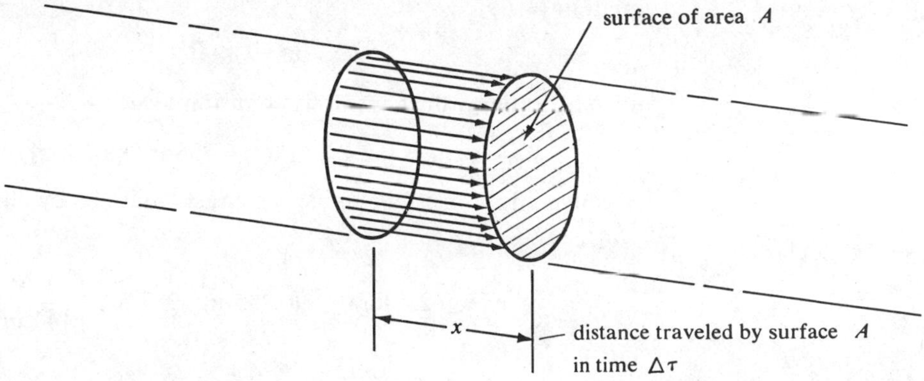

Figure 4–2 Illustration of mass flow.

or, since we have denoted $m/\Delta\tau$ by the notation $\dot{m}$, then

$$\dot{m} = \rho A\bar{V} \tag{4–6}$$

Example 4.1

Kerosene is pumped into an aircraft fuel tank through a hose which has an inside diameter of 4 centimeters. If the velocity of the kerosene is 8 m/s through the hose, determine the mass flow rate. Assume kerosene has a density of 800 kg/m^3.

Solution

The mass flow rate is obtained from equation (4–6). The area of the hose is given by a circular section of 2 centimeters radius, thus

$$A = \pi(2\text{ cm})^2 = 12.6\text{ cm}^2$$

or

$$A = 0.00126\text{ m}^2$$

The mass flow rate is then

$$\dot{m} = (800\text{ kg/m}^3)(0.00126\text{ m}^2)(8\text{ m/s})$$
$$= 8.06\text{ kg/s}$$ *Answer*

Example 4.2

Water flows from a 1-in diameter faucet with a velocity of 8.7 ft/s. Determine the mass flow rate of the water leaving the faucet.

Solution

The water is crossing the circular area of the faucet outlet, equal to pi (π) times the radius squared. Then the area A is calculated by

$$A = \pi\left(\frac{1}{2}\right)^2\text{ in}^2$$

or

$$A = 0.785\text{ in}^2$$

The water is here assumed to be at 78°F where the density is approximately 62.4 lbm/ft^3. The mass flow rate is then determined by using equation (4–6):

$$\dot{m} = \rho A\bar{V}$$

and substituting values into this equation gives us

$$\dot{m} = (62.4\text{ lbm/ft}^3)(0.785\text{ in}^2)(8.7\text{ ft/s})$$

To convert to consistent units, we must multiply by the factor 1/144 ft^2/in^2, so

$$\dot{m} = (62.4\text{ lbm/ft}^3)(0.785\text{ in}^2)(8.7\text{ ft/s})\left(\frac{1\text{ ft}^2}{144\text{ in}^2}\right)$$

Thus

$$\dot{m} = 2.96\text{ lbm/s}$$ *Answer*

From these examples we see that mass flow rate can have units of kilograms per second (kg/s), pounds-mass per second (lbm/s), or any other unit equivalent to *mass per unit time*. In some instances the weight flow rate is specified or required. This quantity is defined as the weight per unit of time and is given by

$$\dot{W} = \frac{W}{\Delta\tau} \tag{4–10}$$

The units of weight flow rate are newtons per second (N/s), pounds-force per second (lbf/s), or any unit of force per unit time. In terms of the mass flow rate we can find that

$$\dot{W} = \dot{m}g \text{ (N/s)}$$

or

$$\dot{W} = \frac{g}{g_c}\dot{m} \text{ (lbf/s)} \tag{4–11}$$

Also,

$$\dot{W} = \gamma A\bar{V} \tag{4–12}$$

where γ is the specific weight as defined in section 2.8.

The volume flow rate, defined as the volume of material crossing an area per unit of time and written

$$\dot{V} = \frac{V}{\Delta\tau} \tag{4–13}$$

is another term for describing flow rate. Since the volume V can be described by Ax we have, from equation (4–13), that

$$\dot{V} = A\bar{V} \tag{4–14}$$

The volume flow rate is described by units of cubic meters per second (m^3/s), cubic feet per minute (ft^3/min), gallons per hour (gal/hr), gallons per min (commonly written gpm), or any other compatible combination of volume per unit time.

Example 4.3 — Determine the volume flow rate for the kerosene in example 4.1.

Solution — Using equation (4–14) we obtain

$$\begin{aligned}\dot{V} &= A\bar{V} \\ &= (0.00126\ m^2)(8\ m/s) \\ &= 0.010\ m^3/s\end{aligned}$$ *Answer*

Example 4.4 Determine the volume flow rate of the water in example 4.2.

Solution We may use equation (4–14) to obtain the volume flow rate. Then

$$\dot{V} = (0.785 \text{ in}^2)(8.7 \text{ ft/s}) \frac{1 \text{ ft}^2}{144 \text{ in}^2}$$

$$= 0.0474 \text{ ft}^3/\text{s} \qquad \textit{Answer}$$

Further, using the conversion of 7.48 gallons (gal) = 1 ft³ we find

$$\dot{V} = (0.0474 \text{ ft}^3/\text{s})(7.48 \text{ gal/ft}^3)$$

$$= 0.355 \text{ gal/s} \qquad \textit{Answer}$$

Recall the preceding discussions of the conservation of mass as described by equations (4–1) or (4–5) for closed and open systems respectively; there is no principle of conservation of weight or volume and caution must be used whenever reference is made to volume flow rate or weight flow rate—a conversion to mass flow rate is a safe approach.

Example 4.5 A cylindrical mixing tank having a diameter of 2 feet and containing 620 lbm of water is being filled from two water lines, one line delivering hot water at a rate of 0.7 lbm/s and a second line with $\frac{5}{8}$-inch diameter delivering cold water at 8 ft/s. If we assume the tank has an exit port of $\frac{3}{4}$-inch diameter from which mixed water discharges at 12 ft/s, determine the rate of change of water level in the tank and the mass of the water in the tank 10 seconds after flow begins. (See figure 4–3.)

Solution For determining the rate of change of water level, the rate of change of mass in the mixing tank must be found. Using the principles of conservation of mass, we write equation (4–5)

$$\dot{m}_{\text{in}} - \dot{m}_{\text{out}} = \dot{m}_{\text{system}}$$

where the system is the mixing tank. Now

$$\dot{m}_{\text{in}} = \text{mass flow from line } A + \text{mass flow from line } B$$

which gives us

$$\dot{m}_{\text{in}} = 0.7 \text{ lbm/s} + \left[\underset{(\text{density})}{\rho_B}\underset{(\text{area})}{A_B}\underset{(\text{velocity})}{\bar{V}_B}\right]$$

The velocity of line B, $\bar{V}_B$, is 8 ft/s; the area of B is $\pi \times (5/16)^2$ in². We assume the density to be 62.4 lbm/ft³. Then the mass flow from line B is

$$\dot{m}_B = (62.4 \text{ lbm/ft}^3)\left[\left(\pi \times \left(\frac{5}{16}\right)^2 \times \frac{1}{144} \text{ ft}^2\right)(8 \text{ ft/s})\right]$$

and then

$$\dot{m}_B = 1.064 \text{ lbm/s}$$

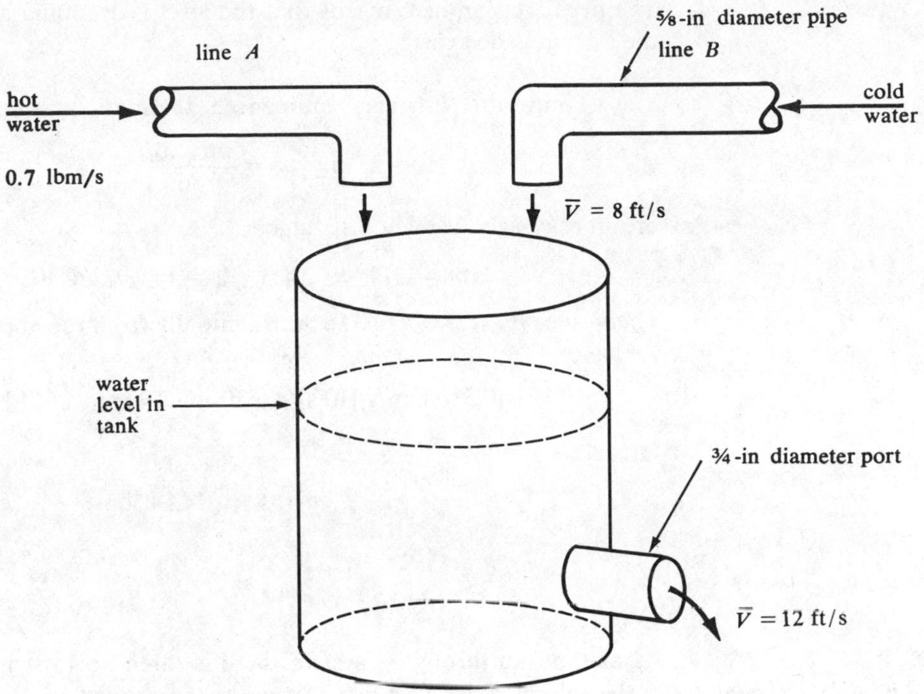

Figure 4–3 Mass flow balance. Example 4.5.

so

$$\dot{m}_{in} = 0.700 \text{ lbm/s} + 1.064 \text{ lbm/s}$$
$$= 1.764 \text{ lbm/s}$$

The mass flow out of the system is just that leaving the port of area $\pi \times (3/4)^2$ in². Assume the density of the mixed water is 62.0 lbm/ft³. We then have

$$\dot{m}_{out} = (62.0 \text{ lbm/ft}^3)\left[\pi \times \left(\frac{3}{8}\right)^2 \text{ in}^2\right]\left(\frac{1}{144} \text{ ft}^2/\text{in}^2\right)(12 \text{ ft/s})$$
$$= 2.28 \text{ lbm/s}$$

The mass rate of change for the system, $\dot{m}_{system}$, is then 1.764 − 2.28 lbm/s or −0.516 lbm/s, from equation (4–5).

The rate of change of the water level in the mixing tank is $\dot{m}_{system}/\rho A_{tank}$:

$$\bar{V}_{level} = \frac{(1.764 - 2.28) \text{ lbm/s}}{(62.0 \text{ lbm/ft}^3)(\pi \times 1 \text{ ft}^2)}$$

or

$$\bar{V}_{level} = \frac{-0.516}{62 \times \pi \times 1 \text{ ft}^2} = -0.00265 \text{ ft/s}$$ *Answer*

The negative sign here means that the level is dropping and the mass of the system is decreasing.

Ten seconds after flow has commenced, the mass is determined from

$$\dot{m}_{\text{system}} = \frac{\Delta m_{\text{system}}}{\Delta \tau}$$

or, more clearly, using a little algebra

$$\Delta \tau \dot{m}_{\text{system}} + m_{\text{system, initial}} = m_{\text{system}} \text{ @ } 10 \text{ s}$$

Then, since $\dot{m}_{\text{system}} = -0.516$ lbm/s initially from the above calculation, we have

$$-(0.516 \text{ lbm/s})(10 \text{ s}) + 620 \text{ lbm} = m_{\text{system}} \text{ @ } 10 \text{ s}$$

or

$$m_{\text{system}} \text{ @ } 10 \text{ s} = 614.84 \text{ lbm} \qquad \textit{Answer}$$

4.2 Steady Flow

As mass flows through a system there is often no loss or gain of mass in the system itself. This tells us then that since $\dot{m}_{\text{system}} = 0$

$$\dot{m}_{\text{in}} - \dot{m}_{\text{out}} = 0$$

or

$$\dot{m}_{\text{in}} = \dot{m}_{\text{out}} \tag{4–15}$$

which indicates that all the mass flows into the system must equal (exactly) the mass flows leaving the system. This condition is called *steady flow* or *steady state* and is the rule rather than the exception in engineering and technological applications. Any engine producing power, refrigerator cooling foods, generator producing electric energy, or any device which is intended to perform for extended periods of time is in steady flow or at the very least, some of the components are in steady flow.

Example 4.6

A nozzle is commonly used to change the velocity of liquids or gases, by changing the cross-sectional area of the flow line. (See section 11.4 for a more complete description of nozzles.) Suppose we have a nozzle with air passing through so that within the nozzle no loss or accumulation of air occurs. The air is entering the nozzle with a velocity of 24 m/s and a density of 1.28 kg/m^3. The density of the air leaving is 1.10 kg/m^3. The nozzle is circular in cross-sectional area and reduces evenly from an entrance diameter of 60 centimeters to an exit diameter of 30 centimeters. Determine the velocity of the air leaving the nozzle. (See figure 4–4.)

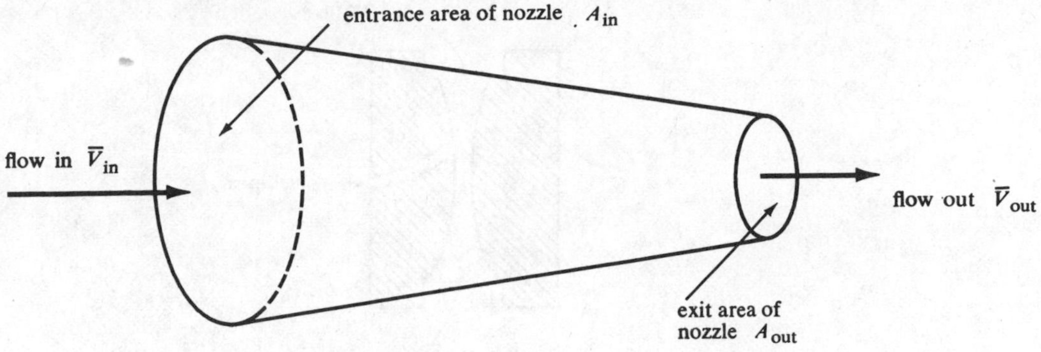

Figure 4–4 Nozzle flow. Example 4.6.

Solution This problem is an example of steady flow, i.e.

$$\dot{m}_{system} = 0$$

so that we may use equation (4–15)

$$\dot{m}_{in} - \dot{m}_{out} = 0$$

If we replace these two terms by using equation (4–6)

$$\rho_{in}A_{in}\bar{V}_{in} - \rho_{out}A_{out}\bar{V}_{out} = 0 \qquad \textbf{(4–16)}$$

we can substitute numbers into this equation. It is an important steady flow relative which we will use often. Into equation (4–16) we substitute values:

$$(1.28 \text{ kg/m}^3)[(\pi)(0.3 \text{ m})^2](24 \text{ m/s}) - (1.10 \text{ kg/m}^3)[(\pi)(0.15 \text{ m})^2](\bar{V}_{out}) = 0$$

which can be solved for $\bar{V}_{out}$:

$$\bar{V}_{out} = 111.7 \text{ m/s}$$ *Answer*

Example 4.7 An automobile carburetor, shown schematically in figure 4–5, mixes air with fuel to provide a combustible mixture for an internal combustion engine. Determine the amount of mixture of fuel and air flowing through the carburetor if 0.01 lbm/s of fuel is consumed and the amount of air per pound-mass of fuel is 20 lbm.

Solution Assuming steady flow through the carburetor we have

$$\dot{m}_a + \dot{m}_f = \dot{m}_{f/a}$$

where

$$\dot{m}_f = 0.01 \text{ lbm/s}$$

and

$$\dot{m}_a = 20 \text{ lbm air/lbm} \times \dot{m}_f = 20 \times 0.01 = 0.2 \text{ lbm/s}$$

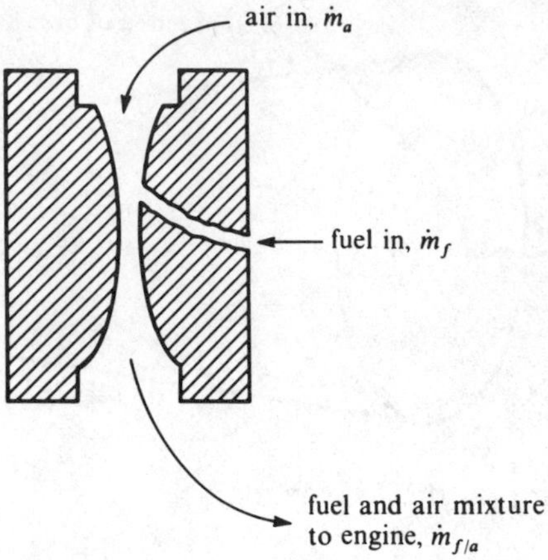

Figure 4–5 Schematic diagram of carburetor of internal combustion engine. Example 4.7.

Thus,

$$\dot{m}_{f/a} = 0.2 \text{ lbm/s} + 0.01 \text{ lbm/s} \qquad \textit{Answer}$$

$$= 0.21 \text{ lbm/s} \qquad \textit{Answer}$$

4.3 Conservation of Energy

As a system goes through a process, some properties of the system are altered and we have postulated that the mass is conserved. If the system is closed or isolated, then the system mass remains unchanged for any process; but if the system is open, the mass will change according to equation (4–5). Similarly, we now postulate that energy is conserved in any process of a system and we write

$$E_{in} - E_{out} = \Delta E_{system} \tag{4–17}$$

where a positive energy change of the system implies an accumulation of energy in the system and a negative value implies loss of system energy. In terms of rates of change we say

$$\dot{E}_{in} - \dot{E}_{out} = \dot{E}_{system} \tag{4–18}$$

in a manner similar to the conservation of mass in section 4.1.

In equation (4–17), E_{in} and E_{out} represent energy crossing the boundary of the system under scrutiny—they are terms of energy in transition. From chapter 3 we recall the definitions of heat and work, which are precisely the energy terms under discussion. We quite arbitrarily assign the following sign conventions:

+*Heat*—implies energy *into* the system.

+*Work*—implies energy *out of* the system.

– *Heat*—implies energy *out of* the system.
– *Work*—implies energy *into* the system.

and, using Q for heat and Wk for work, we get the following from equation (4–17):

$$Q - Wk = \Delta E_{\text{system}} \tag{4–19}$$

Notice that we arrived at equation (4–19) by saying

$$E_{\text{in}} = +Q \quad \text{or} \quad -Wk$$

and

$$E_{\text{out}} = -Q \quad \text{or} \quad +Wk$$

With a positive heat transfer, we "heat up" the system; and with negative heat, we "cool down" the system. Similarly, work gained from a system is positive; work put into the system is negative.

Recalling from chapter 3 that a rate of heat transfer $\dot{Q}$ is a form of the rate of energy transition and defining here $\dot{Wk}$ as the rate of work or *power*, we then obtain the following from equation (4–19):

$$\dot{Q} - \dot{Wk} = \dot{E}_{\text{system}} \tag{4–20}$$

The remainder of this text is devoted to clarification and examples of equations (4–19) and (4–20).

Remember that the energy being used in these equations is any of the forms of energy in the static condition referred to in chapter 3, that is, kinetic, potential, or internal energy. Other forms or adjectives are equally acceptable such as *strain, electromagnetic*, or *chemical*, but only the three common ones will generally be considered in this text.

Example 4.8 A fuel tank is filled with propane gas and heat is transferred from the surroundings at a rate of 30 Btu per hour. Determine the increase in energy of the propane gas over a 24-hr period. (See figure 4–6.)

Solution We first identify the system to be analyzed—the propane gas inside the fuel tank. Assuming the tank is rigid and no gas or air is entering or leaving the tank, we have no mass change of the system and no reversible work since there is no change in volume. We then write the first law as

$$\Delta E = Q - Wk$$

or

$$\Delta E = Q$$

since no work is present, that is,

$$Wk = 0$$

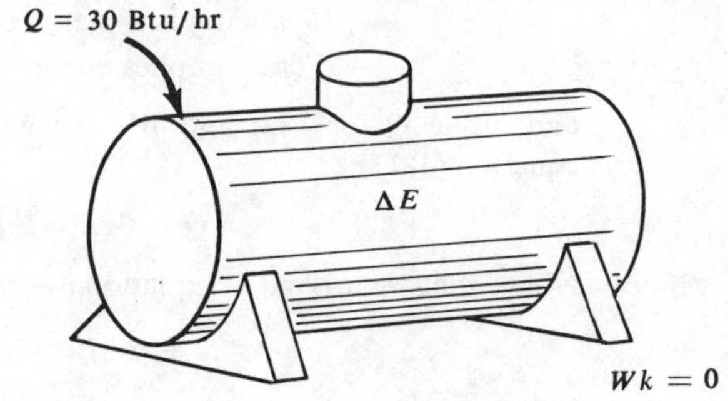

Figure 4–6 Propane tank subject to heat transfer. Example 4.8.

Since the rate of heat transfer was given, we can write

$$\dot{E} = \dot{Q}$$

so that, since $\dot{Q}$ is 30 Btu/hr,

$$\dot{E} = 30 \text{ Btu/hr}$$

The definition of the rate of energy change is

$$\dot{E} = \frac{\Delta E}{\Delta \tau}$$

where $\Delta\tau$ is the time during which the energy changes an amount ΔE. We then have

$$\Delta E = (30 \text{ Btu/hr})\Delta\tau$$

and

$$\Delta\tau = 24 \text{ hr}$$

so

$$\Delta E = 720 \text{ Btu}$$ *Answer*

Example 4.9 A piston-cylinder device does 7800 ft-lbf of work while losing 3.7 Btu of heat. What is the change of energy of the contents of the piston-cylinder? Give the answer in SI units as well as English units.

Solution Here we have a closed system where energy is conserved so we write equation (4–19)

$$\Delta E = Q - Wk$$

Since 7800 ft-lbf of work is done, work is a positive quantity in the equation. The amount of heat lost is a negative quantity. We then obtain

$$\Delta E = -3.7 \text{ Btu} - 7800 \text{ ft-lbf}$$

Since 778 ft-lbf is equal to 1 Btu we have

$$\Delta E = -3.7 \text{ Btu} - \frac{7800}{778} \text{ Btu}$$

$$= -13.7 \text{ Btu}$$ *Answer*

In SI units we have, from Table B.15, the conversion 1054 joules = 1 Btu, so

$$\Delta E = -13.7 \times 1054 \text{ J}$$

$$= -14{,}400 \text{ J} = -14.4 \text{ kJ}$$ *Answer*

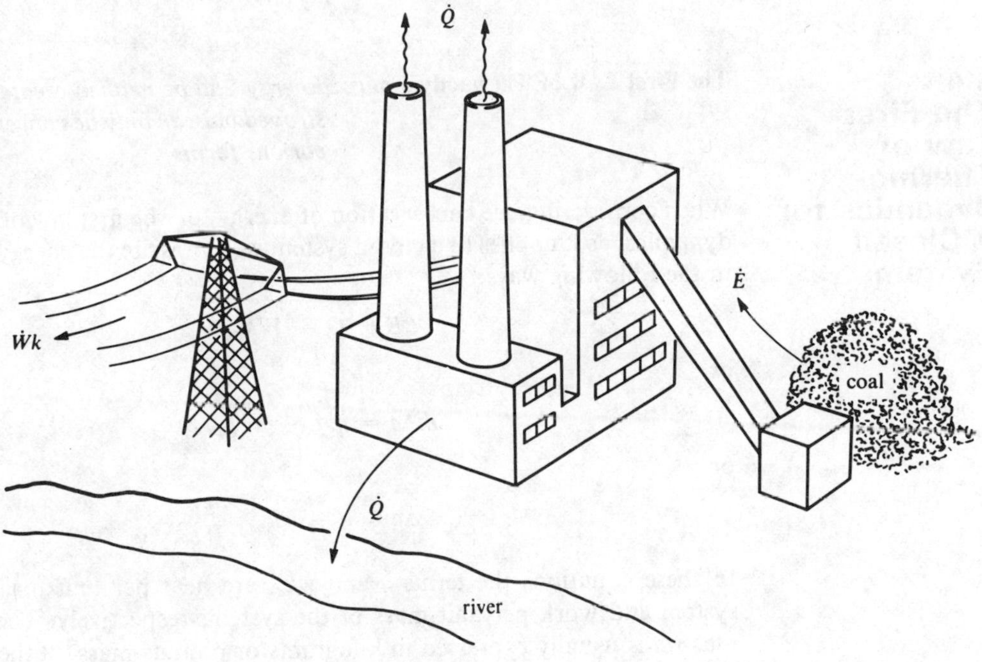

Figure 4–7 Power station as a system. Example 4.10.

Example 4.10 An electric power generating station, sketched in Figure 4–7, produces 100 megawatts (MW) of power. If the coal releases 900×10^6 kJ/hr of energy, determine the rate at which heat is rejected from the power plant.

Solution The system is the power station and the coal pile adjacent to the facility. From the conservation of energy we get

$$\dot{Q} = \dot{E} + \dot{W}k$$

and

$$\dot{Q} = -900 \times 10^6 \text{ kJ/hr} + 100 \text{ MW}$$
$$= -900 \times 10^3 \text{ MJ/hr} + 100 \text{ MW}$$
$$= -250 \text{ MJ/s} + 100 \text{ MW}$$
$$= -250 \text{ MW} + 100 \text{ MW} = -150 \text{ MW} \qquad \textit{Answer}$$

The negative sign given the rate of energy release $\dot{E}$ indicates the condition of decreasing energy whereas the negative sign for the heat release $\dot{Q}$ indicates that the heat was rejected by the system to the river or atmosphere.

4.4 The First Law of Thermodynamics for a Closed System

The First Law of Thermodynamics: *Energy can be neither created nor destroyed but can only be converted to its various forms.*

When considering the conservation of energy, or the first law of thermodynamics, as it applies to a closed system we can write the energy balance in the following ways:

$$\Delta E = Q - Wk_{cs}$$

or

$$m\Delta e = mq - mwk_{cs} \tag{4–21}$$

or

$$\Delta e = q - wk_{cs} \tag{4–22}$$

In these equations the terms q and wk_{cs} are heat per unit mass of the system and work per unit mass of the system, respectively. The system mass m is usually expressed in kilograms or pounds-mass. If the process is *reversible* we can invoke equation (3–8)

$$Wk_{cs} = \int_{V_1}^{V_2} p\,dV$$

and then have

$$\Delta E = Q - \int_{V_1}^{V_2} p\,dV \tag{4–23}$$

or

$$\Delta e = q - \int_{v_1}^{v_2} p\,dv \tag{4–24}$$

We can also consider time as a variable, so equation (4–21) becomes

$$\dot{E} = \dot{Q} - \dot{W}k_{cs} \tag{4–25}$$

Using the same terms as for equation (4–21) we have

$$\frac{d}{d\tau}me = m\frac{d}{d\tau}e + e\frac{d}{d\tau}m \tag{4–26}$$

but

$$\frac{dm}{d\tau} = 0 = \dot{m} \quad \text{(closed system)} \tag{4–27}$$

so

$$\dot{E} = \frac{d}{d\tau}me = m\dot{e} \tag{4–28}$$

Similarly

$$\dot{Q} = m\dot{q} \tag{4–29}$$

and

$$\dot{W}k_{cs} = m\dot{w}k_{cs} \tag{4–30}$$

which then gives us, for the first law of thermodynamics applied to closed systems

$$\dot{e} = \dot{q} - \dot{w}k_{cs}$$

or

$$m\dot{e} = \dot{Q} - \dot{W}k_{cs} \tag{4–31}$$

We have at this time one form of equations (4–23) or (4–24) which can be solved arithmetically. This is the *constant pressure process* considered in example 3.4 where we arrived at the following:

$$\begin{aligned} Wk_{cs} &= p\Delta V \\ &= p(V_{\text{final}} - V_{\text{initial}}) \end{aligned}$$

For *constant pressure reversible process* only, equation (4–21) becomes

$$\Delta E = Q - p(V_{\text{final}} - V_{\text{initial}}) \tag{4–32}$$

4.5 The First Law of Thermodynamics for an Isolated System

Considerations of isolated systems eliminate any heat or work or mass transfer. All we can say about an isolated system then is

$$\Delta E = 0$$

Since $\Delta m = 0$ for isolated systems and $E = me$ then

$$m\Delta e = 0 \quad \text{or} \quad \Delta e = 0 \tag{4–33}$$

for a system having mass m and energy e. All that happens here is that energy is converted from one form to another, and interestingly there is no indication outside of the isolated system as to what is transpiring inside. Some consider our universe as an isolated system, although it has not been established as such and although no actual isolated systems have been identified in the strictest sense.

Example 4.11 In Figure 4–8 is shown a closed chamber whose walls do not allow heat transfer (called adiabatic walls), and which contains 9 kilograms of dust uniformly distributed. The kinetic energy of the dust is 100 joules as the dust is moving about in the chamber. Determine the changes in energy as the dust settles to the bottom.

Solution Since the chamber does not allow heat transfer ($Q = 0$) and since it is closed off, it cannot convey mechanical work ($Wk = 0$); thus we write

$$\Delta E = 0$$

or

$$\Delta \text{KE} + \Delta \text{PE} + \Delta U = 0$$

The initial kinetic energy is 100 joules and the final is zero. Then

$$\begin{aligned}\Delta \text{KE} &= \text{KE}_{\text{final}} - \text{KE}_{\text{initial}}\\ &= 0 - 100\text{ J}\\ &= -100\text{ J}\end{aligned}$$

Answer

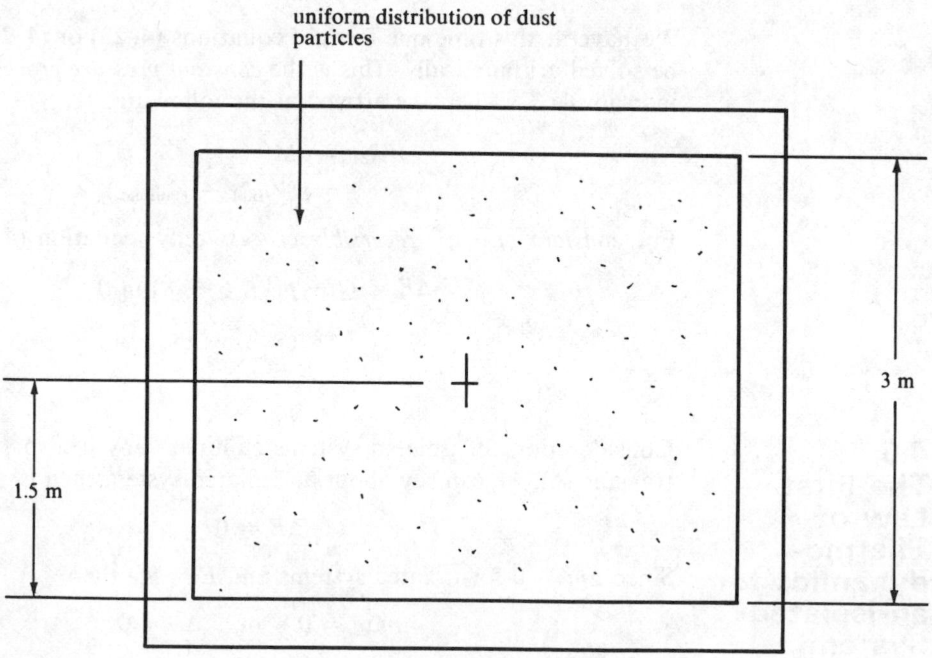

Figure 4–8 Adiabatic chamber containing dust. Example 4.11.

Since the dust is uniformly distributed, the initial potential energy of the dust is simply the total weight of the dust particles times the average height. Thus

$$\text{PE}_{\text{initial}} = mgz = (9\ \text{kg})(9.8\ \text{m/s}^2)(1.5\ \text{m})$$
$$= 132\ \text{kg} \cdot \text{m}^2/\text{s}^2 = 132\ \text{N} \cdot \text{m}$$
$$= 132\ \text{J}$$

where the gravitational acceleration is assumed to be 9.8 m/s². The final potential energy is zero if the bottom of the chamber is assigned as the level of no potential energy. Then

$$\Delta\text{PE} = \text{PE}_{\text{final}} - \text{PE}_{\text{initial}} = 0 - 132\ \text{J}$$
$$= -132\ \text{J}$$ *Answer*

and

$$\Delta U = -\Delta\text{KE} - \Delta\text{PE}$$

or

$$\Delta U = U_{\text{final}} - U_{\text{initial}} = -(-100\ \text{J}) - (-132\ \text{J})$$
$$= 232\ \text{J}$$

If we arbitrarily say

$$U_{\text{initial}} = 0$$

then

$$U_{\text{final}} = 232\ \text{J}$$ *Answer*

In this example we used the term *adiabatic wall* to denote no heat transfer. Throughout the text we will quite often refer to an *adiabatic process* which is a process where there is no heat transfer (Q or $\dot{Q}$ is zero), whether the process applies to open, closed, or isolated systems. Example 4.11 is an adiabatic process.

4.6 Flow Energy and Enthalpy

Thermodynamics has been shown to apply quite generally to closed and isolated systems. No mass transfer is involved and only energy transfer in the form of work and heat is of concern. But most of the systems encountered in technology are open systems, and we should hope to use the thermodynamic approach for these as well. Let us then look carefully at the unique mechanism of the open system—mass moving across the system boundary. In figure 4–9, we have defined an open system and for simplistic purposes provided only two pipes, *A* and *B*, through which mass may enter or leave the system. Now consider some mass moving

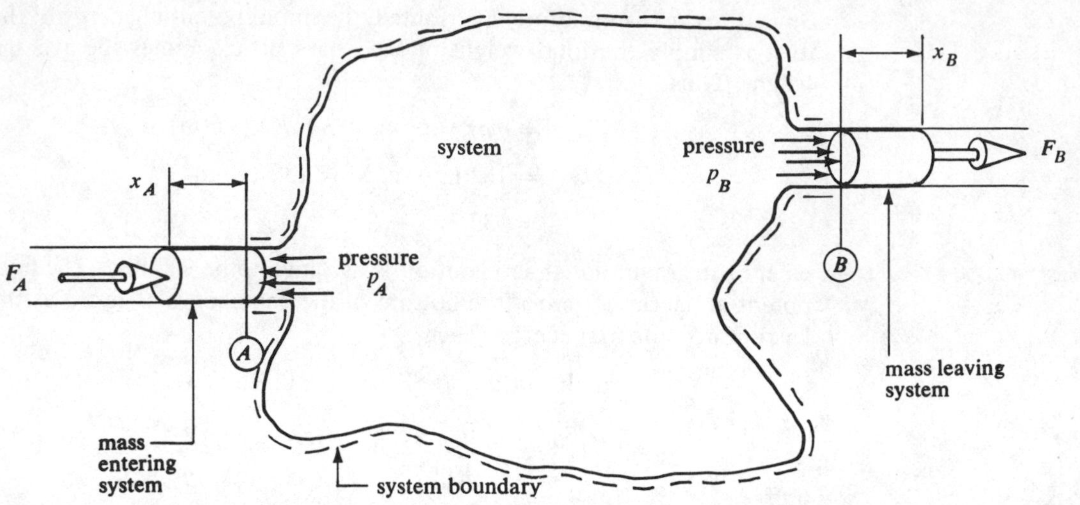

Figure 4–9 Open system.

into the system on the left boundary of the system in figure 4–9. In order that the mass fully enter the system, it must move the distance x_A along pipe A. But for it to move this distance *someone or something external must exert a force* F_A through the distance x_A. We then say there is flow energy required to "flow the mass" into the system:

$$\text{Flow energy} = F_A x_A$$

Also, F_A is equal to the pressure p_A of the system, times the cross-sectional area of the pipe, A. Then

$$\text{Flow energy at } A = p_A A x_A = p_A V_A \tag{4–34}$$

since

$$\begin{aligned} A x_A &= \text{Volume of the mass entering at } A \\ &= V_A \end{aligned}$$

Frequently the term flow work is used for flow energy but for this development let us use the term flow energy.

The flow energy per unit mass at A is

$$p_A \frac{V_A}{m_A} = p_A v_A \tag{4–35}$$

At pipe B we consider that mass is leaving the system, so the mass will move out of its own, propelled by a back pressure p_B, and, associated with it, the flow energy out of the system. At B, the flow energy per unit-mass is $p_B v_B$ in a completely analogous development as for pipe A. Keep in mind, though, that while work (or flow energy) was required

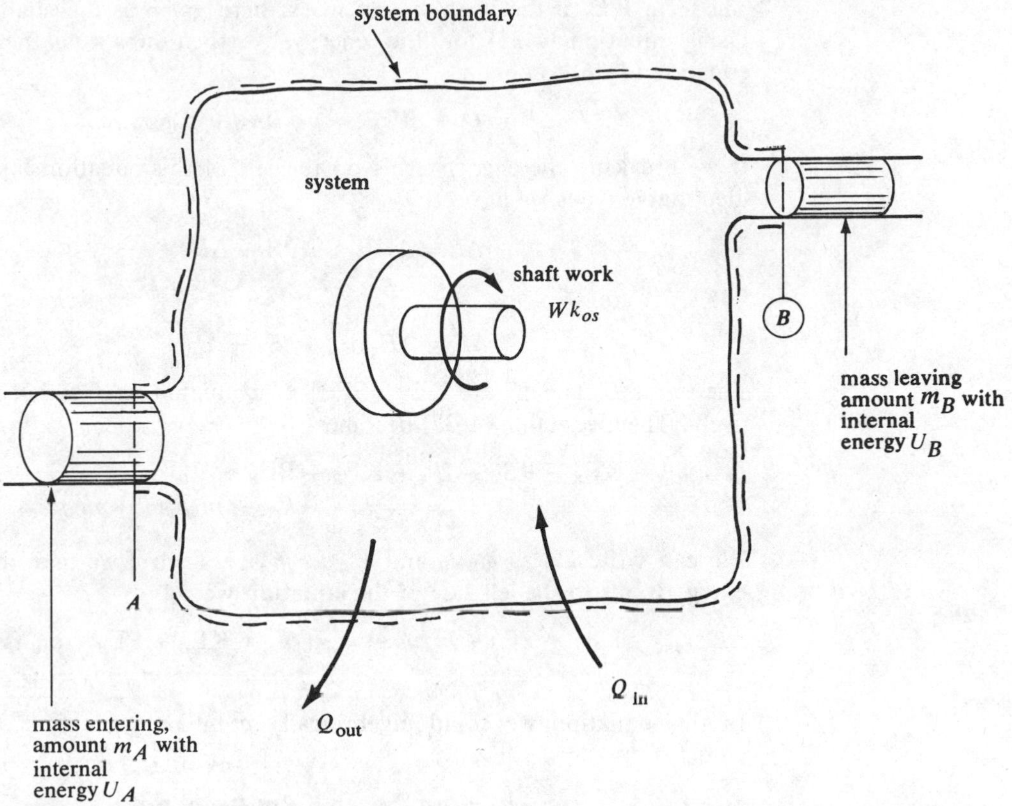

Figure 4–10 Open system.

external to this system at A to shove the mass in, work is done by the system in ejecting mass at B.

Now let us look at the energy balance or first law for this system. In figure 4–10 is our open system with some mechanical additions to account for the fact that work and heat may occur. We write equation (4–21)

$$\Delta E = Q - Wk$$

or, since in this illustration

$$Wk = \text{shaft work} - \text{flow energy at } A + \text{flow energy at } B$$

we have, for our energy balance

$$\Delta E = Q - Wk_{os} - \text{flow energy at } B + \text{flow energy at } A \quad \textbf{(4–36)}$$

The flow energy at A was *into* the system, or done *on* the system, and therefore negative. The flow energy at B was *out* and therefore positive. Upon substituting into the energy balance both signs change and therefore the equation (4–36) results.

The term Wk_{os} is the open system work, here given as the shaft work. Using equation (4–35) for flow energy, we then substitute these into equation (4–36) to obtain

$$\Delta E = Q - Wk_{os} - p_B v_B m_B + p_A v_A m_A \tag{4–37}$$

If we break up the energy terms on the left of this equation into their identifiable types we have

$$\Delta E = \Delta \mathrm{KE} + \Delta \mathrm{PE} + \Delta U \tag{4–38}$$

but

$$\Delta E = \Delta E_{\text{system}} - E_A + E_B$$

The terms E_A and E_B are KE + PE + U evaluated at A and B respectively. Then, equation (4–37) becomes

$$\Delta E_{\text{system}} - \mathrm{KE}_A - \mathrm{PE}_A - U_A + \mathrm{KE}_B + \mathrm{PE}_B + U_B = Q - Wk_{os} - m_B p_B v_B + m_A p_A v_A \tag{4–39}$$

We can write U_B as $m_B u_B$ and U_A as $m_A u_A$. Then if we take the flow energy terms to the left side of the equation we get

$$\Delta E_{\text{system}} - \mathrm{KE}_A - \mathrm{PE}_A - m_A u_A - m_A p_A v_A + \mathrm{KE}_B + \mathrm{PE}_B + m_B u_B + m_B p_B v_B = Q - Wk_{os}$$

In this equation we could algebraically combine the terms $u + pv$, namely

$$m_A(u_A + p_A v_A) \quad \text{and} \quad m_B(u_B + p_B v_B)$$

The quantities inside the brackets here are defined as the specific enthalpy and are denoted by h. Then $h_A = u_A + p_A v_A$, and $h_B = u_B + p_B v_B$, or generally

$$h = u + pv \tag{4–40}$$

Specific enthalpy is specified by units of energy per unit mass, and if it is multiplied by the mass, the resulting property is called *enthalpy*, H. We have

$$\begin{aligned} H &= mh \\ &= mu + mpv \\ &= U + pV \end{aligned} \tag{4–41}$$

Total enthalpy is specified by energy units. Remember that enthalpy and total enthalpy are merely mathematical combinations which may or may not have physical significance in a given problem.

Example 4.12 An air pump shown in figure 4–11 takes in air at 14.7 psia, 78°F, and with a density of 0.075 lbm/ft³ at station (1). The pump compresses the air

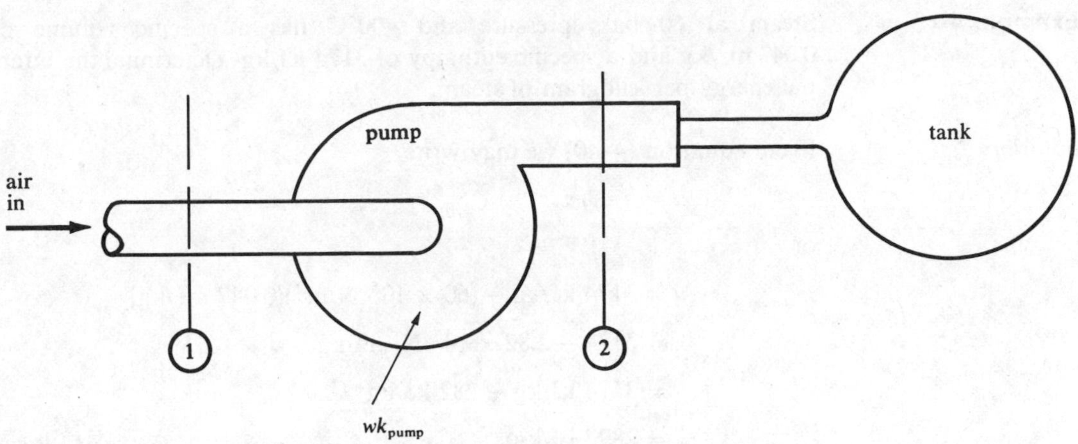

Figure 4–11 Air pump. Example 4.12.

flowing to the tank so that at station (2) the density is found to be 0.050 lbm/ft³ and the pressure is 100 psia. Determine the flow energy and specific enthalpy of the air at stations (1) and (2), if the internal energy of the air is 60 Btu/lbm at (1) and 180 Btu/lbm at (2).

Solution The flow energy at station (1) is negative, and is equal to $(p_1 v_1)$ or (p_1/ρ_1) so that flow work at (1) = 14.7 lb/in² × 1 ft³/0.075 lbm. Converting to compatible units, we get

$$-14.7 \text{ lb/in}^2 \times 144 \text{ in}^2/\text{ft}^2 \times 1 \text{ ft}^3/0.075 \text{ lbm} = 28{,}224 \text{ ft-lbf/lbm} \quad \textit{Answer}$$

At station (2)

$$\text{flow energy} = +100 \text{ lb/in}^2 \times 144 \text{ in}^2/\text{ft}^2 \times 1 \text{ ft}^3/0.050 \text{ lbm} = 288{,}000 \text{ ft-lbf/lbm} \quad \textit{Answer}$$

To calculate the specific enthalpy we use equation (4–40) and at station (1) we obtain

$$h_1 = u_1 + p_1 v_1$$

or

$$h_1 = 60 \text{ Btu/lbm} + 28{,}224 \text{ ft-lbf/lbm}$$

$$= 96.3 \text{ Btu/lbm} \quad \textit{Answer}$$

Then, at station (2) we have

$$h_2 = 180 \text{ Btu/lbm} + 288{,}000 \text{ ft-lbf/lbm}$$

$$= 550 \text{ Btu/lbm} \quad \textit{Answer}$$

Example 4.13 Steam at 60 bars pressure and 400°C has a specific volume of 0.047 m^3/kg and a specific enthalpy of 3174 kJ/kg. Determine the internal energy per kilogram of steam.

Solution From equation (4–40) we may write

$$u = h - pv$$

or

$$\begin{aligned} u &= 3174 \text{ kJ/kg} - (60 \times 10^5 \text{ N/m}^2)(0.047 \text{ m}^3/\text{kg}) \\ &= 3174 - 2.82 \times 10^5 \text{ N} \cdot \text{m/kg} \\ &= 3174 \text{ kJ/kg} - 282 \text{ kJ/kg} \\ &= 2892 \text{ kJ/kg} \end{aligned}$$

Answer

4.7 The First Law of Thermodynamics for an Open System

One way to consider the energy balance or first law of thermodynamics is as the bookkeeping rule of balancing the "energy budget" of a system. All energy must be accounted for in a given process as is done in the closed and isolated systems. Here let us take an accounting of the energy of an open system process—very generally and intuitively—in the same manner as in section 4.6. We could generalize the system as anything, such as a pump, fan, radiator, cooling tower, boiler, turbine, or a biological system. In figure 4–12 is shown a general open system which allows heat transfer, work, and mass flow into and out of the region. Assuming mass enters at station (1) and leaves at station (2), then from $\Delta E = Q - Wk$ we write, per unit of time

$$\dot{KE}_2 + \dot{PE}_2 + \dot{H}_2 - \dot{KE}_1 - \dot{PE}_1 - \dot{H}_1 + \dot{KE}_s + \dot{PE}_s + \dot{U}_s = \dot{Q}_{in} - \dot{Q}_{out} - \dot{Wk} - \dot{Wk}_s \quad \textbf{(4–42)}$$

Combining $\dot{Wk}$ and $\dot{Wk}_s$ into the term called $\dot{Wk}_{os}$, and combining $\dot{Q}_{in}$ and $\dot{Q}_{out}$ into $\dot{Q}$ and merely recalling the convention of signs for heat and work (section 4.3), then

$$\dot{KE}_2 + \dot{PE}_2 + \dot{H}_2 - \dot{KE}_1 - \dot{PE}_1 - \dot{H}_1 + \dot{KE}_s + \dot{PE}_s + \dot{U}_s = \dot{Q} - \dot{Wk}_{os} \quad \textbf{(4–43)}$$

This equation is generally too cumbersome to use, so if we neglect kinetic and potential energy changes of the system, we have

$$\dot{KE}_2 + \dot{PE}_2 + \dot{H}_2 - \dot{KE}_1 - \dot{PE}_1 - \dot{H}_1 + \dot{U}_s = \dot{Q} - \dot{Wk}_{os} \quad \textbf{(4–44)}$$

Even this equation can be extremely difficult to use under completely general conditions of variable flow, mass change of system, or variations in heat or work. For most common engineering problems, the assump-

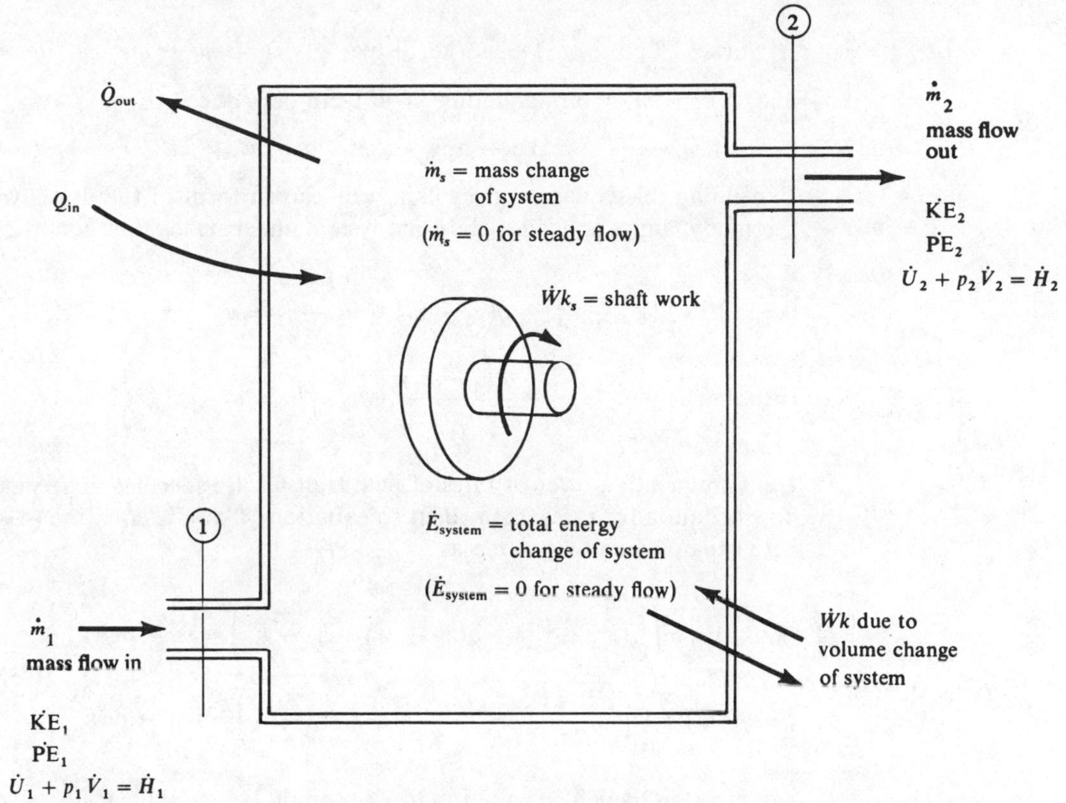

Figure 4–12 General open system.

tion of steady flow is sufficiently general. Steady flow means that the mass flow rates in and out are constant in time and that there is no change in system mass with time. This is equivalent then to writing

$$\text{System mass} = \text{constant} \tag{4–45}$$

and

$$\dot{m}_2 = \dot{m}_1 = \text{constant} \tag{4–46}$$

The mass flow rates in and out are the same from conservation of mass. If steady state is assumed (the state of the system remains constant in time) then

$$\dot{E}_s = \dot{U}_s = 0$$

and the energies evaluated at stations (1) and (2) are constant in time. The steady flow-steady state energy equation (hereafter referred to as the *steady flow energy equation*) is then written

$$\dot{\mathrm{KE}}_2 + \dot{\mathrm{PE}}_2 + \dot{H}_2 - \dot{\mathrm{KE}}_1 - \dot{\mathrm{PE}}_1 - \dot{H}_1 = \dot{Q} - \dot{W}k_{os} \tag{4–47}$$

or

$$\dot{m}_2(\text{ke}_2 + \text{pe}_2 + h_2) - \dot{m}_1(\text{ke}_1 + \text{pe}_1 + h_1) = \dot{m}q - \dot{m}wk_{os} \quad \textbf{(4–48)}$$

and, if $\dot{m} = \dot{m}_1 = \dot{m}_2$, equation (4–48) can be reduced to

$$\Delta\text{ke} + \Delta\text{pe} + \Delta h = q - wk_{os} \quad \textbf{(4–49)}$$

Expanding this equation gives us a well-known form of the first law of thermodynamics applied to the open system under steady flow conditions,

$$\left.\begin{aligned} &\text{(kJ/kg):} \quad \frac{1}{2}(\bar{V}_2^2 - \bar{V}_1^2) + g(z_2 - z_1) + h_2 - h_1 = q - wk_{os} \\ &\text{(Btu/lbm):} \quad \frac{\bar{V}_2^2 - \bar{V}_1^2}{2g_c} + \frac{g(z_2 - z_1)}{g_c} + h_2 - h_1 = q - wk_{os} \end{aligned}\right\} \quad \textbf{(4–50)}$$

Quite frequently power or rates of heat transfer are specified. The steady flow equation for rates, equivalent to equation (4–50), is equation (4–48) here expanded and rewritten as

$$\left.\begin{aligned} &\text{(kJ/kg):} \quad \dot{m}\left(\frac{1}{2}(\bar{V}_2^2 - \bar{V}_1^2) + g(z_2 - z_1) + h_2 - h_1\right) = \dot{m}q - \dot{m}wk_{os} \\ &\text{(Btu/lbm):} \quad \dot{m}\left(\frac{\bar{V}_2^2 - \bar{V}_1^2}{2g_c} + \frac{g(z_2 - z_1)}{g_c} + h_2 - h_1\right) = \dot{m}q - \dot{m}wk_{os} \end{aligned}\right\} \quad \textbf{(4–51)}$$

All terms on the left in equation (4–51) should be easily identifiable with some physical quantity. On the right side of the equation, heat or heat transfer rates can be calculated from those equations discussed in chapter 3, but Wk_{os} is without a defining equation. We say for a closed system and a reversible process that

$$Wk_{cs} = \int p\,dV$$

or Wk_{cs} is the area under a curve in a p-V diagram.

Here we note that since Wk_{os} is the same *work* we have referred to *minus* the flow energy, then we can write

$$Wk_{os} = Wk_{cs} - \Delta pV \quad \textbf{(4–52)}$$

where ΔpV is the representation of the two flow energy terms, namely,

$$\Delta pV = p_2V_2 - p_1V_1$$

Mathematically this is the same as

$$\int_{p_1V_1}^{p_2V_2} d(pv)$$

and since for reversible processes

$$Wk_{cs} = \int_{V_1}^{V_2} pdV$$

Then from equation (4–52)

$$Wk_{os} = \int_{V_1}^{V_2} pdV - \int_{p_1V_1}^{p_2V_2} d(pV)$$

But, recalling that the derivative of a product such as pV, written $d(pV)$ is $pdV + Vdp$, then

$$-\int Vdp = \int pdV - \int dpV$$

We can then see that

$$Wk_{os\ rev} = -\int Vdp \tag{4–53}$$

Equation (4–53) is true for a reversible steady flow process without kinetic or potential energy changes and is shown geometrically in figure 4–13. We have stated that $Wk_{cs\ rev}$ is equivalent to the area under a curve in a p-V diagram; the $Wk_{os\ rev}$ is equal to the area to the left of the curve in the p-V diagram as can be seen in figure 4–13. We will use equation (4–53) as the general definition for open system work when kinetic, potential, and chemical energy changes are neglected.

Per unit mass the reversible work of the open system is

$$wk_{os} = -\int vdp \tag{4–54}$$

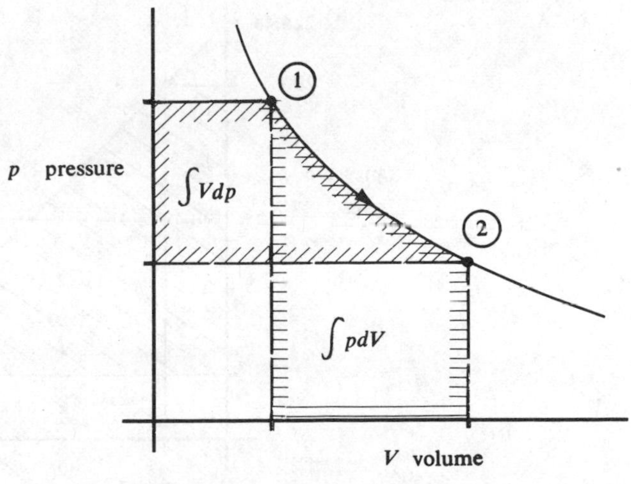

Figure 4–13 Graphical comparison of *pdV* and *Vdp* terms.

and the power for an open system under steady flow conditions is

$$\dot{W}k_{os} = -\dot{m}\int v\,dp \tag{4-55}$$

If kinetic or potential energy changes are not negligible, then the work must be gotten from

$$wk_{os} = -\int v\,dp - \Delta\text{ke} - \Delta\text{pe} \tag{4-56}$$

Example 4.14

A steam turbine running under reversible steady flow conditions takes in steam at 200 psia and exhausts it at 15 psia. Assuming the steam has a specific volume of 4.0 ft³/lbm at the inlet and the pressure-volume relation is

$$p = 228.48 - 7.12v$$

where p is in psia units and v is in ft³/lbm units, determine the work done per pound-mass of steam flowing through the turbine. Neglect kinetic and potential energy changes of the steam.

Solution

If we construct the *p-v* diagram for the expanding steam, we have the curve shown in the graph of figure 4–14. The work done per unit mass is the shaded area of the figure, or $wk_{os} = -\int v\,dp$. This area is a rectangle and a triangle, i.e.

$$\begin{aligned} wk_{os} &= (200 - 15)\ \text{lbf/in}^2 \times 144\ \text{in}^2/\text{ft}^2 \times 4\ \text{ft}^3/\text{lbm} \\ &\quad + (200 - 15)(144)\ \text{lbf-ft} \times \frac{1}{2} \times (29.98 - 4.00)\ \text{ft}^3/\text{lbm} \\ &= 106{,}560\ \text{ft-lbf/lbm} + 346{,}054\ \text{ft-lbf/lbm} \end{aligned}$$

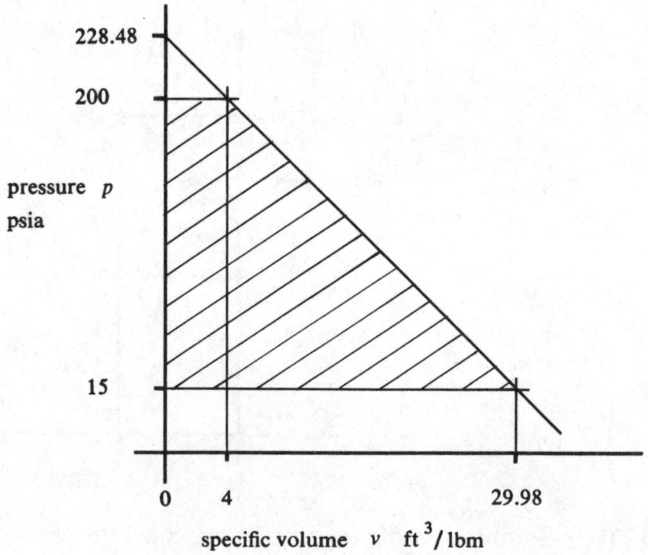

Figure 4–14 *p-v* diagram for process of example 4.14.

or

$$wk_{os} = 452{,}614 \text{ ft-lbf/lbm}$$

or

$$wk_{os} = 581.8 \text{ Btu/lbm} \qquad \textit{Answer}$$

We may also calculate this answer by using equation (4–54).

We have from the given pressure-volume relation that

$$-v = (p - 228.48)\frac{\text{lbm}^2/\text{ft}^3}{7.12 \text{ lb/in}^2 \times \text{lbm}}$$

If we substitute this into equation (4–54) we get

$$\begin{aligned} wk_{os} &= \int_{p_1}^{p_2} (p - 228.48)\frac{1}{7.12}\,dp \\ &= \frac{1}{7.12}\left(\frac{1}{2}p^2 - 228.48p\right)_{p_1}^{p_2} \\ &= \frac{1}{7.12}\left(\frac{1}{2}(15^2 - 200^2) - 228.48(15 - 200)\right) \\ &= \frac{1}{7.12}\left(-19887.5 + 42268.8\right) \\ &= \frac{144 \text{ in}^2/\text{ft}^2}{7.12 \text{ lbf lbm/in}^2 \text{ ft}^3}(22381.3 \text{ lbf}^2/\text{in}^4) \\ &= 452{,}656 \text{ ft-lbf/lbm} \\ &= 581.8 \text{ Btu/lbm} \end{aligned} \qquad \textit{Answer}$$

This answer agrees with the geometric one within 1%.

Example 4.15 A water pump, shown schematically in figure 4–15, delivers 1.0 m^3/min at 0.5 bars gage pressure through a 2.5-cm-diameter pipe. If the water is at 1.01 bars pressure, is at 27°C, and has negligible velocity at the inlet station (1) and there is no heat transfer or friction in the system, determine the power required by the pump, $\dot{W}k_{\text{pump}}$.

Solution The pump is an open system and is assumed to have steady flow. If the boundary is then drawn around the pump as shown in figure 4–15, we can apply equation (4–51) to the system

$$\dot{m}\left(\frac{1}{2}(V_2^2 - V_1^2) + g(z_2 - z_1) + h_2 - h_1\right) = \dot{Q} - \dot{W}k_{\text{pump}}.$$

Since no heat is transferred, $\dot{Q}$ is zero and V_1 is nearly zero. The elevation of the water is increased by 25 meters, so that $(z_2 - z_1) = 25$ meters. If the internal energy change of the water is neglected, then $u_2 \doteq u_1$ and the enthalpy change becomes

$$h_2 - h_1 = p_2 v_2 - p_1 v_1$$

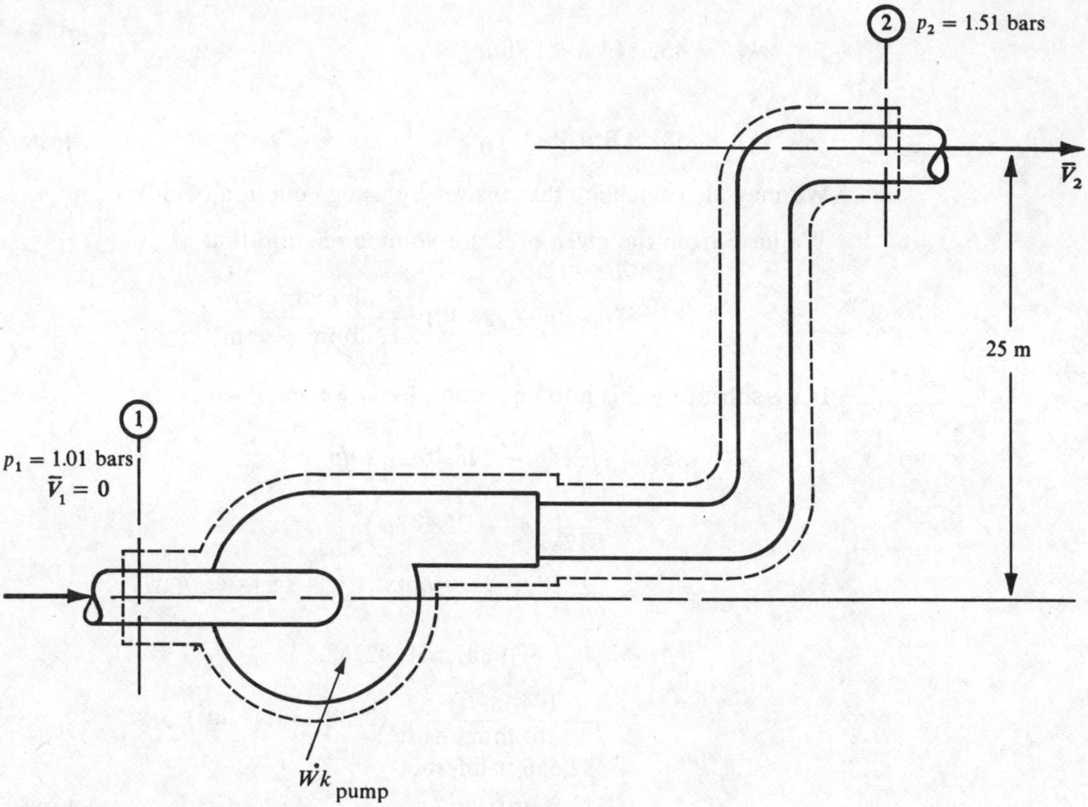

Figure 4–15 Example 4.15.

For water we may use a value of 1000 kg/m³ for the density, so the specific volume v becomes 0.001 m³/kg. The volume flow through the pump is 1.0 m³/min, or 0.0167 m³/s, and the mass flow is simply

$$\dot{m} = \dot{V}\rho$$

or

$$\dot{m} = (0.0167 \text{ m}^3/\text{s})(1000 \text{ kg/m}^3) = 16.7 \text{ kg/s}$$

The velocity at station (2) is, from equation (4–6)

$$\overline{V}_2 = \dot{m}_2/A_2\rho_2$$

where A_2 is the cross-sectional area at (2). We find

$$A_2 = \pi(1.25 \text{ cm})^2 = 4.91 \text{ cm}^2 = 4.91 \times 10^{-4} \text{ m}^2$$

Then the velocity is

$$\overline{V}_2 = \frac{(16.7 \text{ kg/s})}{(4.91 \times 10^{-4} \text{ m}^2)(1000 \text{ kg/m}^3)} = 34.0 \text{ m/s}$$

Substituting the above values into the energy equation yields

$$(16.7\ \text{kg/s})\left(\frac{1}{2}(34.0\ \text{m/s})^2 + (9.8\ \text{m/s}^2)(25\ \text{m}) + 1.51\ \text{bars} \times 0.001\ \text{m}^3/\text{kg} - 1.01\ \text{bars} \times 0.001\ \text{m}^3/\text{kg}\right) = -\dot{W}k_{\text{pump}}$$

or

$$(16.7\ \text{kg/s})(578\ \text{m}^2/\text{s}^2 + 245\ \text{m}^2/\text{s}^2 + 151\ \text{N}\cdot\text{m/kg} - 101\ \text{N}\cdot\text{m/kg}) = -\dot{W}k_{\text{pump}}$$

Notice that the units of m^2/s^2 are equivalent to the units of $N \cdot m/kg$. This can be seen by recalling that $1\ N = 1\ kg \cdot m/s^2$, and by substituting this into the $N \cdot m/kg$ units we obtain $kg \cdot m \cdot m/s^2 \cdot kg$ or m^2/s^2. Continuing now, we find

$$\dot{W}k_{\text{pump}} = -14600\ \text{kg}\cdot\text{m}^2/\text{s}$$

or

$$\dot{W}k_{\text{pump}} = -14{,}600\ \text{N}\cdot\text{m/s} = -14{,}600\ \text{J/s}$$

$$= -14{,}600\ \text{W} = -14.6\ \text{kW} \qquad \textit{Answer}$$

Many pumps are still rated in horsepower (hp) units and from table B.15 we find 1.34 hp = 1 kW, so

$$\dot{W}k_{\text{pump}} = -14.6 \times 1.34 = 19.6\ \text{hp} \qquad \textit{Answer}$$

4.8 Summary

In thermodynamics, the conservation principles are at the foundation of all applications. The first conservation law considered is that of mass,

$$m_{\text{in}} - m_{\text{out}} = \Delta m_{\text{system}} \tag{4–2}$$

or

$$\dot{m}_{\text{in}} - \dot{m}_{\text{out}} = \dot{m}_{\text{system}} \tag{4–5}$$

for steady flow we have

$$\dot{m}_{\text{in}} = \dot{m}_{\text{out}} \tag{4–15}$$

or

$$\rho A\overline{V} = \text{constant} \tag{4–16}$$

The conservation of energy, here considered the first law of thermodynamics, is the principle upon which this text will be primarily built. For the system we have

$$\Delta E = Q - Wk \tag{4–19}$$

or

$$\dot{E} = \dot{Q} - \dot{W}k \tag{4–20}$$

and for isolated systems these reduce to

$$\Delta E = 0 = \dot{E}$$

For the closed system

$$\Delta E = Q - Wk_{cs} \qquad \textbf{(4–21)}$$

where, for reversible processes we can use

$$Wk_{cs} = \int pdv$$

The closed system energy balance per unit mass is

$$\Delta e = q - wk_{cs}$$

For the open system, flow energy is identified as the energy associated with motion of mass across system boundaries. This term is evaluated from the product pV, or pv per unit mass, and together with internal energy is called *enthalpy* H, i.e.

$$H = U + pV \qquad \textbf{(4–41)}$$

or

$$h = u + pv \qquad \textbf{(4–40)}$$

The open system conservation of energy for steady flow condition is, per unit mass,

$$\frac{1}{2}(V_2^2 - V_1^2) + g(z_2 - z_1) + h_2 - h_1 = q - wk_{os}$$

where wk_{os} is the work of the open system and is evaluated from the equation

$$wk_{os} = -\int vdp - \Delta\text{ke} - \Delta\text{pe} \qquad \textbf{(4–56)}$$

Practice Problems

Problems using SI units are indicated with (M) under the problem number, and those using the English units are indicated with (E). Mixed unit problems are listed with (C) under the problem number.

Sections 4.1 and 4.2

4.1. **(M)** Water having a density of 1000 kg/m^3 is flowing with a velocity of 3 m/s through a round pipe. There is a restriction within the pipe where the diameter is one-half the normal diameter. Determine the water velocity at the restriction.

4.2. **(M)** Methyl alcohol at 20°C flows through a plastic tube and there are required 1 kg/s delivered. If the plastic tube limits the velocity to 5 m/s or less, what is the minimum acceptable diameter of the tube?

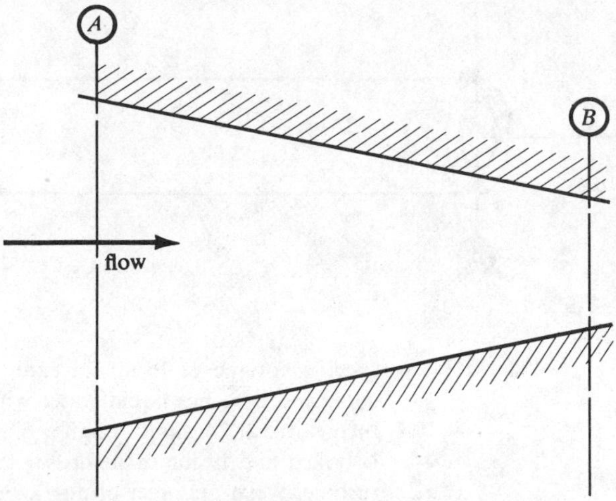

Figure 4–16

4.3. **(M)** Air flows through a converging nozzle shown in figure 4–16 with a velocity of 240 m/s at station *A*. The density of the air is found to be 0.48 kg/m^3 at *A* and 1.12 kg/m^3 at *B*. If the area of the nozzle at *A* is 0.1 m^2 and at *B* it is 0.05 m^2, determine

(a) Mass flow rate of the air through the nozzle.

(b) Velocity of the air at station *B*.

4.4. **(M)** Water at 20°C is delivered through a pipe at 1.0 m^3/min. If the pipe has an inside diameter of 4 centimeters, determine the specific kinetic energy of the water.

4.5. **(M)** Pipes *A* and *B* are joined together to supply pipe *C* as shown in figure 4–17. Pipe *A* is required to convey 1.5 m^3/min of water and pipe *B* 2.5 m^3/min. If the pipes restrict the maximum velocities to 6 m/s, determine the diameters of pipes *A*, *B*, and *C*.

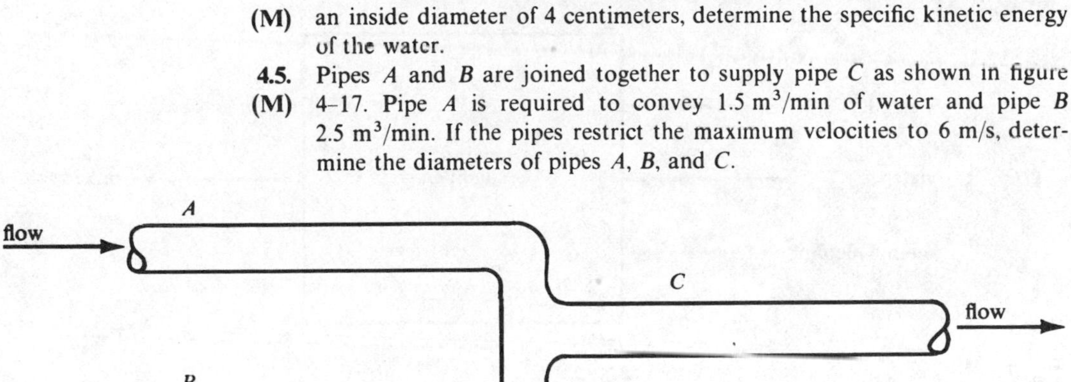

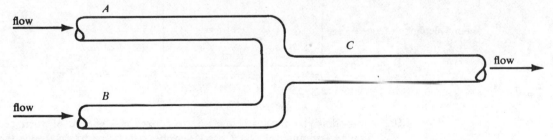

Figure 4–17

4.6. **(M)** In figure 4–18, air flows through a 2.5-cm-diameter duct which is being heated from below. If the entering air has a density of 1.2 kg/m^3 at station *A* and 0.64 kg/m^3 at *B*, find the entering velocity of the air if it is 30 m/s at *B*.

4.7. **(M)** 40,000 kg/hr of steam is exhausted from a steam turbine at a nuclear power station. The steam, flowing through a 5-cm-diameter pipe with a

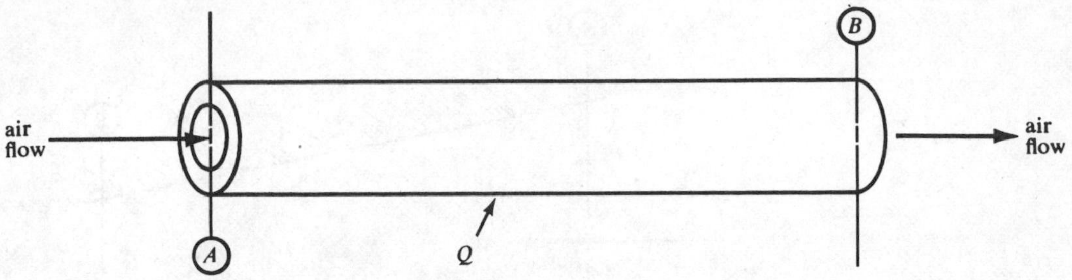

Figure 4–18

specific volume of 15 m^3/kg enters a closed condenser where the steam cools and becomes liquid water with a density of 1,000 kg/m^3. This water then flows at 24 m/s through a pipe back to the nuclear reactor where it is boiled and becomes steam for yet another passage through the steam turbine. What diameter of pipe would you select for conveying the water from the condenser?

4.8. **(M)** A sprinkler consisting of an open tank 60 centimeters square discharges 1.0 kg/s of water through holes in the tank bottom. If the tank is half full at a certain time, how great a flow must be provided from an external faucet to fill the tank to the top with water in 2 minutes if the discharge is constant at 1.0 kg/s? The tank is 60 centimeters tall.

4.9. **(M)** Into a mixing chamber flows 13 kg/s of water at 12°C, 9 kg/s of water at 85°C, and 20 kg/min of methyl alcohol as shown in figure 4–19. If 23 kg/s of mixed solution is flowing out, determine the rate of accumulation or decline of mass in the mixing chamber.

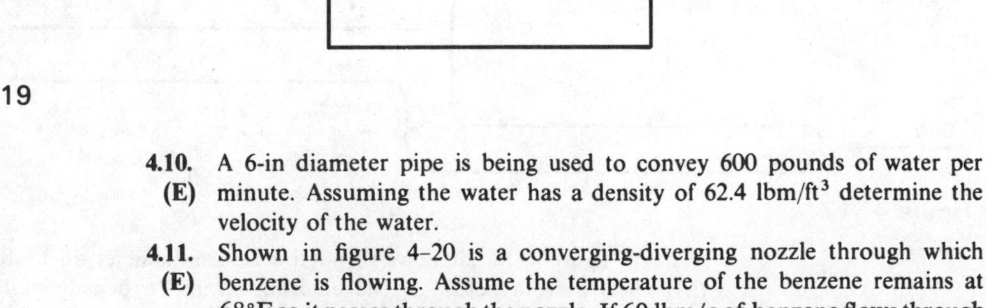

Figure 4–19

4.10. **(E)** A 6-in diameter pipe is being used to convey 600 pounds of water per minute. Assuming the water has a density of 62.4 lbm/ft^3 determine the velocity of the water.

4.11. **(E)** Shown in figure 4–20 is a converging-diverging nozzle through which benzene is flowing. Assume the temperature of the benzene remains at 68°F as it passes through the nozzle. If 60 lbm/s of benzene flows through the nozzle, determine

(a) Mass flow at the throat, station *B*, and at the exit, station *C*.
(b) Velocity of benzene at station *B*.
(c) Velocity at station *C*.

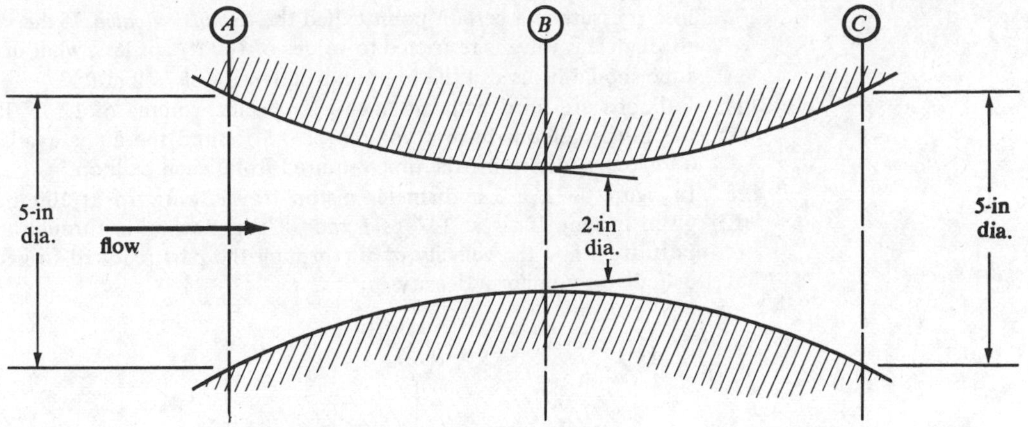

Figure 4–20

4.12. **(E)** Air flows through the converging-diverging nozzle of problem 4.11. The density of the air is found to be 0.045 lbm/ft³ at station *A*, 0.060 lbm/ft³ at station *B*, and 0.050 lbm/ft³ at station *C*. If the velocity of air entering at station *A* is 400 ft/s, determine
(a) Mass flow rate of air.
(b) Velocity of air at station *B*.
(c) Velocity of air at station *C*.

4.13. **(E)** Shown in figure 4-21 is a system diagram for a simple carburetor used to mix air and fuel. Under ideal conditions assume the carburetor mixes 0.04 lbm of fuel for every lbm of air where the density of air is 0.08 lbm/ft³. Assume the fuel has a density of 60 lbm/ft³ and the required air fuel mixture is 2 lbm/min. Determine the mass flow rates of fuel and air.

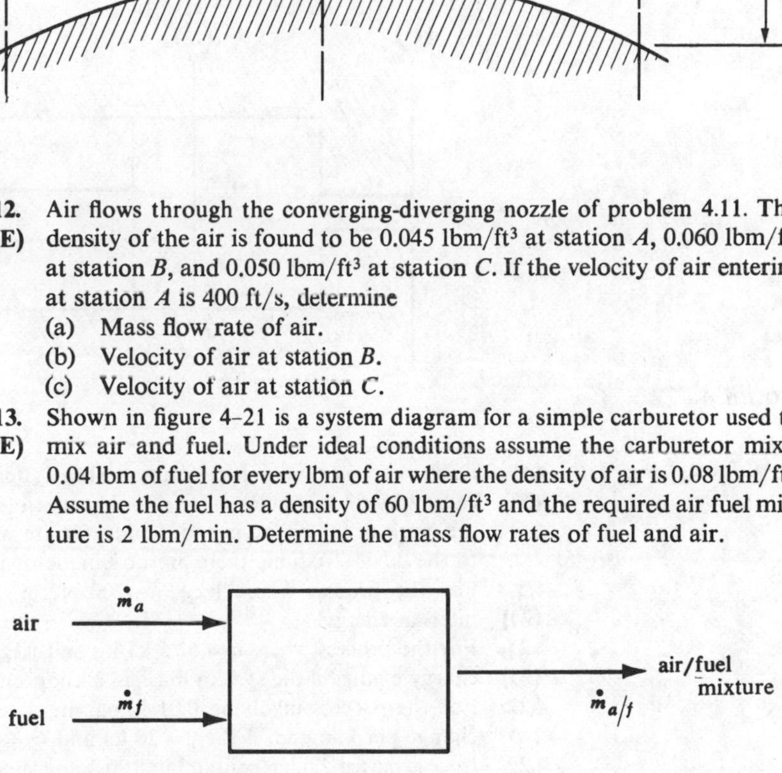

Figure 4–21

4.14. **(E)** Water flows through $\frac{3}{8}$-inch I.D. boiler tubes. Before being heated the water in the pipes has a density of 62.5 lbm/ft³ and a velocity of 10 ft/s. What would you expect the velocity of the water to be downstream, after it has been heated such that the water density is 61.8 lbm/ft³?

4.15. **(E)** In a combustor of a jet engine 30,000 ft³/min of air having a density of 0.06 lbm/ft³ enters through a cross-sectional area of 1 ft². In the combustor 0.02 lbm of fuel is mixed and burns with every 1 lbm of air. If the burned gases exit through a 1-ft² area, determine their velocities. Assume the burned gases have a density of 0.03 lbm/ft.³

4.16. **(E)** A refrigeration system rated at 60 tons uses 260 lbm/min of Freon-12. During the flow cycle of Freon through the refrigerator, it is required to lose pressure at a certain point called the *expansion valve*. If the velocity through this valve is restricted to values of 100 ft/s or less, what diameter tube should be used if the Freon has a density of 78 lbm/ft^3?

4.17. (E) Balloons are filled with air having a specific volume of 12 ft^3/lbm. If the filled balloons have a volume of 0.5 ft^3 and the air is available at 0.01 lbm/s, determine the time required to fill each balloon.

4.18. (E) In figure 4–22, a 3-in-diameter piston travels outward at 100 ft/s at a given instant. If air at 14.7 psia and 78°F is flowing in through a 1-in^2 port at *A*, find the velocity of air through the part required to keep the cylinder at a uniform density.

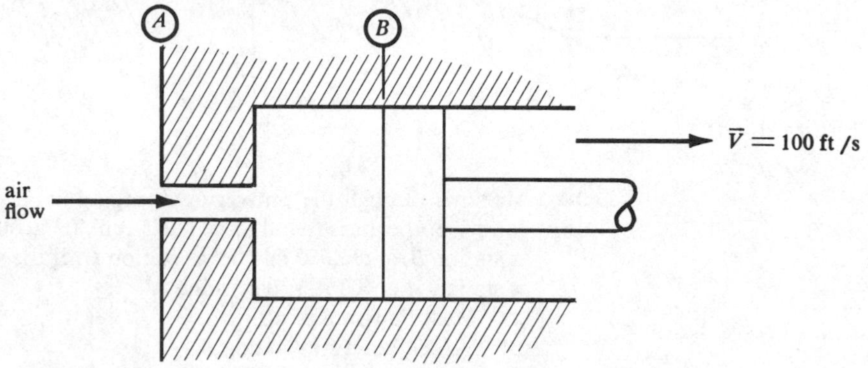

Figure 4–22

Sections 4.3 and 4.4

4.19. (M) Gases enclosed in a frictionless piston-cylinder increase their internal energy by 30 kilojoules when 40 kilojoules of heat are added. Determine the work of the piston-cylinder and indicate whether it is output or input to the device. Assume there are no kinetic or potential energy changes.

4.20. (M) For the process where $wk_{cs} = -200$ N · m, $Q = 0$, find the change in internal energy.

4.21. (M) For the process where $q = 62.5$ kJ/kg and $wk_{cs} = 60$ kJ/kg, determine the energy change if the system mass is 2 kilograms.

4.22. (M) For the process involving 0.01 kilogram of air, find the internal energy change per kilogram if $Wk_{cs} = 20$ kJ and $Q = -10$ kJ.

4.23. (M) Into a closed boiler containing 100 kilograms of water there are transferred 150 kJ/s of heat. Determine the rate of change of energy of the water and the specific energy change; that is, find $\dot{U}$ and $\dot{u}$.

4.24. (E) For the process where $\Delta U = -20$ Btu, $Q = -20$ Btu, find Wk_{cs}.

4.25. (E) For the process where $wk_{cs} = 778$ ft-lbf/lbm and $\Delta u = 0.75$ Btu/lbm, determine the heat transferred per pound-mass.

4.26. (E) For a process involving no heat transfer, find the output if the internal energy change is -16.8 Btu.

4.27. (E) For the process where $q = 8$ Btu/lbm and $wk_{cs} = 6224$ ft-lbf/lbm, find the energy change.

4.28. (E) During a reversible process, 10 horsepower are being delivered external to the system. If the system energy is changing by -10 Btu/s, determine the rate of heat transfer.

4.29. A battery supplies 100 watt-hours of electric energy. If 10 Btu are lost by the battery during this process, what was the total decline in energy of the battery?

Section 4.4

4.30. **(M)** Three kilograms of methanol are contained in a closed, perfectly insulated jar (no heat transfers allowed). A paddle is inserted through the top as shown in figure 4–23 and, driven by an electric motor, agitates the methanol so that its internal energy increases by 24 kilojoules. Determine

(a) Wk_{cs}.

(b) Irreversible work Wk_{irr} and reversible work $Wk_{cs\ rev}$.

(c) Q.

(d) wk_{cs}, wk_{irr}, and q.

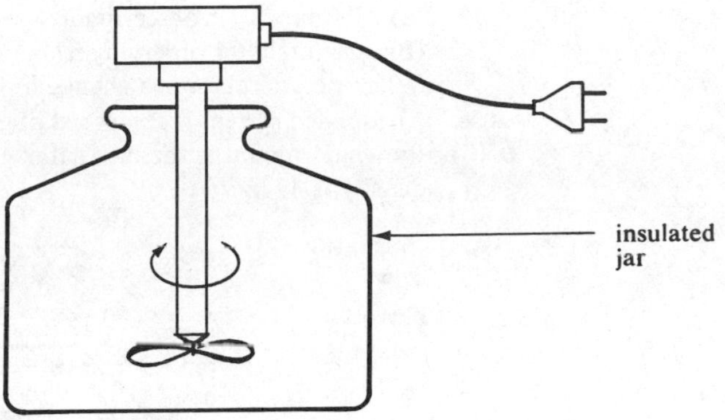

Figure 4–23

4.31. **(M)** In figure 4–23 the electric motor supplies 20 kilojoules of paddle work (or irreversible work) to stir up 3 kilograms of methanol contained in the insulated jar. Assuming the jar is not perfectly insulated so that 1 kilojoule of heat is transferred out during the stirring, find the change in energy of the methanol.

4.32. **(E)** One pound-mass of air at 78°F and 75 psia is contained in a piston-cylinder device. During a reversible process where the volume of the cylinder increases by 2 ft^3 due to the piston moving out, the pressure of the air remains constant. Determine the work produced during this process.

4.33. A heat engine drives an electric generator, thereby producing 100,000 watts of power. If this process is completely reversible and the heat engine has no change in its internal energy, determine the rate of heat transfer required to the heat engine.

Section 4.5

4.34. **(M)** Two 1-kg balls, each with 20 joules of kinetic energy in the positions shown in figure 4–24, constitute an isolated system along with the insulated container. The balls bounce around and come to rest at the bottom. By what amount has the internal energy increased in the system due to this dissipation process?

4.35. **(E)** An isolated system composed of 30 lbm and with a total energy of 3000 Btu exists. What is its energy and mass

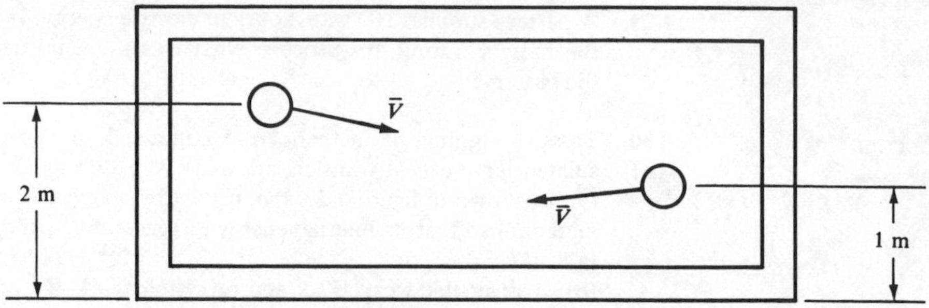

Figure 4–24

(a) 2 hours after observing it?
(b) 2 years after observing it?
When do you observe a change in its mass and/or energy?

4.36. **(E)** An isolated system is composed of grain in a box as shown in figure 4–25. By what amount is the internal energy capable of increasing if the grain density is 38 lbm/ft^3?

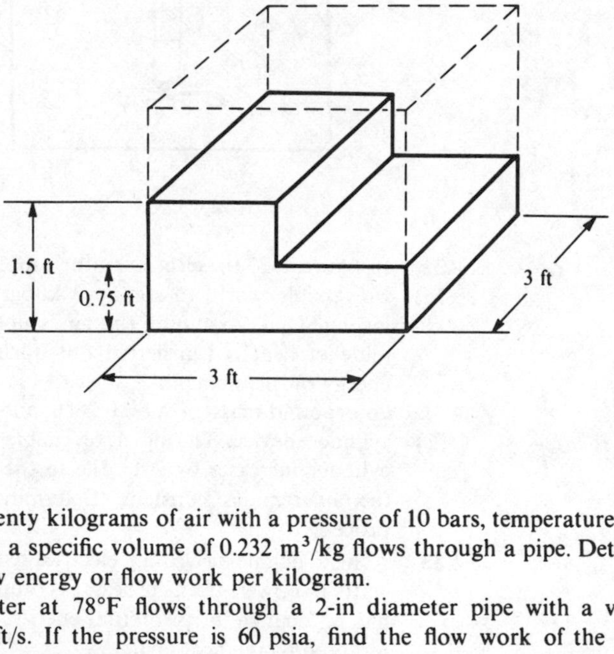

Figure 4–25

Section 4.6

4.37. **(M)** Twenty kilograms of air with a pressure of 10 bars, temperature of 250°C, and a specific volume of 0.232 m^3/kg flows through a pipe. Determine its flow energy or flow work per kilogram.

4.38. **(E)** Water at 78°F flows through a 2-in diameter pipe with a velocity of 30 ft/s. If the pressure is 60 psia, find the flow work of the water per pound-mass.

4.39. **(E)** A fuel pump is used to convey gasoline from a tank to a mixing chamber. At the entrance to the pump, the gasoline has a pressure of 14 psia and a density of 42 lbm/ft^3. As the gasoline leaves the pump, it has a pressure of 14.8 psia and its density is 42 lbm/ft^3. If 0.1 lbm/s of fuel is pumped, determine the rate of change in total flow work of the gasoline per unit of time due to the pump.

4.40. **(C)** Calculate the rate of flow energy, $(\dot{pV}) = \frac{d}{dt}pV$, for
(a) Problem 4.37.
(b) Problem 4.38.

For problems 4.41 through 4.46, calculate flow energy per unit-mass, enthalpy, and total enthalpy.

4.41. **(M)** Two kilograms of ammonia at 2.0 bars, 0.755 m^3/kg, and internal energy of 1405.6 kJ/kg.

4.42. **(M)** Seven kilograms of steam having a pressure of 2.0 bars, specific volume of 1.36 m^3/kg, and specific internal energy of 2839 kJ/kg.

4.43. **(M)** Seventy-five kilograms of air at 1.01 bars density of 1.3 kg/m^3, and internal energy of 180 kJ/kg.

4.44. **(E)** One pound-mass of nitrogen at 20 psia, 120 Btu/lbm of internal energy, and density of 0.1 lbm/ft^3.

4.45. **(E)** One pound-mass of nitrogen at 200 psia, 1000 Btu/lbm of internal energy, and 0.1 lbm/ft^3 density.

4.46. **(E)** Ten pounds-mass of Freon-22, used as a refrigeration medium, under pressure of 112 psia with a specific volume of 0.7 ft^3/lbm and internal energy of 122.93 Btu/lbm.

Section 4.7

4.47. **(M)** Steam having a specific enthalpy of 160 kJ/kg and pressure of 0.01 bar enters an adiabatic pump. Upon leaving the pump the steam is at a pressure of 12.0 bars and has a specific enthalpy of 170 kJ/kg. No kinetic or potential energy changes occur through the pump. Determine
(a) Pump work, wk_{os}, per kilogram of steam.
(b) Average density of steam, assuming the pump is reversible in nature.

4.48. **(M)** Steam with a specific enthalpy of 3278 kJ/kg enters a nozzle at station *A* shown in figure 4–26 with a velocity of 15 m/s. The exit area of the nozzle at station *B* is one-third the size of the inlet area at *A* and the sides are assumed to be adiabatic, allowing no heat transfer. Determine the enthalpy per kilogram of steam leaving the nozzle if the steam is incompressible.

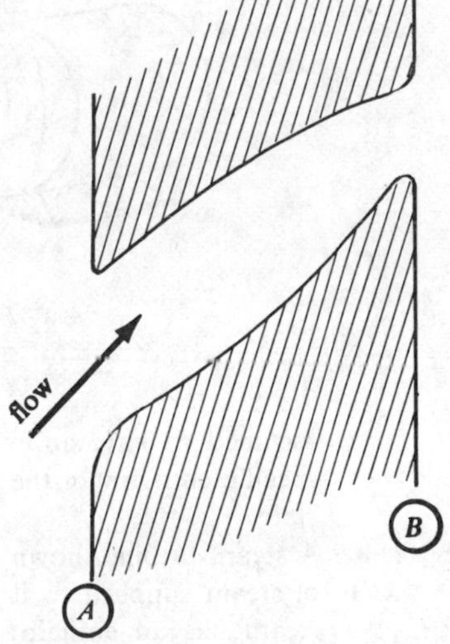

Figure 4–26

4.49. **(M)** A simplified single-cylinder internal combustion engine which produces 20 horsepower is shown in figure 4–27. It loses 40 kJ/min in radiated and conducted heat while using 0.73 kg/min of fuel and air mixture which is

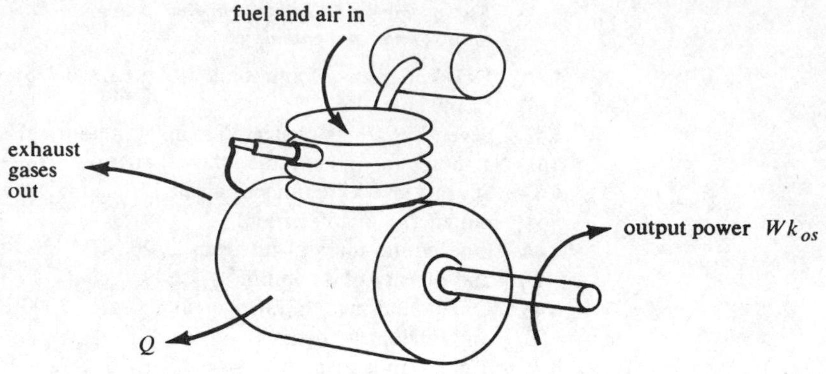

Figure 4–27

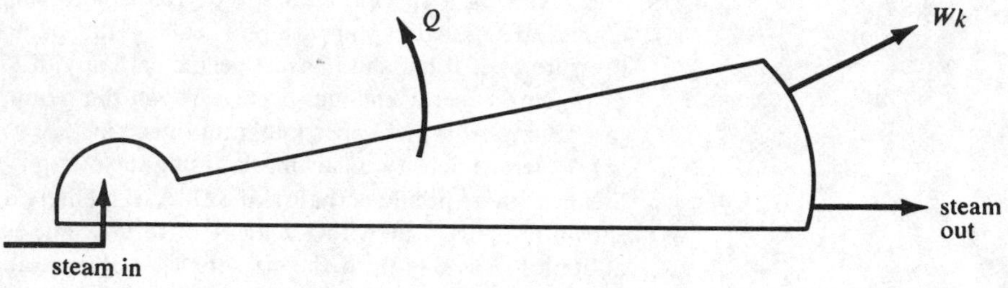

Figure 4–28

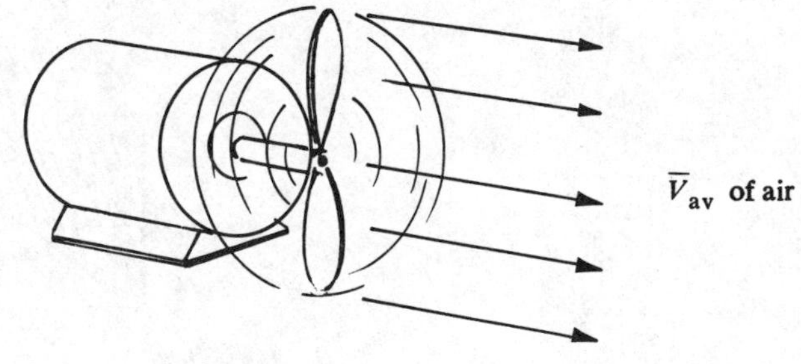

Figure 4–29

assumed to have an enthalpy of 2600 kJ/kg of mixture. Determine the specific enthalpy of the exhaust gases, assuming the engine is in a steady state.

4.50. (E) A steam turbine shown in figure 4–28 produces 290 Btu per pound-mass of steam supplied to it while it loses 8 Btu/lbm of heat. If the entering steam has an enthalpy value of 1530 Btu/lbm and there is negligible kinetic energy loss through the turbine, determine the enthalpy of exiting steam.

4.51. (E) A compressor used in a gas turbine engine increases the pressure of 3000 lbm/min of air from 15 psia to 150 psia without changing the kinetic or potential energy of the air. If the enthalpy of entering air is 118 Btu/lbm and of the compressed air is 230 btu/lbm, determine the power required to drive the compressor in horsepower if it is assumed to be reversible and adiabatic.

4.52. **(E)** In figure 4–29, a fan driven by a $\frac{1}{4}$-hp electric motor moves 40 lbm/min of air. Assuming this process is done without heat transfer and the fan is able to divert the flow of air in a uniform direction, determine the average velocity V_{av} of air leaving the fan if it has negligible velocity before passing around the fan. (Note: Assume the air has no enthalpy change.)

4.53. **(E)** A parallel flow heat exchange transfers energy from one stream to another. In figure 4–30, 2 lbm/s of sodium flows through pipe A and 10 lbm/s of water flows through B. If the sodium loses 85 Btu/lbm during its passage through the exchanger, what is the change in specific enthalpy of the water if 2 Btu/s of heat are lost to the surroundings?

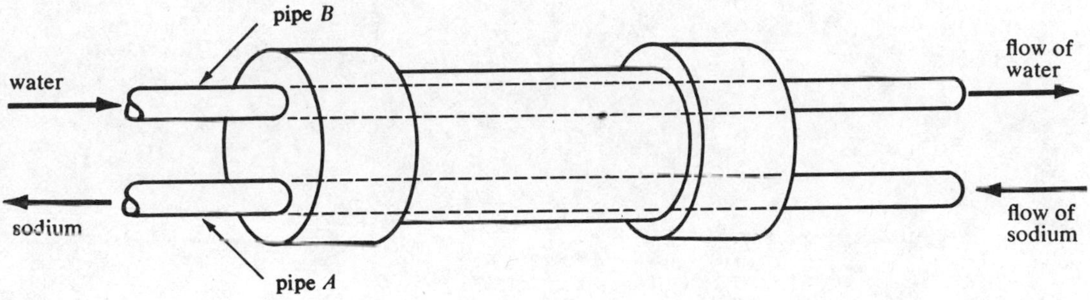

Figure 4–30

5

Equations of State and Calorimetry

The concepts of energy exchanges (heat and work) and the conservation of energy during such exchanges or processes were presented in the previous two chapters. In the discussions of these concepts, it was inferred that the description of the system or its state was sufficiently complete. In this chapter, elementary attempts at satisfactorily describing the state of the system are presented. These involve the use of *equations of state* to relate known system properties to the unknown properties at a given location and time.

The equations of state relating pressure, volume (or density), and temperature for gases will be introduced with particular attention being given to the relationship resulting from the assumption of a *perfect gas*. Solids and liquids will be treated very lightly due to the difficulty of the mathematics required to relate satisfactorily their properties at a given state.

Joule's experiment, which proved that the internal energy of a perfect gas is related only to the temperature, will be presented. Then the *caloric equations of state* will be introduced for describing gases, liquids, and solids. These equations relate the temperature to the internal energy through the concept of specific heats.

Calorimetry, or the methods for measuring internal energy and enthalpy, will be presented, precipitating an introduction to the common thermodynamics tables (a condensed collection of which appears in the text appendix) and their uses in engineering problem solving.

5.1 Equations of State – The Perfect Gas

There are many properties, such as pressure, volume, temperature, mass, density, energy, and shape, which help to describe the thermodynamic system. In order to define the system state and make the first law of thermodynamics usable for processes, we need a complete description of the system, which means we need an infinite number of data points to identify fully all the system properties. We need this description of the system state immediately when the process begins and at least once more at the conclusion of the process — then we can apply the concepts of the first law. Fortunately, we are assuming homogeneous and simple systems which means that only a few property data need to be determined to represent the system state.

In addition, some of the properties of the system, such as mass, density, and volume, are related at given states so that somewhat fewer properties need to be measured in the physical situation. Here we would like to develop some other relations which can still further reduce the number of properties which need to be measured to determine the system state.

Gases, such as air, have long been observed to change density as the pressure changed. Leonardo da Vinci remarked in the late 1400s that "Air does not resist unless it grows denser." This was an observation that the density of air increases as the pressure increases, which is exactly the relation commonly referred to as *Boyle's law*, namely

$$\frac{p}{\rho} = \text{constant} = B \qquad \textbf{(5–1)}$$

Boyle's law is correct only for light or thin gases and for those cases where temperature remains constant. Equation (5–1) is generally written as

$$pv = B \qquad \textbf{(5–2)}$$

In conjunction with this law, another relationship called *Charles' law* has been found to describe the volume and temperature of a gas at constant pressure. This is

$$V = CT \qquad \textbf{(5–3)}$$

where C is another constant. It can be shown that during an arbitrary process of a gas which obeys Boyle's and Charles' laws independently, the gas also obeys the relationship

$$pv = RT \qquad \textbf{(5–4)}$$

We can also write this relationship, without any loss of generality as

$$pV = mRT \qquad \textbf{(5–5)}$$

and call this the *perfect gas law*. Equations (5–4) and (5–5) are also referred to as the *perfect gas equations* since any gas obeying them is defined as a perfect or an ideal gas. Note that Boyle's and Charles' laws are special relations resulting from restrictions imposed on equations (5–4) and (5–5).

In these equations, the constant R is denoted as the gas constant, values for which are tabulated in the appendix table B.4. From this table it can also be seen that the gas constant is related to the molecular weight of a gas, MW, through the relationship

$$R = \frac{8.3143 \text{ kJ/kg} \cdot \text{mole} \cdot \text{K}}{\text{MW}} \tag{5–6}$$

where the constant 8.3143 kJ/kg · mole · K is called the *universal gas constant*, R_u. In the English system R_u is 1544 ft-lbf/lbm · mole · °R. Values for molecular weights of all the natural elements are listed in appendix table B.18.

The perfect gas relationship is the most common and easily used equation of state, but caution must be exercised in its application. Materials which obey the perfect gas equations have the following qualitative characteristics:

1. Sufficiently rarefied so that no attractive or repulsive forces exist between any of the atoms.
2. Sufficiently dense so that the material can be considered a continuous, uniform medium without voids or vacuums.
3. Atoms with perfectly elastic collisions (reversible collisions) between themselves and the container enclosing the material.

These assumptions are quite difficult to check, but it is generally helpful to "have a feel" for these characteristics. Those materials which obey the perfect gas law include air, most light gases, and free electrons in a solid, electrically conducting material. Obviously, liquids, solids, and dense gases such as moist steam do not obey the perfect gas law and the relationships (5–4) or (5–5) should not be used for defining these materials.

One of the earliest attempts at deriving a more general equation of state was van der Waal's relationship

$$\left(p + \frac{a}{v^2}\right)\left(v - b\right) = RT \tag{5–7}$$

where a and b are constants. This relationship was introduced to describe real gases better than the perfect gas law does, but it still has many short-comings. A better general approach to describing gases by an algebraic relationship is the *virial equation of state:*

$$\frac{pv}{RT} = 1 + \frac{b}{v} + \frac{0.625b^2}{v^2} + \frac{0.2896b^3}{v^3} + \ldots \tag{5–8}$$

Though equations (5–7) and (5–8) represent more precise relationships between gas properties, the perfect gas relationship is still much more widely used due to its relative simplicity.

The equations of state for solids are generally in the realm of pure research, and any reference to an equation for solids would be misleading in this text. For liquids and dense gases, one of the equations of state used with some success is written

$$\frac{pV}{RT} = \frac{2(V/N)^{1/3}}{2(V/N)^{1/3-2} - 2d} \quad \textbf{(5–9)}$$

where N is the number of molecules in the volume V and d is the molecular diameter. Again, this equation is not as simple as the perfect gas law, and consequently, is generally used only if tables of properties are not available.

Example 5.1 Oxygen gas is in a 3 m^3-tank at 20°C and 60 bars gage pressure. How much gas is in the tank in kilograms?

Solution We assume that oxygen acts like a perfect gas and then we may use the perfect gas relation, equation (5–5). We may determine the gas constant for oxygen from equation (5–6) where the molecular weight (MW) is 32 kg/kg · mole. Thus

$$R = (8.3143 \text{ kJ/kg} \cdot \text{mole} \cdot \text{K})/(32 \text{ kg/kg} \cdot \text{mole})$$

$$= 0.260 \text{ kJ/kg} \cdot \text{K}$$

We also could have gotten R from table B.4.

The pressure is 60 bars gage and if the atmospheric pressure is 1.01 bars, we have that $p = 61.01$ bars. Then, using equation (5–5) we get

$$m = \frac{pV}{RT}$$

or

$$m = \frac{(61.01 \text{ bars})(3 \text{ m}^3)}{(0.260 \text{ kJ/kg} \cdot \text{K})(293 \text{ K})}$$

Converting the units of bars and kilojoules to their base units we have

$$m = \frac{(61.01 \times 10^5 \text{ N/m}^2)(3 \text{ m}^3)}{(260 \text{ N} \cdot \text{m/kg} \cdot \text{K})(293 \text{ K})}$$

and

$$m = 240 \text{ kg}$$ *Answer*

Example 5.2 A perfect gas at 10 psia and 40°F is enclosed in a rigid tank of 3 ft³. This gas is heated to 540°F and 20 psia at which point its density is 0.1 lbm/ft³. Determine the mass and the gas constant for this material.

Solution

Since the gas is contained in a container of constant volume (3 ft³), we can use the relation between density, volume, and mass to determine the mass. Thus,

$$\rho = \frac{m}{V}$$

and from which we have

$$m = \rho V$$

Substituting values into this equation yields

$$m = (0.1 \text{ lbm/ft}^3)(3 \text{ ft}^3)$$

or

$$m = 0.3 \text{ lbm} \qquad \textit{Answer}$$

The gas constant can now be found from the perfect gas equation (5–5)

$$pV = mRT$$

and the gas constant can be determined at either state. Let us first calculate it at the state with a pressure of 10 psia and 30°F. We can rearrange equation (5–5) to read

$$R = \frac{pV}{mT}$$

and substituting values into this equation obtain

$$R = \frac{(10 \text{ lbf/in}^2)(3 \text{ ft}^3)(144 \text{ in}^2/\text{ft}^2)}{(0.3 \text{ lbm})(40 + 460°\text{R})}$$

or

$$R = 28.8 \text{ ft-lbf/lbm} \cdot °\text{R} \qquad \textit{Answer}$$

At the second state, as a check on this last answer, we get

$$R = \frac{(20)(3)(144)}{(0.3)(540 + 460)}$$
$$= 28.8 \text{ ft-lbf/lbm} \cdot °\text{R}$$

and this agrees with the result calculated at the first state, as it should.

5.2 Joule's Experiment

Before we proceed with a detailed investigation of how internal energy is related to the common properties of temperature, pressure, and volume, let us here see how a simple experiment can indicate some powerful conclusions regarding internal energy and its relation to other properties. This experiment, named after James Joule who first performed it in

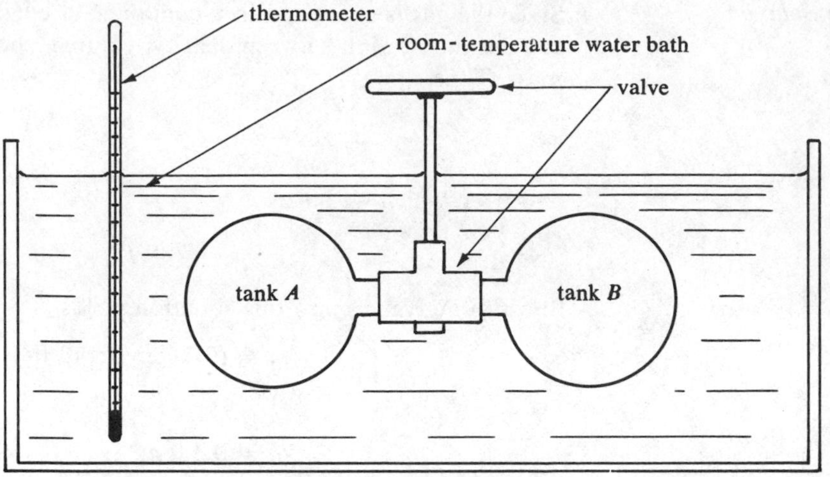

Figure 5–1 Apparatus for Joule's experiment.

1843, involves the use of the apparatus shown in figure 5–1. Tanks A and B are submerged in a water bath at room temperature. Tank A is filled with air at some high pressure, say 300 psia, while tank B is empty or near minus (−) 14.7 psig. The two tanks are connected by a valve which is opened quickly. The temperature is recorded before and after the valve is opened and is found to be the same. During this process, the air expanded to fill both tanks, the pressure reached an equilibrium in A and B, and the volume of air increased from that in tank A initially to that in both tanks after expanding. The temperature remained constant, however, and since no work or heat was done across the boundary of the two tanks A and B, the internal energy remained fixed or constant — this from the first law of thermodynamics, i.e.

$$\Delta U = Q - Wk$$

but

$$Q = 0 \quad \text{and} \quad Wk = 0$$

so

$$\Delta U = 0$$

From these observations it is seen that internal energy of the air cannot be a function of pressure or volume of the air, and consequently, must be a function of temperature only. This is true because pressure and volume changed, while temperature and internal energy remained unchanged during the expansion. We then write for air

$$U = f(T) \tag{5–10}$$

and this has been found true for all gases which can be considered as perfect gases. The significance here is that the internal energy of a perfect gas is *only* a function of temperature and this result is a helpful simplification in studying gases.

5.3 The Specific Heats

In section 5.2 was introduced a result which is very helpful; that is, internal energy is only a function of temperature for perfect gases. This result must be expanded to be useful. For instance, there are many functions which satisfy equation (5–10), a few of which are

$$U = CT + B \tag{5–11}$$

$$U = CT^2 + B \tag{5–12}$$

and

$$U = CT^3 + BT \tag{5–13}$$

where C and B are constants. Keep in mind that these three equations are attempts at describing internal energy for perfect gases. Quite generally, for any gas or liquid, we define the change in internal energy per unit temperature change during a constant volume process as the total specific heat at constant volume C_v and write

$$C_v = \left(\frac{dU}{dT}\right)_{v=\text{constant}} \tag{5–14}$$

The notation here of bracketing the differential and subscripting with "v = constant" indicates that the differential is only with respect to temperature, and not volume. Alternately, the specific heat at constant volume, c_v, is related to the total, C_v, by

$$mc_v = C_v$$

or

$$c_v = \frac{C_v}{m} \tag{5–15}$$

and then

$$c_v = \left(\frac{du}{dT}\right)_{v=\text{constant}}$$

The units for total specific heat are easily seen to be energy per unit temperature, that is Btu/°R, kJ/K, or other similar units. The units for specific heat at constant volume are then in terms of energy per unit mass per unit temperature; kJ/kg · K or BTU/lbm · °R.

For perfect gases, U is only a function of temperature and in this case, equation (5–14) becomes

$$C_v = \frac{dU}{dT} \tag{5–16}$$

or, for changes in internal energy,

$$dU = C_v dT \tag{5–17}$$

where now no restriction is placed on the differential regarding constant volumes. Equation (5–17) is correct for gases or liquids which do not behave like perfect gases, but correct only if the process during which the internal energy and temperature changes occur is one which has no volume change. Again, for perfect gases, no such restriction is placed on equation (5–17).

Now let us integrate equation (5–17) to arrive at a *caloric equation of state;* first we will assume C_v is constant and then

$$\int dU = C_v \int dT$$

from which we obtain the caloric equation of state

$$U = C_v T + U_0 \tag{5–18}$$

where U_0 is a constant of integration. Obviously, equation (5–11) is a good description of internal energy of a perfect gas with constant C_v since it agrees with equation (5–18), provided that C and C_v are identified as one and the same. The value of U_0 cannot be determined but is arbitrarily assigned some value, generally zero. For U_0 assigned the value of zero, internal energy has the value of zero when absolute temperature is zero. This may or may not be true physically, but the absolute value is not as important as is the change of value in internal energy. This point can easily be checked by noting that in the statements of the first law of thermodynamics, only changes in energy were considered, not single absolute values.

From equation (5–17) or (5–18) the change in internal energy for constant specific heats is

$$\Delta U = C_v \Delta T = mc_v \Delta T$$

and

$$\Delta u = c_v \Delta T \tag{5–19}$$

For any substance, we define now the total specific heat at constant pressure, C_p, as the change in enthalpy per unit change in temperature when pressure is kept constant, and write

$$C_p = \left(\frac{dH}{dT}\right)_{p=\text{constant}} \tag{5–20}$$

The specific heat at constant pressure c_p is related to C_p by

$$mc_p = C_p \tag{5–21}$$

or

$$c_p = \frac{C_p}{m} = \left(\frac{dh}{dT}\right)_{p=\text{constant}} \tag{5–21}$$

Recalling the definition of enthalpy

$$H = U + pV$$

we see that for a perfect gas, U is a function of temperature only and pV is equivalent to mRT which is also only a temperature function. For a perfect gas, then, enthalpy is a function of temperature only and equation (5–20) can be written

$$C_p = \frac{dH}{dT} \tag{5–22}$$

or

$$dH = C_p dT \tag{5–23}$$

From equation (5–23) we can get another caloric equation of state by assuming C_p is constant. Then

$$\int dH = C_p \int dT \tag{5–24}$$

and from this we obtain

$$H = C_p T + H_0 \tag{5–25}$$

Here H_0 denotes an integration constant in the same manner as U_0 for equation (5–18). We see then that like internal energy, enthalpy cannot be determined absolutely; only changes from one state to another can be measured with some precision. The change in enthalpy for constant specific heats is obtained from integrating equation (5–24) between limits, so that

$$\Delta H = C_p \Delta T = mc_p \Delta T \tag{5–26}$$

and

$$\Delta h = c_p \Delta T$$

From equation (5–22), substituting the terms $U + pV$ for H, we get

$$C_p = \frac{dU}{dT} + \frac{dpV}{dT}$$

and since $pV = mRT$ for perfect gases

$$C_p = \frac{dU}{dT} + mR\frac{dT}{dT}$$

or

$$C_p = \frac{dU}{dT} + mR \tag{5-27}$$

Now, from equation (5–16), the first term on the right side of equation (5–27) is identified as C_v or mc_v, so

$$C_p = mc_v + mR \tag{5-28}$$

and consequently, since $C_P = mc_p$, we obtain

$$c_p = c_v + R \tag{5-29}$$

The equations (5–28) and (5–29) are true only for perfect gases but represent useful tools in problem solving.

Another notation which will be used in subsequent analyses is the ratio of specific heats, k, defined as

$$k = \frac{C_p}{C_v} = \frac{c_p}{c_v} \tag{5-30}$$

Under most conditions, k has a value near 1.4 and unless it is stated otherwise the student can safely use this value for a perfect gas. Values of k for various gases are listed in table B.4 of the appendix.

5.4 Measurement of Internal Energy

We have no energy meters available with which to measure internal energy of a system — but we have thermometers, and armed with them and the caloric equations of state as given by (5–18) and (5–25), we can obtain internal energy and enthalpy data for perfect gases. We do, however, need to know values for the specific heats, and the following example should give you the basis for determining specific heats at constant volume.

Example 5.3

In figure 5–2, air at 27°C is contained in a rigid tank which is placed on an electric burner. The sides and top of the tank are insulated while the bottom allows free heat transfer. A thermometer is inserted into this device to record the temperature while the air is heated. After 20 minutes, it was found that the air temperature inside the tank was 138°C and the amount of electric energy required to heat up the air was 1.0 watt-hour.

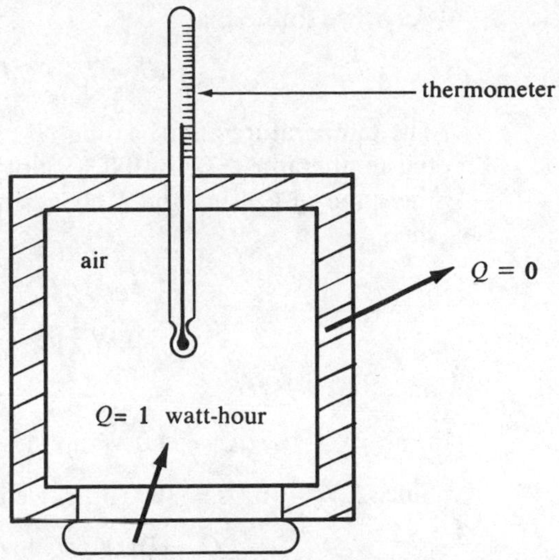

Figure 5–2 Example 5.3.

Determine C_v and c_v for the air. Assume the mass of air contained is 0.045 kilogram.

Solution

We can assume air behaves as a perfect gas over the range of 27°C to 138°C and even if it does not, the process is one of constant volume, so we can use

$$dU = C_v dT \tag{5–17}$$

and assume C_v will be constant. Then we can use the caloric equation of state (5–18), calling state (2) the final and state (1) the initial, we have

$$U_2 = C_v T_2 + U_0$$
$$U_1 = C_v T_1 + U_0$$

and the change of energy is

$$U_2 - U_1 = C_v(T_2 - T_1) = \Delta U$$

This result agrees with a result obtainable from equation (5–19) for changes in internal energy.

Stipulating the system as the air inside the tank, we see that no work was done; but heat was transferred into this system, the amount of which was 1.0 watt-hour. We assumed no heat losses during the process, so using the first law of thermodynamics

$$\Delta U = Q - Wk = Q$$

since $Wk = 0$. Then

$$\Delta U = 1.0 \text{ W} \cdot \text{hr}$$

Also, we know that

$$\Delta U = U_2 - U_1 = C_v(T_2 - T_1)$$

The temperature after heating, T_2, is 138°C + 273 = 411 K and the initial temperature, T_1, is 300 K. Notice that the change in temperature is the same in Celsius and Kelvin scales. The change in internal energy is then

$$\Delta U = C_v(411 - 300)$$
$$1.0 \text{ W} \cdot \text{hr} = C_v(111 \text{ K})$$

and

$$C_v = 1.0 \text{ W-hr}/111 \text{ K} = 0.009 \text{ W} \cdot \text{hr/K}$$

Since 1 W = 1 J/s = 3600 J/hr, we obtain

$$C_v = 0.009 \text{ w} \cdot \text{hr/K} \times 3600 \text{ J/W} \cdot \text{hr}$$
$$= 32.4 \text{ J/K} = 0.0324 \text{ kJ/K}$$ *Answer*

The specific heat at constant volume per unit mass is obtained from the relationship $c_v = C_v/m$ where m is 0.045 kilogram in this example. Thus

$$c_v = \frac{0.0324 \text{ kJ/K}}{0.045 \text{ kg}}$$
$$= 0.72 \text{ kJ/kg} \cdot \text{K}$$ *Answer*

Table B.4 lists values for specific heats of various substances, including air.

Calorimetry: *The science and technology concerned with precisely measuring energy and enthalpy.*

From it the values of specific heats listed in table B.4 were determined in much the same manner as described by example 5.3. In addition, some gases, as well as liquids, have been found to have specific heat values which vary with the temperature of the substance — not an unexpected turn of events. In table B.14 are equations which show that specific heats at constant pressure, c_p, are functions of temperature, and utilizing the relation for perfect gases

$$c_p - c_v = R \qquad \textbf{(5–27)}$$

we can see that c_v is also a function of temperature for those gases listed in the table. For variable specific heats we must use, provided we still have a perfect gas,

$$dU = C_v dT \qquad \textbf{(5–17)}$$

in order that a change in internal energy may be found. To obtain this change in energy, the equation must be integrated, so we have

$$\int_{U_1}^{U_2} dU = U_2 - U_1 = \Delta U = \int_{T_1}^{T_2} C_v dT \qquad \textbf{(5–31)}$$

or, per pound-mass,

$$\int du = u_2 - u_1 = \Delta u = \int c_v dT \qquad \textbf{(5–32)}$$

Either of these two equations can be used for determining internal energy data for gases and liquids deviating from a perfect gas. From tests using empirical methods of calorimetry, temperature and power are recorded. By assigning a value of zero for u_1 in equation (5–32) when T is zero degrees absolute, u_2 is identified then as the power required to elevate the material temperature above zero. In measuring the power data, we bring into use much sophisticated instrumentation — suffice it here to say that from an apparatus as described in example 5.3, watt-meters could measure the power. From other methods of supplying heat to the container of test substances, such as Bunsen burners, heat pipes, or heat exchangers, the power data can also be determined; and from some of the later application chapters we will see how this is done.

Air is not precisely a perfect gas, nor does it have a constant value of its specific heat at constant volume. In table B.6 are listed data of internal energy as functions of temperature for air — data which were determined empirically in a manner just discussed. Table B.13 contains internal energy-temperature data for other gases. In those thermodynamic tables where internal energy data are missing but enthalpy is listed, use of the equation

$$u = h - pv$$

can by made to obtain the data.

Example 5.4 Determine the change in internal energy of 2 lbm of oxygen gas as it changes temperature from 70°F to 90°F.

Solution We assume that oxygen is a perfect gas and also that the specific heats are constant over the specific temperature range. Then we have

$$\Delta U = mc_v \Delta T \qquad \textbf{(5–19)}$$

for which we identify the following terms

$$m = 2 \text{ lbm}$$

$$\Delta T = 90°\text{F} - 70°\text{F} = 20°\text{F} = 20°\text{R}$$

and

$$c_v = 0.157 \text{ Btu/lbm} \cdot {}^\circ\text{R} \quad \text{(from table B.4 of the appendix)}$$

Substituting these values into the caloric equation of state (5–17) yields

$$\Delta U = (2 \text{ lbm})(0.157 \text{ Btu/lbm} \cdot {}^\circ\text{R})(20^\circ\text{R})$$

$$= 6.28 \text{ Btu} \qquad \textit{Answer}$$

and we have determined the change in internal energy.

5.5 Measurement of Enthalpy

We determine enthalpy by using equation (5–23), provided the substance is a perfect gas. Even then, we must have a value for c_p in order that we obtain the desired values. The following example should provide the background for understanding the origin of values for c_p.

Example 5.5

Twelve grams of air are contained in a frictionless piston-cylinder device. The piston weighs 2 newtons and the walls of the cylinder and the piston are perfectly insulated (adiabatic surfaces) allowing no heat transfer. This device is shown in figure 5–3 where a paddle, driven by an electric motor, is inserted through the bottom. The paddle shaft is frictionless and, after the electric motor has expended 847 watt-seconds of energy to drive the paddle, the air temperature in the cylinder is found to have increased from 10°C to 80°C. Determine the total specific heat at constant pressure, C_p, and the specific heat at constant pressure, c_p.

Solution

Here we have an addition of energy to our system (the air inside the cylinder) which is completely irreversible. The paddle wheel stirred up the air in 847 watt-seconds, and thereby increased the air temperature through internal friction of the air itself. In example 5.3, the addition of energy to the air could have been made in this same way rather than by heat transfer. The heat for this process is zero so we write the first law

$$\Delta U = Q - Wk = Wk$$

The work term is composed of a reversible and an irreversible part. The irreversible work is, as we have mentioned, equal to 847 watt-seconds, while the reversible work is the motion of the piston. Since the air is elevated in temperature, it also will tend to increase pressure; but the piston will be raised instead, to increase the volume of air. The process here is conducted at constant pressure since the piston weight is constant as it moves up. Since we have a frictionless piston, the reversible work done by the system is given by

$$Wk_{cs\ \text{rev}} = \int p dV = p \int dV$$

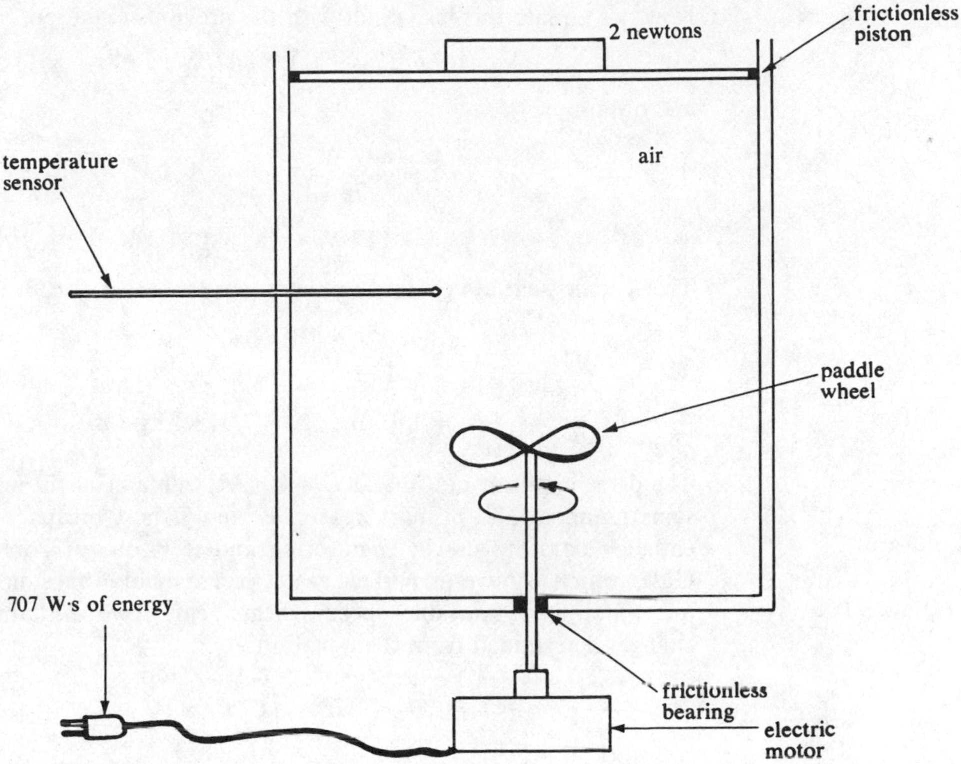

Figure 5–3 Device for determining specific heat values.

and then

$$Wk_{cs\ \mathrm{rev}} = p(V_2 - V_1)$$

The first law then becomes

$$U_2 - U_1 = -(-847\ \mathrm{W \cdot s}) - p(V_2 - V_1)$$

or

$$U_2 - U_1 + pV_2 - pV_1 = 847\ \mathrm{W \cdot s}$$

Now, using the definitions for enthalpy, we see the left side here is $H_2 - H_1$ and then using equation (5–24), we see that

$$\int dH = H_2 - H_1 = \int C_p dT$$

We assume that air is a perfect gas and that the specific heat is a constant. This gives us

$$H_2 - H_1 = C_p(T_2 - T_1)$$

Now, we equate this last result with the previous equation,

$$C_p(T_2 - T_1) = 847 \text{ W} \cdot \text{s}$$

and obtain

$$C_p = \frac{847 \text{ W} \cdot \text{s}}{70 \text{ K}}$$

$$= 12.1 \text{ W} \cdot \text{s/K} = 12.1 \text{ J/K} \qquad \textit{Answer}$$

The specific heat at constant pressure per unit mass is $c_p = C_p/m$, so

$$c_p = \frac{12.1 \text{ J/K}}{12 \text{ g}}$$

$$= 1.01 \text{ J/g} \cdot \text{K} = 1.01 \text{ kJ/kg} \cdot \text{K} \qquad \textit{Answer}$$

The determination of C_p and C_v for various materials is pursued in the same manner. The primary reason is that it is a means of obtaining enthalpy data for use in engineering and technology problem solving. Gases which behave as perfect gases, or reasonable facsimilies thereof, but which have variable specific heats, can have enthalpy property changes determined from the equations

$$\Delta H = \int C_p dT \qquad \textbf{(5–33)}$$

$$\Delta h = \int c_p dT \qquad \textbf{(5–34)}$$

with the value of C_p obtained from tables such as B.4 or B.14.

Example 5.6

Determine the change in enthalpy per pound-mass of methane gas as it goes through a temperature change from 600°F to 200°F.

Solution

Assuming that methane gas behaves as a perfect gas with constant values for the specific heats over the temperature range, we have

$$\Delta h = c_p \Delta T \qquad \textbf{(5–24)}$$

for which

$$\Delta T = 200°\text{F} - 600°\text{F}$$
$$= -400°\text{F}$$

and

$$c_p = 0.5318 \text{ Btu/lbm} \cdot °\text{R} \quad \text{(from table B.4 of the appendix)}$$

Substituting values into the above equation gives us

$$\Delta h = (0.5318 \text{ Btu/lbm} \cdot °\text{R})(-400°\text{R})$$

$$= -212.72 \text{ Btu/lbm} \qquad \textit{Answer}$$

Example 5.7 Determine the change in enthalpy per pound-mass for carbon monoxide (CO) as the gas is cooled from 1500°F to 500°F, assuming the gas does not have a constant specific heat.

Solution We will assume the gas, CO, is a perfect gas, so

$$\Delta h = \int c_p dT \tag{5–34}$$

and we must find the relation c_p has to temperature. From table B.14 we can see that there is a choice of relations for CO, namely

$$c_p = \left(9.46 - \frac{3.29(10^3)}{T} + \frac{1.07(10^6)}{T^2}\right) \text{ Btu/lbm mole} \cdot {}^\circ\text{R}$$

or

$$c_p = [a + b(10^{-3})T + c(10^{-6})T^2 + d(10^{-9})T^3] \text{ cal/g} \cdot \text{mole} \cdot \text{K}$$

where a, b, c, and d can take on two different sets of values. Let us use the second equation with the following set of constants:

$$a = 6.480$$
$$b = 1.566$$
$$c = -0.2387$$
$$d = 0$$

Then

$$c_p = (6.48 + 1.566T \times 10^{-3} - 0.2387T^2 \times 10^{-6}) \text{ cal/g} \cdot \text{mole} \cdot \text{K}$$

and the initial and final temperatures are

$$T_1 = 1500^\circ\text{F} + 460^\circ = 1960^\circ\text{R}$$
$$= \frac{5}{9} \times 1960^\circ\text{R} = 1089 \text{ K}$$

and

$$T_2 = 500^\circ\text{F} + 460 = 960^\circ\text{R}$$
$$= \frac{5}{9} \times 960 = 533 \text{ K}$$

respectively. We now integrate equation (5–34),

$$\Delta h = \int_{1089\text{ K}}^{533\text{ K}} (6.48 + 1.566T \times 10^{-3} - 0.2387T^2 \times 10^{-6})dT$$
$$= \left(6.48T + 0.783T^2 \times 10^{-3} - 0.0796T^3 \times 10^{-6}\right)_{1089\text{ K}}^{533\text{ K}}$$
$$= [6.48(533 - 1089) + 0.783 \times 10^{-3}(533^2 - 1089^2) - 0.0796 \times 10^{-6}(533^3 - 1089^3)] \text{ cal/g} \cdot \text{mole}$$
$$= \{-3603 - 706 + 91\} \text{ cal/g} \cdot \text{mole}$$
$$= -4218 \text{ cal/g} \cdot \text{mole}$$

Per gram, the change in enthalpy is found by using the molecular weight of CO, 28 g/g · mole:

$$\Delta h = -4218 \text{ cal/g-mole} \times \frac{1}{28} \text{ g/g-mole CO}$$
$$= -150.6 \text{ cal/g}$$

Using the conversions listed in table B.15 we obtain

$$\Delta h = (-150.6 \text{ cal/g})(454 \text{ g/lbm})/(252 \text{ cal/Btu})$$
$$= -271 \text{ Btu/lbm}$$

Answer

5.6 The Thermodynamic Tables

For perfect gases we can use the equations (5–17) and (5–23) in integrated form,

$$\int dU = \int C_v dT = \Delta U$$
$$\int dH = \int C_p dT = \Delta H$$

and with constant values C_v and C_p, obtain equations (5–19) and (5–26)

$$\Delta U = C_v \int dT = C_v \Delta T$$
$$\Delta H = C_p \int dT = C_p \Delta T$$

to determine changes in internal energy and enthalpy. All we need here are temperature data and numerical values for the specific heats. These equations are also reasonably accurate for nonperfect gases, particularly if we utilize variable specific heats and integrate equations (5–17) and (5–23) with the according complication as shown in example 5.6. However, for the vast bulk of materials subject to engineering usage — liquids, dense gases, and solids — we have tables of thermodynamic properties at our disposal with which to determine enthalpy and internal energy. Tabular values of these properties are also recommended over caloric equations of state like (5–19) and (5–26) for gases approximating perfect gases if the desired results must be accurate to more than 5%.

Tables B.1–B.3, B.6, B.8–B.13 in appendix B list enthalpy and internal energy value for various substances at representative temperatures and pressure, and using interpolation of data we can solve all the problems presented in this text. More extensive data of thermodynamics properties are published in various literature, and much effort is being expended in calorimetry to compile more precise and expanded data treating more and more materials. The following example should be helpful if you are not familiar with *interpolation*, a technique giving detailed scope to tabular data.

Example 5.8 Determine the enthalpy of saturated mercury vapor h_g at 4.5 bars pressure.

Solution From table B.10, which lists the properties of mercury vapor, we see that the enthalpy of the saturated vapor h_g is 345.4 kJ/kg at 4.0 bars and 346.3 kJ/kg at 5.0 bars pressure. The difference between the two values is 0.9 kJ/kg over a pressure difference of 1.0 bars. We may then use a method of proportions or a pressure-enthalpy plot as shown in figure 5–4. From the figure we can see that the value for the enthalpy at 4.5 bars must be exactly midway between the values at 4.0 and 5.0 bars pressure. Thus

$$h = 345.4 + 0.45 = 345.9 \text{ kJ/kg} \qquad \textit{Answer}$$

Using the method of proportions we write

$$\frac{4.5 \text{ bars} - 4.0 \text{ bars}}{5.0 \text{ bars} - 4.0 \text{ bars}} = \frac{h - 345.4 \text{ kJ/kg}}{346.3 \text{ kJ/kg} - 345.4 \text{ kJ/kg}}$$

and then

$$\frac{0.5 \text{ bar}}{1.0 \text{ bar}} = \frac{h - 345.4 \text{ kJ/kg}}{0.9 \text{ kJ/kg}}$$

$$(0.9)(0.5) = h - 345.4 \text{ kJ/kg}$$

$$345.4 \text{ kJ/kg} + 0.45 \text{ kJ/kg} = h$$

$$h = 345.9 \text{ kJ/kg} \qquad \textit{Answer}$$

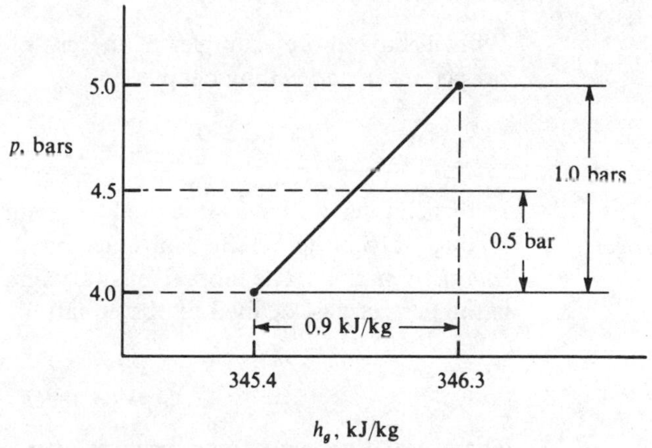

Figure 5–4 Graphical illustration of interpolation in Example 5.8.

Example 5.9 Determine the enthalpy and specific volume of steam at 560°C and 30 bars. Also determine the enthalpy and specific volume of ammonia at 100°C and 14 bars pressure.

Solution The properties of steam can be found from table B.2. We read

$$h = 3592 \text{ kJ/kg} \qquad \textit{Answer}$$

$$v = 0.126 \text{ m}^3\text{/kg} \qquad \textit{Answer}$$

The properties of ammonia are found from table B.12. We read

$$h = 1653.8 \text{ kJ/kg} \qquad \textit{Answer}$$

$$v = 0.1217 \text{ m}^3\text{/kg} \qquad \textit{Answer}$$

Example 5.10 Determine the specific volume of saturated steam in the vapor state , v_g, at 50 psia.

Solution From table B.1 we find that values are listed for 49.2 psia but not for 50 psia. If we then interpolate between the values at 40 and 60 psia we can find v_g at 50 psia. Here we have 50 psia being midway between 40 and 60, so

$$\frac{50 \text{ psia} - 40 \text{ psia}}{60 \text{ psia} - 40 \text{ psia}} = \frac{v_g - 10.497 \text{ ft}^3\text{/lbm}}{7.174 \text{ ft}^3\text{/lbm} - 10.497 \text{ ft}^3\text{/lbm}}$$

$$\frac{10}{20} = \frac{v_g - 10.497 \text{ ft}^3\text{/lbm}}{-3.323 \text{ ft}^3\text{/lbm}}$$

$$(-3.323 \text{ ft}^3\text{/lbm})\left(\frac{1}{2}\right) = v_g - 10.497 \text{ ft}^3\text{/lbm}$$

$$8.8355 \text{ ft}^3\text{/lbm} = v_g \qquad \textit{Answer}$$

We will have more examples of the use of the tables of thermodynamic properties in succeeding chapters.

5.7 Summary

In this chapter we have attempted to relate the properties at given states through algebraic relationships, called *equations of state* and *caloric equations of state*. The most elementary and usable material to be studied is the perfect gas, defined by the equations of state,

$$pv = RT \qquad \textbf{(5–4)}$$

$$pV = mRT \qquad \textbf{(5–5)}$$

There are other equations, such as van der Waal's equation and the virial equation, which are attempts to relate properties for dense gases and liquids, but which will not generally be used here.

The caloric equations of state for perfect gases with variable specific heats are

$$\Delta U = \int C_v dT \tag{5-31}$$

or

$$\Delta u = \int c_v dT \tag{5-32}$$

and

$$\Delta H = \int C_p dT \tag{5-33}$$

or

$$\Delta h = \int c_p dT \tag{5-34}$$

which follow from definitions of the specific heats. The total specific heats, C_v and C_p, were defined for any material as

$$C_v = \left(\frac{dU}{dT}\right)_{v=\text{constant}} \tag{5-14}$$

and

$$C_p = \left(\frac{dH}{dT}\right)_{p=\text{constant}} \tag{5-20}$$

From Joule's experiment it was concluded that for perfect gases, internal energy and enthalpy are functions of temperature only. Equations (5–14) and (5–20) then reduce to (5–31) and (5–33) respectively.

The specific heats, more widely used than the total specific heats, are given by

$$c_v = \frac{C_v}{m} = \left(\frac{du}{dt}\right)_{v=\text{constant}}$$

and

$$c_p = \frac{C_p}{m} = \left(\frac{dh}{dT}\right)_{p=\text{constant}} \tag{5-21}$$

for any substance. For perfect gases, the following two useful relations hold:

$$C_p - C_v = mR \tag{5-28}$$

and

$$c_p - c_v = R \tag{5-29}$$

Equations (5–31), (5–32), (5–33), and (5–34) are used, along with the assumption of constant values for C_v and C_p, to give us

$$\Delta U = C_v \Delta T \qquad (5\text{–}19)$$

$$\Delta H = C_p \Delta T \qquad (5\text{–}26)$$

$$\Delta u = c_v \Delta T \qquad (5\text{–}19)$$

$$\Delta h = c_p \Delta T \qquad (5\text{–}26)$$

These specific equations allow us to calculate changes in internal energy and enthalpy for *perfect gases* with constant specific heats. Tables of thermodynamic properties developed from calorimetric techniques are used to obtain enthalpy and internal energy of nonperfect gases and liquids.

Practice Problems

Problems designated with an asterisk * preceding the number should only be attempted by those having a background which includes the knowledge of integral calculus.

Given following conditions, determine the unknown property of a perfect gas, for problems 5.1–5.12.

Section 5.1

5.1. (M) Pressure is 1.4 bars gage, volume is 0.085 m^3, mass is 0.7 kilogram, and the gas constant is 300 J/kg · K.

5.2. (M) Density is 1.5 kg/m^3, the gas constant is 600 J/kg · K, and the temperature is 80°C.

5.3. (M) Oxygen gas at 12 bars and 400°C.

5.4. (M) A gas whose molecular weight is 13.5 kg/kg · mole and which is at a gage pressure of 2 bars. The temperature is 800 K.

5.5. (M) Two hundred seventy grams of argon at a pressure of 1.6 bars and a volume of 0.13m^3.

5.6. (M) A gas in a 3-liter container at a pressure of 3 bars, a temperature of 700°C, and with a mass of 0.66 gram.

5.7. (E) A gas having a specific volume of 7 ft^3/lbm, a pressure of 120 psia, and a temperature of 1000°F.

5.8. (E) Determine the gage pressure if the atmospheric pressure is 14.7 psia, the gas constant is 96 ft-lbf/lbm · °R, the temperature is 700°R, and the specific volume is 10 ft^3/lbm.

5.9. (E) Carbon dioxide gas (CO_2) at a pressure of 15 psia and 90°F.

5.10. (E) Seventy pounds-mass of a gas are contained in a rigid container at 200 psia and 80°F. The gas is then expanded to fill a 2000-ft^3 volume at a pressure of 20 psia and a temperature of 70°F. Determine the volume of the rigid container.

5.11. (E) During a constant pressure process, 0.05 lbm of hydrogen gas increases in temperature from 70°F to 200°F. Determine the final volume of the gas if its density is initially 0.09 lbm/ft^3.

5.12. **(E)** A constant temperature process is executed during which the pressure of an ideal gas increases by a ratio of 10 to 1. If the gas was initially enclosd in 2 ft^3, determine the final volume.

5.13. **(E)** Determine the temperature of carbon monoxide gas at 20 psia and 6 ft^3/lbm using van der Waal's equation of state and using the values

$$a = 375 \text{ atm} \cdot \text{ft}^6/\text{mole}^2$$

$$b = 0.63 \text{ ft}^3/\text{mole}$$

Compare your answer to that derived from using the perfect gas law.

Use table B.4 for gas constants when needed.

Sections 5.2, 5.3, and 5.4

5.14. **(M)** Determine the increase in internal energy of 2 kilograms of argon if its temperature increases from 30°C to 130°C.

5.15. **(M)** Six hundred thirty grams of propane gas are cooled from 38°C to 13°C. Determine the total internal energy change and the specific internal energy change.

5.16. **(M)** Helium gas is cooled from 80°C to 20°C in 60 minutes. Determine the internal energy change and the rate of change of energy per gram.

5.17. **(M)** Neon gas exhibits a change in temperature of 1000°C. Determine its change in internal energy per gram.

***5.18.** **(M)** Determine the change in internal energy of ammonia (NH_3) as its temperature increases from 700 K to 800 K. (Hint: Use table B.14 and equation $c_v = c_p - R$.)

5.19. **(E)** Calculate the change in internal energy per pound-mass of sulfur dioxide as its temperature increases by 50°R.

5.20. **(E)** Three pounds-mass of a perfect gas, having constant specific heat, can absorb 70 Btu of heat while increasing temperature by 80°R. Determine C_v and c_v of the gas.

***5.21.** **(E)** Determine, within 1.1% error, the change of internal energy of oxygen gas between 1000°R and 2000°R. (Hint: Use table B.14 and equation $c_v = c_p - R$.)

***5.22.** **(E)** The specific heat, c_v, of a certain gas obeys the relation

$$c_v = [(0.25) + 0.01\,T] \text{ Btu/lbm} \cdot {}^\circ\text{R}$$

Determine the change in internal energy of this gas as its temperature increases from 200°F to 300°F.

Section 5.5

5.23. **(M)** A perfect gas increases temperature from 70°C to 130°C. Determine the change in specific enthalpy if the specific heat is 1.0 kJ/kg · K.

5.24. **(M)** Determine the increase in the enthalpy of 100 kilograms of sulfur dioxide gas (SO_2) between 30°C and 110°C.

5.25. **(M)** A perfect gas is known to have a gas constant of 14.3 kJ/kg · K and a ratio of specific heats, k, of 1.405. Determine c_p and c_v.

5.26. **(M)** Assuming variable specific heats, determine the enthalpy change of propane gas (C_3H_8) as it is cooled from 98°C to 20°C. Check your answer against the answer you get by using constant specific heat values.

5.27. **(E)** Determine the enthalpy change of helium as it is cooled from 120°F to 60°F.

5.28. **(E)** Determine the values of k and c_p of a perfect gas which has c_v of 0.225 Btu/lbm · °R and a gas constant of 66 ft-lbf/lbm · °R.

5.29. **(E)** A perfect gas is known to have a gas constant value of 78.5 ft-lbf/lbm · °R and a ratio of specific heats, k, of 1.28. Determine c_p and c_v.

Use the tables in the appendix for problems 5.30–5.41.

Section 5.6

5.30. Determine the enthalpy of dry air at 1150°R and 5400 K.

5.31. **(M)** Determine the enthalpy and internal energy of saturated mercury vapor at 3.0 bars.

5.32. **(M)** Determine the enthalpy of saturated refrigerant R-22 vapor at 1.5 bars.

5.33. **(M)** Determine the enthalpy of saturated steam liquid at 40°C.

5.34. **(M)** Determine the enthalpy of superheated steam at 3.0 bars and 720°C.

5.35. **(M)** Determine the enthalpy of saturated steam vapor at 1.013 bars.

5.36. **(E)** Determine the internal energy for the following gases at 3400°R: O_2, N_2, CO, and H_2O.

5.37. **(E)** Determine enthalpy and internal energy of saturated mercury vapor at 50 psia.

5.38. **(E)** Determine the enthalpy and internal energy of saturated refrigerant R-22 vapor at 120°F.

5.39. **(E)** Determine the enthalpy and internal energy of saturated refrigerant R-22 liquid at 160°F.

5.40. **(E)** Determine the enthalpy of superheated steam at 200 psia and 1000°F.

5.41. **(E)** Determine the enthalpy of saturated steam vapor at 14.696 psia.

6

Processes

In chapter 5, those equations relating system properties at one particular state were introduced. Here we introduce equations called *process equations*, which determine and relate a system's properties at various states during a process. The advantage of an accurately descriptive process equation is that it gives additional information necessary to determine the amount of work done during a reversible process. The reversible process is given heaviest emphasis in this text, but some attempt is made to show how irreversibilities might affect predicted properties and work.

Tables which list the various forms of the relationships resulting from process equations are presented at the end of the chapter. These tabulations can circumvent much of the mathematical manipulations which students may find too difficult, and they should be useful references for subsequent problem solving.

6.1 The Constant Pressure Process

The most common process occurring in our mechanized, technological society is probably the constant pressure process — any closed system executing a change in volume against the atmosphere is involved in a constant pressure process. There are, of course, many other situations which will produce this type of process; we will now investigate a general one.

For a closed reversible system, the work is calculated from $\int p dV$. In particular, if the pressure is constant, we have

$$Wk_{cs} = p\int_{V_1}^{V_2} dV = p(V_2 - V_1)$$

or

$$Wk_{cs} = p(\Delta V) = p(V_2 - V_1) \tag{6–1}$$

Here V_2 is the final volume of the closed system, and V_1 is the initial volume. This result has been reached previously, but is repeated here for completeness. It is correct for any material — liquid, imperfect gas, or perfect gas.

For perfect gases we have other results which are quite useful. For the constant pressure process we have

$$p_1 = p_2 = \text{constant} \tag{6–2}$$

where the subscripts (1) and (2) represent initial and final states. For the perfect gas this implies

$$\frac{V_1}{T_1} = \frac{V_2}{T_2}$$

or

$$\frac{V_1}{V_2} = \frac{T_1}{T_2} \tag{6–3}$$

Using the relationship $v = V/m$ we also have

$$\frac{v_1}{v_2} = \frac{T_1}{T_2} \tag{6–4}$$

and from the relation $\rho = 1/v$ we obtain

$$\frac{\rho_2}{\rho_1} = \frac{T_1}{T_2} \tag{6–5}$$

From these results, and given enough data, we can calculate properties of perfect gases at different states. These relationships hold for either open or closed systems.

For the open system, we have indicated that reversible work can be found from the equation

$$Wk_{os} = -\int V dp$$

During the constant pressure process this integral vanishes, i.e.

$$dp = 0$$

so that

$$Wk_{os} = 0$$

and for an open system, the only contributions to work when pressure remains constant must be in the form of kinetic or potential energy changes. Frequently, the constant pressure process is referred to as the *isobaric process.*

Example 6.1 In figure 6–1, a perfect gas flows through a chamber at constant pressure, with no change in velocity. If the density of the gas increases by a factor of 2.5, and the inlet temperature is 650°C, determine the work and the exhaust temperature.

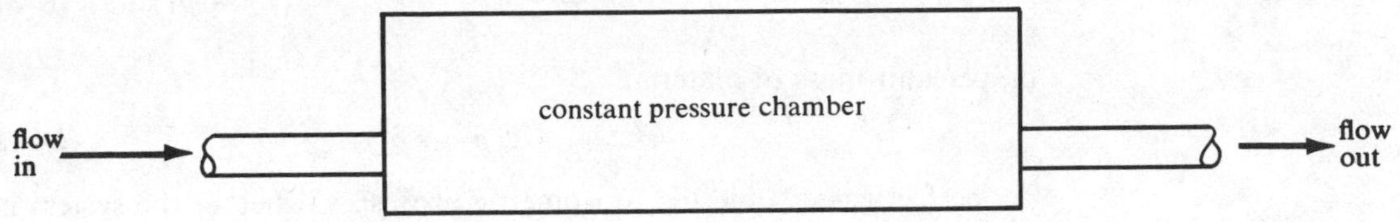

Figure 6–1 Example 6.1.

Solution Since we are here concerned with a system through which gas flows, that is, an open system, we can directly find the work done. We have, for this isobaric process

$$Wk_{os} = 0$$

The exhaust temperature we seek from equation (6–5). We then have

$$\frac{\rho_2}{\rho_1} = 2.5 = \frac{T_1}{T_2} = \frac{(650 + 273)\ \text{K}}{T_2}$$

yielding, for the exhaust temperature

$$T_2 = \frac{923\ \text{K}}{2.5} = 369\ \text{K} \qquad \textit{Answer}$$

6.2 The Constant Volume Process

The *constant volume process*, commonly called the *isometric process*, is approximated quite frequently in engineering situations. Even solid materials can be conveniently considered if we assume that the materials retain their initial volume throughout a twisting, stretching, or shearing process. This then is the meaning of a constant volume process — no volume change.

For a closed, simple system the work done during an isometric process is zero; at least the reversible work due to volume change is zero since

$$Wk_{cs} = \int p dV$$

and obviously here $dV = 0$, so $Wk_{cs} = 0$.

We can, of course, have irreversible work or reversible work manifested by changes in kinetic, potential, or strain energy. (See example 6.4.) Reversible work in the open system, neglecting kinetic or potential energy changes, is

$$Wk_{os} = -\int_{p_1}^{p_2} Vdp = -V\int_{p_1}^{p_2} dp = -V(p_2 - p_1) \qquad \textbf{(6-6)}$$

or, per unit-mass of material

$$wk_{os} = -v(p_2 - p_1) \qquad \textbf{(6-7)}$$

For perfect gases subjected to isometric processes, whether the system is open or closed, we have

$$\frac{p_1}{p_2} = \frac{T_1}{T_2} \qquad \textbf{(6-8)}$$

Remember that though they appear very similar, equations (6–3) and (6–8) are in no way related to each other — they are each descriptive of entirely different processes.

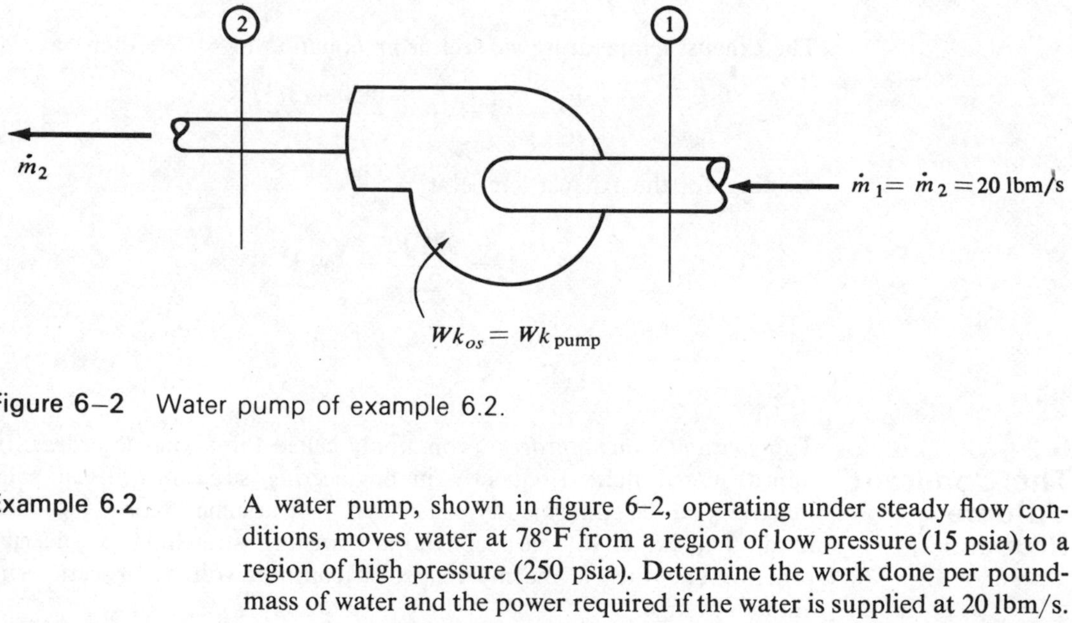

Figure 6–2 Water pump of example 6.2.

Example 6.2 A water pump, shown in figure 6–2, operating under steady flow conditions, moves water at 78°F from a region of low pressure (15 psia) to a region of high pressure (250 psia). Determine the work done per pound-mass of water and the power required if the water is supplied at 20 lbm/s.

Solution The system involved here is a water pump whose boundaries cross an inlet and an exhaust water pipe. Since we are here treating water, we assume it is incompressible and therefore of constant volume. In this circumstance

$$wk_{os} = -\int vdp$$

$$= -v\int dp$$

$$= -v\Delta p = -v(p_2 - p_1)$$

so that, using a value of 0.016 ft³/lbm for the specific volume of water at 78°F

$$wk_{os} = -(0.016 \text{ ft}^3/\text{lbm})(250 - 15 \text{ lbf/in}^2)(144 \text{ in}^2/\text{ft}^2)$$

$$= -541.4 \text{ ft-lbf/lbm}$$ *Answer*

Now, the power is determined from

$$\dot{W}k_{os} = \dot{m}wk_{os}$$

and

$$\dot{W}k_{os} = 20 \text{ lbm/s } (-541.4 \text{ ft-lbf/lbm})$$

$$= -10829 \text{ ft-lbf/s} = -19.7 \text{ hp}$$ *Answer*

For the irreversible case, this power requirement must be increased; but we could not determine exactly how much by using only thermodynamics.

Example 6.3 Inside the closed chamber of a piston-cylinder device, fuel and air burn at constant volume with a releasing of 350 kilojoules of energy. If the volume is 100 cm³, the temperature is 280°C before burning, and the pressure is 6.0 bars, determine the temperature and pressure of the gases after burning if the gases behave like air.

Solution The process is one of a closed system, so we may write the energy equation as follows:

$$\Delta U = Q \qquad (Wk_{cs} = 0)$$

for constant volume, closed systems. Using specific heats and temperatures, we have

$$\Delta U = mc_v T = Q = 350 \text{ kJ}$$

For air, $c_v = 0.719$ kJ/kg · K and $R = 0.287$ kJ/kg · K from table B.4. We need to know the gas constant R in order that we may solve for the mass from the perfect gas equation. We find

$$m = \frac{pV}{RT} = \frac{(6.0 \text{ bars})(10^5 \text{ N/m}^2 \cdot \text{bar})(100 \text{ cm}^3)}{(0.287 \text{ kJ/kg} \cdot \text{K})(553 \text{ K})(10^6 \text{ cm}^3/\text{m}^3)}$$

$$= 0.378 \text{ kg}$$

Then, from the above

$$\Delta T = \frac{Q}{mc_v} = \frac{350 \text{ kJ}}{0.378 \text{ kg} \times 0.719 \text{ kJ/kg} \cdot \text{K}}$$

or

$$\Delta T = 1288\ \text{K} = T_2 - T_1$$

Since $T_1 = 280°\text{C} = 553$ K, we find

$$T_2 = 1841\ \text{K}$$ *Answer*

The pressure after burning is found from equation (6–8):

$$p_2 = p_1\left(\frac{T_2}{T_1}\right) = (6.0\ \text{bars})\left(\frac{1841\ \text{K}}{553\ \text{K}}\right)$$

$$= 20.0\ \text{bars}$$ *Answer*

Example 6.4

A cylindrical bar 10 inches long and 1 inch in diameter is subjected to an axial load of 3000 lbf, as shown in figure 6–3. The bar is composed of steel having a modulus of elasticity of 30×10^6 lbf/in² where the modulus of elasticity, Y, is defined as

$$Y = \frac{\text{stress}}{\text{strain}} \tag{6–8}$$

The stress and strain are

$$\text{Stress} = \frac{\text{force}}{\text{area}} = \frac{F}{A}$$

and

$$\text{Strain} = \xi = \frac{\Delta l}{l}$$

where Δl is the change in length of a given length l bar. Determine the work done for this process, if the bar does not change volume when stretched and if no heat is transferred.

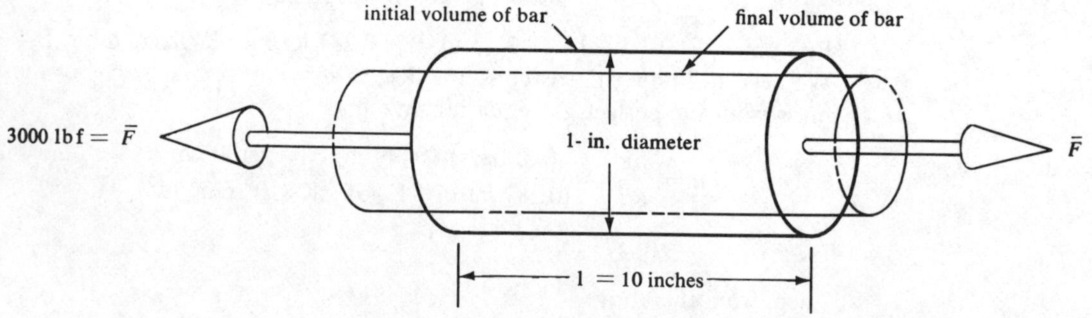

Figure 6–3 Strain energy application. Example 6.4.

Solution

This problem has been included in this text to demonstrate the scope of the thermodynamic approach in analyzing a diverse set of technical problems. We have here a constant volume process for a closed system, but the system is not simple. We are assuming the material is perfectly elastic (which means it will return to its initial shape when external forces are removed) and, therefore, reversible. However, we have interactions between the constituent atoms and our first law becomes

$$\Delta E = Q - Wk_{cs}$$

We assume no heat transfer or changes in kinetic and potential energy or enthalpy, but we must account for strain energy, SE. Strain energy results from the interaction of the atoms or molecules. We define the change in strain energy by

$$\Delta \text{SE} = \int V \times \frac{F}{A} \times d\left(\frac{\Delta l}{l}\right) = \int V \times \frac{F}{A} \times d\xi \qquad \textbf{(6–9)}$$

Note that $F/A = Y\xi$ from equation (6–8). Then the strain energy change is

$$\Delta \text{SE} = \int VY\xi d\xi \qquad \textbf{(6–10)}$$

From our first law equation applied to the bar, since $Q = 0$

$$\Delta \text{SE} = -Wk_{os} \qquad \textbf{(6–11)}$$

Since we seek the work required to stress the bar under 3000 lbf load, we set $\xi_1 = 0$ and from the definition of the modulus of elasticity, $Y = \dfrac{F/A}{\xi}$, obtain

$$\begin{aligned}\xi_2 &= \frac{3000 \text{ lbf}}{YA}\\ &= \frac{3000 \text{ lbf}}{(\pi \times \frac{1}{4}) \text{ in}^2(30 \times 10^6 \text{ lbf/in}^2)}\\ &= 1.27 \times 10^{-4} \text{ in/in}\\ &= 1.27 \times 10^{-4} = \xi_2\end{aligned}$$

Then

$$\begin{aligned}Wk_{cs} = -\Delta \text{SE} &= -\int_{\xi_1}^{\xi_2} VY\xi d\xi\\ &= \left[-VY\left(\frac{1}{2}\xi^2\right)\right]_{\xi_1}^{\xi_2}\\ &= -VY\left(\frac{1}{2}\right)(\xi_2^2 - \xi_1^2)\end{aligned}$$

We now substitute numbers into this equation. First, the volume is just the volume of the cylindrical bar, i.e.

$$V = (10 \text{ in})\left(\pi \times \frac{1}{4} \text{ in}^2\right)$$

and then

$$Wk_{cs} = -\left(\frac{10\pi}{4} \text{ in}^3\right)(30 \times 10^6 \text{ lbf/in}^2)\left(\frac{1}{2}\right)[(1.27 \times 10^{-4})^2 - 0]$$

or

$$Wk_{cs} = 1.93 \text{ in-lbf}$$ *Answer*

This represents the energy expanded in applying 3000 lbf to the described bar and allowing equilibrium to be slowly achieved. In equilibrium the bar will have been stretched 1.27×10^{-4} inches/inch of bar length or a total of $1.27 \times 10 \times 10^{-4}$ or 1.27×10^{-3} inches.

Notice that for a modulus of elasticity, Y, which remains constant during a stress process the work can be written

$$Wk_{cs} = \frac{1}{2}VY(\xi_2^2 - \xi_1^2) \quad \textbf{(6–12)}$$

also. If the initial strain, ξ_1, is zero then

$$Wk_{cs} = \frac{1}{2}VY\xi_2^2 \quad \textbf{(6–13)}$$

This relationship holds true for those materials having no prestress.

6.3 The Constant Temperature Process

We now treat the *constant temperature* or *isothermal* process which has many applications. This type of process will frequently involve heat transfer and work, as well as energy changes. If the system undergoing an isothermal process can be described as a perfect gas, we see that the enthalpy and internal energy changes are zero. Thus, since $\Delta T = 0$ and enthalpy and internal energy are functions of temperature only

$$\Delta H = 0$$

and

$$\Delta U = 0$$

Let us consider the closed, reversible, isothermal system. Assuming the system is composed of a perfect gas we have

$$pV = mRT = \text{constant} = C \quad \textbf{(6–14)}$$

or

$$p_1V_1 = p_2V_2 = C \quad \textbf{(6–15)}$$

As a consequence of this, let us consider the reversible work for the closed system. Here

$$Wk_{cs} = \int p dV$$

and from equation (6–14)

$$Wk_{cs} = \int \frac{C dV}{V} \quad \textbf{(6–16)}$$

We will integrate this between the limits of V_1 and V_2, i.e.

$$Wk_{cs} = C \int_{V_1}^{V_2} \frac{dV}{V}$$

The integration produces the result

$$Wk_{cs} = \Big(C \ln V\Big)_{V_1}^{V_2} = C(\ln V_2 - \ln V_1)$$

or

$$Wk_{cs} = C \ln \frac{V_2}{V_1} \quad \textbf{(6–17)}$$

where the constant C has the value calculated from equation (6–15). Remember that equation (6–17) is correct for perfect gases only and represents the area under the curve

$$pV = \text{constant}$$

on a p-V diagram, as indicated in figure 6–4.

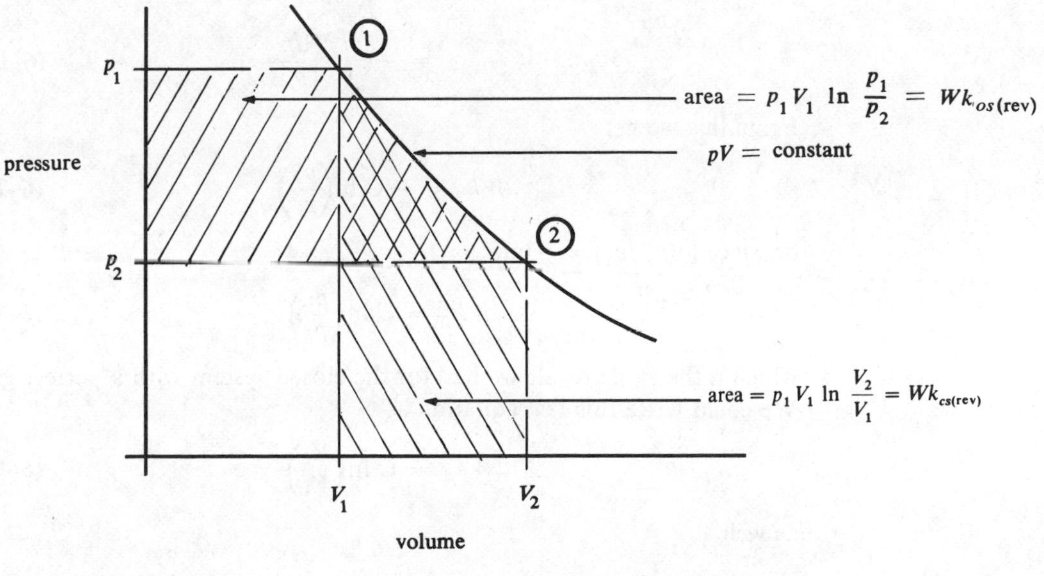

Figure 6–4 p-V diagram for isothermal process with a perfect gas.

For closed systems without strain, kinetic, or potential energy, the internal energy change is zero for isothermal processes of a perfect gas and, from the first law

$$\Delta U = 0 = Q - Wk_{cs}$$

We then have

$$Q = Wk_{cs} = C \ln \frac{V_2}{V_1}$$

In addition, by using equation (6–15) we have

$$\frac{p_1}{p_2} = \frac{V_2}{V_1}$$

so

$$\ln\left(\frac{V_2}{V_1}\right) = \ln\left(\frac{p_1}{p_2}\right)$$

and substituting this result into equation (6–17) yields

$$Wk_{cs} = C \ln\left(\frac{p_1}{p_2}\right) \tag{6–18}$$

Let us now consider the isothermal process in an open system. Again, as for the closed system, we assume a perfect gas and obtain

$$p_1V_1 = p_2V_2 = \text{constant} = C$$

and observe that the reversible work is given by

$$Wk_{os} = -\int V dp$$
$$= -C\int_{p_1}^{p_2} \frac{dp}{p} \tag{6–19}$$

From this we get

$$Wk_{os} = -C \ln\left(\frac{p_2}{p_1}\right) \tag{6–20}$$

or, since $\ln(p_2/p_1) = -\ln(p_1/p_2)$

$$Wk_{os} = C \ln\left(\frac{p_1}{p_2}\right) \tag{6–21}$$

which is the same result we had for the closed system with a perfect gas. We could write this last equation as

$$Wk_{os} = C \ln\left(\frac{V_2}{V_1}\right) \tag{6–17}$$

as well.

For those gases not obeying the perfect gas relation, the above results must be altered, and generally with an increase in algebraic difficulty and complication. At this point we will not consider isothermal processes for other than perfect gases.

Example 6.5 During an expansion of a perfect gas from 12.0 bars pressure and a specific volume of 1.1 m^3/kg to 1.01 bars pressure, the temperature remains constant. Determine the constant C in the process equation $pv = C$ describing the process and determine the final specific volume.

Solution The constant C is equal to the product of the pressure and specific volume at any state during the process. Thus, at the initial condition we have

$$C = (12.0 \text{ bars})(1.1 \text{ m}^3/\text{kg})$$
$$= (12.0 \text{ bars} \times 10^5 \text{ N/m}^2 \cdot \text{bars})(1.1 \text{ m}^3/\text{kg})$$
$$= 1320 \text{ kJ/kg} \qquad \textit{Answer}$$

Also, we may write the process equation as

$$p_1 v_1 = p_2 v_2$$

yielding

$$v_2 = v_1 \left(\frac{p_1}{p_2}\right)$$

Substituting values into this relationship gives

$$v_2 = (1.1 \text{ m}^3/\text{kg})\left(\frac{12 \text{ bars}}{1.01 \text{ bars}}\right)$$
$$= 13.1 \text{ m}^3/\text{kg} \qquad \textit{Answer}$$

Example 6.6 During the compression of 0.01 lbm of air in a cylinder (see figure 6–5), heat is transferred through the cylinder walls to keep the air at a constant

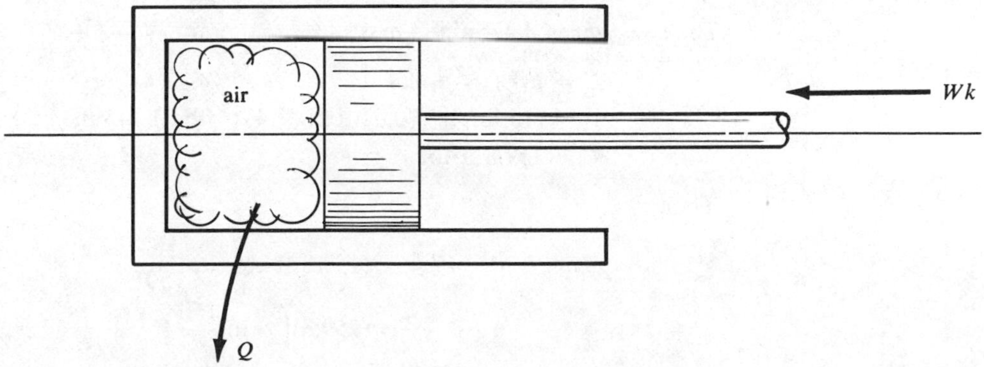

Figure 6–5 Isothermal process of a piston-cylinder device.

temperature. The air pressure increases from 15 psia to 150 psia after the air is fully compressed. The initial specific volume of the air is 7.4 ft^3/lbm. Determine the operating air temperature, the change in internal energy and in enthalpy, the work done, and the heat transferred during this process.

Solution This is an isothermal process and we will assume it to be reversible as well. If the air is behaving like a perfect gas, which we assume, then the operating temperature can be found from

$$T = \frac{pV}{mR} = \frac{pv}{R}$$

or, initially

$$T_1 = \frac{p_1 v_1}{R} = \frac{(15 \text{ lbf/in}^2)(7.4 \text{ ft}^3\text{/lbm})(144 \text{ in}^2\text{/ft}^2)}{53.3 \text{ ft-lbf/lbm} \cdot {}^\circ\text{R}}$$

so that

$$T_1 = 300^\circ\text{R}$$

and then

$$T_2 = 300^\circ\text{R}$$

The change in internal energy is

$$\Delta U = mc_v \Delta T = 0$$

and for the enthalpy change we have

$$\Delta H = mc_p \Delta T = 0$$

The work done is reversible and this we obtain from equation (6–17)

$$Wk_{cs} = C \ln\left(\frac{V_2}{V_1}\right)$$

or, more conveniently, from equation (6–18)

$$Wk_{cs} = C \ln\left(\frac{p_1}{p_2}\right)$$

The constant is determined first:

$$\begin{aligned} C &= p_1 V_1 = p_1 m v_1 \\ &= (15 \text{ lbf/in}^2)(0.01 \text{ lbm})(7.4 \text{ ft}^3\text{/lbm})(144 \text{ in}^2\text{/ft}^2) \\ &= 159.8 \text{ ft-lbf} \end{aligned}$$

Then

$$\begin{aligned} Wk_{cs} &= (159.8 \text{ ft-lbf}) \ln \frac{15}{150} \\ &= (159.8)\left(-\ln \frac{150}{15}\right) \\ &= 159.8\,(-\ln 10) \end{aligned}$$

$$= -368 \text{ ft-lbf} \qquad \textit{Answer}$$

The heat transferred is equal to the work done, so

$$Q = -368 \text{ ft-lbf} \qquad \textit{Answer}$$

and Q is, as the sign indicates, removed from the system. For the irreversible isothermal process, the internal energy change can still be zero; but the work and heat increase in absolute values; that is, more work is required and more heat transfer is demanded to retain constant temperature.

6.4 The Adiabatic Process

During the preceding sections we have considered processes where an important property is fixed or constant, that is, constant pressure, volume, or temperature. Here we will consider the condition when heat transfer is zero — called the *adiabatic process.* While no real process is completely adiabatic, there are physical conditions when it is approximated, such as a well-insulated system, or a rapidly occurring process.

The assumption of a perfect gas system has proved to be profitable before, so we will assume this condition again. Let us then write the first law for the adiabatic process with a perfect gas.

$$\Delta U = Q - Wk_{cs} \tag{6–22}$$

but $Q = 0$, so

$$\Delta U = m\int c_v dT = -Wk_{cs} \tag{6–23}$$

Now let us assume the process is reversible as well. Then the work term can by replaced by $\int pdV$ and we get

$$m\int c_v dT = -\int pdV \tag{6–24}$$

For the perfect gas, $p = mRT/V$, so

$$m\int c_v dT = -\int \frac{mRTdV}{V}$$

and dividing both sides by mT yields

$$\int c_v \frac{dT}{T} = -\int \frac{RdV}{V} \tag{6–25}$$

We now assume that c_v is a constant and recall that R can be replaced by $c_p - c_v$. Then

$$c_v\int \frac{dT}{T} = -(c_p - c_v)\int \frac{dV}{V}$$

or

$$\int \frac{dT}{T} = -\left(\frac{c_p}{c_v} - 1\right)\int \frac{dV}{V} = (1 - k)\int \frac{dV}{V} \qquad \textbf{(6–26)}$$

Integrating both sides of this equation gives us

$$\ln T = (1 - k)\ln V + \text{constant}$$

or, since $T = pV/mR$

$$\ln \frac{pV}{mR} = (1 - k)\ln V + \text{constant}$$

This can be written

$$\ln p + \ln V = \ln V - k \ln V + C_0 \qquad \textbf{(6–27)}$$

where C_0 is a new constant containing all the other constants. Now we have, from equation (6–27)

$$\ln p + k \ln V = C_0 \qquad \textbf{(6–28)}$$

or

$$pV^k = C \qquad \textbf{(6–29)}$$

The curve of this is plotted in figure 6–6. In this curve, C is a newer constant to account for the antilog operation between equations (6–28) and (6–29). For reversible adiabatic processes, equation (6–29) is correct only if the system involved is composed of a perfect gas. Let us see what this has done for us.

The reversible work of a closed system indicated as the area under a curve described by pV^k = constant in a p-V diagram is

$$Wk_{cs} = \int p\,dV$$

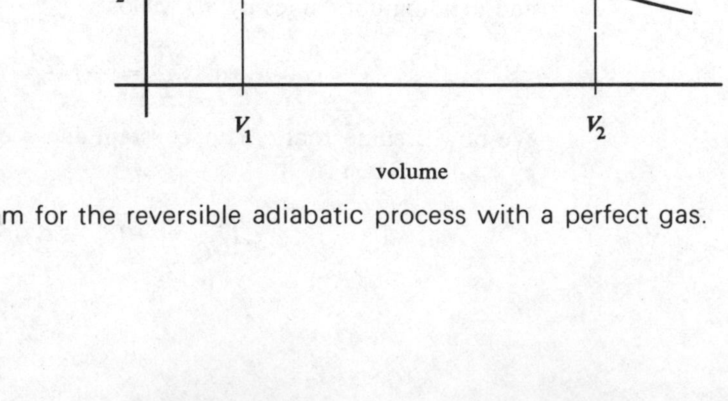

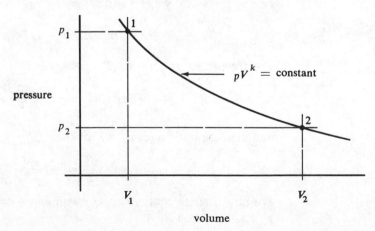

Figure 6–6 p-V diagram for the reversible adiabatic process with a perfect gas.

From equation (6–29) we write

$$p = \frac{C}{V^k}$$

and then

$$Wk_{cs} = \int_{V_1}^{V_2} \frac{C\,dV}{V^k} \tag{6–30}$$

Integrating between limits of V_1 and V_2 yields

$$Wk_{cs} = C\left(\frac{1}{1-k}\right)(V_2^{1-k} - V_1^{1-k}) \tag{6–31}$$

This can be condensed somewhat by substituting equation (6–29)

$$C = p_1 V_1{}^k = p_2 V_2{}^k$$

into equation (6–31)

$$Wk_{cs} = \frac{1}{1-k}(p_2 V_2{}^k V_2{}^{1-k} - p_1 V_1{}^k V^{1-k})$$

to obtain

$$Wk_{cs} = \frac{1}{1-k}(p_2 V_2 - p_1 V_1) \tag{6–32}$$

Since this is an equation for perfect gases it could easily be written as

$$Wk_{cs} = \frac{mR}{1-k}(T_2 - T_1) \tag{6–33}$$

It should not be surprising that, for the reversible adiabatic process, the change in internal energy and enthalpy are found from the equations

$$\Delta U = mc_v \Delta T$$

and

$$\Delta H = mc_p \Delta T$$

assuming constant specific heats for the perfect gas.

For open systems, the reversible work (neglecting kinetic and potential energy changes) is gotten from the relation

$$Wk_{os} = \frac{k}{1-k}(p_2 V_2 - p_1 V_1) \tag{6–34}$$

which can easily be derived in the manner of getting equation (6–32). This is left as an exercise for the student. (See problem 6.36.) Also, we can revise equation (6–34) to read

$$Wk_{os} = \frac{(mRk)}{1-k}(T_2 - T_1) \tag{6–35}$$

by a simple algebraic manipulation, assuming we still have a perfect gas.

Example 6.7

A perfect gas is allowed to expand reversibly and adiabatically in a piston-cylinder from 15.0 bars and 0.05 m³/kg to 1.01 bars pressure. If the gas has a specific heat ratio, k, of 1.38 determine the work done during this process.

Solution

For a reversible adiabatic process we may use equation (6–29)

$$pv^k = C$$

from which we derived equation (6–32)

$$wk_{cs} = \frac{1}{1-k}(p_2 v_2 - p_1 v_1)$$

To find the final specific volume v_2 we write the process equation in the form

$$p_1 v_1^k = p_2 v_2^k$$

and then

$$v_2 = v_1 \left(\frac{p_1}{p_2}\right)^{1/k}$$

Substituting values into this relationship gives

$$v_2 = (0.05\ \text{m}^3/\text{kg})\left(\frac{15.0\ \text{bars}}{1.01\ \text{bars}}\right)^{1/1.38}$$

$$= 0.353\ \text{m}^3/\text{kg}$$

Using this result, we can now compute the work from equation (6–32)

$$wk_{cs} = \frac{1}{1-1.38}[(1.01 \times 10^5\ \text{N/m}^2)(0.353\ \text{m}^3/\text{kg})$$

$$- (15 \times 10^5\ \text{N/m}^2)(0.05\ \text{m}^3/\text{kg})]$$

$$= (-2.63)(-0.393 \times 10^5\ \text{N}\cdot\text{m/kg})$$

$$= 103\ \text{kJ/kg}$$ *Answer*

Example 6.8

A gas turbine receives air at 2000 K, allows reversible adiabatic expansion of the air past the turbine buckets, and exhausts this air at 2 lbm/s to the atmosphere at 1000 K. Determine the heat transferred, the work produced per lbm of air, and the power produced if there are negligible kinetic and potential energy changes for the air.

Solution

Our system is here specified as the gas turbine which is a device converting energy from a flowing stream (such as air) to a rotational energy of a

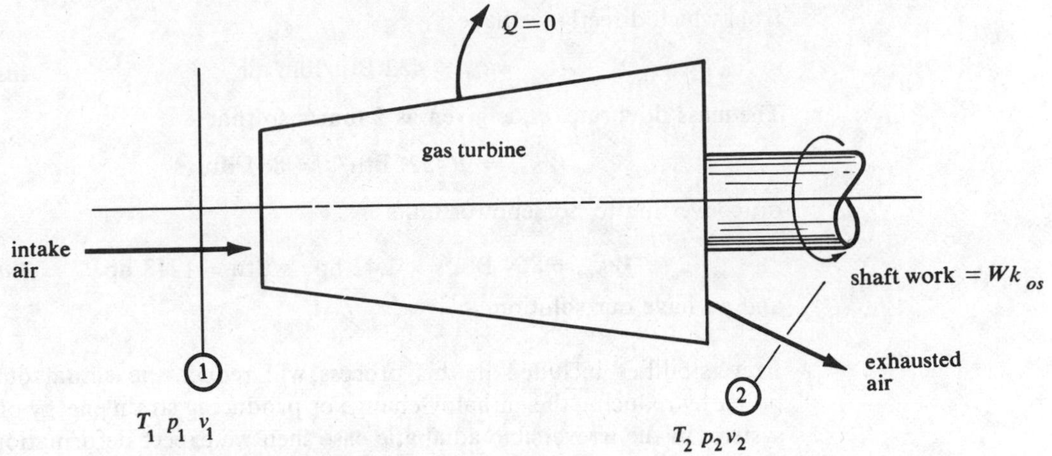

Figure 6–7 Sketch of system for adiabatic expansion of air in a gas turbine.

wheel or shaft. The system is an open one as indicated in figure 6–7, so we write the first law as

$$\Delta H = Q - Wk_{os}$$

In terms of rates per unit time, this becomes

$$\dot{H} = \dot{Q} - \dot{W}k_{os} \quad \textbf{(6–36)}$$

which can be rewritten as

$$\dot{m}\Delta h = \dot{Q} - \dot{m}wk_{os} \quad \textbf{(6–37)}$$

The enthalpy change, Δh, is obtained from

$$\Delta h = c_p \Delta T$$

or

$$\Delta h = h_2 - h_1 = c_p(T_2 - T_1)$$

after assuming the air is a perfect gas with constant specific heat. From table B.4 of appendix B we find c_p to be 0.24 Btu/lbm · °R, or

$$c_p = 0.24 \text{ Btu/lbm} \cdot {}^\circ\text{R} \times \frac{9\ {}^\circ\text{R}}{5\ \text{K}} = 0.432 \text{ Btu/lbm} \cdot \text{K}$$

Then

$$\Delta h = 0.432 \text{ Btu/lbm} \cdot \text{K} \times (1000\ \text{K} - 2000\ \text{K}) = -432 \text{ Btu/lbm}$$

and since the process is adiabatic, $\dot{Q} = 0$. Then equation (6–37) can be written

$$\dot{m}(-432 \text{ Btu/lbm}) = -\dot{m}wk_{os} \quad \textbf{(6–38)}$$

from which directly we have

$$wk_{os} = 432 \text{ Btu/lbm air} \qquad \textit{Answer}$$

The mass flow rate, $\dot{m}$, is given as 2 lbm/s so that

$$\dot{W}k_{os} = 2(432) \text{ Btu/s} = 864 \text{ Btu/s}$$

or, converting to horsepower units

$$\dot{W}k_{os} = 864 \text{ Btu/s} \times 1.41 \text{ hp} \cdot \text{s/Btu} = 1218 \text{ hp} \qquad \textit{Answer}$$

and we have our solution.

Irreversibilities included in this process will reduce the actual output power by reducing the enthalpy change or producing strain energy of the system. In the irreversible adiabatic case then we expect deformation or wear in the system.

Before leaving the reversible adiabatic process of perfect gases let us return to the equation

$$pV^k = C \qquad \textbf{(6–29)}$$

and seek relations between p, V, and T. From this equation we can easily write

$$p_1V_1^k = p_2V_2^k$$

or

$$\frac{p_1}{p_2} = \left(\frac{V_2}{V_1}\right)^k \qquad \textbf{(6–39)}$$

By using the perfect gas relation we have $p_1 = mRT_1/V_1$ and $p_2 = mRT_2/V_2$ which yields for equation (6–39) the result

$$\frac{mRT_1/V_1}{mRT_2/V_2} = \left(\frac{V_2}{V_1}\right)^k$$

and

$$\frac{T_1}{T_1} = \left(\frac{V_1}{V_2}\right)\left(\frac{V_2}{V_1}\right)^k$$

which reduces then to

$$\frac{T_1}{T_2} = \left(\frac{V_2}{V_1}\right)^{k-1} \qquad \textbf{(6–40)}$$

Also we could revise this last equation to read

$$\frac{V_2}{V_1} = \left(\frac{T_1}{T_2}\right)^{1/k-1} \qquad \textbf{(6–41)}$$

or

$$\left(\frac{V_2}{V_1}\right)^k = \left(\frac{T_1}{T_2}\right)^{k/k-1} \qquad \textbf{(6–42)}$$

which, substituted back into equation (6–39) yields

$$\left(\frac{p_1}{p_2}\right) = \left(\frac{T_1}{T_2}\right)^{k/k-1} \qquad \textbf{(6–42)}$$

We will need these relations in succeeding problem solutions

6.5 The Polytropic Process

In the discussions of the processes considered in this chapter, we have always arrived at an equation relating pressure to volume — an equation which can be obtained from the general equation

$$pV^n = C \qquad \textbf{(6–43)}$$

which is called the *polytropic equation.* For the isobaric process we set n equal to zero in equation (6–43) and find

$$pV^0 = p = c$$

Similarly, for the isometric process we revise (6–43) to read

$$p^{1/n}V = C^{1/n} = C' \quad \text{(a new constant)}$$

and set n equal to infinity. Then

$$p^0V = V = C'$$

as we required for this process.

For the constant temperature process of a perfect gas we set n equal to unity (1) and obtain from the polytropic equation

$$pV^1 = pV = C$$

For the reversible adiabatic process we arrived at equation (6–29) which is identical to the polytropic equation but with $k = n$.

The most general process then we call the *polytropic process,* which is any process of a system which can be described by equation (6–43), where n can take any value desired. We therefore consider n an empirical constant (it could also be a very complicated variable if desired!) which fits equation (6–43) to actual data.

Table 6–1 summarizes the previous discussion, and in figure 6–8 are plotted the curves characterizing the five common processes considered here.

Table 6–1
The Polytropic Exponent

Processes	*Value of n in polytropic equation* $pV^n = C$
Isobaric	0
Isometric	∞
Isothermal	1
Reversible adiabatic	k
Polytropic	n

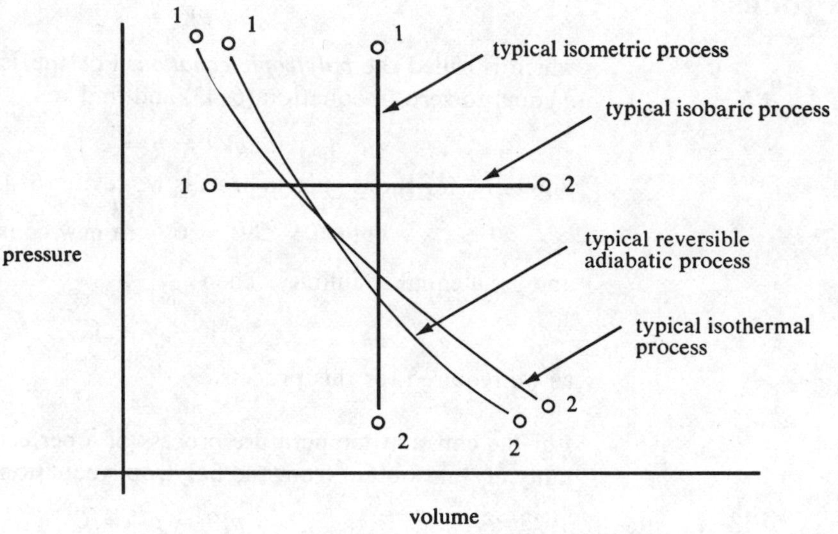

Figure 6–8 The various polytropic processes, shown on the p-V plane, for perfect gases.

For the closed system, the reversible work of a polytropic process can readily be shown to be

$$Wk_{cs} = \frac{1}{1 - n}(p_2V_2 - p_1V_1) \tag{6–44}$$

by a derivation much like that for the work of a reversible adiabatic process. The open system reversible work, likewise can be obtained from the equation

$$Wk_{os} = \frac{n}{1 - n}(p_2V_2 - p_1V_1) \tag{6–45}$$

which is closely related to the corresponding reversible adiabatic case. Similarly, the relations between pressure, volume, and temperature at various states along a polytropic process are given by

$$\frac{p_1}{p_2} = \left(\frac{V_2}{V_1}\right)^n = \left(\frac{v_2}{v_1}\right)^n \tag{6–46}$$

$$\frac{T_1}{T_2} = \left(\frac{V_2}{V_1}\right)^{n-1} = \left(\frac{v_2}{v_1}\right)^{n-1} \tag{6–47}$$

and

$$\frac{p_1}{p_2} = \left(\frac{T_1}{T_2}\right)^{n/(n-1)} \tag{6–48}$$

Example 6.9 Air is polytropically expanded through a reversible turbine from 15 bars to 1.0 bar pressure. The inlet air temperature is 1200 K, the mass flow rate is 1.0 kg/s, and the polytropic exponent n is 1.5. Determine the power produced and the rate of heat transfer.

Solution Since the system is an open one and is reversible, the work can be computed from the equation

$$wk_{os} = \frac{n}{1-n}(p_2 v_2 - p_1 v_1) \tag{6–49}$$

and assuming a perfect gas, this becomes

$$wk_{os} = \frac{nR}{1-n}(T_2 - T_1) \tag{6–50}$$

To compute an answer from this equation we must first find the final temperature T_2. Using equation (6–42) we have

$$T_2 = T_1\left(\frac{p_2}{p_1}\right)^{(n-1)/n}$$

or

$$T_2 = (1200\ \text{K})(1\ \text{bar}/15\ \text{bars})^{0.5/1.5}$$

$$= 487\ \text{K}$$

The work is then obtained from the computation

$$wk_{os} = \frac{1.5}{1-1.5}(0.287\ \text{kJ/kg}\cdot\text{K})(487 - 1200)\ \text{K}$$

$$= 614\ \text{kJ/kg}$$

The power can be found from the relationship with the mass flow rate

$$\dot{W}k_{os} = \dot{m}(wk_{os})$$

to give

$$\dot{W}k_{os} = (1.0 \text{ kg/s})(614 \text{ kJ/kg})$$
$$= 614 \text{ kJ/s} = 614 \text{ kW} \qquad \textit{Answer}$$

Using the conversion in table B.15 to horsepower units we get

$$\dot{W}k_{os} = 823 \text{ hp} \qquad \textit{Answer}$$

If we neglect kinetic and potential energy changes through the turbine, we may write the energy equation for the turbine as

$$\dot{m}h = \dot{Q} - \dot{W}k_{os}$$

or

$$\dot{Q} = \dot{m}h + \dot{W}k_{os}$$

from which we may now compute the heat transfer. Thus

$$\dot{Q} = \dot{m}c_p(T_2 - T_1) + Wk_{os}$$

Using a value for the specific heat of 1.007 kJ/kg · K we have

$$\dot{Q} = (1.0 \text{ kg/s})(1.007 \text{ kJ/kg} \cdot \text{K})(-713 \text{ K}) + 614 \text{ kW}$$
$$= -104 \text{ kW} \qquad \textit{Answer}$$

Example 6.10

During a reversible polytropic compression of an ideal gas, the following data are found:

$$p_1 = 14.7 \text{ psia}$$
$$v_1 = 14.5 \text{ ft}^3/\text{lbm}$$
$$p_2 = 165 \text{ psia}$$
$$v_2 = 2.42 \text{ ft}^3/\text{lbm}$$

Determine the polytropic exponent n.

Solution

Since the process is reversible polytropic, for a perfect or ideal gas we write the process equation

$$p_1 v_1^n = p_2 v_2^n$$

or

$$\left(\frac{v_1}{v_2}\right)^n = \frac{p_2}{p_1}$$

and

$$(n)\ln\left(\frac{v_1}{v_2}\right) = \ln\left(\frac{p_2}{p_1}\right)$$

Then

$$n = \frac{\ln(p_2/p_1)}{\ln(v_1/v_2)}$$

and we compute n

$$n = \frac{\ln(165 \text{ psia}/14.7 \text{ psia})}{\ln(14.5 \text{ ft}^3/\text{lbm}/2.42 \text{ ft}^3/\text{lbm})}$$

$$= 1.35$$ *Answer*

6.6 Summary

We have considered the most common processes encountered in thermodynamics or energy transfer problems. The numerous equations accumulated during these last two chapters are listed in tables 6–2 and 6–3 to provide ready references. Note that table 6–2 lists the general

Table 6–2

General Process

(Excluding Processes Involving Phase Change)

Term	*Reversible Isobaric*	*Reversible Isometric*	*Reversible Isothermal*	*Reversible Adiabatic*	*Reversible Polytropic*
p-V relation	$p = c$	$V = c$	$T = c$	$s = c$	$pV^n = C$
$^*Wk_{cs}$	$p(V_2 - V_1)$	0	$\int pdV$	$\int pdV$	$\frac{1}{1-n}(p_2V_2 - p_1V_1)$
$^*Wk_{os}$	0	$V(p_1 - p_2)$	$-\int Vdp$	$-\int Vdp$	$\frac{n}{1-n}(p_2V_2 - p_1V_1)$
Q_{os}	$m\Delta h$	$m\Delta u$	$-\int Vdp$	0	$m\Delta h + Wk_{os}$
Q_{cs}	$m\Delta h$	$m\Delta u$	$\int pdV$	0	$m\Delta U + Wk_{cs}$
Δh	$\int c_p dT$	$\int c_p dT$	0	$\int c_p dT$	$\int c_p dT$
Δu	$\int c_v dT$	$\int c_v dT$	0	$\int c_v dT$	$\int c_v dT$
Δs	$\int \frac{dh}{T}$	$\int \frac{du}{T}$	$\frac{q}{T}$	0	$\int \frac{dq}{T}$

* Assuming there are no kinetic, potential, or strain energy changes.

Table 6–3

Perfect Gas Process

Term	Reversible Isobaric	Reversible Isometric	Reversible Isothermal	Reversible Adiabatic	Reversible Polytropic
p-v relation	$p = c$	$V = c$	$T = c$	$pV^k = c$	$pV^n = c$
p-v-T relations	$\frac{V_1}{V_2} = \frac{T_1}{T_2}$	$\frac{p_1}{p_2} = \frac{T_1}{T_2}$	$\frac{p_1}{p_2} = \frac{V_2}{V_1}$	$\frac{p_1}{p_2} = \left(\frac{V_2}{V_1}\right)^k$ $\frac{p_1}{p_2} = \left(\frac{T_1}{T_2}\right)^{k/k-1}$	$\frac{p_1}{p_2} = \left(\frac{V_2}{V_1}\right)^n$ $\frac{p_1}{p_2} = \left(\frac{T_1}{T_2}\right)^{n/n-1}$
$^*Wk_{cs}$	$p(V_2 - V_1)$	0	$p_1V_1 \ln\left(\frac{V_2}{V_1}\right)$	$\frac{1}{1-k}(p_2V_2 - p_1V_1)$	$\frac{1}{1-n}(p_2V_2 - p_1V_1)$
$^*Wk_{os}$	0	$V(p_1 - p_2)$	$p_1V_1 \ln\left(\frac{V_2}{V_1}\right)$	$\frac{k}{1-k}(p_2V_2 - p_1V_1)$	$\frac{n}{1-n}(p_2V_2 - p_1V_1)$
Q_{os}	$m\Delta h$	$m\Delta u$	$p_1V_1 \ln(V_2/V_1)$	0	$m\Delta h + Wk_{os}$
Q_{cs}	$m\Delta h$	$m\Delta u$	$p_1V_1 \ln(V_2/V_1)$	0	$m\Delta u + Wk_{cs}$
Δh	$c_p \Delta T$	$c_p \Delta T$	0	$c_p \Delta T$	$c_p \Delta T$
Δu	$c_v \Delta T$	$c_v \Delta T$	0	$c_v \Delta T$	$c_v \Delta T$
Δs	$c_p \ln\left(\frac{T_2}{T_1}\right)$	$c_v \ln\left(\frac{T_2}{T_1}\right)$	$R \ln(V_2/V_1)$	0	$c_v \ln(T_2/T_1)$ $+ R \ln(V_2/V_1)$

* Assuming there are no kinetic, potential, or strain energy changes.

equations which are applicable to any material but which are frequently too cumbersome; table 6–3 lists the specific equations pertinent to perfect gases only. In table 6–3 are listed only the reversible cases and in addition a new property, *entropy*, is tabulated as a change in its value, Δs. We will speak more of entropy in the next chapter but its formulas are presented here for completeness.

Practice Problems

Problems designated with an asterisk * preceding the number should only be attempted by those having a background which includes the knowledge of integral calculus.

Problems using SI units are indicated with (M) under the problem number, and (E) indicates the use of English units.

Section 6.1

6.1. **(M)** Determine the work of a process when the volume changes from 3 m^3 to 5 m^3. The pressure has a constant value of 3 bars.

6.2. **(M)** A piston-cylinder contains 30.0 gram of air at 1.2 bars and 40°C. The piston is then compressed at constant pressure so that the volume of air is one-half the original value. Determine the final temperature, final volume, and work required for this process.

6.3. **(E)** Air contained in a frictionless piston-cylinder expands from 1 in^3 to 10 in^3. The mass of air is 0.02 lbm at a temperature of 140°F, and the air has a volume of 1 in^3. If the pressure remains constant during the process, calculate the final air temperature and the work.

6.4. **(E)** Ammonia gas flows through a chamber at constant pressure. The specific volume of the entering ammonia is 8 ft^3/lbm and 9.5 ft^3/lbm when leaving. Assuming no kinetic or potential energy changes, determine the temperature of the ammonia leaving if the initial temperature is 200°F.

6.5. A water pump transfers an incompressible liquid without kinetic or potential energy changes. If the pressure also remains constant, what work is required to drive the pump? Assume the process to be reversible.

Section 6.2

6.6. **(M)** A rigid container is filled with a perfect gas having a gas constant of 200 J/kg · K. The volume is 1.0 m^3, the gas pressure is 1.5 bars, and the temperature is 30°C. If the gas is heated to 200°C, determine the final pressure and the work.

6.7. **(M)** For the container and contents described in problem 6.6, assuming the specific heat at constant volume is 1.0 kJ/kg · K, determine the reversible work and the irreversible work if the temperature of the gas is increased from 30°C to 200°C by a paddle stirring the gas instead of by heat transfer.

6.8. **(M)** Methyl alcohol having a density of 0.80 g/cm^3 is pumped with no kinetic or potential energy changes from a tank at 1.0 bar to a pressurized container at 2.0 bars. Determine the work required for a reversible process.

6.9. **(M)** A round rod 50 centimeters long and 5 centimeters in diameter is subjected to a load of 1200 kilonewtons in an axial direction as shown in figure 6–9. If the rod is made of aluminum having a modulus of elasticity of 1.2×10^9 kN/m^2, determine the work or energy absorbed by the rod.

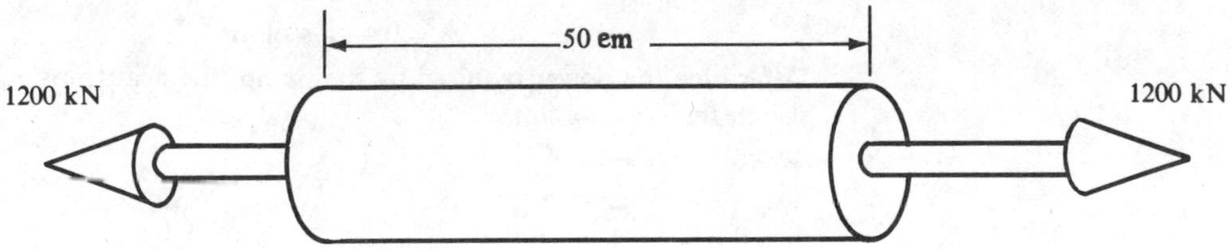

Figure 6–9

6.10. **(E)** Kerosene having a density of 51.0 lbm/ft^3 is pumped from a tank at 14.7 psia to a reservoir 30 feet above the tank where the pressure is 20 psia. If 8 lbm/s are transferred, determine the power required by the pump if we assume complete reversibility.

6.11. **(E)** A block of wood is subjected to a uniform load of 800 lbf, as shown in figure 6–10. If the modulus of elasticity of the wood is 6×10^5 lbf/in^2, determine the energy absorbed by the wood for this loading.

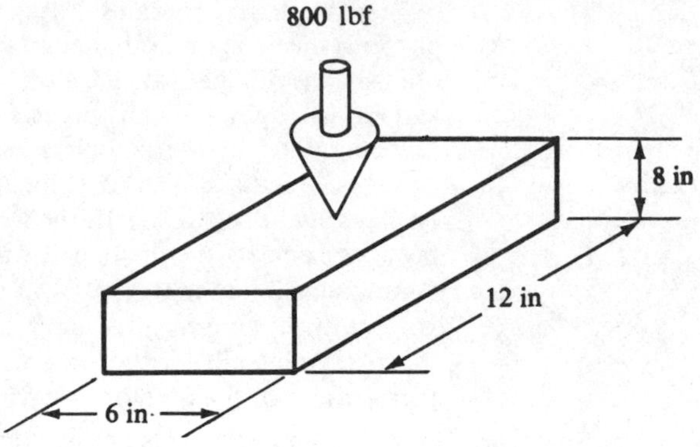

Figure 6–10

6.12. **(E)** In a piston-cylinder device, 0.2 lbm of air and a fuel burn, releasing 45 Btu of heat at constant volume. If the pressure is 80 psia and the temperature is 300°F before burning, determine the final pressure and temperature.

Section 6.3

6.13. **(M)** Air in a piston-cylinder is subjected to a reversible isothermal process where the pressure increases from 1.02 bars to 3.06 bars. Determine the work per unit mass of air for this process if the air temperature is 32°C.

6.14. **(M)** Ammonia gas is compressed in a piston-cylinder at a constant temperature of 130°C from a pressure of 1 bar (= 10^6 dyne/cm^2) to 3.5 bars. Determine the work required, the heat transferred, and the change in internal energy. Also plot the *p-V* diagram of this process.

6.15. **(M)** The following data are obtained from the pump shown in figure 6–11:

$$T_1 = T_2 = 150°\text{C}$$

$$p_1 = 1.1 \text{ bars}$$

$$p_2 = 2.2 \text{ bars}$$

$$\dot{m} = 0.4 \text{ kg/s of air}$$

Determine the power required by the pump, the heat transfer rate, and sketch the *p-v* diagram.

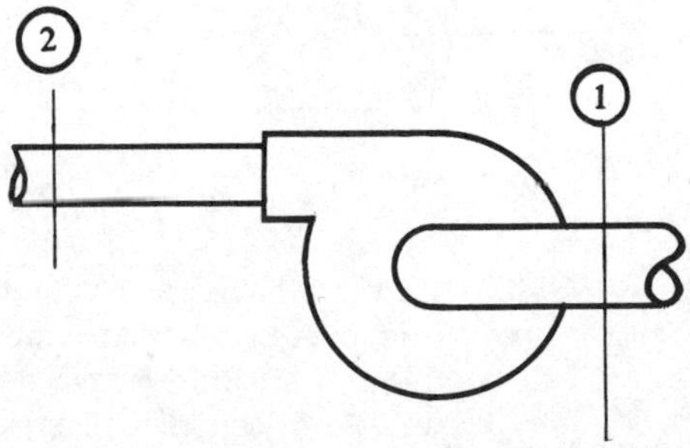

Figure 6–11

6.16. **(E)** One pound-mass of air is compressed in a piston-cylinder from 15 psia to 100 psia. If the air temperature remains constant at 85°F, determine the work required.

6.17. **(E)** Determine the heat transfer during an isothermal expansion of 2 lbm of air from 15 ft³ to 30 ft³ if the temperature is 110°F and the process is carried out in a closed container.

6.18. **(E)** Two cubic feet of air are expanded in a piston-cylinder from 600 psia and 3400°F to a final volume of 6 ft³. Assuming the process is isothermal, determine the final pressure, the work done, the heat transferred, and the internal energy change. Also sketch the p-V diagram of this process.

6.19. **(E)** A fan draws 2 lbm/s of air through the wind tunnel shown in figure 6–12. The following data are given:

$$T_1 = T_2 = 85°\text{F}$$
$$p_1 = 14.7 \text{ psia}$$
$$p_2 = 15.2 \text{ psia}$$
$$\mathcal{V}_1 = \mathcal{V}_2$$

Determine the fan motor power requirement and the heat transfer through the wind tunnel walls.

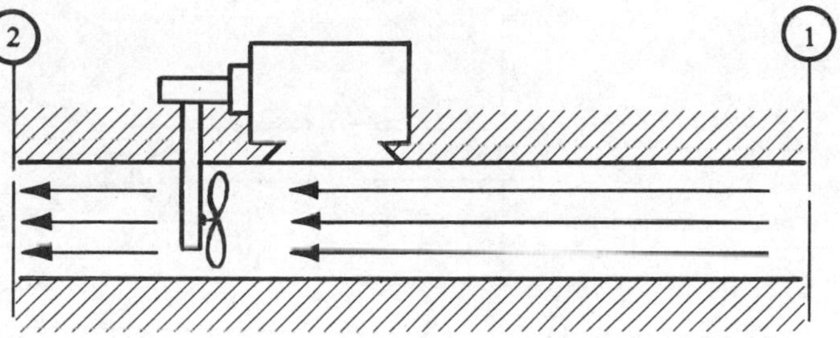

Figure 6–12

Section 6.4

6.20. **(M)** In a piston-cylinder device, 10 grams of air are compressed reversibly and adiabatically. The air is initially at 27°C and 1.1 bars. After being compressed, the air is at 450°C. Determine the final pressure, heat transferred, increase in internal energy, and work required.

6.21. **(M)** A rotary compressor handles 400 g/s of air, increasing the pressure, reversibly and adiabatically, from 1 bar to 10 bars. If the air temperature initially is 10°C and $k = 1.4$, determine the power required by the compressor. Also plot the p-v diagram and indicate the work per unit mass on this diagram.

6.22. **(M)** Two kilograms per second of helium at 300 K are expanded through an adiabatic nozzle and turbine so that the pressure drops to one-half the initial value. Assuming the kinetic energy decreases by 30 kJ/s during this expansion, determine the power produced if the process is irreversible and the following process equation holds:

$$pv^k = C$$

Also determine the change in internal energy per unit time.

6.23. **(E)** During a reversible adiabatic process of 1 lbm of an ideal gas in a piston-cylinder, the pressure decreases from 183 psig to 2 psig. The atmosphere pressure is 13.8 psia, the gas constant is 37.7 ft-lbf/lbm · °R, and the initial temperature of the gas is 863°F. Determine the initial and final gas density, the heat transferred, the work produced, and the change in internal energy. Assume the specific heat at constant pressure is 0.22 Btu/lbm · °R.

6.24. **(E)** In a gas turbine 3000 lbm/min of air are expanded in a reversible adiabatic manner. If we neglect kinetic and potential energy changes of the air flow, and if we find the entering air to have a temperature of 2100°R and a pressure of 230 psia, determine the power developed by the turbine. Assume the exhaust air pressure is 15 psia.

6.25. **(E)** One pound-mass per second of oxygen gas (O_2) is expanded through a nozzle reversibly and adiabatically. The dimensions of the nozzle are shown in figure 6–13 and the temperature and pressure of the O_2 are 200°F, and 89 psia entering the nozzle. Calculate

(a) Final temperature.
(b) Final pressure.
(c) Heat transfer rate.
(d) Final density of O_2.

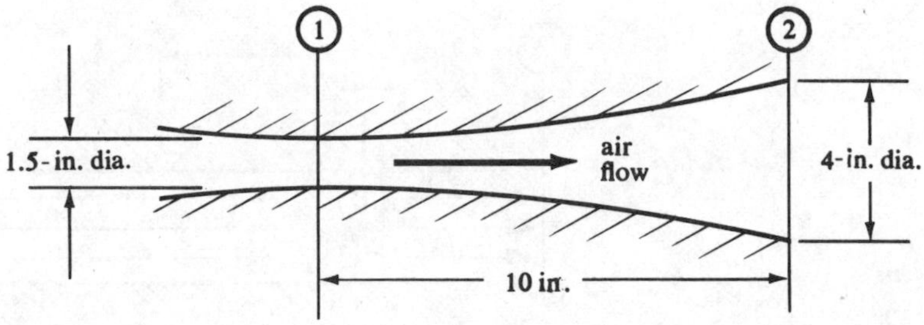

Figure 6–13

6.26. Prove the identity

$$c_v = \frac{-R}{1-k}$$

6.27. Prove the identity

$$c_p = \frac{kR}{k-1}$$

Section 6.5

6.28. **(M)** One kilogram of carbon dioxide (CO_2) is expanded reversibly from 30°C and 2 bars to 1.0 bar pressure. If the expansion is polytropic with $n = 1.27$, determine the work, heat transferred, and energy change. Assume a closed system with no kinetic or potential energy changes.

6.29. **(M)** A gas at 10 bars and 150°C expands to 2.0 bars in a closed container. If the final temperature is 50°C, determine the value for n and the heat transferred. Assume R + 0.28 kJ/kg·K and $c_v = 0.7$ kJ/kg·K.

6.30. **(M)** Seventy grams per second of air flows through a gas turbine, expanding from 15.0 bars to 1.01 bars in a reversible polytropic manner. If the inlet air temperature is 2400 K and $n = 1.5$, determine the power generated, the rate of heat transfer, and the rate of enthalpy change. Assume that kinetic and potential energy changes are negligible.

6.31. One pound-mass of air at 82°F and 14.7 psia is compressed reversibly
(E) and polytropically ($n = 1.25$) to 6.5 atmospheres in a piston-cylinder. Determine the work required.

6.32. Ammonia is compressed polytropically in a piston-cylinder from 15 psia
(E) and 3 ft^3 to 150 psia and 0.4 ft^3. Determine the value for n and the heat transferred.

6.33. A compressor handles 3500 lbm/min of air in a steady flow polytropic
(E) process ($n = 1.36$). The entering air is at 11.5 psia, 85°F, and has negligible velocity, while the compressed air is at 115 psia and has a velocity of 30 ft/s. Determine the power required of the compressor, the heat transfer rate, and the rate of enthalpy change of the air.

6.34. Three pounds-mass per second of air are reversibly and polytropically
(E) ($n = 1.48$) expanded in a gas turbine from 2100°F to 900°F. If the exhaust pressure is 15 psia, determine the work produced per pound-mass of air, the power generated, and the rate of heat transfer.

***6.35.** Derive an expression for Wk_{cs} in terms of n, pressure, and volume for a reversible polytropic process.

***6.36.** Derive equation (6–34) for a reversible adiabatic process of an open system, neglecting kinetic and potential energy changes.

***6.37.** Derive an expression for Wk_{os} (in terms of n, pressure, and volume for a reversible polytropic process) if kinetic and potential energy changes are neglected.

7

Entropy, Heat, and Temperature

One of the most important properties, *entropy*, is presented in this chapter. It is introduced as the property that relates temperature and heat — not like the caloric equations of state, which related temperature and energy — but like volume, which related pressure and work. Analogies are presented to give the student a broad concept for this very important property. Emphasis is placed, however, on presenting tools for evaluating entropy and its changes during processes.

The *third law of thermodynamics* is shown to be a limiting condition for the value of entropy, a law which probably has more qualitative value than quantitative.

7.1 Entropy

In chapter 3 we arrived at equations for calculating heat transfer rates. These equations were concerned primarily with conditions at the boundary of a system, that is, the system's outer surface temperature and the heat transfer at that boundary. In this section let us see if a work-heat analogy is profitable. For the reversible work of a closed system we have written

$$Wk_{cs} = \int_{V_1}^{V_2} p\,dV$$

which looks like an intensive, surface property (pressure) mechanically motivating a change in an extensive, bulk property (volume). In a like manner, temperature induces heat transfer and changes some extensive system property, i.e.

$$\text{Reversible } Q = Q_{\text{rev}} = \int_{S_1}^{S_2} T\,dS \qquad \textbf{(7–1)}$$

where we quite arbitrarily call entropy S. It is defined from a rearrangement of the equation,

$$dQ_{\text{rev}} = T\,dS$$

and then

$$dS = \frac{dQ_{\text{rev}}}{T}$$

or

$$\Delta S = \int_{T_1}^{T_2} \frac{dQ_{\text{rev}}}{T} \qquad \textbf{(7–2)}$$

Entropy is an extensive property, dependent on the mass of the system, and is described by the units of energy per unit absolute temperature. It can, however, be converted into an intensive property—specific entropy—by the operation

$$\frac{S}{m} = s = \text{specific entropy} \qquad \textbf{(7–3)}$$

where s has units of energy per unit absolute temperature per unit mass. Notice from equation (7–2) that as heat is transferred into the system, entropy will increase and as heat is transferred out of the system, it will decrease.

Entropy is a measure of unavailable energy in a system; the greater the entropy of a system, the less available is that system's energy for doing work or transferring heat. In mechanical systems we have seen that as a system increases in volume, it can perform work, but thereby also has a reduced capability to perform other work. It seeks an equilibrium in pressure with its surroundings and when it expands in volume so that the pressures are equal, the system cannot perform additional work. Thus, a system at high pressure (with respect to its surroundings) has a lower entropy than does the same system upon equilibrating its pressure with the surroundings. Also, if we have a system at a high temperature, it will tend to reach a temperature equal to the surroundings. Here the volume may not change, but entropy will The high temperature system can perform work or heat processes and has a low entropy; but after reducing its temperature, it does not have the same capability — it has greater entropy.

In more general terms, entropy is a measure of disorder or randomness. Suppose we have ten red marbles and ten green marbles in separate boxes.

All the red are in one box and all the green in another—we thus have an ordered system of two boxes. We place the boxes on a high shelf so that we can withdraw the marbles, but cannot look in the boxes. Thus, if we wish to get a red marble, we reach up and remove a marble from one of the boxes. If it is green, then we are assured of a red marble by selecting a second from the other box. Here we have a system that is ordered and has a low entropy. But we now mix the red and green marbles in one box and return the box to the shelf. The system is well mixed, disordered, and has a high entropy. If we want to get a red marble, we are never assured that we will pick one. Perhaps, if we are lucky on the first try, we will get a red marble, but we could also get ten green ones before we would get one red one. In a less abstract manner, suppose we have a system composed of pure hydrogen gas (H_2) and pure oxygen gas (O_2), each in a separate container. The system has a low entropy, it is well ordered or well arranged, and can easily be mixed to produce water. Also, we could get work or electrical energy from it if we wished. However, if we mix the hydrogen and oxygen and produce water (H_2O), we immediately have a higher entropy. The system is less ordered or arranged (we no longer know exactly where the oxygen atoms are) and it is only water. We cannot get work or electrical energy from it anymore. We can thus say that a low entropy implies a broader capability for energy systems, and if we perform reversible processes only, the world will retain this low entropy. The world, as we are reminded by environmentalists, is increasing its entropy, and the choices of power supplies are indeed not as many today as some persons would have us believe.

The property entropy gives us the tools (theoretically, at least) to calculate heat transfer by measuring only the system's properties. Previously, in the heat transfer relationships of conduction, convection, and radiation, we needed to know the properties of the system *and* the surroundings in order to calculate heat transfer. By plotting entropy versus temperature on a diagram, called a *T-S diagram,* we can then represent heat as the area under a curve, as shown in figure 7–1, which is reminiscent of work in an area in a *p-V* diagram. Of course, the heat determined is the *reversible* heat; but under any circumstance, the entropy change will *be at least as great as that determined from* equation (7–2). We can have an irreversible entropy change, ΔS_{irr}, and for any process

$$\Delta S_{actual} = \Delta S_{irr} + \int \frac{dQ_{rev}}{T} \qquad \textbf{(7–4)}$$

For a reversible process, ΔS_{irr} is zero and the irreversible entropy change for a given process can only, at the present state of science, be determined empirically; that is, measure it or find the entropy values from tables. The appendix tables contain listings of specific entropy, values which, in general, were calculated using statistical thermodynamic concepts and computer techniques which we will not discuss.

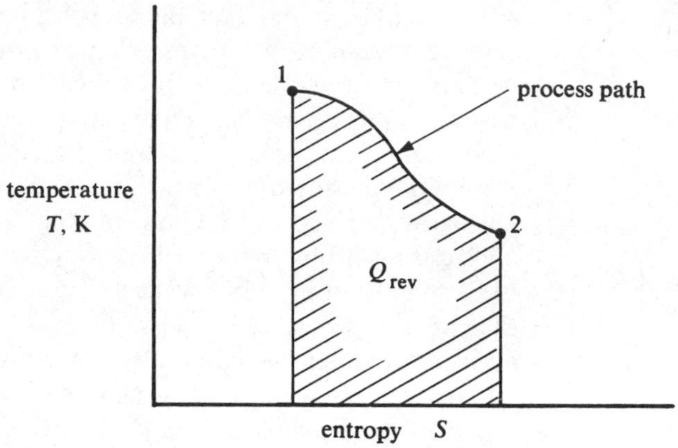

Figure 7–1 Characteristics of a process on a *T-S* plane.

In the following sections, we will introduce ways of calculating entropy changes, with particular emphasis on perfect gases.

7.2 Isothermal Entropy Change

For those reversible processes where temperature is fixed (isothermal), we can readily calculate changes in entropy by

$$\Delta S = \int \frac{dQ_{rev}}{T} = \frac{1}{T}\int dQ_{rev} = \frac{Q_{rev}}{T} \tag{7–5}$$

The heat transfer is then

$$Q_{rev} = T\Delta S \tag{7–6}$$

as long as we treat reversible cases. While most processes are not isothermal, equations (7–5) and (7–6) can be used in some of these other processes. If the heat transfer is carried on with constant temperature surroundings, for instance, the properties T and S of the surroundings will suffice to calculate Q_{rev}, even if the system is exhibiting a temperature change. We see this in the following problem.

Example 7.1

In figure 7–2 a system has a loss of 300 kilojoules of energy in the form of reversible heat transfer to the surroundings. If the temperature of the surroundings is 5°C, determine the entropy change of the system.

Solution

The entropy change of the surroundings can easily be determined. It is $\Delta S_{surr} = Q/T$ and T is 5°C + 273 K or 278 K. The entropy change is, therefore

$$\Delta S_{surr} = 300 \text{ kJ}/278 \text{ K} = 1.08 \text{ kJ/K}$$

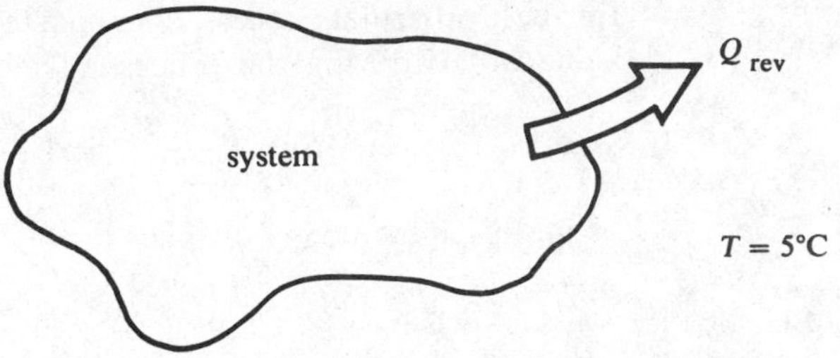

Figure 7–2 Example 7.1.

For a reversible process it can be shown that the sum of all entropy changes resulting from the process is zero. That is,

$$\Sigma\Delta S = 0 = \Delta S_{surr} + \Delta S_{sys} \tag{7–7}$$

and then

$$\Delta S_{sys} = -\Delta S_{surr} = -1.08 \text{ kJ/K} \qquad \textit{Answer}$$

The entropy has, therefore, decreased in the system and increased in the surroundings. Equation (7–7) was utilized to arrive at this result, but notice that this result is true for reversible processes only. For irreversible processes, the entropy must increase or remain constant, i.e.

$$\Sigma\Delta S \geq 0 \tag{7–8}$$

or

$$\Delta S_{surr} + \Delta S_{sys} \geq 0 \tag{7–9}$$

and

$$\Delta S_{sys} \geq -\Delta S_{surr}$$

Then for an irreversible process for example 7.1, equation (7–9) demands that

$$\Delta S_{sys} \geq -1.08 \text{ kJ/K}$$

We have previously inferred that the relation $\int dQ/T$ is equal to or less than the actual ΔS. We must, therefore, never assume that entropy is conserved; it is conserved for reversible processes only. It is continually being increased, and rather effectively, by our technological mills in irreversible processes.

For a perfect gas, we can obtain another relation for calculating entropy change during an isothermal process for a closed system. We first note that, since $\Delta U = 0$

$$Q = Wk_{cs} = mC \ln\left(\frac{V_2}{V_1}\right)$$

for the isothermal process. The constant C is equal to p_2V_2 or any other value of pV during the process. Then from equation (7–5), we obtain

$$\Delta S = \left(\frac{mp_2V_2}{T_2}\right)\ln\left(\frac{V_2}{V_1}\right) = mR\ln\left(\frac{V_2}{V_1}\right) \qquad \textbf{(7–10)}$$

or for specific entropy changes

$$\Delta s = \Delta\frac{S}{m} = R\ln\left(\frac{V_2}{V_1}\right) \qquad \textbf{(7–11)}$$

Example 7.2

During a reversible isothermal compression of 0.5 lbm of air, the pressure ratio is 10. Determine the entropy changes and the specific entropy changes for this process.

Solution

Since the pressure ratio is 10, we know that

$$\frac{p_2}{p_1} = 10 = \frac{V_1}{V_2}$$

Then the entropy change is calculated from equation (7–10)

$$\Delta S = (0.5 \text{ lbm})(53.3 \text{ ft-lbf/lbm} \cdot {}^\circ\text{R})\ln\left(\frac{1}{10}\right)$$

$$= 61.36 \text{ ft-lbf/}^\circ\text{R} = 0.0787 \text{ Btu/}^\circ\text{R}$$ *Answer*

The specific entropy change is obtained from equation (7–11)

$$\Delta s = \frac{\Delta S}{m}$$

$$= \frac{(0.0787 \text{ Btu/}^\circ\text{R})}{(0.5 \text{ lbm})}$$

$$= 0.1577 \text{ Btu/lbm}^\circ\text{R}$$ *Answer*

7.3 Adiabatic Entropy Change

For the adiabatic process, no heat is transferred and consequently there is no change in entropy due to reversible heating. The only entropy change that can occur must be due to irreversible internal or external effects. Then

$$\Delta S = \Delta S_{\text{irrev}} \qquad \textbf{(7–12)}$$

and entropy changes can only be determined by knowing the initial state and final state. From this information, the entropy values can then be extracted from thermodynamic tables. For reversible adiabatic processes, of course, equation (7–12) reduces to

$$\Delta S = 0 \qquad \textbf{(7–13)}$$

and since this implies a constant value for entropy, the reversible adiabatic process is frequently called the *isentropic process.* The reversible adiabatic process, however, is a more restricted process than the isentropic one. We can allow an irreversible process to progress, yet still retain a constant entropy by proper heat transfers; consequently we will generally use the term *reversible adiabatic* instead of *isentropic.*

Example 7.3

Steam at 20.0 bars and 560°C is expanded reversible adiabatically through a steam turbine until the pressure is 2.0 bars. Determine the final steam temperature.

Solution

The entropy change for this process is zero, so

$$s_2 = s_1$$

and from table B.2, we find that steam at 20.0 bars and 560°C has a specific entropy s of 7.596 kJ/kg · K. Therefore, the final specific entropy has this same value and at 2.0 bars the temperature must be between 160°C and 240°C as indicated in table B.2. We will interpolate to determine the precise value:

$$s = 7.324 \text{ kJ/kg} \cdot \text{K at } 160°\text{C}$$
$$= 7.663 \text{ kJ/kg} \cdot \text{K at } 240°\text{C}$$

and then,

$$S = 0.339 \text{ kJ/kg} \cdot \text{K with } t = 80°\text{C}$$

Since $s_2 = 7.596$ kJ/kg · K, we write

$$\frac{7.596 \text{ kJ/kg} \cdot \text{K} - 7.324 \text{ kJ/kg} \cdot \text{K}}{0.339 \text{ kJ/kg} \cdot \text{K}} = \frac{t_2 - 160°\text{C}}{80°\text{C}}$$

$$\frac{0.272}{0.339} \times 80°\text{C} + 160°\text{C} = t_2$$

yielding the result

$$t_2 = 224°\text{C}$$ *Answer*

7.4 Polytropic Entropy Change

With the introduction of the entropy function, the first law can now be written in the following manner:

$$dU = TdS - pdV \tag{7–14}$$

or,

$$du = Tds - pdv \tag{7–15}$$

In addition, if we replace pdv by the sum $dpv - vdp$ we have, for equation (7–15) that

$$du = Tds - d(pv) + vdp$$

or,

$$du + d(pv) = Tds + vdp$$

But, $du + d(pv)$ equals dh, so

$$dh = Tds + vdp \tag{7–16}$$

What we are seeking is a general relation for the property entropy involving the more common properties of pressure, volume, and temperature. From equation (7–15) we find

$$ds = \frac{du}{T} + \frac{pdv}{T}$$

or, integrating both sides

$$\Delta s = \int \frac{du}{T} + \int \frac{pdv}{T} \tag{7–17}$$

From equation (7–16) we obtain, similarly, that

$$\Delta s = \int \frac{dh}{T} - \int \frac{vdp}{T} \tag{7–18}$$

These last two equations are general; they can be used for open or closed systems and for any material. If we assume a perfect gas with constant specific heats, then equation (7–17) can be reduced to

$$\Delta s = c_v \int_{T_1}^{T_2} \frac{dT}{T} + R \int_{v_1}^{v_2} \frac{dv}{v} \tag{7–19}$$

or

$$\Delta s = c_v \ln\left(\frac{T_2}{T_1}\right) + R \ln\left(\frac{v_2}{v_1}\right)$$

and equation (7–18) can become

$$\Delta s = c_p \ln\left(\frac{T_2}{T_1}\right) - R \ln\left(\frac{p_2}{p_1}\right) \tag{7–20}$$

These two equations then are correct for determining the change in entropy of any polytropic process, reversible or irreversible, involving a perfect gas.

The isobaric (constant pressure) process implies $p_2 = p_1$ and then in equation (7–20)

$$R \ln\left(\frac{p_2}{p_1}\right) = R \ln 1 = 0$$

so that

$$\Delta s = c_p \ln\left(\frac{T_2}{T_1}\right) \tag{7–21}$$

For the isometric process, v_2 is equal to v_1, reducing equation (7–19) to

$$\Delta s = c_v \ln\left(\frac{T_2}{T_1}\right) \tag{7–22}$$

A process involving constant temperature perfect gases implies that $\ln T_2/T_1$ is zero so that

$$\Delta s = R \ln\left(\frac{v_2}{v_1}\right) \tag{7–23}$$

from equation (7–19) or

$$\Delta s = -R \ln\left(\frac{p_2}{p_1}\right) = R \ln\left(\frac{p_1}{p_2}\right) \tag{7–24}$$

from equation (7–20). For the isothermal process we found $p_1/p_2 = v_2/v_1$ so equations (7–23) and (7–24) are identical. We could also use the equation $\Delta S = Q/T$ as we did in section 7.2. If we note that $v_2 = V_2/m$ and $v_1 = V_1/m$ so that

$$\frac{v_2}{v_1} = \frac{mV_2}{mV_1} = \frac{V_2}{V_1}$$

then equation (7–11) is identical to equation (7–23), as it should be.

Example 7.4 At constant pressure 0.8 lbm of carbon dioxide is expanded from 10 ft^3 to 30 ft^3. Determine the entropy change and the specific entropy change.

Solution The process is an isobaric one and the entropy can therefore be determined from equation (7–23) in the form

$$\Delta S = m\Delta s = mR \ln \frac{v_2}{v_1} \tag{7–25}$$

The specific volumes are

$$v_2 = \frac{V_2}{m} = \frac{30 \text{ ft}^3}{0.8 \text{ lbm}}$$

$$v_1 = \frac{V_1}{m} = \frac{10 \text{ ft}^3}{0.8 \text{ lbm}}$$

and the gas constant is 35.12 ft-lb/lbm · °R from table B.4. Then

$$\Delta S = (0.8 \text{ lbm})(35.12 \text{ ft-lb/lbm} \cdot \text{°R})\ln\left(\frac{30 \text{ ft}^3/0.8 \text{ lbm}}{10 \text{ ft}^3/0.8 \text{ lbm}}\right)$$

$$= 28.096 \text{ ft-lbf/°R} \ln 3 = 30.87 \text{ ft-lbf/°R}$$

$$= +\,0.0397 \text{ Btu/°R}$$ *Answer*

Also,

$$\Delta s = \frac{\Delta S}{m} = \frac{0.0397}{0.8} = 0.0496 \text{ Btu/lbm} \cdot {}^\circ\text{R} \qquad \textit{Answer}$$

Example 7.5

Air expands polytropically through a nozzle such that the exponent n is 1.45. The exhaust pressure of the air is 1.02 bars and the temperature is 90°C. If the inlet pressure is 4.0 bars gage, determine the change in specific entropy of the air as it passes through the nozzle.

Solution

For any polytropic process we can use equation (7–20)

$$\Delta s = c_p \ln \frac{T_2}{T_1} - R \ln \frac{p_2}{p_1}$$

and from table B.4 find that

$$c_p = 1.007 \text{ kJ/kg} \cdot \text{K}$$

The gas constant R is found to be

$$R = 287 \text{ J/kg} \cdot \text{K} = 0.287 \text{ kJ/kg} \cdot \text{K}$$

Also, the pressure ratio p_2/p_1 is 1.02 bars/(4.0 + 1.01 bars), assuming an atmospheric pressure of 1.01 bars. Then

$$\frac{T_2}{T_1} = \left(\frac{p_2}{p_1}\right)^{(n-1)/n}$$

from equation (6–47). This gives us

$$\frac{T_2}{T_1} = \left(\frac{1.02}{5.01}\right)^{0.45/1.45} = 0.610$$

Consequently, the specific entropy change can now be determined from equation (7–20)

$$\begin{aligned}\Delta s &= (1.007 \text{ kJ/kg} \cdot \text{K})\ln(0.610) - (0.287 \text{ kJ/kg} \cdot \text{K})\ln(0.204) \\ &= -0.498 \text{ kJ/kg} \cdot \text{K} + 0.456 \text{ kJ/kg} \cdot \text{K} \\ &= -0.042 \text{ kJ/kg} \cdot \text{K} \qquad \textit{Answer}\end{aligned}$$

7.5 The Third Law of Thermodynamics

In calculating entropy changes due to processes, temperature directly affects the entropy function. A decrease in temperature will induce a decrease in entropy and an increase in temperature will induce increased entropy. We might ask, how low can entropy be? The answer is given by the third law of thermodynamics.

Third Law of Thermodynamics: *Entropy tends to a minimum constant value as temperature tends to absolute zero. For a pure element this minimum value is zero, but for all other substances it is not less than zero, but possibly more.*

This third statement or law is a result of experimentation in the temperature regime near absolute zero and has not been violated—therefore it is considered a "law." From a practical viewpoint, it tells us that it is impossible to reach an absolute zero temperature by other than a reversible process since near the zero point (as illustrated in figure 7–3), the change in entropy is zero and the only irreversible means of lowering entropy further is to have a surrounding which is cooler yet than absolute zero. This is impossible, so the final approach to absolute zero for cooling any material must be reversible and adiabatic (isentropic). Another manner of observing this is to take the definition, equation (7–2), and set ΔS equal to zero.

$$\Delta s = \int \frac{dQ}{T} = 0$$

In figure 7–3, for substance A near absolute zero, point (1), the entropy s_1 is equal to the entropy when the temperature will be zero, s_0. Then

$$\Delta s = s_0 - s_1 = 0$$

and for equation (7–2) to be satisfied dQ must be zero (adiabatic), else the term dQ/T will approach infinity, which is impossible. The process from 1 to 0 must therefore be adiabatic and reversible.

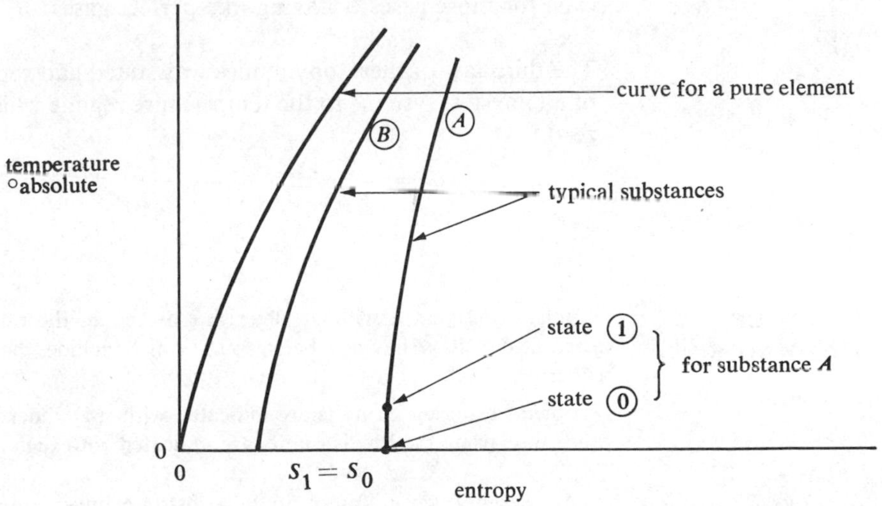

Figure 7–3 Behavior of entropy near absolute zero of temperature.

The point in this discussion is that the state of absolute zero temperature is a highly desirable one for making energy available, but it is impossible to achieve. Materials close to zero temperature are, however, attractive as sources of heat or work.

7.6 Summary

The property entropy, S, has been introduced, along with equations for calculating changes in it. These equations are listed in tables 6–2 and 6–3 for ready references. In these tables are listed specific entropy changes s but the relation

$$\Delta S = m\Delta s$$

can readily allow conversion from specific to total entropy if desired.

Entropy is defined mathematically in terms of equation (7–2)

$$dQ_{rev} = TdS$$

as a tool for evaluating heat transfer. Entropy is a measure of the order of a system or the unavailability of thermal energy.

If we wish to know the entropy change for a given process, we may calculate such change from various equations developed in this chapter (depending on what type of process is occurring), or we may determine the change from thermodynamic tables listing entropy values. Care must be exercised, however, when using equations to calculate entropy changes. For instance, unless a perfect gas (or a substance reasonably like a perfect gas) represents the system, we should not use those equations developed from the perfect gas assumption. Generally, greater accuracy can be achieved by using entropy values from thermodynamic tables, even for those gases behaving like perfect gases.

The third law of thermodynamics was stated and represents a restriction of all physical systems to the temperature regime which excludes absolute zero.

Practice Problems

Problems designated with an asterisk * preceding the number should only be attempted by those having a background which includes the knowledge of integral calculus.

Problems using the SI units are indicated with (M) under the problem number, and those using the English units are indicated with (E).

Sections 7.1 and 7.2

7.1. **(M)** Seventy-two grams of air in a piston-cylinder are heated by 135 joules while going through a reversible isothermal process at 70°C. Determine the entropy change and the specific entropy change.

7.2. Steam is expanded from 20.0 bars to 5.0 bars at 400°C. Determine the
(M) specific entropy change.

7.3. The entropy function of a certain gas is plotted in figure 7–4. Determine
(M) the energy required in the form of heat transfer to heat the material to 300°K from absolute zero temperature.

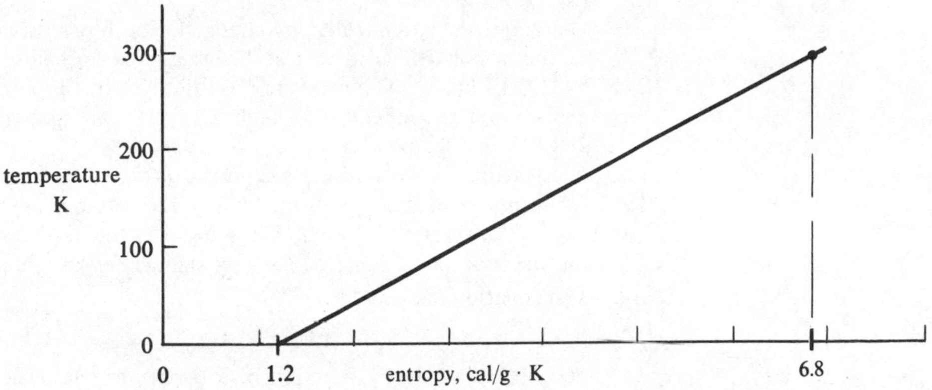

Figure 7–4

7.4. Helium gas is increased in pressure from 15 psia to 35 psia. Determine
(E) the specific entropy changes of the helium if the compression is done reversibly under constant temperature conditions.

7.5. During a reversible, isothermal process, 700 Btu of heat transfer are
(E) directed into a system at 80°F. Determine the entropy change.

7.6. On a summer day when the temperature is 80°F, a trash fire emits
(E) 30,000 Btu to the atmosphere. By how much has the entropy increased?

Section 7.3

7.7. Superheated steam is reversibly and adiabatically expanded from 30.0
(M) bars and 480°C to 5.0 bars. Determine the final temperature.

7.8. During an adiabatic process argon gas exhibits an increase of specific
(M) entropy of 0.12 kJ/kg · K. What is the irreversible entropy increase of 20 grams of argon?

7.9. Determine the final temperature of 70 lbm of steam at 75 psia and
(E) 1200°F contained in an insulated chest. The chest expands isentropically until the steam is at 10 psia.

Section 7.4

7.10. During an isometric process of 0.1 kilogram of nitric oxide (NO), the
(M) specific entropy decreases by 0.30 kJ/kg · K. If the NO was initially at −10°C, determine the final temperature.

7.11. In a rigid stainless steel tank, 540 grams of propane (C_3H_3) are heated
(M) from −20°C to 5°C. Determine the entropy change in kJ/K.

7.12. During a constant pressure process, 1.5 kilograms of sulfur dioxide (SO_2)
(M) are heated to 70°C from an initial temperature of 10°C. Determine the total and specific entropy changes.

7.13. Mercury vapor is heated from a saturated liquid to a saturated vapor.
(M) Determine the specific entropy change if the pressure is

(a) 0.06 bar.
(b) 1.4 bars.
(c) 10.0 bars.

7.14. **(M)** A drop in entropy of 1.0 J/g·K is calculated for propylene gas as it is cooled to 20°C in an isobaric process. Determine the initial temperature.

7.15. **(M)** Refrigerant R-22 is heated from saturated vapor at −20°C to superheated vapor at

(a) −5°C at 3.93 bars.

(b) 10°C at 2.18 bars.

Determine the specific entropy change for both processes.

7.16. **(M)** During a polytropic process involving air, the specific entropy decreases by 0.03 kJ/kg · K. The pressure ratio (final to initial pressure) is 12.1 to 1 and the initial temperature was 80°C. Obtain the polytropic exponent n and the final temperature.

7.17. **(M)** A polytropic process with an exponent n of 1.43 is found to describe an expansion through a gas turbine where the entrance pressure is 12.0 bars and the exhaust is 1.2 bars. Assuming that the gas flowing through the turbine is air, determine the entropy change per kilogram.

*** 7.18.** **(M)** The equation

$$S = (15.0\ T^3 + 325)\ \text{kJ/K}$$

describes the entropy function of a particular gas. Determine the heat transfer of a reversible process involving this gas if the temperature decreases from 425 K to 25°C.

7.19. **(E)** A perfect gas contained in a rigid container in 90°F surroundings exhibits a pressure drop from 28 psia to 20 psia. Assuming the gas constant is 63 ft-lbf/lbm · °R and the specific heat, c_p, is 0.30 Btu/lbm · °R, calculate the temperature change and the specific entropy change for the gas.

7.20. **(E)** Acetylene gas is contained in a rigid steel tank while its pressure increases from 10 psig to 15 psig. Determine the change in specific entropy of the gas during this process.

7.21. **(E)** Determine the specific entropy change of superheated steam during an isometric process where the initial state is 1500°F and 150 psia and the final is 700°F.

7.22. **(E)** In a flexible container, 1.8 lbm of air are expanded at constant pressure from an initial volume of 20 ft³ to 40 ft³ finally. If the initial temperature is 195°F, determine the final pressure and the change in total and specific entropy.

7.23. **(E)** Determine the specific entropy change for refrigerant R-22 cooled from 50°F to 20°F at

(a) 5 psia.

(b) 30 psia.

7.24. **(E)** Steam is heated from saturated liquid at 15 psia to superheated steam at 200 psia and 600°F. Determine the specific entropy change.

7.25. **(E)** Steam is expanded from 300 psia and 800°F to a saturated vapor in a reversible adiabatic manner. Determine the final pressure of the steam.

7.26. **(E)** A gas having the same thermodynamic properties as air is contained in the combustion chamber of an internal combustion engine. The gas is at 200°F and 200 psia from which it expands polytropically to 15 psia. If the polytropic exponent is 1.51, determine the specific entropy change of the gas during this process.

***7.27.** **(E)** A perfect gas is found to have an entropy function given by the equation

$$S = (0.006T^2 + 0.031T)\ \text{Btu/°R}$$

Determine the heat transferred during a reversible process of this gas when the temperature increases from 100°F to 300°F.

8

The Heat Engine and the Second Law of Thermodynamics

In this chapter, cyclic devices which have heat exchanges with the surroundings, called *heat engines*, are described. The *Carnot engine* is cited and discussed as an example of a heat engine.

The general definition of *thermodynamic efficiency* is given, and a development is made of the *Clausius inequality* which provides a criterion for a comparison of actual to ideal reversible power cycles. The *second law of thermodynamics* is stated as a physical limitation on real and ideal cyclic devices. *Cyclic heat* and *work* are then discussed, leading into the differentiation of *closed* and *open cycles* — synonymous with the closed and open systems.

Two rather extended example problems are presented at the close of this chapter; one treats the *Carnot cycle* which produces power, and the other treats a *reversed Carnot cycle* which is the basis for a type of refrigerator or heat pump.

8.1 Cyclic Devices – System Diagrams

A thermodynamic system which progresses through a set of processes and periodically returns to its initial or beginning state, we have defined as a *cyclic device*. One complete set of processes which allows the first return of a system to its beginning state we call a *cycle*. There are many examples of cycling devices, some of which are as follows: electrical power plants which use coal or nuclear fuel; electric motors; internal combustion engines using gasoline; jet and steam engines; refrigerators; and air conditioners. Items which are *not* cyclic devices include the solid or liquid rocket engines; electrical batteries; fuel cells; and most biological systems, for example, trees.

A particular type of cyclic device which transfers heat to its surroundings we call a *heat engine*. A heat engine and its important thermodynamic characteristics are visualized concisely in a *system diagram* in figure 8–1, where two types of cyclic devices are presented, the open system and the closed system. While these diagrams contribute no information regarding the details of how the heat engine mechanically functions, they do provide a complete and clear picture of the thermodynamic operation of the devices. Thus, a system diagram can be a worthwhile exercise in the analysis or synthesis of a cycling heat engine. Keep in mind that the term *heat engine* does not necessarily require mechanical components such as gears, shafts, or piston-cylinders, but it is any device which involves heat and work.

Notice in figure 8–1 that heat is transferred from some high temperature source (which may physically be within the system boundary or may be more than one source) to the heat engine. The heat engine then converts the transferred energy to mechanical work and dumps the excess into a

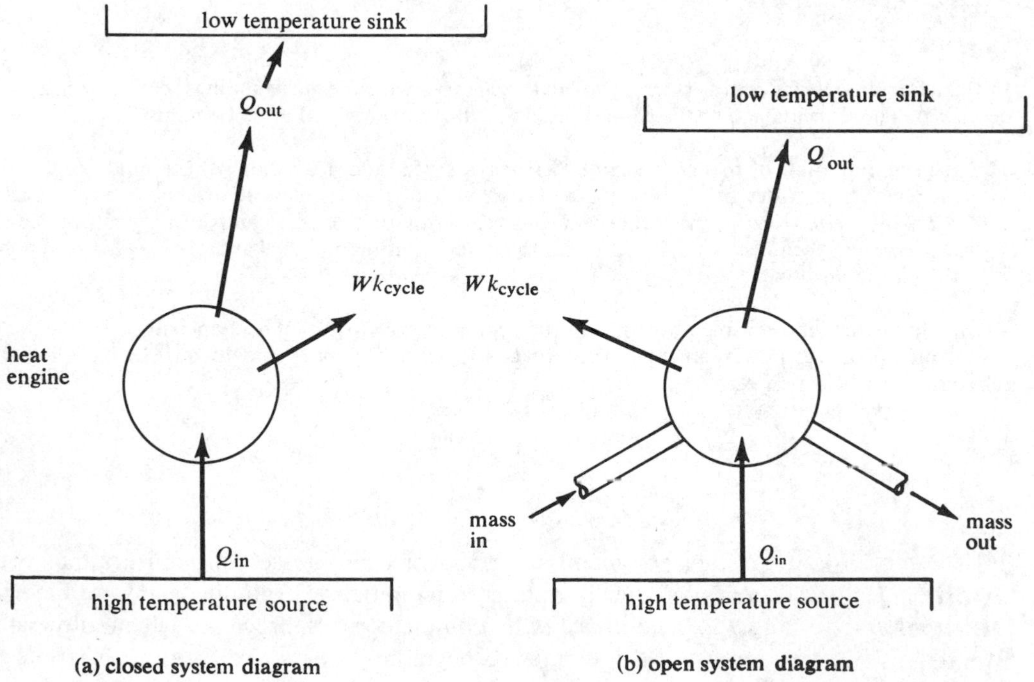

Figure 8–1 The system diagram.

low temperature sink. This is the general operation of all heat engines, and where these phenomena occur in a particular device can be found from an initial study of a system diagram. Note, also, that in the cycling device no energy is accumulating or being drained from the engine itself — all energy is taken from some foreign source and dumped in another place.

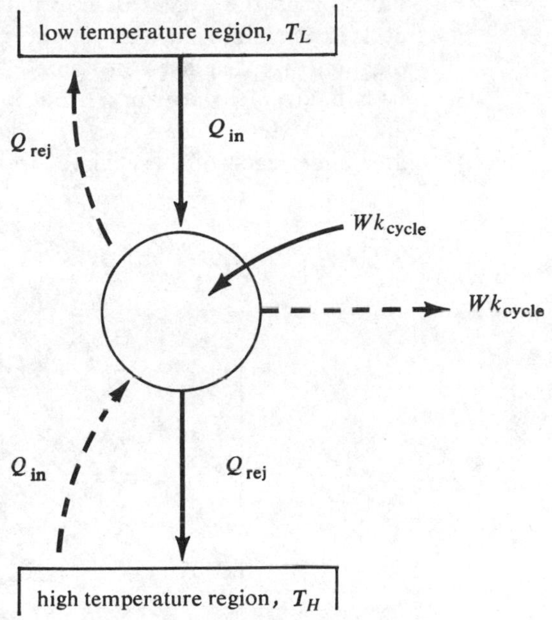

Figure 8–2 Reversible heat engine.

If we consider the heat engine to be reversible, then we can visualize the energy transfers of heat and work to take either direction. In figure 8–2 is shown such a reversible heat engine where the discontinuous arrows represent the set of transfers which would occur reversed from the set represented by solid arrows. It can be seen that the arrangement of energy transfers diagrammed with the solid arrows is just a device which will provide refrigerating capability; that is, it pumps energy from a lower temperature into a higher temperature region by expending work. This type of heat engine is generally referred to as a *heat pump*.

8.2 The Carnot Engine

In 1824, Sadi Carnot published a treatise on thermodynamics in which he invented a cycle composed of four special processes. The heat engine which would run on this cycle (but which no one has yet built) has since been called the *Carnot engine* and the cycle, the *Carnot cycle*. The four processes, in order, which constitute this cycle are as follows:

(1–2) Reversible isothermal compression at temperature T_L.
(2–3) Reversible adiabatic compression from the low temperature T_L to a higher temperature T_H.
(3–4) Reversible isothermal expansion at temperature T_H.
(4–1) Reversible adiabatic expansion from temperature T_H to T_L.

Notice that the above descriptions of the processes indicate that the cycle is reversible since all the individual processes are. The *p-V* and *T-s* diagrams are displayed in figures 8–3 and 8–4 respectively, and we see from figure 8–3 that work is put into the engine during processes (1–2) and (2–3) while processes (3–4) and (4–1) produce work. If we add up the area under each of these curves we have

$$\int_{V_1}^{V_2} pdV + \int_{V_2}^{V_3} pdV + \int_{V_3}^{V_4} pdV + \int_{V_4}^{V_1} pdV$$

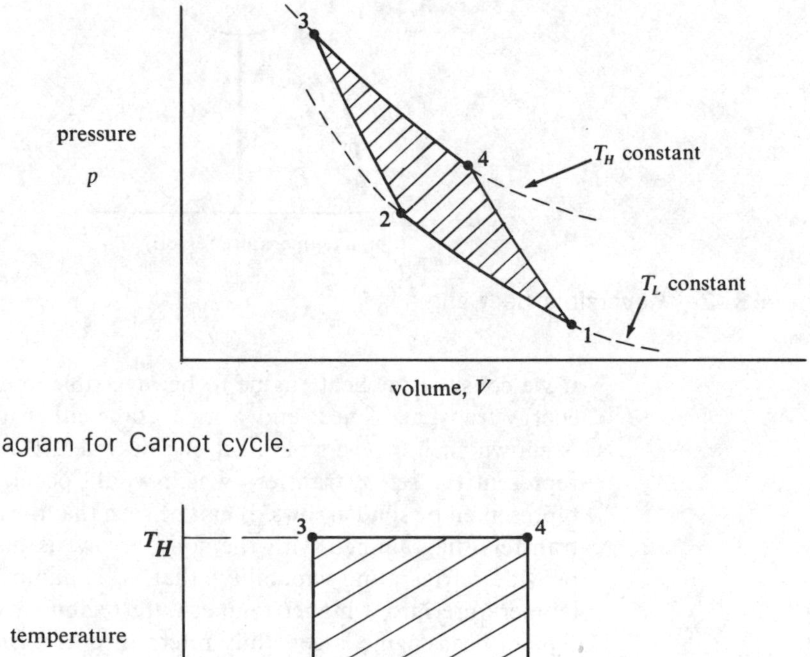

Figure 8–3 *p-V* diagram for Carnot cycle.

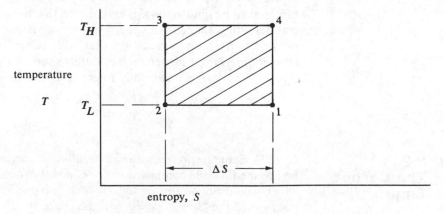

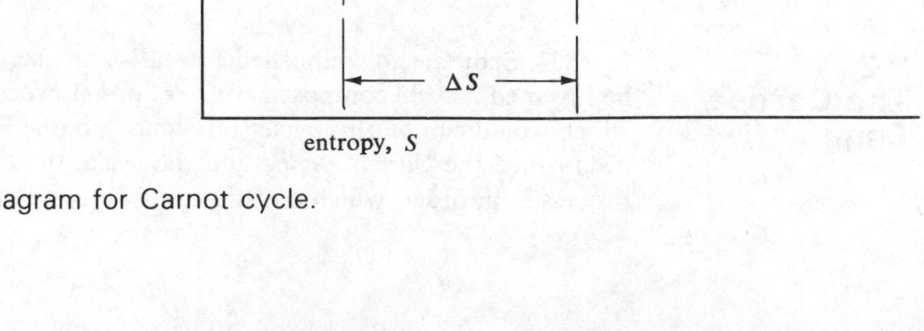

Figure 8–4 *T-s* diagram for Carnot cycle.

and this sum we denote by

$$\oint p\,dV = \oint dWk$$

We will use the sign $\oint$ to represent a cyclic sum or sum of the integrals during one complete cycle. We do not have the details here to arithmetically add up the area under curves, but the shaded area in figure 8–3 should be recognized as just this sum we have discussed, i.e.

$$\oint dWk = \text{net work of a Carnot cycle} = Wk_{\text{rev cycle}}$$

In a similar manner, in figure 8–4 the shaded area enclosed by the curves, a simple rectangle, represents the sum of the four $\int Tds$ terms which together represent the net heat transferred into the engine. Then

$$\oint Tds = \oint dQ_{\text{rev}} = \text{net heat of a cycle} = Q_{\text{rev cycle}}$$

Only two of the four processes contribute to the sum of the heat transfer: processes (1–2) and (3–4). Process (1–2) is a heat rejection one, so

$$Q_{\text{rej}} = T_{\text{L}} \Delta s_{12}$$

and process (3–4) is a heat addition one, so

$$Q_{\text{add}} = T_{\text{H}} \Delta s_{34}.$$

As we shall see in the following example problems, Δs_{12} is a negative quantity since the entropy is reduced in the process whereas Δs_{34} is positive. The magnitudes of these two quantities are equal, however, so that we may write, referring to figure 8–4

$$|\Delta s_{12}| = |\Delta s_{34}| = \Delta s$$

Then

$$\oint dQ_{\text{rev}} = Q_{\text{add}} + Q_{\text{rej}}$$

and for the Carnot cycle, denoting $\oint dQ_{\text{rev}}$ as Q_{Carnot}, we have

$$Q_{\text{Carnot}} = (T_H - T_L)\Delta S \qquad \textbf{(8–1)}$$

Since the Carnot engine is a cycle, we may say that $\oint dU$ is zero. That is, the sum of internal energy changes for one cycle is zero, as it should be for any property. We have

$$\oint dU = \int_1^2 dU + \int_2^3 dU + \int_3^4 dU + \int_4^1 dU$$

or

$$\oint dU = U_2 - U_1 + U_3 - U_2 + U_4 - U_3 + U_1 - U_4 = 0$$

and we see that this statement is true; in fact it is true for *all cycles*, reversible or not. The first law then can be written

$$\oint dU = \oint dQ - \oint dWk \qquad \textbf{(8–2)}$$

but

$$\oint dU = 0$$

so

$$\oint dQ = \oint dWk \tag{8–3}$$

or

$$Q_{\text{cycle}} = Wk_{\text{cycle}} \tag{8–4}$$

For the Carnot cycle, then

$$(T_H - T_L)\Delta S = Wk_{\text{Carnot}}$$

from the result in equation (8–1).

We should note that a result of equation (8–4) is that the shaded areas of the diagrams in figures 8–3 and 8–4 must be equal.

It was stated that no one has yet built a Carnot engine; there are at least two reasons:

1. Reversible cycles can be approached but never fully realized.
2. While reversible adiabatic processes can be closely approximated in technology, the reversible isothermal process is difficult to put into practice and still have sufficient amounts of heat transfer in a reasonable time period.

The Carnot cycle/engine concepts are theoretical tools which have added immensely to the thermodynamic method; we will refer to them frequently as a standard of comparison.

Example 8.1

Assume an internal combustion engine operates on a reversible Carnot cycle between 20°C and 2000°C. If this engine generates 100 kilowatts of power, determine the amount of heat added and the heat rejected by the engine.

Solution

We know that the cyclic work equals the cyclic heat from equation (8–4), so

$$\dot{Q}_{\text{cycle}} = \dot{W}k_{\text{cycle}} = 100\ \text{kW} = \dot{Q}_{\text{Carnot}}$$

Also,

$$\dot{Q}_{\text{Carnot}} = (T_H - T_L)(\dot{S})$$

where $\dot{S} = \Delta S/\Delta\tau$ is the entropy generation. The heat added is given by the equation

$$\dot{Q}_{\text{add}} = T_H\dot{S}$$

so, from the above, we have

$$\dot{S} = \frac{\dot{Q}_{\text{Carnot}}}{T_H - T_L}$$

Since the temperatures are given, we may compute the entropy generation

$$\dot{S} = \frac{100 \text{ kW}}{2273 \text{ K} - 293 \text{ K}}$$

$$= 0.0505 \text{ kW/K}$$

The heat addition and heat rejected may then be found:

$$\dot{Q}_{add} = (2273 \text{ K})(0.0505 \text{ kW/K})$$

$$= 115 \text{ kW}$$ *Answer*

and

$$\dot{Q}_{rej} = (293 \text{ K})(0.0505 \text{ kW/K})$$

$$= 14.8 \text{ kW}$$ *Answer*

8.3 Thermodynamic Efficiency

In this section we will consider the methods of determining the effectiveness of heat engines; that is, we will define the efficiencies of cyclic devices. As we will see, one efficiency is not an accurate description of an engine. Rather, we will have many efficiencies with precise meanings for each. For the heat engine analysis, the term we will use is *thermodynamic efficiency*, η_T. It is defined as

$$\eta_T = \left(\frac{Wk_{cycle}}{Q_{add}}\right)(100) \tag{8-5}$$

and it gives us the ratio of the *net* work *out* to the heat *in*. For some of the other information we might want, such as the reversibility of the cycle, the unnecessary degradation of available energy, or the rate at which work can be furnished, we must seek other efficiency terms.

We can obtain some idea of the thermodynamic irreversibility from the *heat engine efficiency* η defined as

$$\eta = \left(\frac{Wk_{actual}}{Wk_{cycle}}\right)(100) \tag{8-6}$$

where Wk_{actual} is the actual work gained from the heat engine and the Wk_{cycle} is the ideal or reversible work.

If we consider systems which have no thermal effects, the *mechanical efficiency* η_{mech} is defined as

$$\eta_{mech} = \left(\frac{Wk_{out}}{Wk_{in}}\right)(100) \tag{8-7}$$

and gives us some idea of the energy dissipated during mechanical conversions of energy. This is probably the most familiar statement of efficiency — output divided by input — but the reciting of all these dimensionless efficiency numbers is not the heart of a true technical analysis.

Example 8.2

A Carnot engine operates in an atmosphere at 343 K, with an addition of heat from combusting gases burning at 1473 K. Determine the thermodynamic efficiency of the engine.

Solution

We are concerned here with a reversible heat engine which operates between high and low temperature regions. We can visualize this Carnot engine better from a system diagram as shown in figure 8–5. In this figure we see that

$$Q_{\text{add}} - Q_{\text{rej}} = Wk_{\text{cycle}}$$

which is a restatement of equation (8–4). The term Q_{add} is equal to $T_H \Delta S$, where T_H is the high temperature region at 1473 K. Similarly, Q_{rej} is $-T_L \Delta S$ where T_L is 343 K. Then

$$(T_H - T_L)\Delta S = Wk_{\text{cycle}}$$

and from the equation

$$\eta_T = \frac{Wk_{\text{cycle}}}{Q_{\text{add}}} \times 100 \qquad \textbf{(8–5)}$$

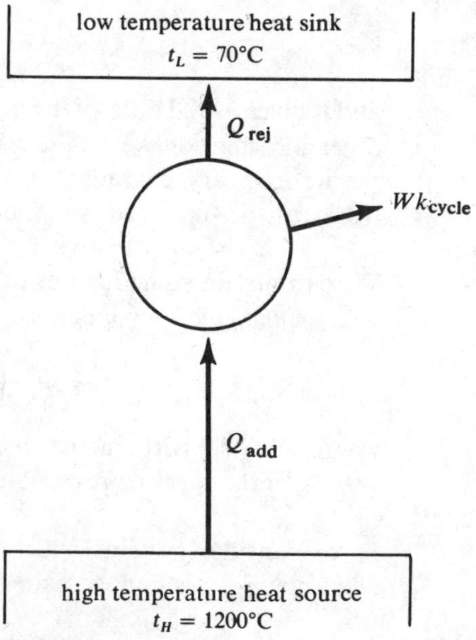

Figure 8–5 Carnot system diagram for example 8.2.

we have, for the general Carnot engine

$$\eta_T = \frac{(T_H - T_L)\Delta S}{T_H \Delta S} \times 100$$

or

$$\eta_T = \left(1 - \frac{T_L}{T_H}\right) \times 100 \tag{8–8}$$

Specifically for this problem, we obtain

$$\eta_T = \left(1 - \frac{343}{1473}\right) \times 100 = 76.7\% \qquad \textit{Answer}$$

Notice that for the Carnot heat engine, the thermodynamic efficiency is only a function of the two temperature regions, and no other cycle can have a higher efficiency.

8.4 Clausius' Inequality

We have considered the property *entropy* and noted that for irreversible processes the entropy change is limited by the inequality

$$\Delta S > \int \frac{dQ}{T} \tag{8–9}$$

For a heat engine which has irreversibilities in it, we say the sum of ΔS is

$$\Sigma \Delta S = \oint dS$$

and

$$\oint dS > \oint \frac{dQ}{T} \tag{8–10}$$

But for any cycle the sum of the property changes for one cycle must be zero so

$$\oint dS = 0 > \oint \frac{dQ}{T} \tag{8–11}$$

This statement is called *Clausius' inequality*. For reversible cycles the equality

$$\oint \frac{dQ}{T} = 0 \tag{8–12}$$

holds; but for real, irreversible cycles, equation (8–11) must hold. This inequality has been the beginning of much conjecture and advanced discussion in thermodynamic analysis. For us, the use of Clausius' inequality is in the inference that the entropy increase of a heat engine cannot be determined solely from the heat transfer data. For instance, for the Carnot reversible engine we found that

$$\frac{Q_{\text{add}}}{T_H} - \frac{Q_{\text{rej}}}{T_L} = 0$$

but for an irreversible heat engine this sum does not add to zero. From Clausius' inequality, we see that

$$\frac{Q_{\text{add}}}{T_H} - \frac{Q_{\text{rej}}}{T_L} < 0$$

or

$$\frac{Q_{\text{add}}}{T_H} < \frac{Q_{\text{rej}}}{T_L}$$

where the left side represents entropy *reduction* of some energy source while the right represents *increases* to a surrounding energy sink. Obviously the entropy has a net increase to the surroundings for irreversible processes.

8.5 The Second Law of Thermodynamics

We have considered the first law of thermodynamics and from it came an elementary restriction on any device we wish to analyze or design — energy cannot be created. It would be a tremendous feat to have an engine that would operate as diagrammed in figure 8–6, but this device is creating energy from nothing and the first law tells us this is impossible. Engines such as that shown in figure 8–6 are called *perpetual motion machines of the first kind* (PMM1) and occasionally we hear pronouncements that someone has invented one. They are impossible to achieve, however, because of the natural restrictions stated in the first law, and a simple system diagram can usually pinpoint the fallacy in any such supposed device.

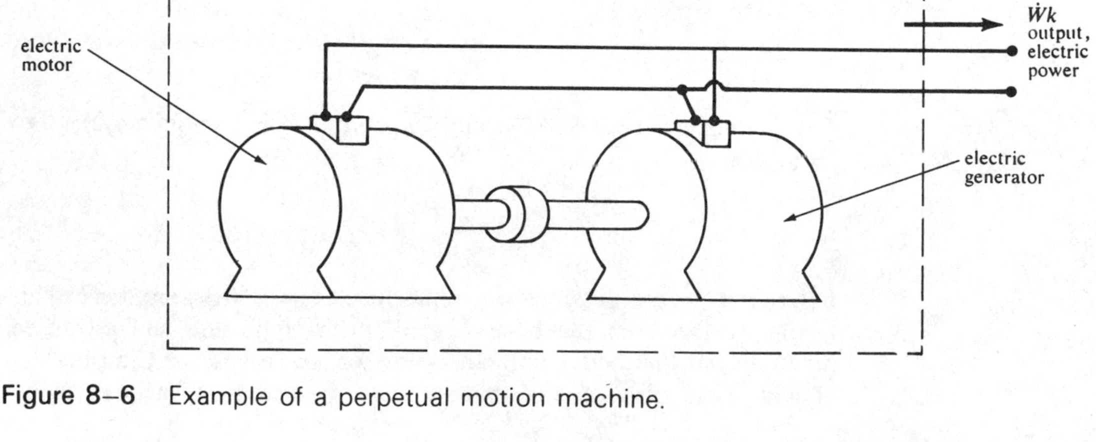

Figure 8–6 Example of a perpetual motion machine.

There is, however, a more subtle restriction we must contend with above and beyond the first law, contained in the second law.

Second Law of Thermodynamics: *No heat engine can produce a net work output by exchanging heat with a single fixed temperature region.*

The effect of this categoric statement is that a restriction is placed on all heat engines, a restriction which requires that both Q_{add} and Q_{rej} be nonzero. You may have considered that to achieve 100% efficiency for the Carnot engine all that was required was for the heat rejected to be zero, or that the heat sink to be at 0 K. That is

$$\eta_T = \left(1 - \frac{T_L}{T_H}\right) \times 100 = 100\% \qquad \textbf{(8–8)}$$

when $T_L = 0$. The second law now tells us it is impossible to reach 100% efficiency for a heat engine since the heat rejected must be greater than zero. For the bulk of technological applications, we must be further restricted to a low temperature sink which is at the atmospheric state. The term T_L in the efficiency equation (8–8) has a value of between 270 K and 320 K or between 460°R and 550°R for many common cycles.

The second law will not provide us with any new computational or bookkeeping techniques as the first law did, but its underlying usefulness cannot be overestimated. Suppose, for instance, that we have a reversible heat engine which is capable of producing net work from only one temperature region. This machine, called a *perpetual motion machine of the second kind* (PMM2), would be converting 100% of the heat into work. In addition, since we need be concerned with only one temperature region, our heat engine can run at any lower temperature and the efficiency of the engine (100%) is independent of everything. This type of device does not in any way violate the first law, and yet it has never been achieved. It has a certain advantage over a PMM1 in that it can provide work and refrigeration by extracting heat from any temperature region. The PMM1 merely provided work from nothing. We thus are led to the conclusion that for a heat engine to operate within the laws of nature, it must be somehow dependent upon the surroundings. The second law provides the answer.

There are other statements of the second law which are equivalent to the one given above. One of these alternatives is attributed to Clausius, who stated that it is impossible for a heat pump to have as its only effect the transfer of heat from a low to a high temperature region. (Thus, a refrigerator cannot operate without work supplied to it.) The equivalence of the two second law statements can easily be shown. Suppose, as indicated in figure 8–7(a), that we have a reversible heat pump requiring

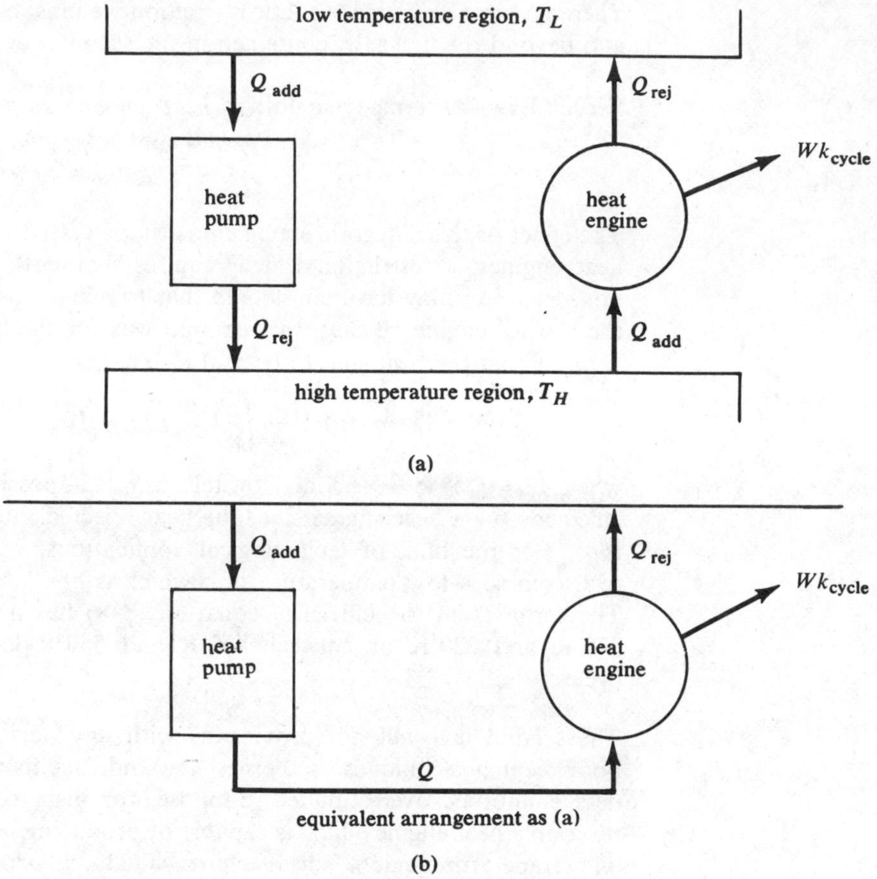

Figure 8–7 Hypothetical heat pump and engine combination.

no work input and a reversible heat engine. The arrangement of (a) is then the same as (b), and in this second system diagram there is a transfer of heat with only one temperature region. This is an obvious violation of the initial statement of the second law, so we see Clausius' statement is its equivalent.

8.6 Heat and Work of a Cycle

During the execution of a set of processes on a system or heat engine, which together constitute a cycle, none of the properties change during the complete cycle; but for each individual process the properties will definitely change. Using the equation of state, caloric equations of state, and process equation, we may obtain values of heat and/or work for each process, and by arithmetic addition the net work and heat of a cycle are determined. These net values of a cycle have been denoted by the cyclic integral sign, $\oint$, and from the first law it was shown

$$\oint dWk = \oint dQ \tag{8–13}$$

for any cycle. This is equivalent to the statement

$$Wk_{\text{cycle}} = Q_{\text{net}} \tag{8–14}$$

where in general

$$Q_{\text{net}} = Q_{\text{add}} + Q_{\text{rej}} \tag{8–15}$$

Most references to the work or heat of cycle mean the net values, as these are the only quantities which affect the surroundings. Caution should be exercised, however, in interpreting this terminology, as there are possibilities for referring to individual work or heat quantities for the separate processes making up the cycle of a heat engine or pump.

Notice that equation 8–15 indicates, correctly, an addition of Q_{add} and Q_{rej}, but Q_{rej} will carry a sign opposite that of Q_{add}.

8.7 Closed Cycle

System identification and classification as to open or closed has been the approach in analyzing a complete thermodynamic problem. In this chapter, we introduced the system as a device which is rotating or in some manner going through a cycle — and called it a *heat engine* or *pump*. Just as we classified systems, we can classify heat engines as open or closed, that is, open or closed cycles. The closed cycle heat engine/pump is a device which allows no mass transfer across the boundaries; the device retains the same mass during the cycle. The system diagram for a closed cycle is shown in figure 8–1(a). Examples of closed cycles are the heat pump system of the refrigerator (as shown in figure 8–8), the steam turbine - electric generating system (shown in figure 8–9), and a piston-cylinder device (shown in figure 8–10).

Notice in the first examples that while the complete devices were closed, there were four components in which each was individually represented as an open system. In figure 8–8, the pump, the condenser, the expansion valve, and the evaporator each contribute a process to the closed cycle of the refrigerator; but each of the processes is carried out in open systems. A similar situation exists in the steam turbine cycle of figure 8–9, but in figure 8–10 we have a closed system contributing all the processes to a complete closed cycle.

It is then apparent that a closed cycle and closed system are not one and the same. Open systems can be a part of a closed cycle and obviously, so can closed systems.

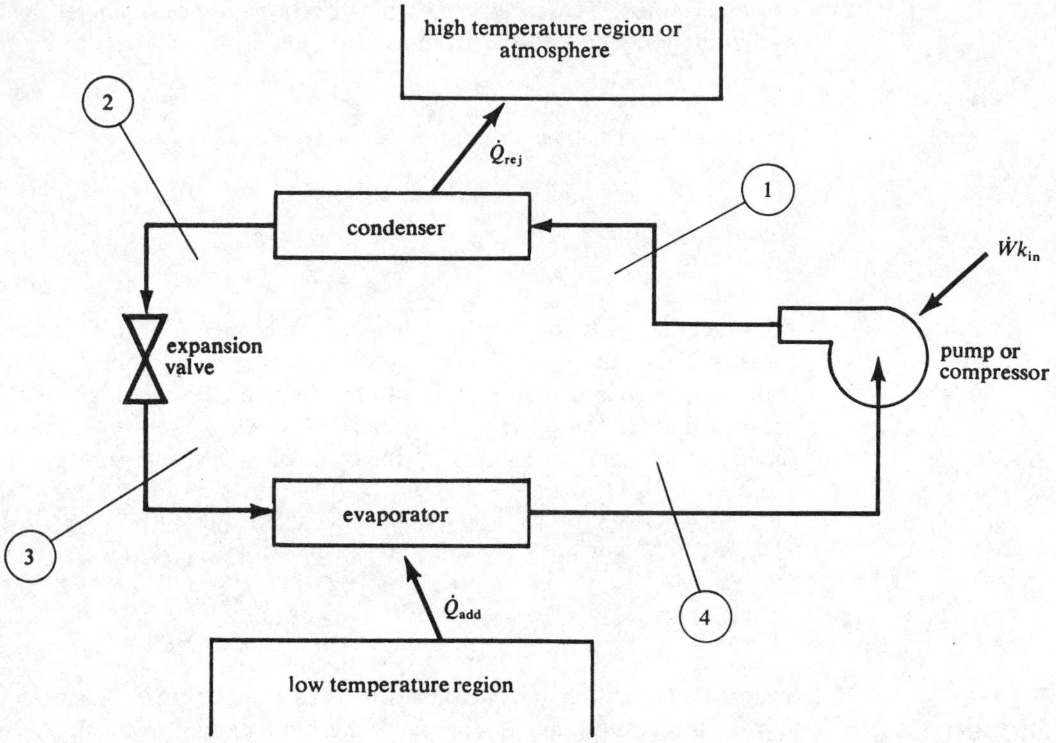

Figure 8–8 Typical refrigerating cycle.

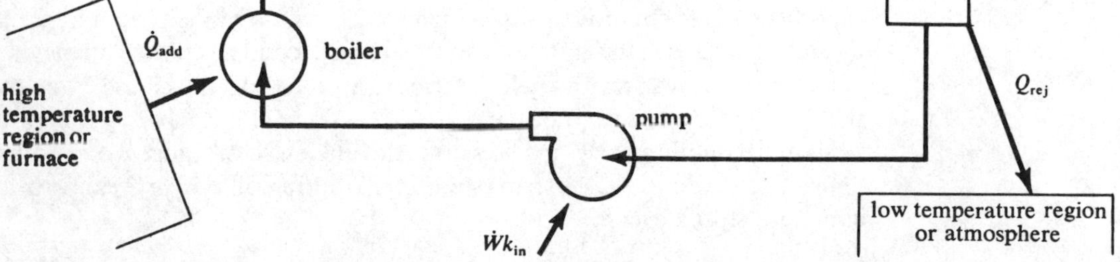

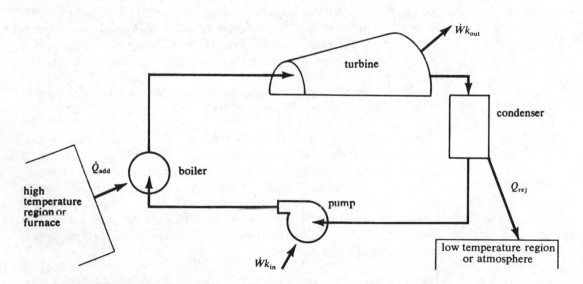

Figure 8–9 Steam turbine cycle.

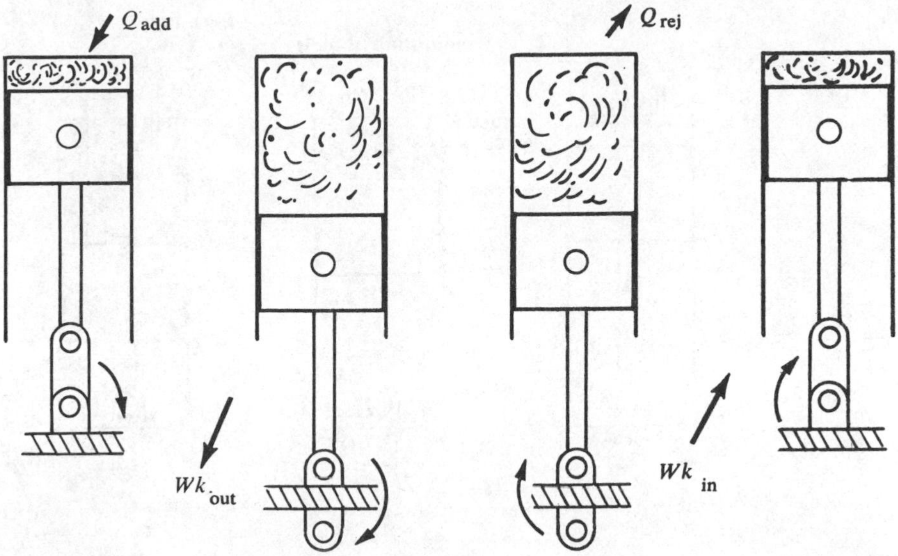

Figure 8–10 Piston-cylinder device executing one cycle of a closed cycle.

8.8 Open Cycle

We consider an open cycle heat engine/pump to be one which involves mass transfer across the device boundary. Since we are considering cyclic devices, there is no change in the amount of mass over a cycle, but the mass may be flowing through continuously or flushed and rinsed periodically during specific cycle processes. The system diagram of a typical, generalized open cycle is shown in figure 8–1(b). Examples of open cycles are given in figures 8–11 and 8–12. Notice that in figure 8–11 during process (1), the mass is being exchanged; this exchange represents the only difference in theory from that shown in figure 8–10.

The gas turbine shown in figure 8–12 is obviously open — all the components are open with air or gases flowing through each of them under steady flow conditions.

For the open cycle then, we may have some processes which are individually closed, or we may have all processes open.

8.9 Carnot Power Cycle

In this section, we will examine the details of a Carnot engine and attempt to devise a machine which would function under the requirements of the Carnot cycle.

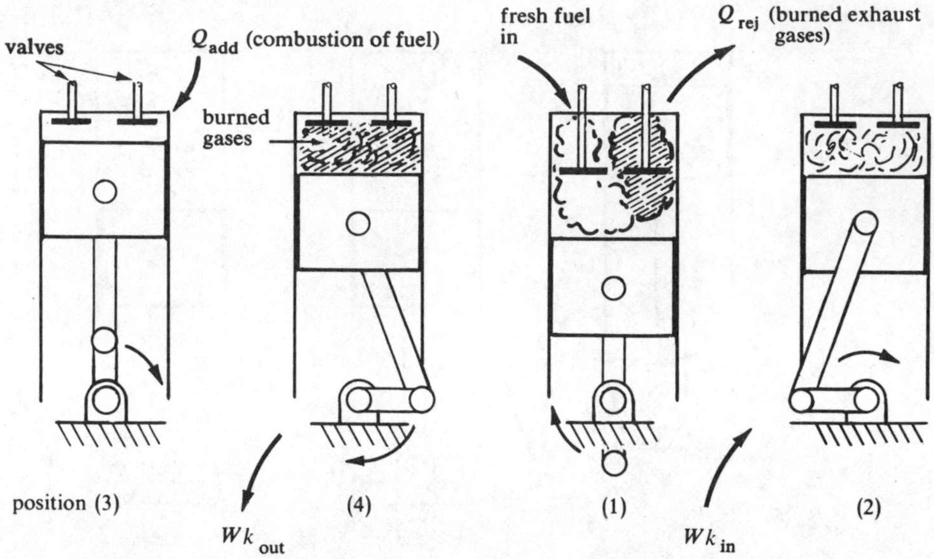

Figure 8–11 Open cycle of a piston-cylinder device exchanging fresh fuel and air mixture for spent exhaust gases.

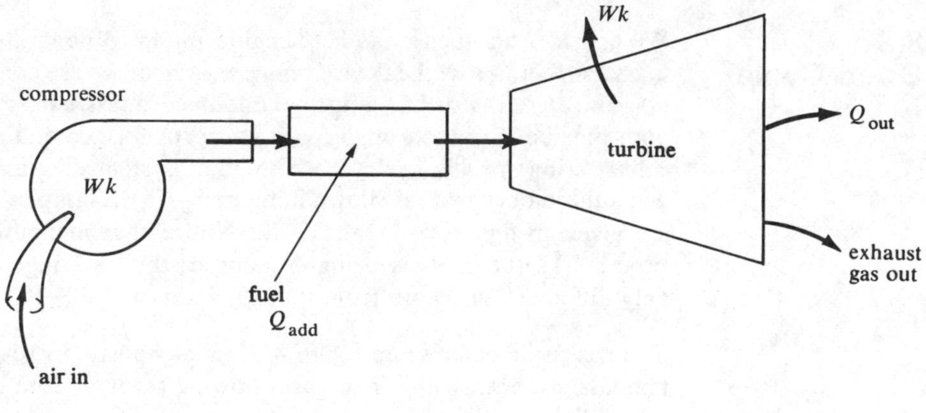

Figure 8–12 Typical gas turbine cycle.

Example 8.3 A Carnot engine is proposed which has the physical characteristics of a piston-cylinder while the gas contained within the cylinder is air having a mass of amount 0.002 lbm. The piston has a diameter of 3 inches and travels a distance of 4 inches, reciprocating from within $\frac{1}{4}$ inch of the cylinder and to $4\frac{1}{4}$ inches. The arrangement is shown in figure 8–13. Heat proceeds to be added when the piston is at the extreme inward position (1) and proceeds to be rejected when the piston is in the extreme outward position (3). If the high temperature region is at 1200°F and the low temperature region is at 80°F, determine the engine's thermal efficiency, net work per cycle, heat added, and heat rejected.

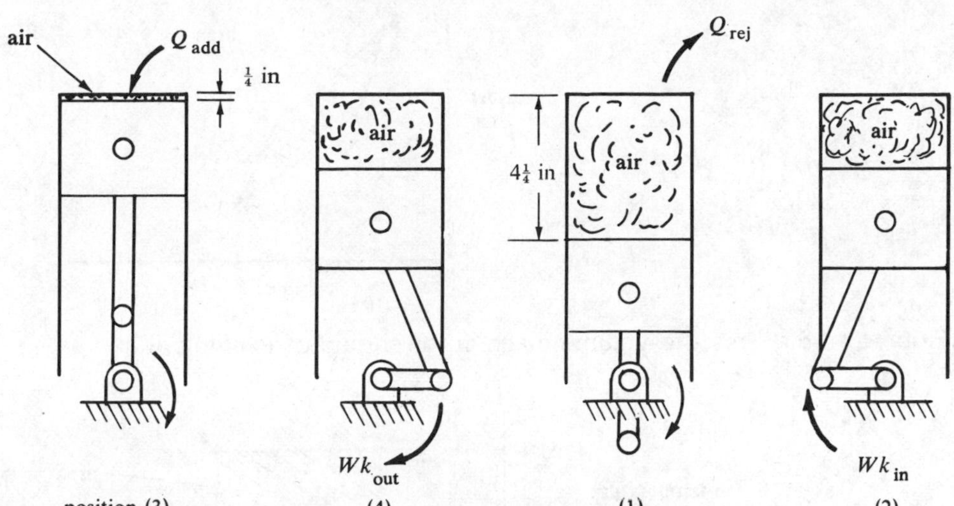

Figure 8–13 Carnot engine using the piston-cylinder mechanism.

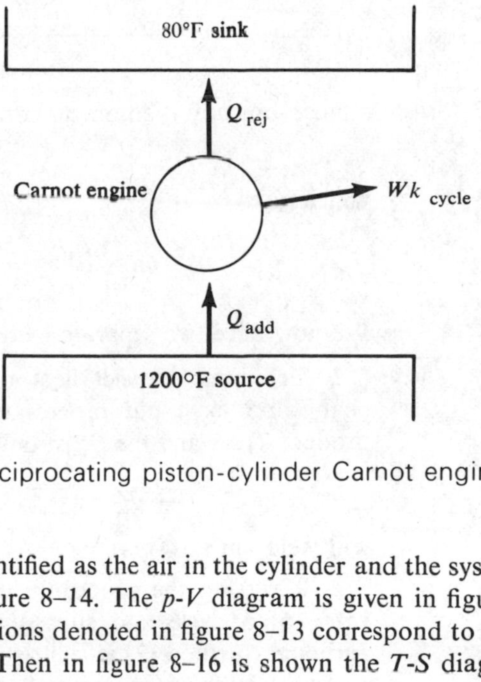

Figure 8–14 System diagram for reciprocating piston-cylinder Carnot engine.

Solution The system is identified as the air in the cylinder and the system diagram is sketched in figure 8–14. The *p-V* diagram is given in figure 8–15 and the physical positions denoted in figure 8–13 correspond to the numbers in this diagram. Then in figure 8–16 is shown the *T-S* diagram for the given Carnot engine.

The thermal efficiency we obtain from

$$\eta_T = \left(1 - \frac{T_L}{T_H}\right) \times 100$$

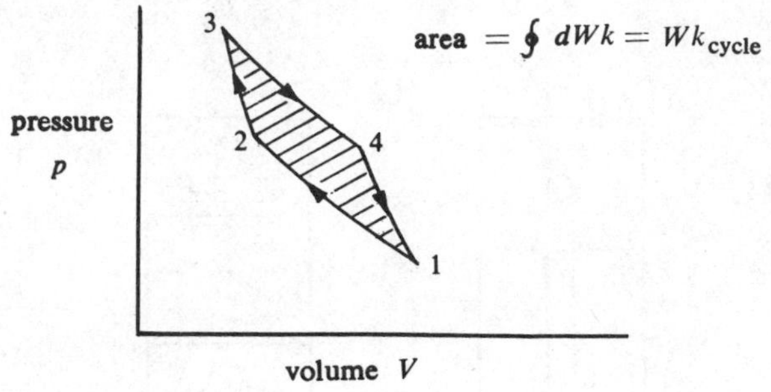

Figure 8–15 Pressure-volume diagram for engine of example 8.3.

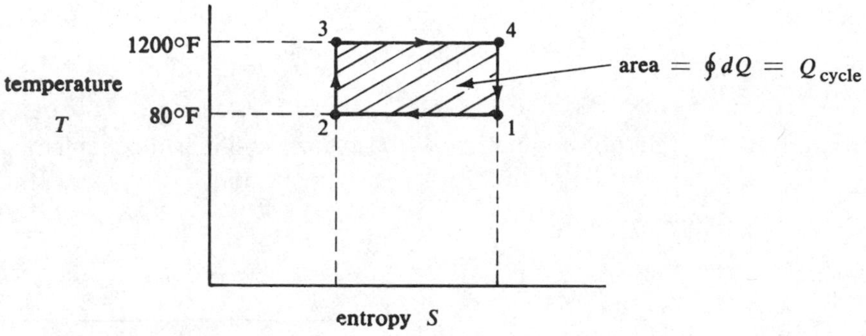

Figure 8–16 Temperature-entropy diagram for engine of example 8.3.

which gives us

$$\eta_T = \left(1 - \frac{540}{1660}\right) \times 100 = 67.5\% \qquad \textit{Answer}$$

We now have two separate methods of obtaining our desired answers:

1. Determine the net heat transfer $\oint dQ$ by obtaining the entropy change ΔS from our process equations. Then we get Q_{add} from the product $T_1\,\Delta S$ and the Q_{rej} from $T_4\,\Delta S$. Using the sum

$$\oint dQ = Q_{add} - Q_{rej} = \oint dWk = Wk_{cycle}$$

will yield our answers.

2. Reversing the procedure of method (1) by first determining the work of a cycle and subsequently finding the heat transfer terms. Although method (2) is a little lengthier than method 1, we will use it also to help clarify the computational techniques we have accumulated.

In method (1), to compute Δs, we observe that it is given by the relations $|\Delta s| = |s_2 - s_1|$ and $|\Delta s| = |s_4 - s_3|$. From the results in chapter 7, listed in table 6–3, for the isothermal processes

$$s_2 - s_1 = \Delta s_{12} = R \ln \frac{V_2}{V_1}$$

and

$$s_4 - s_3 = \Delta s_{34} = R \ln \frac{V_4}{V_3}$$

From table B.4 we have $R = 53.3$ ft-lbf/lbm · °R, and the volumes are

$$V_1 = (\text{piston face area})(4\tfrac{1}{4} \text{ inches})$$

$$V_3 = (\text{piston face area})(\tfrac{1}{4} \text{ inch})$$

or, since the face area is circular, we obtain

$$V_1 = (\pi)\left(\frac{3^2}{4} \text{ in}^2\right)(4.25 \text{ in}) = 30.04 \text{ in}^3$$

$$V_3 = (\pi)\left(\frac{3^2}{4} \text{ in}^2\right)(0.25 \text{ in}) = 1.767 \text{ in}^3$$

Also, since processes (2–3) and (4–1) are reversible adiabatic, we have that

$$\frac{V_1}{V_4} = \frac{T_4}{T_1}^{1/(k-1)} = \frac{1660°\text{R}}{540°\text{R}}^{1/0.4} = 16.6$$

Thus

$$V_4 = \frac{V_1}{16.6} = 1.81 \text{ in}^3$$

and

$$\frac{V_2}{V_3} = \frac{T_3}{T_2}^{1/(k-1)} = 16.6$$

so

$$V_2 = 16.6 V_3 = 29.33 \text{ in}^3$$

Notice that for the Carnot cycle we have

$$\frac{V_1}{V_4} = \frac{V_2}{V_3}$$

Now, we can find the entropy changes

$$S_2 - S_1 = m\Delta s_{12} = mR \ln \frac{V_2}{V_1}$$

$$= (0.002 \text{ lbm})(53.3 \text{ ft-lbf/lbm} \cdot °\text{R})\left(\ln \frac{29.33}{30.04}\right)$$

$$= -0.0025 \text{ ft-lbf/°R}$$

and

$$S_4 - S_3 = mR \ln \frac{V_4}{V_3}$$

$$= (0.002 \text{ lbm})(53.3 \text{ ft-lbf/lbm} \cdot {}^\circ\text{R})\left(\ln \frac{1.81}{1.767}\right)$$

$$= +0.0025 \text{ ft-lbf/}^\circ\text{R}$$

The net or cyclic work is

$$Wk_{\text{cycle}} = (T_H - T_L)(\Delta S)$$

$$= (1660^\circ\text{R} - 540^\circ\text{R})(0.0025 \text{ ft-lbf/}^\circ\text{R})$$

$$= 2.80 \text{ ft-lbf/cycle}$$ *Answer*

The heat added is

$$Q_{\text{add}} = T_H \Delta S = (1660^\circ\text{R})(0.0025 \text{ ft-lbf/}^\circ\text{R})$$

$$= 4.15 \text{ ft-lbf/cycle}$$ *Answer*

and the heat rejected is

$$Q_{\text{rej}} = T_L \Delta S = (540^\circ\text{R})(0.0025 \text{ ft-lbf/}^\circ\text{R})$$

$$= -1.35 \text{ ft-lbf/cycle}$$ *Answer*

Using method (2) we write the work of the cycle as the sum

$$\oint dWk = Wk_{\text{cycle}} = Wk_{(1-2)} + Wk_{(2-3)} + Wk_{(3-4)} + Wk_{(4-1)}$$

The first term of this sum we find from equation (6–17)

$$Wk_{(1-2)} = p_1 V_1 \ln \frac{V_2}{V_1} = p_2 V_2 \ln \frac{V_2}{V_1}$$

since process (1–2) is isothermal. Similarly

$$Wk_{(3-4)} = p_4 V_4 \ln \frac{V_4}{V_3} = p_3 V_3 \ln \frac{V_4}{V_3}$$

for the reason that process (3–4) is isothermal. Processes (2–3) and (4–1) are both reversible adiabatic (or isentropic as well) and the work is gotten from equation (6–33)

$$Wk_{(2-3)} = \frac{mR}{1-k}(T_3 - T_2)$$

and

$$Wk_{(4-1)} = \frac{mR}{1-k}(T_1 - T_4)$$

respectively. But we see that T_3 equals T_4, and T_1 equals T_2, so $Wk_{(2-3)} = -Wk_{(1-4)}$, and we need only worry about the two isothermal work terms.

The volumes are computed as they were in method 1, so we have $V_1 = 30.04\ \text{in}^3$, $V_2 = 29.33\ \text{in}^3$, $V_3 = 1.767\ \text{in}^3$, and $V_4 = 1.81\ \text{in}^3$. The work then is obtained from

$$Wk_{\text{cycle}} = p_1V_1 \ln \frac{V_2}{V_1} + p_3V_3 \ln \frac{V_4}{V_3}$$

and since $pV = mRT$, we have

$$Wk_{\text{cycle}} = mR\left(T_1 \ln \frac{V_2}{V_1} + T_3 \ln \frac{V_4}{V_3} \right)$$

Substituting values in, we obtain

$$\begin{aligned} Wk_{\text{cycle}} &= (0.002\ \text{lbm})(53.3\ \text{ft-lbf/lbm} \cdot {}^\circ\text{R}) \\ &\quad \times \left(1660^\circ\text{R} \ln \frac{1.81}{1.767} + 540 \ln \frac{29.33}{30.04} \right) \\ &= (0.1066\ \text{ft-lbf/}^\circ\text{R})[1660^\circ\text{R} \ln 1.024 - 540^\circ\text{R} \ln 1.024] \\ &= 2.83\ \text{ft-lbf} \end{aligned}$$

Answer

This answer represents the work produced by the Carnot engine per cycle. In heat units it is

$$Wk_{\text{cycle}} = \frac{2.83}{778}\ \text{Btu} = 0.0036\ \text{Btu}$$

and the heat added can be quickly determined from the efficiency

$$\eta_T = 67.5\% = \left(\frac{Wk_{\text{cycle}}}{Q_{\text{add}}} \right)(100)$$

so

$$Q_{\text{add}} = \frac{0.0036}{0.675} = 0.0053\ \text{Btu}$$

Answer

The heat rejected can be calculated from the relation

$$\oint dQ = \oint dWk = 0.00289 = Q_{\text{add}} - Q_{\text{rej}}$$

so

$$Q_{\text{rej}} = 0.0053 - 0.0036 = 0.0017\ \text{Btu}$$

Answer

This problem, while somewhat lengthy, should illustrate the drawing together of the various ideas and equations so necessary in solving thermodynamics problems in engineering.

8.10 Reversed Carnot Cycle

Consider the Carnot cycle operating in reverse. This is the heat pump which, from the second law of thermodynamics, is required to have a net work or power put into the device to make it practical. With the following example problem we introduce the full significance of its operation.

Example 8.4

A refrigerator operates on a reversed Carnot cycle. If the heat is taken in at $-13°C$ and rejected to $27°C$ and the entropy change is 0.1 J/K, determine the work required to drive the refrigerator and the amount of cooling it will have per cycle.

Solution

For a reversed Carnot cycle we refer to figure 8–2 and see that we add heat at the low temperature T_L and reject heat at T_H. The cyclic work is

$$Wk_{\text{cycle}} = Q_{\text{add}} + Q_{\text{rej}}$$
$$= T_H \Delta S - T_H \Delta S$$
$$= (T_L - T_H)\Delta S$$

Converting the temperatures to absolute values we obtain

$$Wk_{\text{cycle}} = (260\text{ K} - 300\text{ K})(0.1\text{ J/K})$$
$$= -4.0\text{ J/cycle}$$ *Answer*

The negative sign indicates that work is put into the system. The cooling or refrigerating effect is the heat added, or

$$Q_{\text{add}} = T_L \Delta S = (260\text{ K})(0.1\text{ J/K})$$
$$= 26.0\text{ J/cycle}$$ *Answer*

Example 8.5

If the refrigerator of example problem 8.4 operates at 300 cycles per minute, determine the power required and the rate of cooling.

Solution

The power is computed from the relationship

$$\dot{W}k = (Wk_{\text{cycle}})(300\text{ cycles/min})$$
$$= -1200\text{ J/min} = -20\text{ J/s}$$
$$= -0.02\text{ kW}$$ *Answer*

The rate of cooling is computed from

$$\dot{Q}_{\text{add}} = (Q_{\text{add}})(300\text{ cycles/min})$$
$$= (26\text{ J/cycle})(300\text{ cycles/min})$$
$$= 7800\text{ J/min} = 130\text{ J/s}$$
$$= 0.13\text{ kW}$$ *Answer*

Example 8.6

A Carnot heat pump operates between the limits of 10°F and 120°F. Its mechanical aspects are diagrammed in figure 8–17, and it operates with

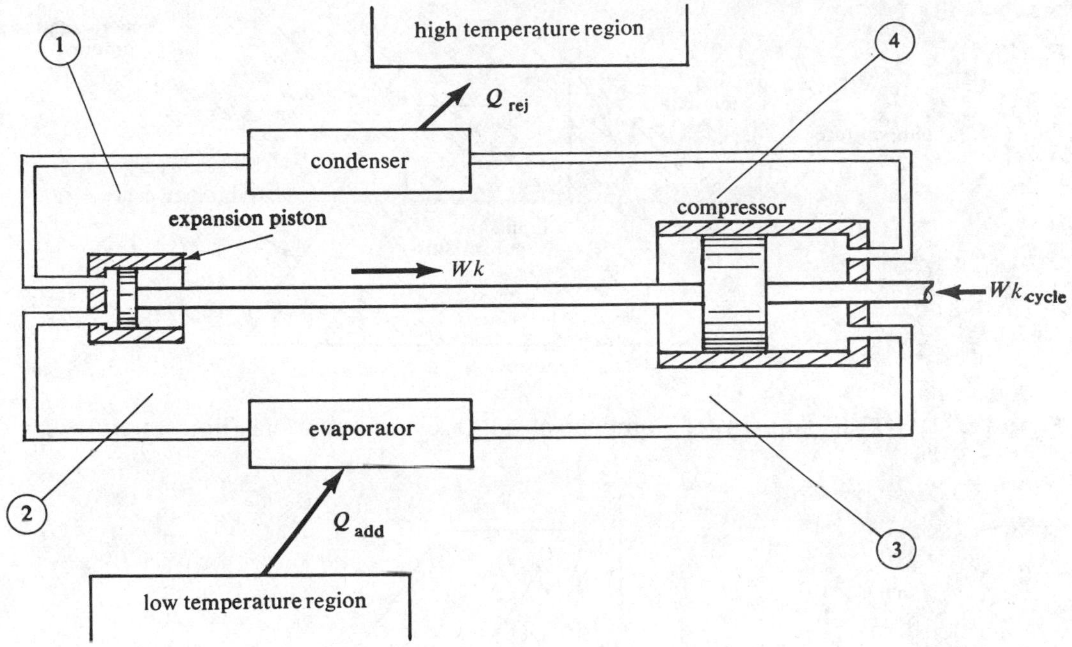

Figure 8–17 An arrangement for a Carnot heat pump.

ammonia as the working fluid. At point (1) in the closed cycle, the ammonia is completely saturated liquid at 120°F while at point (4) the ammonia is saturated vapor at 120°F. The ammonia expands adiabatically from (1) to (2), furnishing some work to help drive the compressor. The operating temperature at point (2) is 10°F. Heat is added to the working fluid in the evaporator and rejected in the condenser. The compressor supplies the energy to the ammonia, which increases in pressure and temperature, making the heat rejection to the high temperature (120°F or less) region possible. Determine Q_{add}, Q_{rej}, Wk_{cycle}, and heat pump efficiency.

Solution

The *T-S* and *p-v* diagrams are shown in figures 8–18 and 8–19 respectively. Notice that a line, called the *saturation curve*, has been included on the *T-S* diagram and the area under this bell-shaped curve is designated as "liquid and vapor mixture." The significance here is that the liquid and vapor mixture represents the state a material (in this case ammonia) must pass through to change from a liquid phase to a gaseous or vapor phase. This area under the saturation curve we call the *saturated region;* a *saturated liquid* is a point on the left side of the saturation curve, and a *saturated vapor* is a point on the right side. These points are tabulated for various temperatures or pressures in the table B.11 and to find states in between, we need to know the percent vapor or the ratio of liquid to vapor. For instance, a point midway between (1) and (4), point (a),

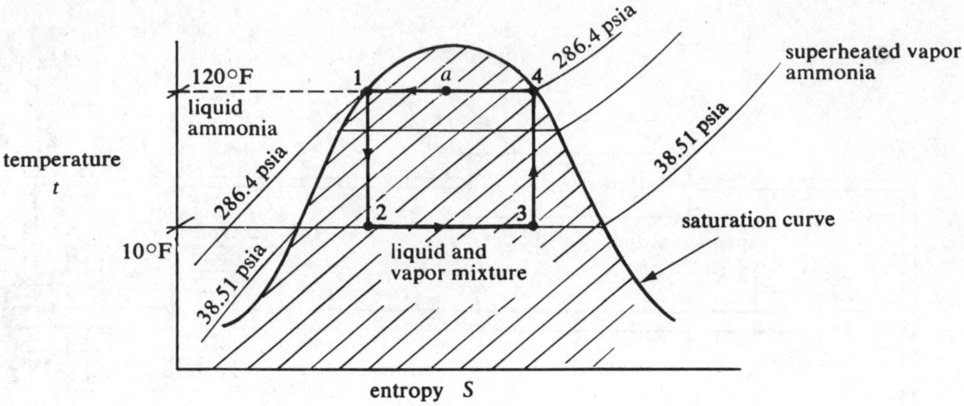

Figure 8–18 *T-S* diagram for Carnot heat pump using ammonia (not drawn to scale).

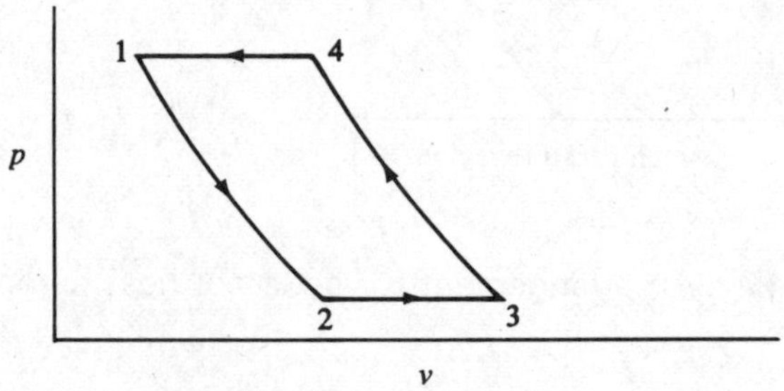

Figure 8–19 *p-v* diagram for Carnot heat pump.

would be 50% quality. We denote quality by the Greek letter chi χ and define it as follows:

$$\chi_a = \left(\frac{\text{mass of vapor at } a}{\text{mass of vapor and liquid at } a}\right) \times 100 \qquad \textbf{(8–16)}$$

which can be reduced to the equation

$$\chi_a = \left(\frac{s_a - s_1}{s_4 - s_1}\right) \times 100 \qquad \textbf{(8–17)}$$

Points (1) and (4) in figure 8–18 correspond to the saturated liquid and vapors respectively, the entropies of which are denoted in the appendix tables as s_f for saturated liquid, and for saturated vapor, s_g. In some tables the difference $s_g - s_f$ is denoted s_{fg}, so equation (8–17) can then be written

$$\chi_a = \left(\frac{s_a - s_f}{s_g - s_f}\right) \times 100 = \left(\frac{s_a - s_f}{s_{fg}}\right) \times 100 \qquad \textbf{(8–18)}$$

or, equivalently

$$s_a = s_{fg}\frac{\chi_a}{100} + s_f \qquad \textbf{(8–18)}$$

and the other properties at state a are as follows:

$$h_a = h_{fg}\left(\frac{\chi_a}{100}\right) + h_f \quad \textbf{(8–19)}$$

$$u_a = u_{fg}\left(\frac{\chi_a}{100}\right) + u_f \quad \textbf{(8–20)}$$

and

$$v_a = v_{fg}\left(\frac{\chi_a}{100}\right) + v_f \quad \textbf{(8–21)}$$

The quantity h_{fg} is equal to the difference of the enthalpy of saturated vapor, h_g, and saturated liquid, h_f, or

$$h_{fg} = h_g - h_f$$

and is called the *latent heat, heat of vaporization*, or *heat of condensation.* A better term would be *energy of vaporization* since it is energy — not heat — which is required to convert a liquid to a vapor, but the initial term has stuck from earlier days.

In figure 8–18 are drawn two typical constant pressure lines (for 286.4 psia and 38.51 psia) which correspond to isothermal lines in the phase change region but which diagonally cross isothermal lines elsewhere. The fact that pressure and temperature both remain constant during a phase change accounts for the slight difference in the p-v diagram in figure 8–19 from the normal Carnot cycle of figure 8–3.

We will consider in further detail the mechanisms of phase change in chapter 12 when we consider steam and water transformations, but now let us return to our reversed Carnot engine. The first law, applied to the condenser, is

$$\Delta h = q_{\text{rej}} - wk$$

since the condenser is an open system and the ammonia flows through without kinetic or potential energy changes. Also, wk is zero in the condenser so we obtain

$$\Delta h = h_1 - h_4 = q_{\text{rej}}$$

but, from table B.11 we have

$$h_1 = h_f \text{ (at 120°F)} = 179.0 \text{ Btu/lbm}$$

and

$$h_4 = h_g = 634.0 \text{ Btu/lbm}$$

We get then

$$q_{\text{rej}} = 179.0 - 634.0 = -455.0 \text{ Btu/lbm} \qquad \textit{Answer}$$

We are using the tabulated values because the equations developed previously are primarily of value for perfect gases and this obviously is not a suitable assumption for this problem. The heat added q_{add} can be gotten from a first law application to the evaporator:

$$\Delta h = h_3 - h_2 = q_{add} - wk$$

Again, wk is zero in an evaporator and we need only determine the enthalpy values from the tables. To find these enthalpy values we need to know the quality. At state (2) this is from equation (8–17)

$$\chi_2 = \frac{s_2 - s_f}{s_g - s_f} \times 100$$

where, at 10°F

$$s_f = 0.1208 \text{ Btu/lbm} \cdot \text{°R}$$

and

$$s_g = 1.3157 \text{ Btu/lbm} \cdot \text{°R}$$

Since the ammonia is condensed, or is a saturated liquid upon leaving the condenser, the entropy of the ammonia at state (1) is $s_1 = s_f$ at 120°F. From Table B.11 we find

$$s_f = 0.3576 \text{ Btu/lbm} \cdot \text{°R}$$

so

$$s_1 = 0.3576 \text{ Btu/lbm} \cdot \text{°R}$$

Then the entropy at (2) is the same as it is at (1) so that in reversible adiabatic or isentropic process (1)

$$s_1 = s_2 = 0.03576 \qquad \text{(from table B.11)}$$

and the quality then can be calculated

$$\chi_2 = \frac{0.3576 - 0.1208}{1.3157 - 0.1208} \times 100 = \frac{0.2368}{1.1949} \times 100 = 19.8\%$$

The enthalpy at point (2) is then obtained from equation (8–19)

$$h_2 = \chi_2 h_{fg} + h_f$$

and the table B.11 data gives us, at 10°F

$$h_g - h_f = h_{fg} = 561.1 \text{ Btu/lbm}$$

$$h_f = 53.8 \text{ Btu/lbm}$$

from which we calculate h_2

$$h_2 = \frac{19.8}{100} \times 561.1 \text{ Btu/lbm} + 53.8 \text{ Btu/lbm}$$

$$= 164.9 \text{ Btu/lbm}$$

In a like manner, we calculate the enthalpy at point (3) from

$$h_3 = x_3 h_{fg} + h_f$$

where the h_{fg} and h_f values are the same as at point (2) since the temperature is the same. The quality, however, is obtained from

$$x_3 = \left(\frac{s_3 - s_f}{s_g - s_f}\right) \times 100$$

where

$$s_f = 0.1208 \text{ Btu/lbm} \cdot {}^\circ\text{R} \quad \text{(at 10°F)}$$
$$= 1.3157 \quad \text{(at 10°F)}$$

and

$$s_3 = s_4 = 1.1427 \quad \text{(at 120°F)}$$

This last equality $s_3 = s_4$ results from the reversible adiabatic compression process. The quality then is

$$x_3 = \frac{1.1427 - 0.1208}{1.3157 - 0.1208} \times 100$$
$$= 85.5\%$$

and the enthalpy at point (3) is calculated

$$h_3 = \frac{85.5}{100} \times 561.1 + 53.8 = 533.5 \text{ Btu/lbm}$$

We then obtain the heat added from the first law equation:

$$q_{\text{add}} = h_3 - h_2$$

or

$$q_{\text{add}} = 533.5 - 164.9 = 368.6 \text{ Btu/lbm} \qquad \textit{Answer}$$

By using the first law for the full cycle

$$\oint dwk = \oint dq$$

we obtain

$$wk_{\text{cycle}} = q_{\text{add}} + q_{\text{rej}}$$
$$= 368.6 - 455.0 = -86.4 \text{ Btu/lbm} \qquad \textit{Answer}$$

The efficiency or measure of effectiveness of heat pumps is given by the coefficient of performance (COP), defined as

$$\text{COP} = \frac{Q_{\text{rej}}}{Wk_{\text{cycle}}} \qquad \textbf{(8–22)}$$

which is the ratio of the heat produced by the cycle to the amount of work required to drive the device. For refrigerators or other cooling devices, the coefficient of refrigeration (COR) is a more appropriate measure of the cycle effectiveness. This is defined as

$$\text{COR} = \frac{-Q_{\text{add}}}{Wk_{\text{cycle}}} \qquad \textbf{(8–23)}$$

where the term Q_{add} can be associated with the "cooling" or amount of heat withdrawn from a region by the refrigerator. The quantities COP and COR are positive and greater than 1 for all heat pump and refrigerating cycles.

For our problem the COP is easily determined

$$\text{COP} = \frac{-455}{-86.4} = 5.27$$

and the coefficient of refrigeration is

$$\text{COR} = \frac{-368.6}{-86.4} = 4.27$$

We will return to other details concerning the heat pump in chapter 14.

Practice Problems

Problems using the SI units are indicated with the notation (M) under the problem number, and those using English units are designated by (E).

Section 8.2

8.1. **(M)** A Carnot engine operates at 200 rpm with 1 revolution/cycle and with an entropy change of 0.03 kJ/K. Determine the high temperature required to generate 100 kilowatts of power. Assume T_L to be 27°C.

8.2. **(M)** Determine the heat added in the engine of problem 8.1.

8.3. **(E)** Determine the work generated by a Carnot engine operation, between 2000°F and 100°F, if the entropy change is 10 Btu/°R.

8.4. **(E)** Determine the net heat and the heat rejected of the Carnot engine of problem 8.3.

Section 8.3

8.5. **(M)** A Carnot engine operates between 1000°C and 200°C. If the entropy change is 0.01 kJ/kg · K, determine the cycle efficiency.

8.6. **(M)** At 1200°C, 400 kJ/s of heat are added to a Carnot engine. If the surroundings are at a temperature of 20°C, determine the cycle efficiency and the output power.

8.7. **(M)** What maximum efficiency may a heat engine have if it can exchange heat by radiation with the sun and outer space? Assume the temperature of the sun is 10,000°C and outer space is −60°C.

8.8. **(E)** A Carnot heat engine is designed to produce 70 horsepower at 3000 rpm. If the energy source is at 1500°F and sink at 90°F, determine the cycle efficiency and the ratio of heat transfer to the cycle $Q_{\text{rej}}/Q_{\text{add}}$.

8.9. **(E)** Determine the rate of heat rejection of the engine in problem 8.8.

8.10. **(E)** An inventor claims an efficiency of 90% for an engine he has built. He claims that the highest temperature is 130°F. Is his claim possible? Discuss why or why not.

Sections 8.4 and 8.5

8.11. An engine is proposed which extracts energy from sea water by heat transfer and then returns the cooled water to the ocean. No other heat transfer occurs with the engine. Does this violate any of our principles or laws of thermodynamics?

8.12. A device is proposed which is essentially a heat pump driven by a heat engine. A fluid is taken from a low temperature area and by means of the heat pump, is deposited in a high temperature sink. This sink is subsequently used to furnish energy to drive the heat engine which in turn rejects heat and fluid to the low temperature region. Does this device violate the second law? Does this device violate any principle of thermodynamics?

Sections 8.8, 8.9, and 8.10

8.13. (M) A Carnot engine operating with a perfect gas between 800°C and 25°C has an operating pressure range between 0.2 bar and 60 bars. Determine the work of the cycle, heat added, heat rejected, and efficiency, if the gas has physical properties equivalent to those of air.

8.14. (M) A Carnot engine composed of a piston-cylinder uses 0.003 kilogram of air per cycle. The minimum volume of the enclosed cylinder is 50 cm^3 and the maximum is 1000 cm^3, and the temperature range is 700°C and 50°C. Determine the following if the engine is running at 400 cycles/min.

(a) Power output.
(b) Rate of heat addition.
(c) Cycle efficiency.

8.15. (M) A Carnot heat pump, diagrammed in figure 8–17, operates between −15°C and 30°C with ammonia as a working fluid. Assume the ammonia is a saturated vapor at point (4) and a saturated liquid at (1). Determine

(a) Heat added per kilogram of ammonia.
(b) Heat rejected per kilogram.
(c) Net work required per kilogram.
(d) COP and COR.

8.16. (M) For the heat pump of problem 8.15, assume that 20 g/s of ammonia flows through the system. Determine

(a) Power required.
(b) Rate of heat added.
(c) Rate of heat rejected.

8.17. (E) A Carnot heat engine operating at 3000 rpm with a perfect gas having the properties listed below has a heat source at 1500°F and a sink at 85°F. Determine

(a) Cycle efficiency.
(b) Work of the cycle.
(c) Heat added and heat rejected.

The gas properties known are

$$R = 48.3 \text{ ft-lbf/lbm} \cdot {}^\circ\text{R}$$
$$c_p = 0.22 \text{ Btu/lbm} \cdot {}^\circ\text{R}$$
$$c_v = 0.157 \text{ Btu/lbm} \cdot {}^\circ\text{R}$$
$$k = 1.396$$
$$p_1 = 800 \text{ psia}$$
$$p_2 = 70 \text{ psia}$$

8.18. (E) A Carnot heat pump uses nitrogen gas (which behaves like a perfect gas), and operates between 376°F and −15°F. The gas is at 15 psia when entering the compression and 270 psia when leaving. Determine the following for this cyclic device:

(a) Work of the compression per unit mass.
(b) COP.
(c) COR.

8.19. (E) A Carnot heat pump uses refrigerant R-22 as a working fluid. The cycle operates between $-10°F$ and 70°F, and the flow rate of refrigerant is 60 lbm/min. The device is similar to that shown in figure 8–17, but the compressor does not utilize the work developed by the expansion of the refrigerant from state (1) to (2). If the refrigerant is saturated liquid at (1) and saturated vapor at (4), determine

(a) Power required.
(b) Rate of heat addition from the low temperature source.
(c) Rate of heat rejection to the high temperature sink.
(d) COP and COR.

9

Availability and Useful Work

Some concepts are presented here which have general value in determining the limitations imposed on thermodynamic systems as they execute processes. From the first law of thermodynamics, a general equation is developed for useful work. We see that useful work is precisely what the name implies — work that we can directly use. From this concept we progress to a definition of *availability* and a discussion of its practical meaning. Emphasis is placed on evaluating the change in availability, which is equal to *useful work*, but theory is not ignored entirely. Example problems have been included which should aid in clarifying the *energy-entropy-work relationships*, including the ways in which energy progresses from an available to an unavailable condition. Available and unavailable energy are introduced as special cases of our availability function, and the free energy functions are given as alternate statements resulting from availability.

9.1 Useful Work

The simple, homogeneous system can increase in volume and thereby provide work to be used external to the system. For a specific system, such as the piston-cylinder depicted in figure 9–1, we found the work was described by the relationship

$$Wk_{cs} = \int pdV \tag{9–1}$$

if the process was reversible. This equation then takes on various forms (see tables 6–2 and 6–3), depending on the type of process. The piston-cylinder is a detailed configuration of a general system, as shown in figure 9–2, which is capable of changing volume. However, almost all of the systems we can visualize will ultimately be used in an atmosphere

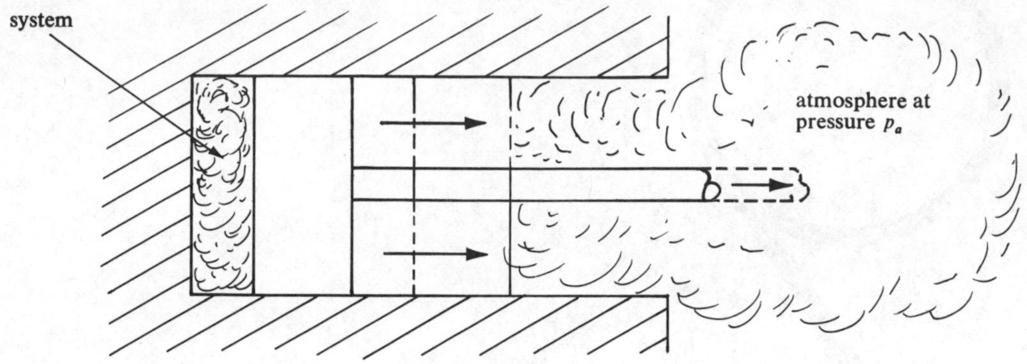

Figure 9–1 The piston-cylinder increasing the system volume.

(the Earth, Mars, or somewhere else, rather than in a vacuum), and as such the system must displace some of the atmosphere, namely the amount by which the system itself changes volume. This requires work, which detracts from the $\int pdV$ term, and consequently we say that useful work, Wk_{use}, is obtained from the equation

$$Wk_{use} = \int_{V_1}^{V_2} pdV - p_a(V_2 - V_1) \qquad \textbf{(9–2)}$$

for a reversible process. The term p_a is the atmospheric or surrounding pressure.

We can also recall the first law for the closed system

$$E_2 - E_1 = Q - \int pdV \qquad \textbf{(9–3)}$$

and substitute the result from equation (9–2), and get

$$E_2 - E_1 = Q - Wk_{use} - p_a(V_2 - V_1) \qquad \textbf{(9–4)}$$

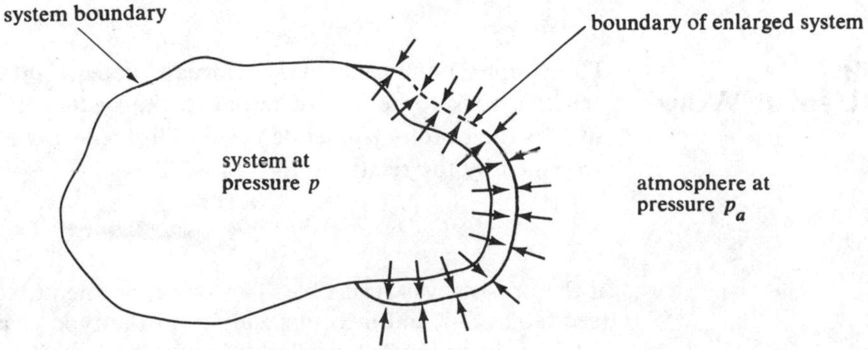

Figure 9–2 System increasing in volume.

Since we are here considering ideal or reversible processes and assuming that heat transfers are conducted with the atmosphere at isothermal processes, we can then write

$$Q = T_a(S_2 - S_1) \tag{9–5}$$

when T_a is the atmospheric temperature. If we substitute this into equation (9–4) and rearrange slightly we have

$$Wk_{\text{use}} = T_a(S_2 - S_1) - p_a(V_2 - V_1) - (E_2 - E_1) \tag{9–6}$$

We can easily include irreversible or real processes by using an inequality sign

$$Wk_{\text{use}} \leq T_a(S_2 - S_1) - p_a(V_2 - V_1) - (E_2 - E_1) \tag{9–7}$$

Of course, we identify the "less than" with an irreversible process and the "equality" with an ideal or a reversible one.

Various information can be drawn from either equation (9–6) or (9–7). For instance, from equation (9–6) we see that a decrease in a system's energy does not, of itself, assure a supply of useful work. The initial energy E_1 might very well be greater than the final energy E_2 but the terms $T_a(S_2 - S_1)$ may be negative and/or $p_a(V_2 - V_1)$ may be large; this could cancel any energy decreases.

Also, it is quite possible to get work out of a system during a single process and not have a decrease in energy—or it is possible to have a heat transfer, as the following example shows.

Example 9.1 A piston-cylinder encloses a perfect vacuum, shown in figure 9–3. The pin holds the piston in the extended position. If the piston diameter is 15 centimeters, determine the useful work obtained from this device when the pin is removed.

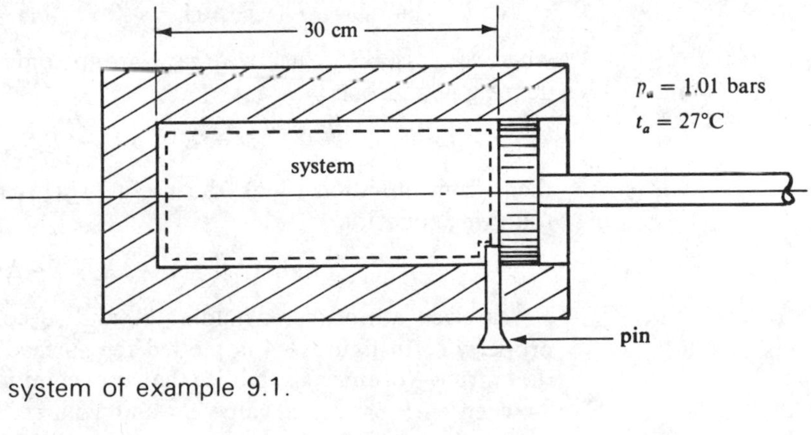

Figure 9–3 Vacuum system of example 9.1.

Solution

We identify the system as the enclosed vacuum and observe that the system has no entropy or energy values before, during, or after the process. The useful work we obtain from

$$Wk_{use} = T_a(S_2 - S_1) - p_a(V_2 - V_1) - (U_2 - U_1) \qquad \textbf{(9–6)}$$

and since

$$U_2 = 0 \qquad U_1 = 0 \qquad S_2 = 0 \qquad S_1 = 0$$

We have

$$Wk_{use} = p_a(V_2 - V_1)$$

The initial volume V_1 is calculated

$$V_1 = (\pi)(30 \text{ cm})(7.5 \text{ cm})^2 = 5300 \text{ cm}^3$$
$$= 5.3 \times 10^{-3} \text{ m}^3$$

The final volume is zero since the atmosphere will push the piston into the cylinder completely. Then

$$Wk_{use} = (1.01 \times 10^5 \text{ N/m}^2)(5.3 \times 10^{-3} \text{ m}^3)$$
$$= 535 \text{ N} \cdot \text{m} = 535 \text{ J} \qquad \textit{Answer}$$

and we see that a significant amount of work can be derived from an absolute vacuum.

9.2 Availability

We have seen that the useful work obtained from a closed system is given by equation (9–6)

$$Wk_{use} = T_a(S_2 - S_1) - p_a(V_2 - V_1) - (E_2 - E_1)$$

or, per unit mass

$$wk_{use} = T_a(s_2 - s_1) - p_a(v_2 - v_1) - (e_2 - e_1) \qquad \textbf{(9–8)}$$

where wk_{use} is Wk_{use}/m. If we now group some properties together and define a new property Λ as

$$\Lambda = E + p_aV - T_aS \qquad \textbf{(9–9)}$$

then, from equation (9–6), the useful work can also be given by the following equation:

$$Wk_{use} = \Lambda_1 - \Lambda_2 = -\Delta\Lambda \qquad \textbf{(9–10)}$$

For a given atmosphere, holding T_a and p_a constant, we can plot the property Λ. In figure 9–4 is plotted the surface of the Λ-function along the entropy-volume axis. Notice that at one unique point, Λ is a minimum — when entropy has a value S_m and volume, a value V_m. This point,

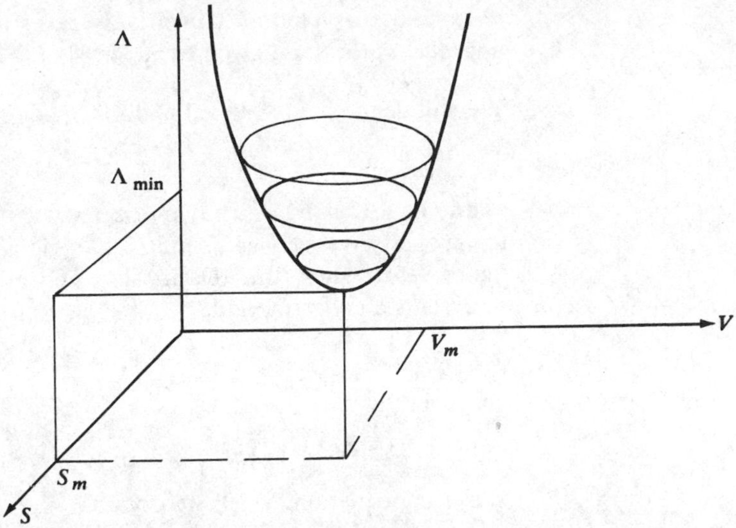

Figure 9–4 Graph of Λ-function in Λ-*V*-*S* space.

Λ_{min}, is a "lowest point" and intuitively we see that a system will gravitate toward the minimum point if it is at some other state. In figure 9–5 is shown the Λ-function for entropy having a constant value S_m. It is indicated that during a process where no work is put into the given system (natural process) and where the process is merely allowed to proceed in

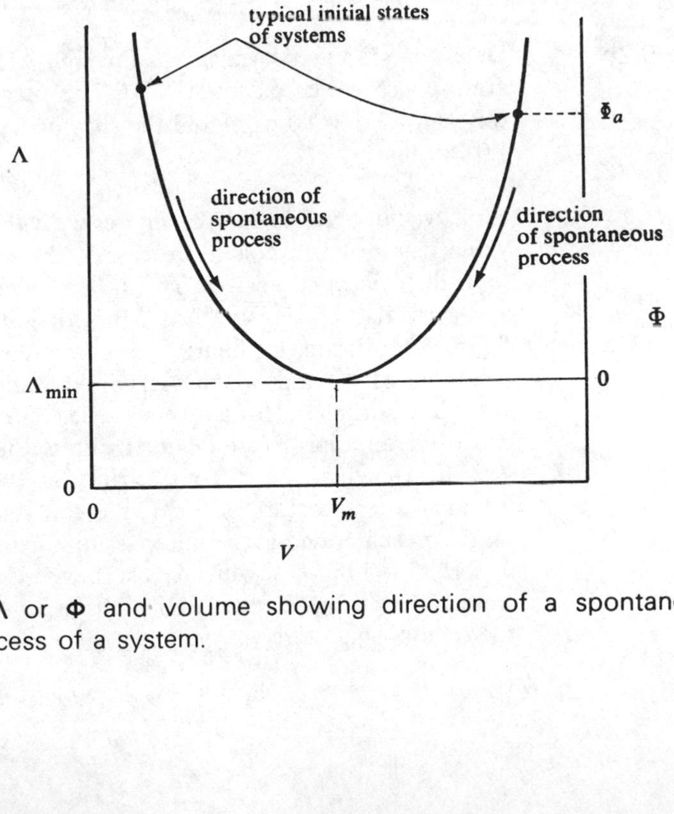

Figure 9–5 Graph of Λ or Φ and volume showing direction of a spontaneous or natural process of a system.

its course, the value of Λ invariably takes a lower value, or the system approaches the state given by Λ_{min}.

We will define the total availability, Φ, as

$$\Phi = \Lambda - \Lambda_{min} \tag{9–11}$$

and remark that Φ has a value of zero when it is at the state specified by the property values S_m, V_m, and E_m. Availability is plotted in the graph of figure 9–5 to show this relationship and it can be seen that since Λ_{min} represents a constant value

$$\Phi_2 - \Phi_1 = \Lambda_2 - \Lambda_1$$

or, generally

$$\Delta\Phi = \Delta\Lambda = (E_2 - E_1) + p_a(V_2 - V_1) - T_a(S_2 - S_1) \tag{9–12}$$

and using equation (9–10) we obtain

$$Wk_{use} = -\Delta\Phi \tag{9–13}$$

Again, we can include all processes, reversible and irreversible, by writing

$$Wk_{use} \leq -\Delta\Phi \tag{9–14}$$

The availability then gives us an upper limit or greatest expected output of a particular system. Let us look at a problem solved by the use of the availability concept.

Example 9.2

Determine the vertical distance through which ice can raise itself in an atmosphere at 14.7 psia and 70°F. The energy required to change ice to water at 32°F is 80 cal/g and the specific heat of water can be taken as 1.0 Btu/lbm · °R.

Solution

First, we must realize that the answer we will get is one that is a maximum value; for real irreversible processes, the vertical distance will be less than the calculated answer. The system is, obviously, the ice (or water) and may also be visualized as cubes of ice — 1-lbm cubes as shown in figure 9–6. Also in this figure we see how the ice is going to lift itself up; it acts like a low temperature reservoir (but ultimately reaches equilibrium with the atmosphere) for a reversible heat engine and the heat engine then can drive an appropriate device to hoist the ice, melting ice, or water. This heat engine can, of course, run only until the ice has melted and attained a temperature of 70°F. We can easily identify the work to be used for each pound-mass of ice, assuming a gravitational acceleration of 32.2 ft/s^2 as 1 lbf $\times$ y, which is just the increase in potential energy of the water. The change in availability of the ice (or water) is obtained from the relationship

$$\Delta\Phi = \Delta\Lambda = \Lambda_a - \Lambda_1$$

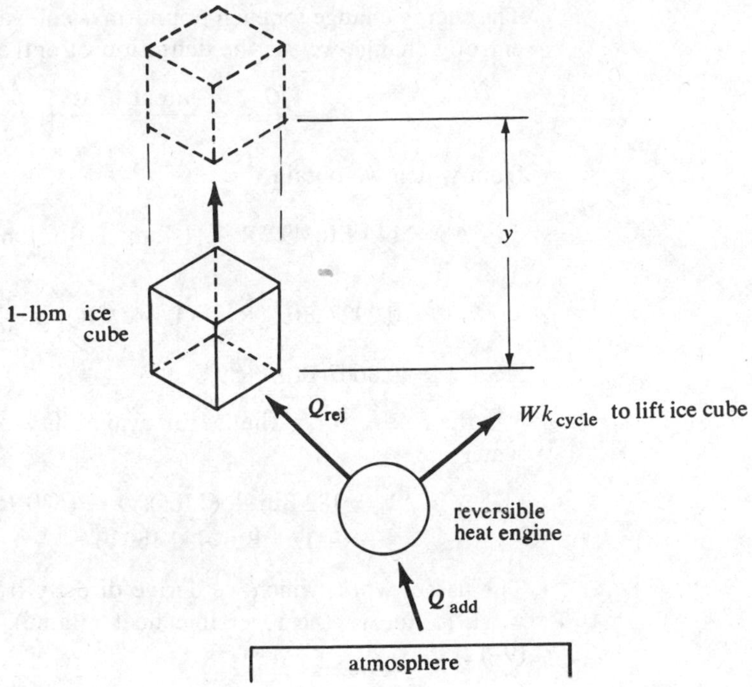

Figure 9–6 The availability of an ice cube.

and then

$$\Delta\Phi = E_a + p_aV_a - T_aS_a - E_1 - p_aV_1 + T_aS_1$$

The energy of the water is in the form of internal energy and potential energy (since it is being hoisted y feet up). We then write

$$\Delta\Phi = U_a - U_1 + W(y) + p_a(V_a - V_1) - T_a(S_a - S_1)$$

Since no appreciable volume change is exhibited by the water in going from ice to 70°F water, we can ignore the $p_a(V_a - V_1)$ term.

As we discussed in problem 8.6, most materials require a proportionately large amount of energy to change phase and this change occurs at constant temperature. This generally is true for the liquid-solid phase change as well as for the vapor-liquid. The energy required to convert a solid to a liquid is frequently called the *latent heat of liquefaction* or just *latent heat*. We know it to be 80 cal/g for water at 32°F, so our energy change can be found from the following:

$$\begin{aligned} U_a - U_1 &= (\text{latent heat}) + mc(T_a - T_1) \\ &= 80 \text{ cal/g} + (1 \text{ lbm})(1 \text{ Btu/lbm} \cdot {}^\circ\text{R})(70^\circ\text{F} - 32^\circ\text{F}) \\ &= 80 \text{ cal/g} \times \frac{454 \text{ g/lbm}}{252 \text{ cal/Btu}} + 38 \text{ Btu} = 144 \text{ Btu} + 38 \text{ Btu} \\ &= 182 \text{ Btu} \end{aligned}$$

The energy change for each pound-mass of water is then 182 Btu. For the entropy change we use the definition of entropy

$$\Delta S = \frac{Q_{\text{rev}}}{T} = \frac{\text{latent heat}}{T_1} + \int_{T_1=32^\circ\text{F}}^{T_2=70^\circ\text{F}} \frac{mcdT}{T}$$

from which we obtain

$$\Delta S = 144 \text{ Btu}/492^\circ\text{R} + (1 \text{ lbm})(1 \text{ Btu/lbm} \cdot {}^\circ\text{R}) \int_{T_1}^{T_2} \frac{dT}{T}$$

$$= 0.2927 \text{ Btu}/^\circ\text{R} + (1 \text{ Btu}/^\circ\text{R})\left(\ln \frac{530}{492}\right) = 0.2927 + 0.074$$

$$= 0.3667 \text{ Btu}/^\circ\text{R}$$

We then have for a change in availability of each pound-mass of the water

$$\Delta\Phi = 182 \text{ Btu} + (1 \text{ lbf})(y) - (530^\circ\text{R})(0.3667 \text{ Btu}/^\circ\text{R})$$
$$= -12.4 \text{ Btu} + 1 \text{ lbf}(y)$$

The useful work, which we derive directly from the ice, is zero (all our work is due to the reversible heat engine), so we get from equation (9–12) that

$$Wk_{\text{use}} = 0 = -\Delta\Phi$$

and

$$\Delta\Phi = -12.4 \text{ Btu} + (1 \text{ lbf})(y)$$

From this we obtain

$$y = 12.4 \text{ Btu}/1 \text{ lbf} = 12.4 \times 778 = 9609 \text{ ft} \qquad \textit{Answer}$$

Notice in this problem that the actual lifting of the water was contingent on a number of mechanical apparatus: a reversible heat engine; a transmission to physically hoist the water; and some type of reversible heat transfer pipe from the heat engine to both atmosphere and water.

We will now concern ourselves with some remarks of the degradation of energy to an unavailable state.

9.3 Energy Degradation

While we have seen that it is not energy but availability which is desirable, we will use the term *energy degradation* to imply a degrading of availability or a moving down the curve in figure 9–5. We lose availability in a real process and in a reversible process it merely remains the same. Loss of availability is subtle and difficult to detect, but predicting irreversibilities, degradation of energy, or increases in entropy in a general process is still more difficult. One of the important areas of research in thermodynamics is the development of more general and accurate equations or theorems which can allow these predictions of irreversibility — it is a crucial area since most of our present tools of thermodynamics can only be used accurately for reversible and static systems which are indeed not the real world.

Example 9.3

We seek a comparison of the maximum useful work obtainable from two systems which begin at identical states, proceed through different processes, and reach states which are not too different.

Solution

System (1), shown in figure 9–7, is an adiabatic chamber enclosing air in the left half. The right half is a vacuum separated from the left by a removable wall. The air pressure is 3.0 bars, the temperature is 40°C, and the volume is 0.5 m^3. The volume of the vacuum is also 0.5 m^3, so when the removable wall is taken away the air will occupy the full 1.0 m^3 volume. When this happens the air pressure will drop to 1.5 bars but the temperature will remain at 40°C. Why?

System (2) in figure 9–7 is a frictionless adiabatic piston-cylinder device which initially encloses 0.5 m^3 of air at 3.0 bars pressure and is at 40°C. The system then progresses through a reversible adiabatic process until the air fills a volume of 1.0 m^3.

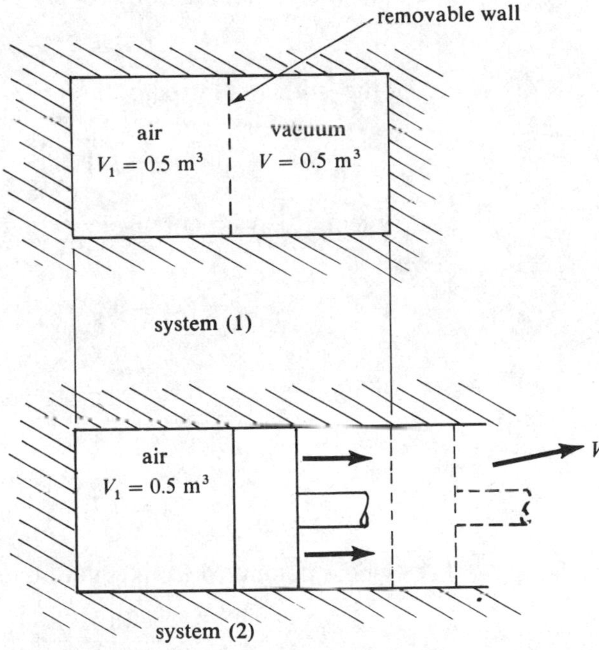

Figure 9–7 Two systems which are going through adiabatic processes with identical volumes but which will reach different states and have different amounts of work produced.

Both systems are at final states that are slightly different; from system (2) we have extracted some useful work while system (1) has its initial energy value. Heat has not been transferred in either system but we can ask, "Have we lost any availability in system (1) by not utilizing the expansion of the air from 0.5 m³ to 1.0 m³?" The answer, as we will subsequently see is "yes."

First let us look at system (1). We assume the air to be a perfect gas so we have the following results:

$$\Delta s = s_2 - s_1 = R \ln \frac{V_2}{V_1} \tag{9–15}$$

$$= (287 \text{ J/kg} \cdot \text{K})\left(\ln \frac{1 \text{ m}^3}{0.5 \text{ m}^3}\right)$$

$$= 199 \text{ J/kg} \cdot \text{K}$$

We have that the work and heat are both zero, so

$$wk = 0 = q \tag{9–16}$$

and from the first law we must then have

$$\Delta u = u_2 - u_1 = 0 \tag{9–17}$$

and since the air acts as a perfect gas

$$\Delta T = T_2 - T_1 = 0 \tag{9–18}$$

so, for the isothermal expansion

$$p_2 = p_1\left(\frac{V_1}{V_2}\right) = 1.5 \text{ bars}$$

For system (2), with air again acting in the system, we have these results:

$$\Delta S = 0$$

$$q = 0$$

$$wk = \frac{R}{1-k}(T_2 - T_1)$$

$$T_2 = T_1\left(\frac{V_1}{V_2}\right)^{k-1} - 237 \text{ K} \tag{9–19}$$

The work is computed from equation (9–19) to give us

$$wk = \frac{287 \text{ J/kg} \cdot \text{K}}{1 - 1.4}(237 \text{ K} - 313 \text{ K})$$

$$= 54{,}500 \text{ J/kg} = 54.5 \text{ kJ/kg}$$

Also we get

$$\Delta u = -wk = -54.5 \text{ kJ/kg}$$

and

$$p_2 = p_1\left(\frac{V_1}{V_2}\right)^k = 1.14 \text{ bars}$$

Obviously, the availability was the same for both systems at their initial states. To determine the availability of the systems at their final states, we need the change in availability. This change we find from equation (9–12)

$$\Delta\Phi = \Phi_2 - \Phi_1 = U_2 - U_1 + p_a(V_2 - V_1) - T_a(S_2 - S_1)$$

For system (1) the change is obtained

$$\Delta\Phi = 0 + p_a(V_2 - V_1) - T_a m(\Delta s)$$

We now assume the surroundings are at a temperature of 27°C and a pressure of 1.01 bars. The system mass is computed from the perfect gas relation:

$$m = \frac{p_1 V_1}{RT_1} = \frac{(3.0 \times 10^5 \text{ N/m}^2)(0.5 \text{ m}^3)}{(287 \text{ J/kg}\cdot\text{K})(313 \text{ K})}$$

$$= 1.67 \text{ kg}$$

Thus, the change in availability is, for system (1)

$$\Delta\Phi = (1.01 \times 10^5 \text{ N/m}^2)(0.5 \text{ m}^3) - (300 \text{ K})(1.67 \text{ kg})(199 \text{ J/kg}\cdot\text{K})$$

$$= -49.1 \text{ kJ}$$

For system (2) the change in availability is

$$\Delta\Phi = m(u_2 - u_1) + p_a(V_2 - V_1) - T_a m(\Delta s)$$

$$= (1.67 \text{ kg})(-54.5 \text{ kJ/kg}) + (1.01 \times 10^5 \text{ N/m}^2)(0.5 \text{ m}^3) - 0$$

$$= -91.0 \text{ kJ} + 50.5 \text{ kJ} = -40.5 \text{ kJ}$$

Notice that since system (2) involved a reversible process, the change in availability is exactly equal in magnitude to the useful work gotten from the piston-cylinder. That is, from equation (9–6) the useful work is just +40.6 kJ. System (1), by going through an irreversible expansion, decreased its availability more than system (2), even though its energy remained constant. The apparent contradiction which gives system (2) a higher final availability than system (1) is due to the fact that the final temperature of system (2) is 237 K, or −36°C, well below the temperature of the surroundings, therefore providing a capability for heat transfer to those surroundings.

One of the practical conclusions to be drawn from this problem is that a gas in a cylinder or any other source of availability must be used in a work process during the actual expansion or else the availability is lost forever. It is exactly like water flowing over a drop; if the potential energy of the water is not immediately converted into some other energy, such as electric power, then the water will flow down and dissipate (quite irreversibly) the kinetic energy and degrade the availability of the energy of that portion of water. Thus, engineers and technologists must be opportunistic in generating power for society by extracting energy from processes at the appropriate times and by better utilizing the naturally occurring processes of nature (such as tides, tornadoes, hurricanes, and winds) for energy sources.

One other task of those who furnish power or availability to society is the storage of such products. In mechanics, a flywheel rotates at high speed, storing kinetic energy for later use. Electric energy is stored in batteries or capacitors, while thermal or internal energy is stored in material which is retained in insulated or adiabatic chambers. When the energy is put into storage, we lose some availability in the mere process of storing. The term *unavailable energy* has traditionally been associated with the amount of energy which is lost during a process and which can never again be put to useful work. *Available energy*, on the other hand, has been associated with that energy which can be reconverted to useful work; it is efficiently stored for future use. These two terms are tied up in the availability concept, but an example problem might provide some enlightenment.

Example 9.4

A heat exchanger is a device which transfers thermal energy from one material to another. In figure 9–8 is shown such a heat exchanger which allows hot exhaust gases to transfer some of their energy to air flowing in the opposite direction. Outside the heat transfer $\dot{Q}$ between the two streams, the system can be assumed to be adiabatic. Assume 6.1 kg/s of hot gases enter at 500°C and leave at 200°C, all at a pressure of 1.01 bars. There are 6.0 kg/s of air flowing into the exchanger at 120°C and 1.01 bars. The specific heats at constant pressure are 0.7 kJ/kg · K for the gases and 1.007 kJ/kg · K for the air; both are assumed to be perfect

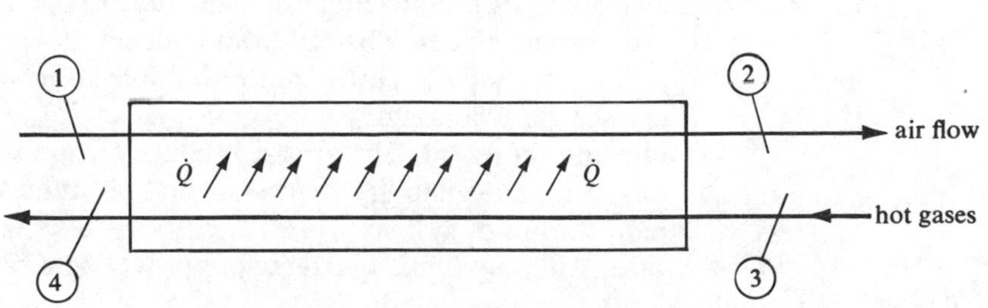

Figure 9–8 Heat exchanger.

gases having R values of 189 J/kg · K and 287 J/kg · K respectively. If the atmospheric conditions are 27°C and 1.01 bars, determine the loss in availability per hour of the system due to the heat exchange.

Solution We recognize that here we have an open system and that our concepts of useful work and availability were adapted to closed systems. It can be shown, however, that for open, steady flow systems, the rate of change in availability, $d\Phi/d\tau$, is obtained from:

$$\dot{\Phi} = \dot{m}\left((h_2 - h_1) + \frac{V_2^2 - V_1^2}{2g_c} + \frac{g}{g_c}(z_2 - z_1) - T_a(s_2 - s_1)\right) \quad \textbf{(9–20)}$$

The total rate of change in availability for this system is given by

$$\dot{\Phi} = \dot{\Phi}_{\text{air}} + \dot{\Phi}_{\text{gas}} \quad \textbf{(9–21)}$$

In figure 9–9 is shown the T-s diagram for the air and gas simultaneously. The total areas under each curve are equal and represent the heat transferred between the gas and air, $\dot{Q}$. That is, the total area under curve (3–4) is equal to the total area under (1–2). Notice also that the areas under the T_a abscissa line are labeled as "unavailable energy." This is a good way of visualizing equation (9–20) or (9–12), but in this graphical representation, the $p_a(V_2 - V_1)$ term is neglected. We see also that the "available energy" areas of figure 9–9 are equal to the availability per unit mass for this steady flow problem. The generation of unavailable energy then is given by the expression

$$\dot{m}(T_a\Delta s) \quad \textbf{(9–22)}$$

For the gases we determine the following properties:

$$h_4 - h_3 = c_p(T_4 - T_3) = (0.7 \text{ kJ/kg} \cdot \text{K})(473 \text{ K} - 773 \text{ K})$$
$$= -210 \text{ kJ/kg}$$

$$s_4 - s_3 = c_p \ln \frac{T_4}{T_3} = (0.7 \text{ kJ/kg} \cdot \text{K})\left(\ln \frac{473 \text{ K}}{773 \text{ K}}\right)$$
$$= -0.344 \text{ kJ/kg} \cdot \text{K}$$

From these results and using equation (9–20), we can calculate the rate of change of availability for the gas, assuming no kinetic or potential energy changes

$$\dot{\Phi}_{\text{gas}} = \dot{m}_{\text{gas}}[(h_4 - h_3) - T_a(s_4 - s_3)]$$
$$= (6.1 \text{ kg/s})[-210 \text{ kJ/kg} + (300 \text{ K})(0.344 \text{ kJ/kg} \cdot \text{K})]$$
$$= -651 \text{ kJ/s}$$

The available energy per unit mass of the gas in the temperature-entropy diagram of figure 9–9 is equal to $h_4 - h_3 - T_a(s_4 - s_3)$, or -106.8 kJ/kg, and the unavailable energy is given by the term $T_a(s_4 - s_3)$ and equals -103.2 kJ/kg.

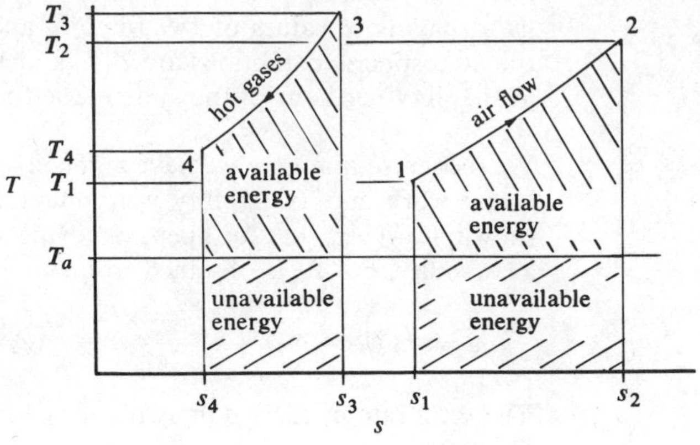

Figure 9–9 *T-s* diagram for heat exchanger.

For the air we have

$$h_2 - h_1 = c_p(T_2 - T_1) = (1.007 \text{ kJ/kg}\cdot\text{K})(T_2 - 393 \text{ K})$$

and the entropy change is given by

$$s_2 - s_1 = c_p \ln \frac{T_2}{T_1}$$

To determine the final air temperature, we notice from the first law of thermodynamics applied to the heat exchanger that the rate of increase in enthalpy of the air is exactly the same magnitude as the decrease of the gas. Therefore

$$\dot{m}_{\text{air}}(h_2 - h_1) = \dot{m}_{\text{gas}}(h_3 - h_4)$$

$$(6.0 \text{ kg/s})(h_2 - h_1) = (6.1 \text{ kg/s})(210 \text{ kJ/kg})$$

and using the relationship

$$h_2 - h_1 = c_p(T_2 - T_1)$$

we obtain

$$T_2 = T_1 + \frac{(6.1 \text{ kg/s})(210 \text{ kJ/kg})}{(6.0 \text{ kg/s})(1.007 \text{ kJ/kg}\cdot\text{K})}$$

Since T_1 is 393 K, we obtain

$$T_2 = 393 \text{ K} + 212 \text{ K} = 605 \text{ K}$$

Then the entropy change is computed

$$s_2 - s_1 = (1.007 \text{ kJ/kg}\cdot\text{K})\left(\ln \frac{605 \text{ K}}{393 \text{ K}}\right) = 0.434 \text{ kJ/kg}\cdot\text{K}$$

Further, the enthalpy change is

$$h_2 - h_1 = (1.007 \text{ kJ/kg}\cdot\text{K})(605 \text{ K} - 393 \text{ K})$$
$$= 213 \text{ kJ/kg}$$

The rate of change of availability, using equation (9–20), is

$$\dot{\Phi} = \dot{m}_{air}[(h_2 - h_1) - T_a(s_2 - s_1)]$$
$$= (6.0 \text{ kg/s})[213 \text{ kJ/kg} - (300 \text{ K})(0.434 \text{ kJ/kg} \cdot \text{K})]$$
$$= 497 \text{ kJ/s}$$

The total rate of change of availability for the heat exchanger is, from equation (9–21)

$$497 \text{ kJ/s} - 651 \text{ kJ/s} = -154 \text{ kJ/s}$$ *Answer*

The available energy of the air per kilogram is given by

$$\frac{\dot{m}_{air}}{\dot{m}_{gas}}(h_2 - h_1) - T_a(s_2 - s_1)$$

or, substituting values, we obtain

$$\frac{6.0 \text{ kg/s}}{6.1 \text{ kg/s}}[(213 \text{ kJ/kg}) - (300 \text{ K})(0.434 \text{ kJ/kg} \cdot \text{K})] = 81.4 \text{ kJ/s}$$

The unavailable energy change per kilogram of gas is

$$\frac{\dot{m}_{air}}{\dot{m}_{gas}}(T_a)(s_2 - s_1) = \frac{6.0 \text{ kg/s}}{6.1 \text{ kg/s}}(300 \text{ K})(0.434 \text{ kJ/kg} \cdot \text{K})$$
$$= 128.1 \text{ kJ/kg}$$

The total available energy change per kilogram of gas is obtained from the difference of the two designated areas under the curves in figure 9–9 or the sum of the available energy changes of the gas and of the air. Thus, it is

$$-106.8 \text{ kJ/kg} + 81.4 \text{ kJ/kg} = -25.4 \text{ kJ/kg}$$

Similarly, the unavailable energy change per kilogram of gas is equal to the sum of the unavailable energy changes for the gas and air, or

$$-103.2 \text{ kJ/kg} + 128.1 \text{ kJ/kg} = 24.9 \text{ kJ/kg}$$

The decrease in available energy here corresponds to the decrease in availability or increase in unavailable energy. The total generation of unavailable energy is given by

$$\dot{m}_{air} T_a \Delta s_{air} + \dot{m}_{gas} T_a \Delta s_{gas} = (6.0 \text{ kg/s})(300 \text{ K})(0.434 \text{ kJ/kg} \cdot \text{K})$$
$$- (6.1 \text{ kg/s})(300 \text{ K})(0.344 \text{ kJ/kg} \cdot \text{K})$$
$$= 152 \text{ kJ/s}$$ *Answer*

9.4 Free Energy

For a system which is in thermal and pressure equilibrium with its surroundings, that is, $T = T_a$ and $p = p_a$, we can write the change in availability in the form

$$\Delta\Phi = (E_2 - E_1) + p(V_2 - V_1) - T(S_2 - S_1) \qquad \textbf{(9–23)}$$

If we have no kinetic and potential energy changes we have

$$E_2 - E_1 = U_2 - U_1$$

and then

$$\Delta\Phi = (U_2 - U_1) + p(V_2 - V_1) - T(S_2 - S_1) \qquad \textbf{(9–24)}$$

The property identified with the quantity $U + pV - TS$ we call the *Gibbs free energy* G′ and write

$$\mathrm{G}' = U + pV - TS \qquad \textbf{(9–25)}$$

The Gibbs free energy has units of energy and the specific Gibbs free energy g′ defined as

$$\mathrm{g}' = \frac{\mathrm{G}'}{m} = u + pv - Ts \qquad \textbf{(9–26)}$$

has units of energy per unit mass. These properties are quite useful in analyzing chemical reactions, in predicting heats of combustion tabulated in table B.7, and, of course, in predicting the useful work obtainable from a system whose pressure and temperature are equal to constant atmospheric conditions. If this equilibrium holds for a system then

$$\Delta\Phi = \Delta\mathrm{G}' \qquad \textbf{(9–27)}$$

and we can directly say

$$Wk_{\mathrm{use}} = -\Delta\mathrm{G}' \qquad \textbf{(9–28)}$$

For systems which are at constant volume we predict the maximum useful work from

$$Wk_{\mathrm{use}} = (E_2 - E_1) - T_a(S_2 - S_1) \qquad \textbf{(9–29)}$$

and, again, if the system is in thermal equilibrium with the atmosphere at constant temperature ($T = T_a$)

$$Wk_{\mathrm{use}} = (E_2 - E_1) - T(S_2 - S_1) \qquad \textbf{(9–30)}$$

We associate with the quantity $U - TS$ a property of the system called the *Helmholtz free energy* H′ written

$$\mathrm{H}' = U - TS \qquad \textbf{(9–31)}$$

and, obviously, for the above conditions of the system with no kinetic or potential energy changes

$$Wk_{\mathrm{use}} = -\Delta\mathrm{H}' \qquad \textbf{(9–32)}$$

We can consider also the specific Helmholtz free energy defined by

$$\mathrm{h}' = \frac{\mathrm{H}'}{m} = u - Ts \qquad \textbf{(9–33)}$$

and thereby get an intensive property of the system.

The Gibbs and Helmholtz free energy properties are frequently used in thermodynamic literature, and this introduction to their derivation should be helpful to the engineer and technologist seeking the full use of thermodynamic tools.

Example 9.5

Compute the specific Gibbs free energy and the specific Helmholtz free energy functions for saturated mercury vapor at 30 psia.

Solution

The specific Gibbs free energy function, g′, can be computed from equation (9–26). First we obtain the properties of saturated mercury vapor from table B.10. At 30 psia we find

$$h_g = 147.2 \text{ Btu/lbm}$$

$$v_g = 2.053 \text{ ft}^3\text{/lbm}$$

$$t = 750.9°\text{F} = 1210.9°\text{R}$$

$$s_g = 0.1331 \text{ Btu/lbm} \cdot °\text{R}$$

so we then have

$$g' = h_g - Ts_g = 147.2 \text{ Btu/lbm} - (1210.9°\text{R})(0.1331 \text{ Btu/lbm} \cdot °\text{R})$$
$$= -13.97 \text{ Btu/lbm}$$ *Answer*

The specific Helmholtz free energy function can be computed from equation (9–33):

$$h' = u_g - Ts_g = g' - pv_g$$
$$= -13.97 \text{ Btu/lbm} - \frac{(30 \text{ psia})(2.053 \text{ ft}^3\text{/lbm})(144 \text{ in}^2\text{/ft}^2)}{(778 \text{ ft-lbf/Btu})}$$
$$= -25.37 \text{ Btu/lbm}$$ *Answer*

Practice Problems

Problems which use the SI units are indicated with the notation (M) under the problem number, and those using the English units are indicated with (E).

Assume the atmospheric conditions are 27°C (80°F) and 1.01 bars (14.7 psia) for the following problems.

Sections 9.1 and 9.2

9.1. **(M)** Determine the Λ-function of 2 kilograms of H_2O at 240°C and
(a) 4.0 bars pressure.
(b) Saturated vapor.
(c) Saturated liquid.

9.2. **(M)** Determine the availability of a vacuum of 2 m³.

9.3. **(M)** Determine the availability of the following gases at 500°C and 1.01 bars, assuming each to be a perfect gas:
(a) Oxygen, O_2.
(b) Methane, CH_4.

(c) Carbon dioxide, CO_2.
(d) Carbon monoxide, CO.

9.4. **(M)** Determine the vertical distance through which 3 kilograms of ice can lift itself at atmosphic conditions.

9.5. **(E)** Determine the Λ-function of 6 lbm of mercury vapor at 90 psia and
(a) Saturated vapor.
(b) Saturated liquid.
Assume $v_f = 0.0013$ ft^3/lbm.

9.6. **(E)** Determine the availability of 3 lbm of refrigerant R-22, at −20°F and
(a) Saturated vapor.
(b) Saturated liquid.

9.7. **(E)** Determine the maximum useful work obtainable from 1 lbm of the following perfect gases at 1000°F and 200 psia:
(a) Air.
(b) Nitrogen, N_2.
(c) Hydrogen, H_2.
(d) Sulfur dioxide, SO_2.

***9.8.** **(E)** Calculate the maximum useful work obtainable for
(a) 1 lbm H_2O at 700°F and 200 psia.
(b) 1 lbm H_2O at 1000°F and 200 psia.
(c) 1 lbm H_2O at 700°F and 300 psia.

Section 9.3

9.9. **(M)** Calculate the generation or rate of change of availability for 200 kg/min of air flowing through a reversible adiabatic turbine from 650°C and 15.0 bars pressure, exhausting to 1.01 bars. (Hint: Assume no kinetic or potential energy changes and that the power generated in the turbine is all useful.)

9.10. **(M)** There is 0.2 kilogram of air contained in a frictionless piston-cylinder. The air is at a pressure of 20.0 bars and a temperature of 600°C. Through a reversible adiabatic process the air pressure is reduced to 1.0 bar. Determine the change in availability of the air and the total change in availability of the universe.

9.11. **(E)** A piston-cylinder contains 0.5 lbm of oxygen gas at 30 psia and 100°F. The device then compresses the gas in a reversible manner to a pressure of 300 psia. Assume $n = 1.45$ and the gas behaves as a perfect gas with constant specific heats. Determine the increase in availability of the oxygen and the total change in availability of the universe due to this compression process.

9.12. **(E)** Calculate the total rate of change of availability for 20 lb/s of air compressed in a reversible adiabatic manner from 100°F and 15 psia to 180 psia. (Hint: Assume no kinetic or potential energy changes. Obviously, the work put into the compressor is useful.)

***9.13.** In the Φ-*S*-*V* space determine expressions for the tangent or slope of the Φ-surface in the plane of
(a) Constant entropy.
(b) Constant volume.
(c) Constant availability.

Section 9.4

9.14. **(M)** Determine the specific Gibbs and the specific Helmholtz free energy functions for H_2O at 320°C and
(a) 10 bars pressure.
(b) 0.1 bar pressure.

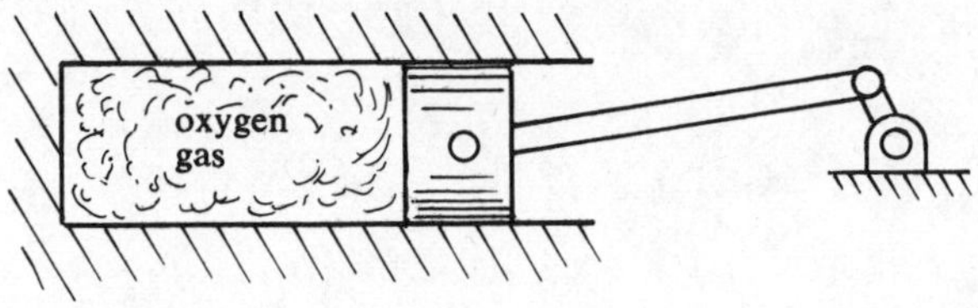

Figure 9–10

(c) Saturated liquid.
(d) Saturated vapor.

9.15. **(E)** Determine the Gibbs free energy for ammonia, NH_3, at 120°F and 20 psia.

9.16. **(E)** Determine the Helmholtz free energy for ammonia at 200°F and 60 psia.

10

The Internal Combustion Engine and the Otto and Diesel Cycles

The *internal combustion* (IC) *engine*, characterized by the reciprocating piston-cylinder gasoline engine, is probably the most common power-producing device in our society. It is utilized in all phases of transportation, for auxiliary electrical power generation and innumerable small, portable power tools. This engine is an example of what we have defined as a *heat engine*, and although the concept of an external heat source may not be clearly compatible with the practicality of an internal combustion (to the system) of fuel, we will see how this ambiguity can be eliminated.

We will define the *ideal Otto cycle* and see how it fits into the analysis of the spark ignited internal combustion engine. Initially we will lean heavily on the assumption that air is the working fluid; but a use of the gas tables in solving Otto cycle problems will help to explain the departures of the air-fuel mixtures and exhaust gases from this assumption.

The analogy between the actual engine and Otto cycle will be sharpened by using the *polytropic process equations*. The actual Otto cycle will then be discussed with some of the more traditional modifications designed to improve the engine power, efficiency, fuel economy, or other engine performance parameters.

We will introduce the *Diesel cycle* and its adaption in the Diesel engine. The *air-standard analysis* will be treated thoroughly and a comparison between the Diesel and Otto cycles will better focus in the reasons for each and their traditional areas of application.

Finally we will discuss some of the design problems inherent in the reciprocating IC engine and the approaches for solving them. The engines described in this chapter have probably absorbed more engineering talent than any other comparable device of man. An overview of the shortcomings and the advantages of these engines should be useful to the student.

10.1 The Ideal Otto Cycle

The ideal Otto cycle is defined as the following set of reversible processes:

(1–2) Adiabatic compression.
(2–3) Constant volume heat addition.
(3–4) Adiabatic expansion.
(4–1) Constant volume heat rejection.

These processes are indicated in figure 10–1 where *p-V* and *T-S* diagrams are used to describe the Otto cycle. Notice that since these processes are all reversible, we can easily identify the enclosed area on the *p-V* diagram as the net work of the cycle and the enclosed area on the *T-S* diagram as the net heat added. To understand how these four processes manage to be descriptive of a real machine operation, let us see how the piston-cylinder can be used with an Otto cycle. In figure 10–2 is shown the sequence of motions of the piston corresponding to those processes. If the piston is reciprocating in a continuous manner, the processes (2–3) and (4–1) must be performed in a zero time period since there is no motion. This is, of course, a deviation the ideal cycle has from the actual case, but we will later see that this is not a large error.

Process (1–2) is a compression of a charge of air and unburned fuel. Upon reaching state (2), the spark plug fires, initiating a chemical reaction between the fuel and air. This is the internal combustion which categorizes the cycle and which releases energy from the fuel (or adds heat to the piston-cylinder), producing a high temperature-pressure gas which drives the piston through an expansion process to state (4). Heat is then quickly removed by opening the exhaust and intake valves, discharging the burned exhaust gases, and just as quickly replacing this volume with a fresh charge of air and unburned fuel to proceed through

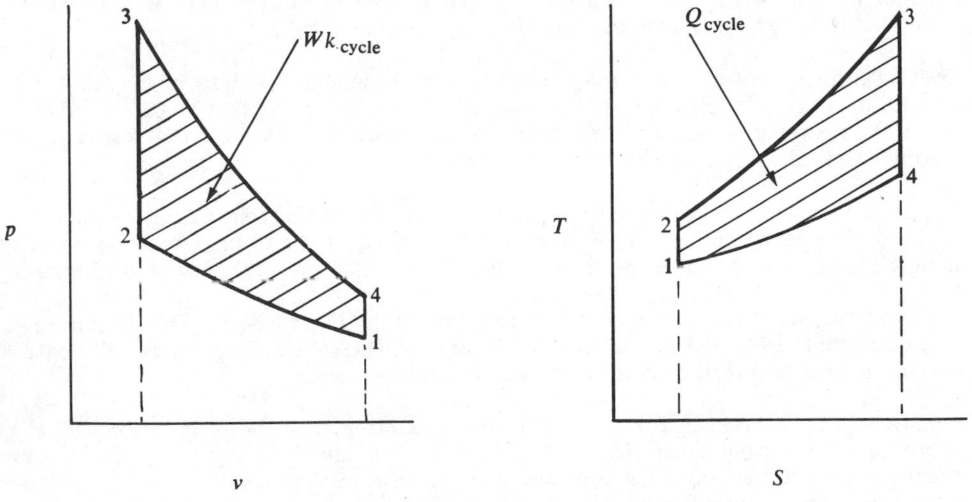

Figure 10–1 Property diagrams for ideal Otto cycle.

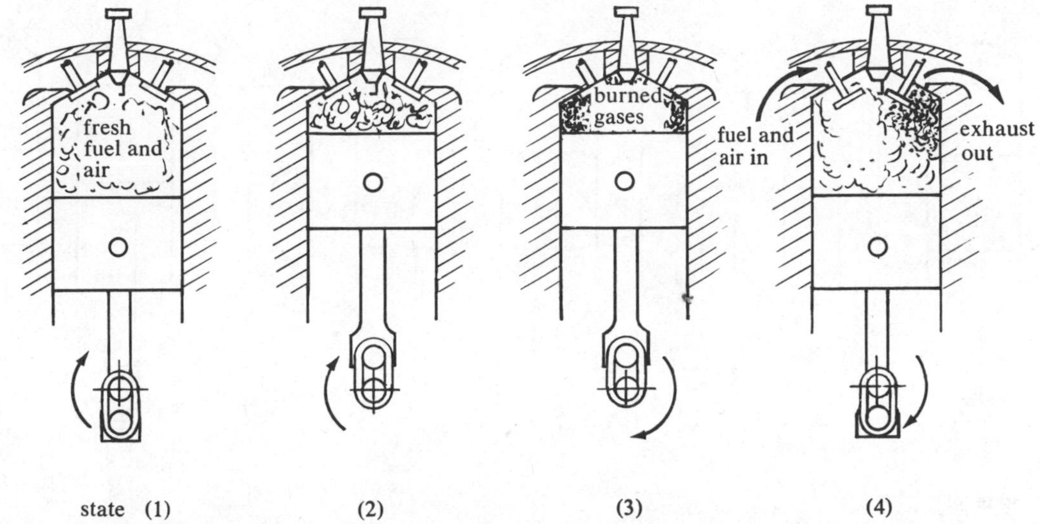

Figure 10–2 Otto cycle application in piston-cylinder device.

the cycle again. This fast shuttling of gases allows us to have power produced on each cycle, called a *two-stroke cycle* or *two-stroke engine*. It is given this name because two strokes complete the cycle (up-down), but the quick gas shuttle requires imaginative design in valves. The configuration of figure 10–2 probably would not give a good exchange of fresh charge for exhaust and much effort is expended in attempting to design a better two-stroke engine which does not waste fuel in discharging the exhaust gases. The prize is one power stroke per revolution of the engine. More often though the engine is changed from the Otto cycle by proceeding through one more revolution to rinse the exhaust gases and add fresh charges more effectively. This variation we call the *four-stroke cycle*, characterized by the opening of the exhaust valve only when the piston-cylinder achieves state (4). This valve remains open through a process (1–5), shown in figures 10–3 and 10–4, during which all the exhaust is pushed out by the piston. At state (5), the exhaust valve closes and the intake valve opens. The fresh fuel-air mixture is then taken in by the retreating piston until state (1) is reached, the intake valve closes, and the normal Otto cycle can proceed. The four-stroke cycle is more commonly used because it allows for better control of the gases than the two-stroke cycle, but there is only one power stroke every other revolution, thus the term "four-stroke" since we have an in-out-in-out motion of the piston for each cycle. Theoretically, the four-stroke engine needs to rotate twice as fast as the two-stroke engine to achieve the same power. This does not generally hold true in practice, however, due to other complications which tend to degrade the attractiveness of the two-stroke cycle.

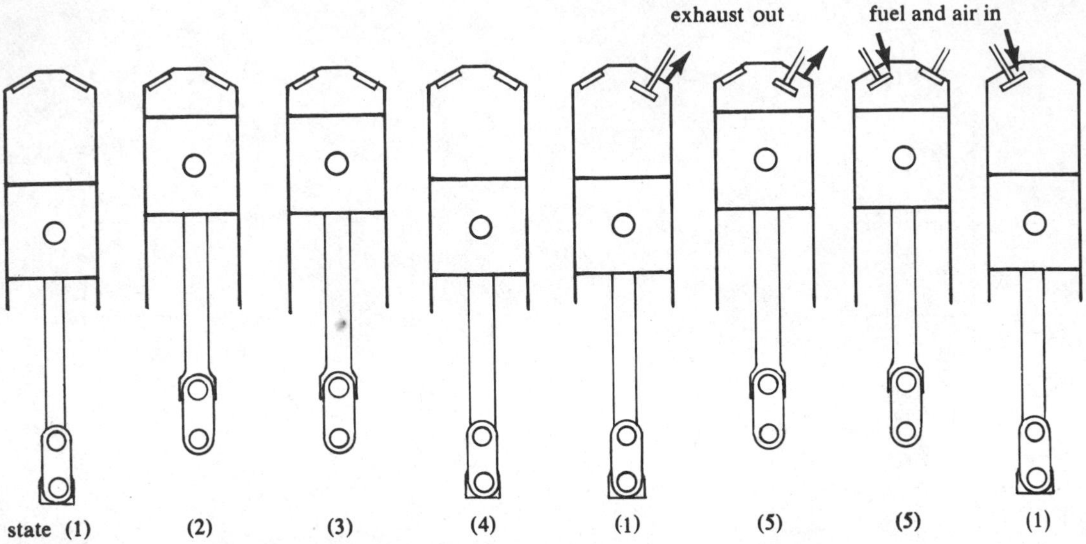

Figure 10–3 Four-stroke Otto cycle.

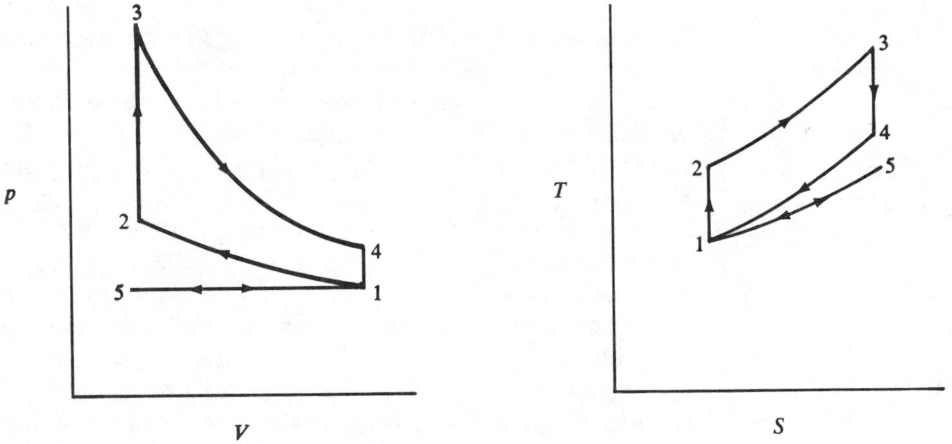

Figure 10–4 Property diagrams for the four-stroke Otto cycle.

The complications of timing, that is, the opening and closing of valves at opportune moments, is a serious deficiency in the whole design for the normal Otto cycle application. In table 10–1, somewhat of a recapitulation, it is indicated when exhaust and intake valves are open or closed. This is done for both the two- and four-stroke cycles and should aid in visualizing the motion of the spark-ignition, internal combustion, piston-cylinder engine.

In figure 10–5 is shown a typical cutaway view of an internal combustion engine which operates on a cycle approximately like the Otto cycle. In

Table 10–1
Valve Positions of Otto Cycles

Process	*4-stroke Cycle* Intake Valve	Exhaust Valve	*2-stroke Cycle* Intake Valve	Exhaust Valve
(1–2)	closed	closed	closed	closed
(2–3)	closed	closed	closed	closed
(3–4)	closed	closed	closed	closed
(4–1)	closed	open	open	open
(1–5)	closed	open		
(5–1)	open	closed		

this figure are shown the major components of the engine, and in figure 10–6 is shown a section of an actual IC engine with water cooling. This view shows the external characteristics of the typical water-cooled IC engine with a portion of it cut away to expose the major workings of the internal parts of the engine. Compare this figure with figure 10–5 and the component parts will be better visualized.

While the Otto cycle is characterized by heat transfer during constant volume, the real engine is continually experiencing heat transfers. Because of this, water is generally directed through cavities in the engine block to keep the cylinder and piston from reaching high temperatures, but the air-cooled engine in figure 10–7 is also effective. In this type of engine, water is not used, but air is forced around the cylinder block to provide convective heat transfer, thus keeping the engine cool.

10.2 The Air-Standard Otto Cycle

Since air is the major constituent of the gases entering the cylinder in the Otto cycle, we shall assume that it is all air. This is the requirement of an air-standard Otto cycle and the analysis will be rather straightforward and familiar. We will use the process equations from chapter 6 in many of the calculating procedures and the student would benefit from referring to chapter 6 to see why certain equations are used. In the following example problems will be shown the thermodynamics of the Otto cycle.

Example 10.1

An internal combustion, spark-ignited (ICSI) four-cylinder engine has a bore (piston diameter) of 3.0 inches, a stroke or piston travel of 3.0

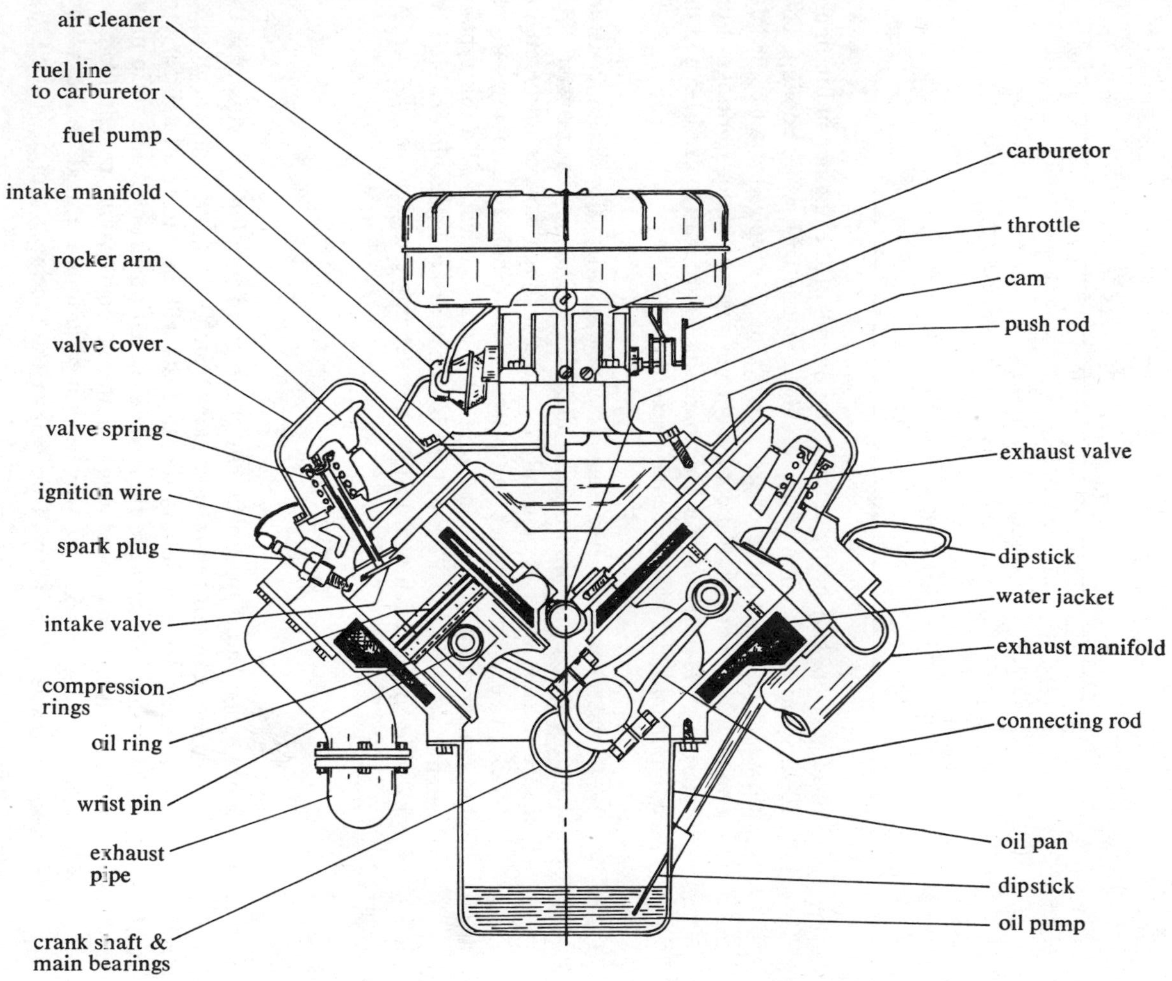

Figure 10–5 Typical internal combustion, spark-ignited engine, V-8 configuration. Cross-sectional view.

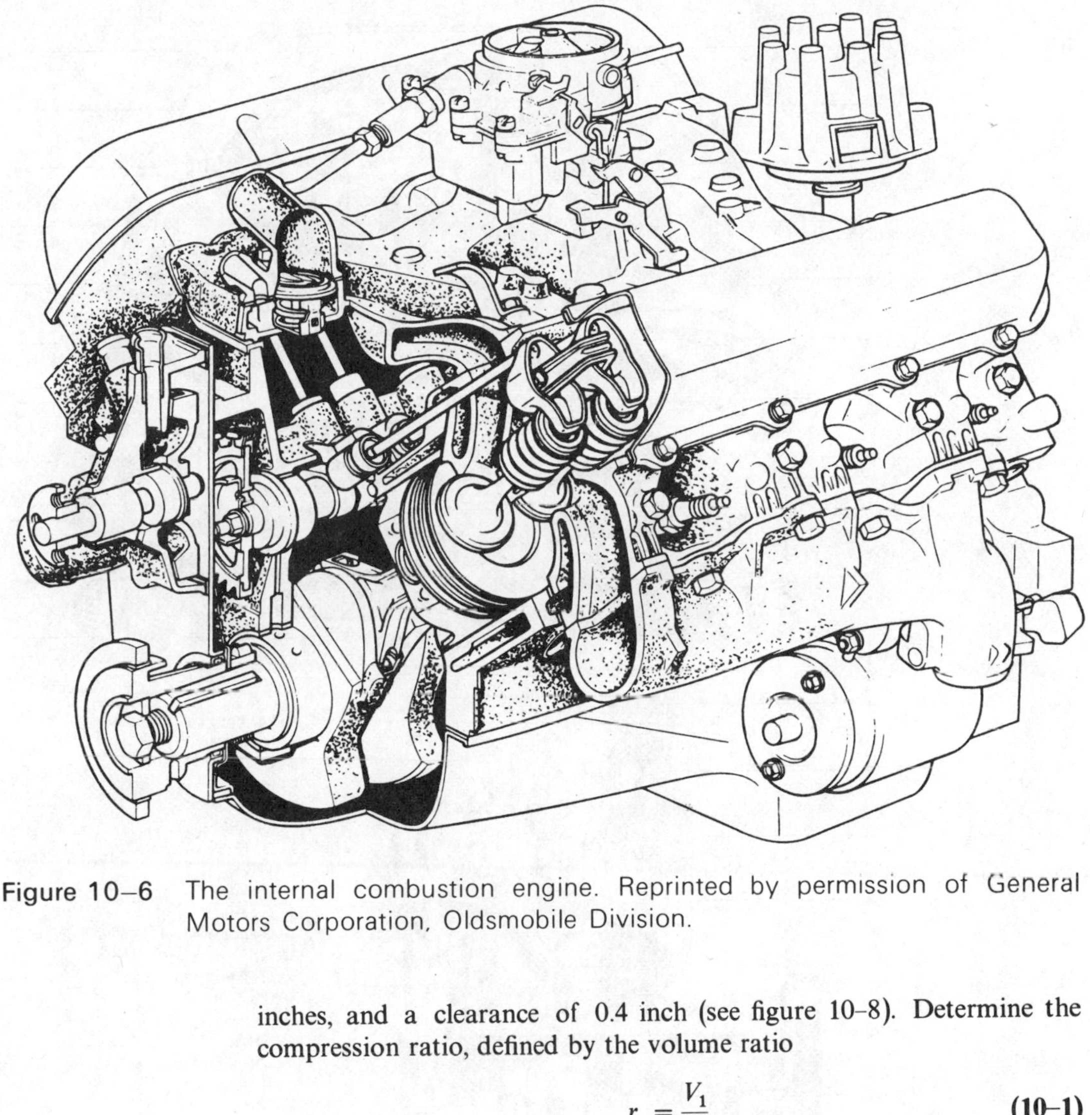

Figure 10–6 The internal combustion engine. Reprinted by permission of General Motors Corporation, Oldsmobile Division.

inches, and a clearance of 0.4 inch (see figure 10–8). Determine the compression ratio, defined by the volume ratio

$$r_v = \frac{V_1}{V_2} \tag{10–1}$$

and the engine displacement, or *displacement*, which is defined as the volume swept or displaced by the piston in traveling from top dead center (TDC) to bottom dead center (BDC).

Solution The compression ratio, computed from equation (10–1), can be determined after the two volumes V_1 and V_2 are computed. Referring to figure 10–8 we find the volume V_1 to be

$$\begin{aligned} V_1 &= (\pi)(\text{bore}/2)^2(\text{clearance} + \text{stroke}) \\ &= (\pi)(1.5\ \text{in})^2(0.4 + 3.0)\ \text{in} \\ &= 24.0\ \text{in}^3 \end{aligned}$$

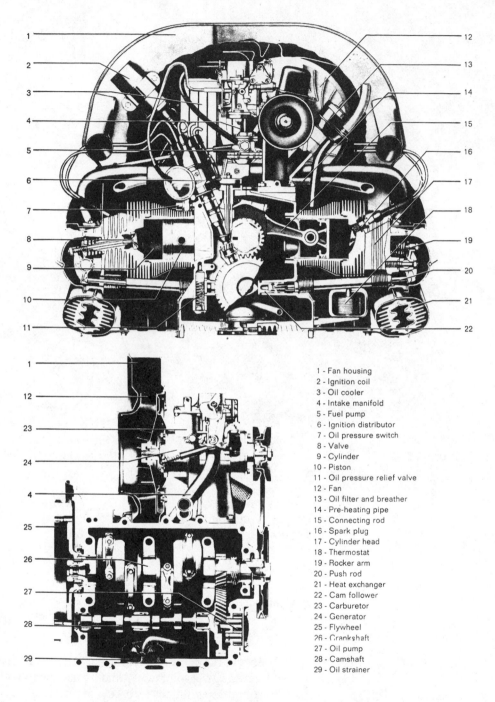

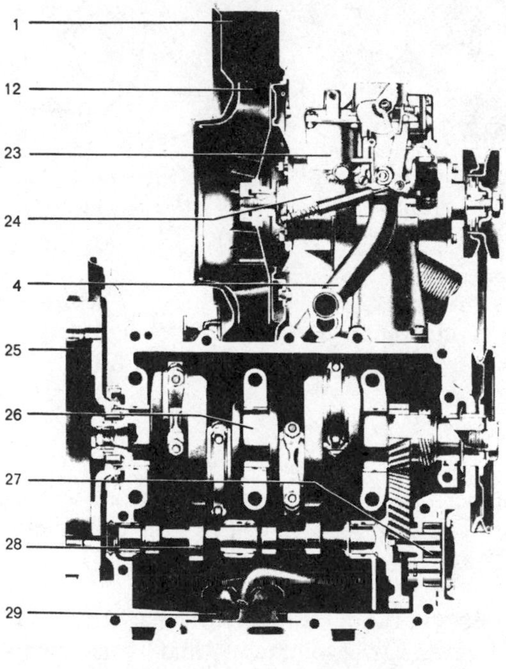

1 - Fan housing
2 - Ignition coil
3 - Oil cooler
4 - Intake manifold
5 - Fuel pump
6 - Ignition distributor
7 - Oil pressure switch
8 - Valve
9 - Cylinder
10 - Piston
11 - Oil pressure relief valve
12 - Fan
13 - Oil filter and breather
14 - Pre-heating pipe
15 - Connecting rod
16 - Spark plug
17 - Cylinder head
18 - Thermostat
19 - Rocker arm
20 - Push rod
21 - Heat exchanger
22 - Cam follower
23 - Carburetor
24 - Generator
25 - Flywheel
26 - Crankshaft
27 - Oil pump
28 - Camshaft
29 - Oil strainer

Figure 10–7 Air-cooled internal combustion engine. Reprinted from Volkswagen Service Manual (VW 1300-1500) (Bielfeld and Berlin, Germany); with permission of Volkswagen of America, Inc., and Delius Klasing and Co.

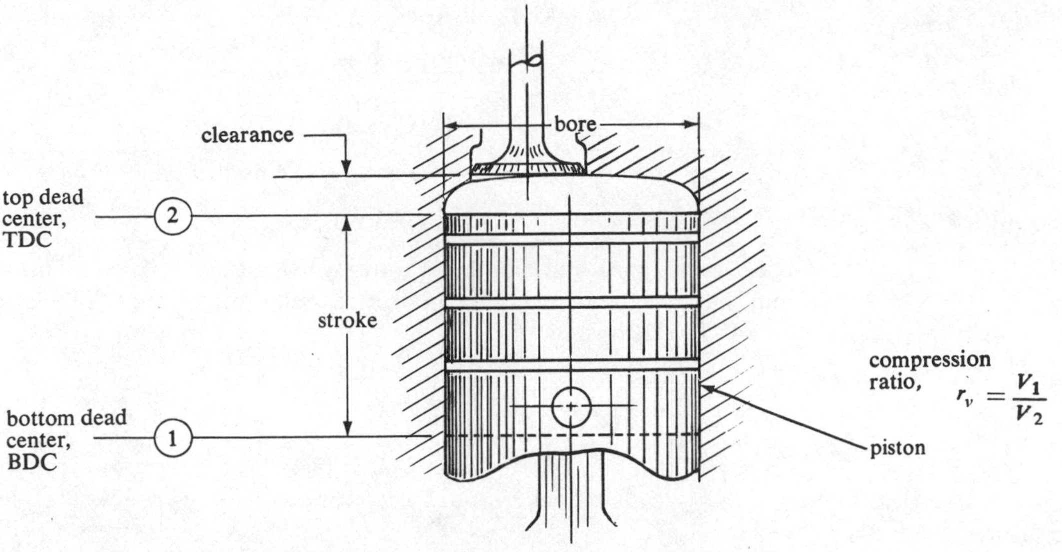

Figure 10–8 Common IC engine parameters.

Similarly, the volume V_2 is

$$V_2 = (\pi)(\text{bore}/2)^2(\text{clearance})$$
$$= (\pi)(1.5 \text{ in})^2(0.4 \text{ in}) = 2.83 \text{ in}^3$$

and the compression ratio r_v is

$$r_v = \frac{24.0 \text{ in}^3}{2.83 \text{ in}^3} = 8.48 \qquad \textit{Answer}$$

STUPID EXAMPLE, WHY NOT USE

Frequently the compression ratio is written 7.99 : 1, or 7.99 to 1.

The displacement V_D can be computed from the equation

$$V_D = (V_1 - V_2)(n) \qquad \textbf{(10–2)}$$

where n is the number of cylinders in the engine. For the engine of this problem we have

$$V_D = (24.0 - 2.83) \text{ in}^3(4) = 84.68 \text{ in}^3 \qquad \textit{Answer}$$

Notice that the displacement is a measure of the volumetric size of the engine.

Example 10.2 For the ICSI engine of example 10.1, assume air is taken into the cylinders at 14.7 psia and 70°F. Determine the pressure and temperature of the air after compression to state (2) if the engine is operating on an Otto cycle.

Solution The compression from state (1) to state (2) is through a reversible adiabatic process, so we may write

$$p_1 v_1^k = p_2 v_2^k$$

or

$$p_2 = p_1\left(\frac{V_1}{V_2}\right)^k = p_1 r_v^k$$

$$= (14.7 \text{ psia})(8.48)^k$$

Since air is the working media, we use $k = 1.4$ and obtain

$$p_2 = 293 \text{ psia} \qquad \textit{Answer}$$

Also, we may now find the temperature by using the p-V-T relationships from chapter 6 or by using the perfect gas relation at state (2). If we use the p-V-T relationship

$$T_2 = T_1\left(\frac{V_1}{V_2}\right)^{k-1} = T_1 r_v^{k-1}$$

$$= (70 + 460°\text{R})(8.48)^{0.4}$$

$$= 1246°\text{R} \qquad \textit{Answer}$$

Notice that the air is very warm after being compressed and would be warm enough to ignite some fuels without external sources such as spark plugs. If this temperature does become too high and ignites the fuel/air mixture, a condition called preignition, or "knock", results. This "knock" causes excessive wear and stresses in the engine and can only be avoided by reducing the compression ratio to thereby reduce the temperature or by using a fuel that will not ignite at the operating temperature.

Example 10.3

An ICSI Otto cycle engine operates on a four-stroke cycle, and has eight cylinders with a total displacement of 1200 cm^3 and a compression ratio of 6 to 1. The air entering the engine is at 27°C and 1.01 bars pressure. The fuel/air mixture during combustion releases 3000 kJ/kg air of heat when the engine is under a load and running at 2200 rpm. Determine the p, V, T properties at the four corners of the cycle and the power produced by the engine.

Solution

First, we notice that the system under consideration is open during one complete cycle. Thus, the system diagram for the heat engine, given in figure 10–9, shows that the engine operates between the temperatures T_3 and T_1. Also, the air and exhaust gases flow in an unsteady fashion through the system, but when observed for an extended period of time, their flows are steady if the engine load and speed are constant.

To determine the properties at each of the four-cycle corners, let us first sketch the p-V and T-s diagrams. (See figure 10–10.) We see that $V_1 = V_4$ and $V_2 = V_3$. Let us then calculate the volumes from the dimensional data.

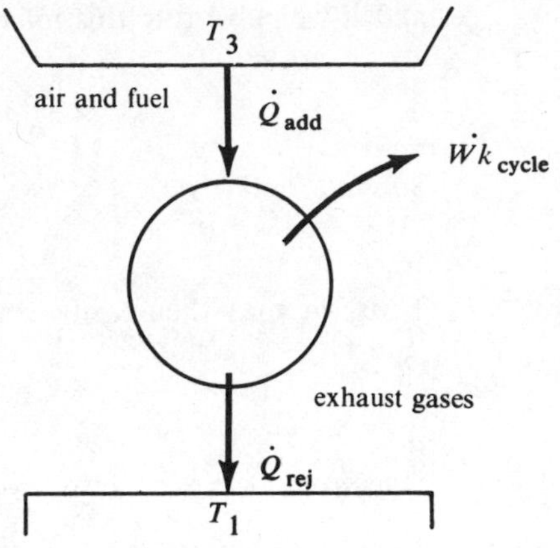

Figure 10–9 System diagram for Otto cycle.

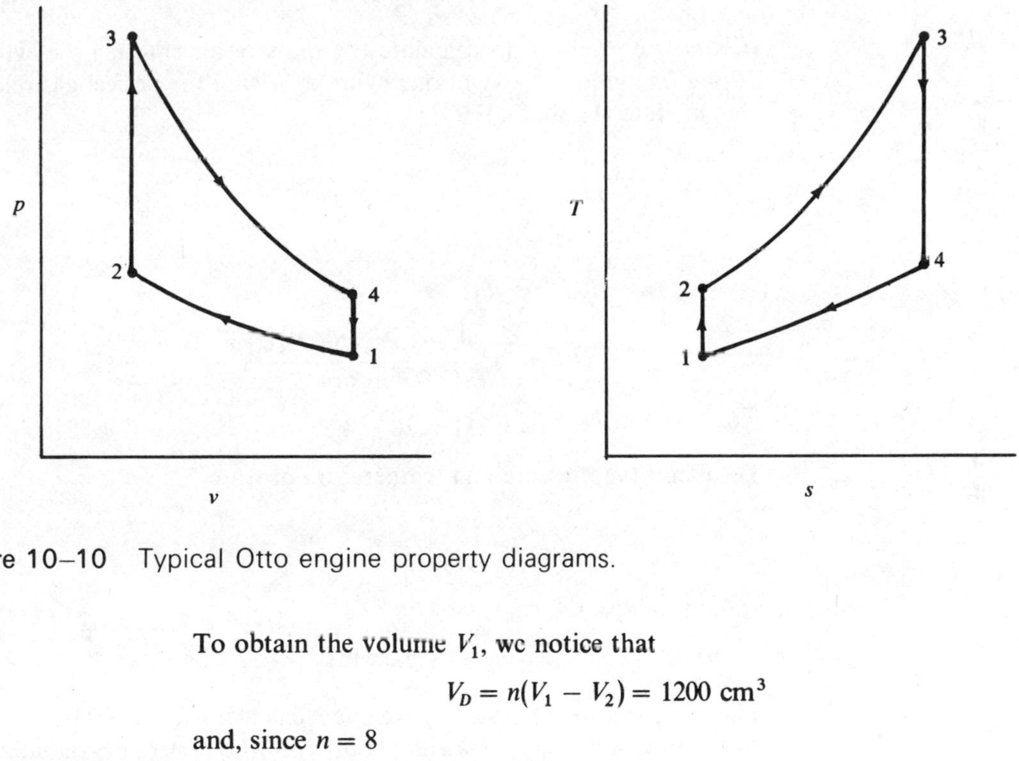

Figure 10–10 Typical Otto engine property diagrams.

To obtain the volume V_1, we notice that

$$V_D = n(V_1 - V_2) = 1200 \text{ cm}^3$$

and, since $n = 8$

$$V_1 - V_2 = \frac{1200}{8} = 150 \text{ cm}^3$$

Further, the compression ratio is

$$r_v = \frac{V_1}{V_2} = 6$$

and if we substitute this into the above relation, we have

$$V_1 - \frac{V_1}{6} = 150 \text{ cm}^3$$

Solving for V_1, we get

$$V_1 = 180 \text{ cm}^3 \qquad \textit{Answer}$$

Thus, we may then easily obtain V_2:

$$V_2 = \frac{V_1}{6} = 30 \text{ cm}^3 \qquad \textit{Answer}$$

Since the cycle is an Otto cycle, we have that

$$V_3 = V_2 = 30 \text{ cm}^3 \qquad \textit{Answer}$$

and

$$V_4 = V_1 = 180 \text{ cm}^3 \qquad \textit{Answer}$$

It is now convenient to calculate the mass of air entering the cylinders during one intake stroke of one cylinder. Using the perfect gas relationship at state (1), we find

$$p_1 V_1 = mRT_1$$

or

$$m = \frac{p_1 V_1}{RT_1}$$

$$= \frac{(1.01 \times 10^5 \text{ N/m}^2)(180 \times 10^{-6} \text{ m}^3)}{(287 \text{ N} \cdot \text{m/kg} \cdot \text{K})(300 \text{ K})}$$

$$= 0.211 \times 10^{-3} \text{ kg}$$

To obtain the pressure and temperature at state (2), we use

$$\frac{p_2}{p_1} = \left(\frac{V_1}{V_2}\right)^k = r_v^k$$

and, using $k = 1.4$, we have

$$p_2 = (1.01 \text{ bars})(6)^{1.4} = 12.4 \text{ bars} \qquad \textit{Answer}$$

The temperature at state (2) we may calculate from either a process equation relating states (2) and (1) or from the perfect gas equation:

$$T_2 = \frac{p_2 V_2}{mR} = \frac{(12.4 \times 10^5 \text{ N/m}^2)(30 \times 10^{-6} \text{ m}^3)}{(0.211 \times 10^{-3} \text{ kg})(287 \text{ N} \cdot \text{m/kg} \cdot \text{K})} = 614 \text{ K} \qquad \textit{Answer}$$

Process (2–3) is a reversible constant volume process involving a heat transfer and no work since the system is closed. Then we can write

$$u_3 - u_2 = q_{\text{add}} = 3000 \text{ kJ/kg air}$$

and since air is a perfect gas with constant specific heats (we assume) we can write

$$u_3 - u_2 = c_v(T_3 - T_2) = 3000 \text{ kJ/kg air}$$

Since c_v is 0.719 kJ/kg · K we get

$$T_3 - T_2 = \frac{(3000 \text{ kJ/kg})}{(0.719 \text{ kJ/kg} \cdot \text{K})}$$

$$= 4172 \text{ K} = T_3 - 614 \text{ K}$$

and

$$T_3 = 4172 \text{ K} + 614 \text{ K} = 4786 \text{ K}$$ *Answer*

Readily then, using the process relations between (2) and (3)

$$\frac{p_3}{p_2} = \frac{T_3}{T_2}$$

and then

$$p_3 = p_2\left(\frac{T_3}{T_2}\right) = (12.4 \text{ bars})\left(\frac{4786 \text{ K}}{614 \text{ K}}\right)$$

$$= 96.7 \text{ bars}$$

The properties at state (4) are found with process equations for the reversible adiabatic process (3–4):

$$\frac{p_4}{p_3} = \left(\frac{V_3}{V_4}\right)^k = \left(\frac{1}{r_v}\right)^k = \frac{1}{6^{1.4}}$$

so

$$p_4 = 7.87 \text{ bars}$$

For the constant volume process (4–1) we have

$$\frac{p_4}{p_1} = \frac{T_4}{T_1}$$

$$T_4 = (300 \text{ K})\left(\frac{7.87 \text{ bars}}{1.01 \text{ bars}}\right) = 2338 \text{ K}$$

We can easily calculate the entropy change per unit mass from equation (7–22):

$$s = s_3 - s_2 = c_v \ln\frac{T_3}{T_2} = (0.719 \text{ kJ/kg} \cdot \text{K})\left(\ln\frac{4786 \text{ K}}{614 \text{ K}}\right)$$

$$= 1.48 \text{ kJ/kg} \cdot \text{K}$$

and

$$s_1 - s_4 = -1.48 \text{ kJ/kg} \cdot \text{K}$$

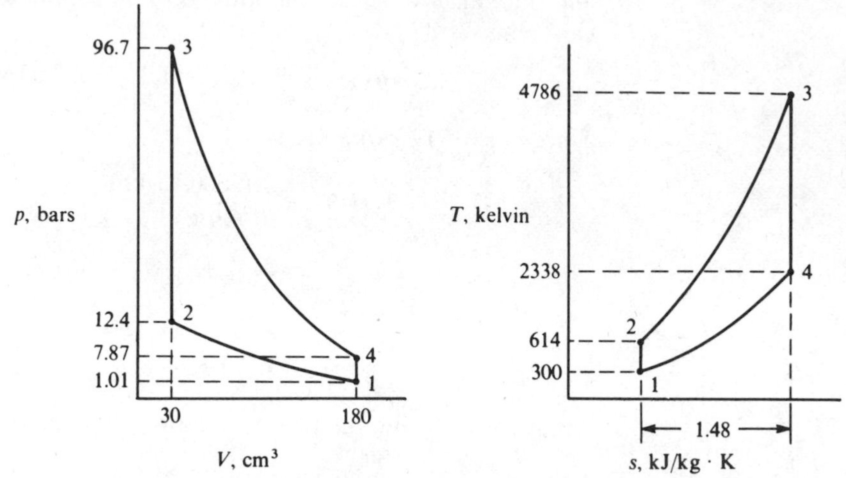

Figure 10–11 Otto engine diagrams with values from calculations.

The net work per cylinder per cycle can be found by determining the enclosed area of the p-V diagram (figure 10–10 or 10–11):

$$Wk_{\text{cycle}} = Wk_{(3-4)} + Wk_{(1-2)}$$

$$= \frac{mR}{1-k}(T_4 - T_3 + T_2 - T_1)$$

$$= \frac{(0.211 \times 10^{-3}\ \text{kg})(287\ \text{N} \cdot \text{m/kg} \cdot \text{K})}{(-0.4)}$$

$$\times (2338\ \text{K} - 4786\ \text{K} + 614\ \text{K} - 300\ \text{K})$$

$$= 323\ \text{N} \cdot \text{m/cycle} \qquad \textbf{(10–3)}$$

The engine will then deliver, from eight cylinders

$$Wk = (8)(323\ \text{N} \cdot \text{m/cycle}) = 2584\ \text{N} \cdot \text{m} \qquad \textit{Answer}$$

The power-produced $\dot{W}k$ will be the work produced per cycle times the number of cycles per unit time. This last term is 1100 cycles/min since the engine speed was 2200 revolutions/min and the engine has a four-stroke cycle. Then

$$\dot{W}k = (2584\ \text{N} \cdot \text{m/cycle})(1100\ \text{cycles/min})$$

$$= 2{,}842{,}400\ \text{N} \cdot \text{m/min} = 47.4\ \text{kJ/s} = 47.4\ \text{kW} \qquad \textit{Answer}$$

This analysis of an Otto cycle was made by assuming reversible processes and perfect gases for the working media. Then the equations derived in chapter 6 were used to compute the properties. Let us look at one more example of the air-standard analysis of the ICSI engine.

Example 10.4

A one-cylinder ICSI engine operating on an ideal two-stroke Otto cycle is to generate 5 horsepower at 4500 rpm with a compression ratio of 10. Assume intake air is at 80°F and 14.7 psia. Determine the engine displacement required and the heat rejected if 1000 Btu/lbm of heat is added when the fuel/air mixture burns.

Solution

Since the engine is a two-stroke design, the work is extracted once each revolution of the engine. Thus,

$$\dot{W}k = Wk_{\text{cycle}} \times 4500 \text{ rpm} = 5 \text{ hp}$$

and the work must be

$$Wk_{\text{cycle}} = \frac{5 \text{ hp}}{4500 \text{ cycles/min}}$$

$$= \frac{(550 \text{ ft-lbf/s} \cdot \text{hp})(5 \text{ hp})}{75 \text{ cycles/s}}$$

$$= 36.7 \text{ ft-lbf/cycle}$$

To compute the displacement of the engine, we must know the two volumes V_1 and V_2. Since the compression ratio is to be 10 we have

$$V_1 = 10V_2$$

To find the volume V_1, we need to know the mass of air entering each cycle so that the perfect gas relation can be used at state (1). The mass is computed from equation (10–2), which requires each temperature to be determined. Proceeding then, we have that $T_1 = 80 + 460 = 540°\text{R}$. Further, we may write

$$\left(\frac{T_2}{T_1}\right)^{k/(k-1)} = \left(\frac{V_1}{V_2}\right)^k$$

or

$$T_2 = T_1\left(\frac{V_1}{V_2}\right)^{k-1}$$

and then

$$T_2 = (540°\text{R})(10)^{0.4}$$

$$= 1356°\text{R}$$

Then, using $c_v = 0.171$ Btu/lbm · °R and the relationship

$$q_{\text{add}} = c_v(T_3 - T_2)$$

we have

$$T_3 = \frac{q_{\text{add}}}{c_v} + T_2$$

$$= \frac{1000 \text{ Btu/lbm}}{0.171 \text{ Btu/lbm} \cdot °\text{R}} + 1356°\text{R}$$

$$= 7204°\text{R}$$

Further, for process (3–4) we have

$$T_4 = T_3\left(\frac{V_3}{V_4}\right)^{k-1}$$

$$= 7204°\text{R}\left(\frac{1}{10}\right)^{0.4} = 2868°\text{R}$$

Then the work is

$$Wk_{\text{cycle}} = \frac{mR}{1-k}(T_4 - T_3 + T_2 - T_1)$$

$$36.7 \text{ ft-lbf/cycle} = (m)\left(\frac{53.3 \text{ ft-lbf/lbm}\cdot°\text{R}}{1-1.4}\right)$$

$$\times (2868°\text{R} - 7204°\text{R} + 1356°\text{R} - 540°\text{R})$$

$$36.7 \text{ ft-lbf/cycle} = (m)(4.69 \times 10^5 \text{ ft-lbf/lbm})$$

and solving for the mass we have

$$m = 7.83 \times 10^{-5} \text{ lbm/cycle}$$

The volume at state (1) can now be found from the perfect gas relation:

$$V_1 = mR\left(\frac{T_1}{p_1}\right)$$

$$= \frac{(7.83 \times 10^{-5} \text{ lbm})(53.3 \text{ ft-lbf/lbm}\cdot°\text{R})(540°\text{R})}{(14.7 \text{ lbf/in}^2)(144 \text{ in}^2/\text{ft}^2)}$$

$$= 0.00106 \text{ ft}^3 = 1.84 \text{ in}^3$$

Then

$$V_2 = \frac{V_1}{10} = 0.184 \text{ in}^3$$

and since there is but one cylinder in the engine, its displacement is

$$V_D = V_1 - V_2 = 1.654 \text{ in}^3$$ *Answer*

Various dimensional arrangements of piston diameter or bore and the stroke could now be selected to provide this particular displacement. The heat rejected per cycle is

$$Q_{\text{rej}} = WK_{\text{cycle}} - Q_{\text{add}}$$

or

$$Q_{\text{rej}} = 36.7 \text{ ft-lbf} - (1000 \text{ Btu/lbm})(778 \text{ ft-lbf/Btu})$$

$$\times (7.83 \times 10^{-5} \text{ lbm})$$

$$= -24.2 \text{ ft-lbf/cycle}$$ *Answer*

Notice that the temperatures we have predicted at combustion are extremely high. At such high temperatures many gases dissociate or ionize.

This means that the gas molecules or atoms lose electrons and thus the gas exhibits quite different characteristics from the normal air. We have not here accounted for such variations.

A method to match theoretical processes to the actual Otto engine is to assume the compression and expansion processes are reversible polytropic. This will account for heat transfers during these processes and the following example should demonstrate the computations involved in this treatment.

Example 10.5

For the engine of example problem 10.3, assume the compression process is polytropic with $n_{12} = 1.3$ and assume that the expansion (3–4) is also polytropic with $n_{34} = 1.25$. If all else is the same as in the previous engine, determine the work of the cycle, the power produced at 2200 rpm, and the heat rejected per cycle.

Solution

The system under analysis is the same as that in example 10.3, so the volumes are the same. The engine can be described by the system diagram as shown in figure 10–8 and the air in the chamber at the beginning of compression, state (1), is at 1.01 bars and 27°C. The compression ratio is 6 to 1 and the volumes are

$$V_1 = V_4 = 180 \text{ cm}^3$$

$$V_2 = V_3 = 30 \text{ cm}^3$$

The pressure at state (2) is computed from the equation

$$p_2 = p_1\left(\frac{V_1}{V_2}\right)^{n_{12}}$$

$$= (1.01 \text{ bars})(6)^{1.3} = 10.4 \text{ bars}$$

and the temperature at state (2) is

$$T_2 = \frac{p_2 V_2}{mR} = \frac{(10.4 \times 10^5 \text{ N/m}^2)(30 \times 10^{-6} \text{ m}^3)}{(0.211 \times 10^{-3} \text{ kg})(287 \text{ N} \cdot \text{m/kg} \cdot \text{K})} = 515 \text{ K}$$

The temperature at state (3) can now be computed from the relationship

$$q_{\text{add}} = c_v(T_3 - T_2) = 3000 \text{ kJ/kg}$$

and then

$$T_3 = (3000 \text{ kJ/kg})/c_v + T_2$$

$$= (3000 \text{ kJ/kg})/(0.719 \text{ kJ/kg} \cdot \text{K}) + 515 \text{ K}$$

$$= 4687 \text{ K}$$

From the process equation

$$\frac{T_4}{T_3} = \left(\frac{V_3}{V_4}\right)^{n-1}$$

We can now calculate the temperature at state (4):

$$T_4 = (4687\ \text{K})\left(\frac{1}{6}\right)^{1.25-1} = 2995\ \text{K}$$

The work is then easily computed from a variation of equation (10–3):

$$\begin{aligned} Wk_{\text{cycle}} &= \frac{mR}{1-n_{43}}(T_4 - T_3) + \frac{mR}{1-n_{21}}(T_2 - T_1) \\ &= (0.211 \times 10^{-3}\ \text{kg})(287\ \text{J/kg}\cdot\text{K}) \\ &\quad \times \left(\frac{2995\ \text{K} - 4687\ \text{K}}{1-1.25} + \frac{515\ \text{K} - 300\ \text{K}}{1-1.3}\right) \\ &= 366.5\ \text{J/cycle} \end{aligned} \tag{10–4}$$

The power produced by the engine will be the work for eight cylinders running at 2200 rpm but at 1100 cycles/min. Thus

$$\begin{aligned} \dot{W}k &= (8)(366.5\ \text{J/cycle} \times 1100\ \text{cycles/min}) \\ &= 323 \times 10^4\ \text{J/min} = 3230\ \text{kJ/min} \\ &= 53.8\ \text{kW} \end{aligned}$$ *Answer*

The heat rejected per cycle we can compute from the balance

$$Q_{\text{rej}} = Wk_{\text{cycle}} - Q_{\text{add}}$$

which gives us

$$\begin{aligned} Q_{\text{rej}} &= (366.5\ \text{J})(8) - (3000\ \text{kJ/kg})(0.211 \times 10^{-3}\ \text{kg})(8) \\ &= -2.13\ \text{kJ} \end{aligned}$$ *Answer*

A comparison of these results with the answers from problem 10.1 indicates a decrease in work or power produced by the engine. This is a typical result of using polytropic processes to represent better the actual engine — the ideal Otto cycle is just not quite as accurate as the polytropic cycle, provided the exponents n are accurate.

In figure 10–12 are shown the property diagrams resulting from problem 10.2. Notice that entropy changes occur in all four of the cycle processes and if we desired, the magnitudes of these changes could be easily calculated from equations listed in tables 6–2 and 6–3.

10.3 Otto Cycle Efficiency

The efficiency of the Otto cycle, as indeed for any cycle, is defined as

$$\eta_T = \frac{(Wk_{\text{cycle}})}{Q_{\text{add}}}(100) \tag{10–5}$$

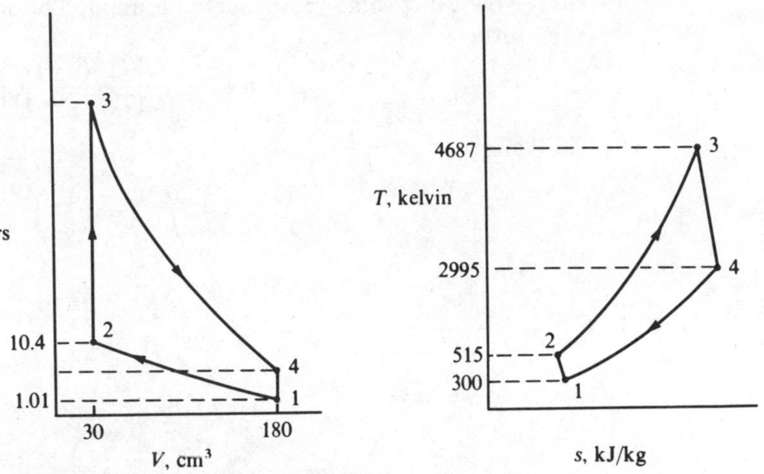

Figure 10–12

For the ideal Otto cycle and assuming a perfect gas working medium, we can reduce this to more specific equations. First we have that

$$Q_{add} = Q_{(2-3)} = U_3 - U_2 = mc_v(T_3 - T_2)$$

and in example 10.3 we saw that

$$Wk_{cycle} = Wk_{(1-2)} + Wk_{(3-4)} = \frac{mR}{(1-k)}(T_2 - T_1 + T_4 - T_3)$$

Since $R = c_p - c_v$ and $k = c_p/c_v$ we can then write

$$Wk_{cycle} = \frac{m(c_p - c_v)}{(1 - c_p/c_v)}(T_2 - T_1 + T_4 - T_3)$$

or

$$Wk_{cycle} = \frac{mc_v(c_p - c_v)}{(c_p - c_v)}(T_1 - T_2 + T_3 - T_4)$$

$$= mc_v(T_1 - T_4) + mc_v(T_3 - T_2)$$

We then obtain the efficiency by substituting these results into equation (10–5):

$$\eta_T = \frac{mc_v(T_1 - T_4) + mc_v(T_3 - T_2)}{mc_v(T_3 - T_2)} \times 100$$

or

$$\eta_T = \left(1 - \frac{T_4 - T_1}{T_3 - T_2}\right) \times 100 \qquad \textbf{(10–6)}$$

Now using some more algebraic manipulations we obtain the following form

$$\eta_T = \left(1 - \frac{T_1(T_4/T_1 - 1)}{T_2(T_3/T_2 - 1)}\right) \times 100$$

and

$$\frac{T_4}{T_3} = \left(\frac{V_3}{V_4}\right)^{k-1} = \left(\frac{V_2}{V_1}\right)^{k-1} = \frac{T_1}{T_2}$$

so

$$\frac{T_4}{T_1} = \frac{T_3}{T_2}$$

and we have

$$\eta_T = \left(1 - \frac{T_1}{T_2}\right) \times 100 \qquad \textbf{(10–7)}$$

Equation (10–7) is frequently reduced further by noting that

$$\frac{T_1}{T_2} = \left(\frac{V_2}{V_1}\right)^{k-1} = \frac{1}{(r_v)^{k-1}}$$

which gives us

$$\eta_T = \left(1 - \frac{1}{(r_v)^{k-1}}\right) \times 100 \qquad \textbf{(10–8)}$$

This result, as mentioned, is good only for ideal Otto cycles using perfect gases with constant specific heats. Notice that the efficiency of the Otto cycle is a function of the compression ratio. Thus, increasing the compression ratio will increase the efficiency of the engine an equivalent amount.

Example 10.6

Determine the efficiency of the engine described in example 10.1.

Solution

The engine in example 10.1 has a compression ratio of 8.48 to 1, so from equation 10.8 we have

$$\eta_T = \left(1 - \left(\frac{1}{8.48}\right)^{0.4}\right)(100) = 57.5\% \qquad \textit{Answer}$$

Example 10.7

Determine the efficiencies of the engines in examples 10.3 and 10.4.

Solution

Again, using equation 10.8 we readily obtain the cycle efficiency for the engine in example 10.3 having a compression ratio of 6 to 1. Thus,

$$\eta_T = \left(1 - \left(\frac{1}{6}\right)^{0.4}\right)(100) = 51.2\% \qquad \textit{Answer}$$

and for the engine of example 10.4 the compression ratio was given as 10 to 1. We compute

$$\eta_T = \left(1 - \left(\frac{1}{10}\right)^{0.4}\right)(100) = 60.2\% \qquad \textit{Answer}$$

It is also interesting to compare the results of these examples to the ideal Carnot cycle. Operating between the high temperature of the engine, T_3, and the low temperature, T_1, the Carnot engine efficiency is calculated using the values from example 10.3:

$$\eta_{\text{Carnot}} = \left(1 - \frac{T_1}{T_3}\right) \times 100 = \left(1 - \frac{300}{4786}\right)(100) = 93.7\%$$

and for example 10.4

$$\eta_{\text{Carnot}} = \left(1 - \frac{300}{4231}\right)(100) = 92.9\%$$

These comparisons are somewhat biased, however, for if we compare *T-s* diagrams of both Otto and Carnot cycles we must conclude that for differential changes in entropy, both cycles have the same efficiency. This we see in figure 10–13 and from this we see that using this criterion, the Carnot efficiency of the differential cycles is, using example 10.3

$$\eta_{\text{Carnot}} = \left(1 - \frac{T_1}{T_2}\right) \times 100 = \left(1 - \frac{300}{614}\right) \times 100$$

$$= 51.7\%$$

which is the same as for the Otto engine.

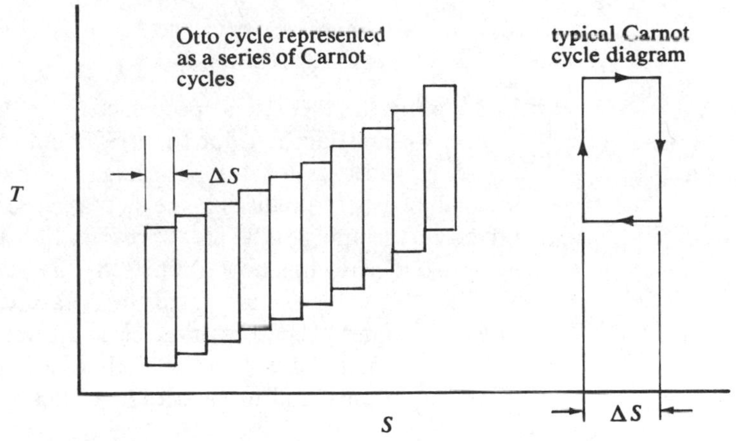

Figure 10–13 Comparison of Carnot and Otto cycles using *T-s* diagram.

One other parameter frequently mentioned with efficiency to describe an Otto engine performance is the mean effective pressure (mep) which is here defined:

$$\text{mep} = \frac{Wk_{\text{cycle}}}{V_D} \qquad \textbf{(10–9)}$$

The mep is a design parameter. It is a theoretical average pressure of the engine gases in the cylinder during the cycle which produces the same work as the actual engine and its variation in gas pressure.

Example 10.8

Determine the mean effective pressure for the engines in examples 10.3, 10.4, and 10.5.

Solution

For the engine of example 10.3 we have that the cyclic work is 22584 N · m. Thus,

$$\text{mep} = \frac{2584 \text{ N} \cdot \text{m}}{1200 \text{ cm}^3 \times 10^{-6} \text{ m}^3/\text{cm}^3}$$

$$= 21.5 \times 10^5 \text{ N/m}^2 = 21.5 \text{ bars} \qquad \textit{Answer}$$

For the engine of example 10.4 we have the cyclic work indicated as 36.7 ft-lbf and the displacement is 1.656 in³. We then find

$$\text{mep} = \left(\frac{36.7 \text{ ft-lbf}}{1.656 \text{ in}^3}\right)(12 \text{ in/ft})$$

$$= 266 \text{ psi} \qquad \textit{Answer}$$

Finally, for the engine in example 10.5 we obtain

$$\text{mep} = \frac{366.5 \text{ N} \cdot \text{m} \times 8 \text{ cylinders}}{1200 \times 10^{-6} \text{ m}^3}$$

$$= 24.4 \text{ bars} \qquad \textit{Answer}$$

10.4 The Gas Tables

In adapting the theoretical processes of the Otto cycle to the actual IC engine, we must account for the irreversible effects as well as the deviations in properties from a perfect gas of the working media. Another method of better predicting the actual processes is the utilizing of gas tables. In the appendix is an abbreviated air table (table B.6) which will serve us in showing how tabulated data can be used in Otto cycle analyses.* In table B.6 are listed temperature, enthalpy, internal energy, and three other properties: the relative pressure p_r; the relative specific volume v_r; and the ϕ-function. All these properties are functions of temperature only, and in particular we define

$$\phi = \int_0^T \frac{c_p dT}{T} \tag{10–10}$$

Notice that this is not the same phi as was defined in chapter 9 (Φ), but when gas tables are used, the definition (10–10) should be understood. The relative pressure p_r we define as

$$p_r = e^{\phi/R} \quad \text{or} \quad \phi = R \ln p_r \tag{10–11}$$

* A more complete air table as well as properties of various gases can be found in the book *Gas Tables*, Keenen and Kaye, John Wiley & Sons Inc. (New York, 1948).

from which the relative volume is written:

$$v_r = \frac{RT}{p_r} \tag{10–12}$$

The general entropy equation resulting from equation (7–18) is

$$\Delta s = \int_{T_1}^{T_2} \frac{c_p dT}{T} - R\int_{p_1}^{p_2} \frac{dp}{p}$$

where we have assumed $dh = c_p dT$ and $V/T = R/p$. Then we can write this as

$$\Delta s = \phi_2 - \phi_1 - R \ln \frac{p_2}{p_1} \tag{10–13}$$

from the definition (10–10). For the reversible adiabatic process, equation (10–13) becomes (since $\Delta s = 0$)

$$\phi_2 - \phi_1 = R \ln\left(\frac{p_2}{p_1}\right)$$

and since we defined relative pressure p_r by the equation (10–11)

$$\phi = R \ln p_r$$

we have

$$R \ln\left(\frac{p_2}{p_1}\right) = R(\ln p_{r_2} - \ln p_{r_1})$$

$$= R \ln \frac{p_{r_2}}{p_{r_1}}$$

Consequently,

$$\frac{p_2}{p_1} = \frac{p_{r_2}}{p_{r_1}} \tag{10–14}$$

Also, from the definition of the relative volume we have

$$\frac{v_2}{v_1} = \frac{v_{r_2}}{v_{r_1}} \tag{10–15}$$

The reader should always remember that equations (10–14) and (10–15) are true only for the reversible adiabatic cases and should not be used for other processes.

Example 10.9 There is 0.2 kilogram of air heated from 127°C and 1.0 bar to 227°C in a rigid container. Determine the entropy change and the enthalpy change.

Solution The system reaches a final pressure p_2 which can be predicted from the equation

$$p_2 = p_1\left(\frac{T_2}{T_1}\right)$$

since we have an isometric process of a perfect gas. Then,

$$p_2 = (1.0 \text{ bar})\left(\frac{500 \text{ K}}{400 \text{ K}}\right) = 1.25 \text{ bars}$$

and from table B.6

$$\phi_1 = 2.805 \text{ kJ/kg} \cdot \text{K as } T_1 = 400 \text{ K}$$

$$\phi_2 = 3.033 \text{ kJ/kg} \cdot \text{K as } T_2 = 500 \text{ K}$$

The entropy change is computed from equation (10–13):

$$\Delta s = 3.033 \text{ kJ/kg} \cdot \text{K} - 2.805 \text{ kJ/kg} \cdot \text{K} - (0.287 \text{ kJ/kg} \cdot \text{K})\left(\ln \frac{1.25}{1.0}\right)$$

$$= 0.164 \text{ kJ/kg} \cdot \text{K}$$ *Answer*

For 0.2 kilogram of air we have

$$\Delta S = m\Delta s = 0.0328 \text{ kJ/kg} \cdot \text{K}$$ *Answer*

The enthalpy change can be determined from the values listed in table B.6. Thus

$$\Delta H = m(h_2 - h_1)$$

$$= (0.2 \text{ kg})(503.0 \text{ kJ/kg} - 401.0 \text{ kJ/kg})$$

$$= 20.4 \text{ kJ}$$ *Answer*

Example 10.10 There is 0.03 lbm of air compressed reversibly and adiabatically from 100°F and 14.8 psia to 120 psia. Determine the final temperature, change in entropy, and change in enthalpy for the process.

Solution Since the system is undergoing a reversible adiabatic process, from equation (10–14) we have

$$\frac{p_{r_1}}{p_{r_2}} = \frac{p_1}{p_2} = \frac{14.8}{120}$$

but from table B.6 we find that $p_{r1} = 1.5742$ and then

$$p_{r_2} = (1.5742)\left(\frac{120}{14.8}\right) = 12.8$$

This is approximately the relative pressure corresponding to a gas temperature of 1000°R, and we therefore say

$$T_2 = 1010°\text{R}$$ *Answer*

The entropy change is obviously zero, but the enthalpy change we determine using tabular values from table B.6:

$$H_2 - H_1 = m(h_2 - h_1) = 0.03 \text{ lbm } (243.47 - 133.86)$$

$$= (0.03)(109.61) \text{ Btu} = 3.29 \text{ Btu}$$ *Answer*

The gas tables (or air tables) can be used to reduce calculations if used with care. They are not meant to replace fully the formulas of chapter 6 but can be used with confidence as additional thermodynamics tools.

10.5 The Actual Otto Cycle

Using our previously developed tools, we will analyze the operation of an Otto engine which could conceivably be used to power an automobile, truck, tractor, or other mechanical device; that is, we will consider the actual or "practical" Otto engine. Before doing this, however, let us introduce additional terminology and cycle parameters, namely, *indicated horsepower* (ihp), *brake horsepower* (bhp), *fuel heating value* (HV), and *brake specific fuel consumption* (bsfc).

The indicated horsepower is determined from a *p-V* diagram of the test engine. This diagram can be obtained by using an engine indicator, which is a mechanism capable of measuring and recording the pressure in a cylinder while concurrently recording the piston position. The engine indicator is a common test device with a history of use dating to before 1900. At any rate, a detailed description of the indicator may be found in various reference literature; here let us look only at the result of the device. In figure 10–14 is shown the typical pressure-volume diagram, of an Otto engine, specifically of an engine with the throttle wide open. The curve can easily be seen to differ somewhat from the typical Otto cycle as shown in figure 10–1. The comparison is made easier by noting figure 10–15 where the ideal Otto cycle is superimposed on an actual Otto cycle. For further analysis of this example we will use the gas

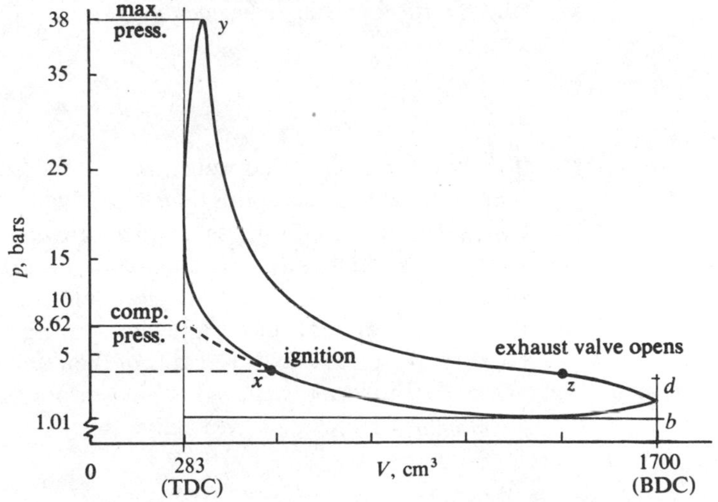

Figure 10–14 Typical *p-V* diagram for spark ignition engines at wide open throttle. Reproduced from E. I. Obert, *Internal Combustion Engines*, 2nd ed. (Scranton, 1959); with permission of International Textbook Co. Revised to SI units by author.

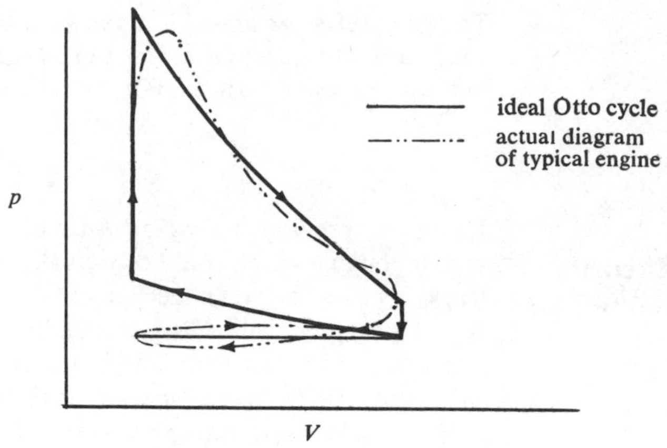

Figure 10–15 p-V diagram of ideal Otto cycle with actual cycle superimposed.

tables to obtain air properties for the Otto cycle curves to calculate the area of the enclosed p-V diagram. This area, however, could be quite authentically calculated using various geometric and mechanical devices (for example, the planimeter), but in any case, the accuracy of the recorded data is probably no less than 2% errors. The concern for accuracy here is, of course, due to the fact that the enclosed area of the p-V diagram is work of the cycle and the rate of this work, power, is referred to as *indicated horsepower*, ihp. It is not the power actually produced by the engine drive shaft but rather the power which could be produced at the piston of a frictionless engine.

The actual power produced by an engine we call *brake horsepower*, bhp, and the mechanically efficiency η_{mech} is then obtained for the actual Otto engine from

$$\eta_{mech} = \frac{\text{bhp}}{\text{ihp}} \times 100 \qquad \textbf{(10–16)}$$

The brake horsepower results are strictly from test data. We have no thermodynamic tools at this time to predict the value of bhp, although we can make some crude guesses from a knowledge of ihp. There are various test devices to measure bhp, all of which utilize some brake or resistance to measure engine output. Dynamometers, prony brakes, and water brakes are common devices for measuring bhp, but we will not here concern ourselves with the details of their workings. Using this measured value bhp, however, we can determine a parameter called the *brake mean effective pressure*, bmep. We define this as

$$\text{bmep} = \frac{\text{bhp}}{V_D \times N} \qquad \textbf{(10–17)}$$

where N is the number of power strokes per unit time. The bmep is generally described in units of pressure, just like mep.

The term *Indicated mean effective pressure* (Imep) is sometimes used to describe an engine's performance and is defined as

$$\text{Imep} = \frac{\text{ihp}}{V_D N} \tag{10–18}$$

The fuel consumed by an engine is of major concern to both designer and user. In particular, the amount of fuel used per bhp per hour is called the *brake specific fuel consumption*, bsfc, and is calculated from

$$\text{bsfc} = \frac{m}{(\text{bhp})\tau} = \frac{\dot{m}_f}{\text{bhp}} \tag{10–19}$$

where the rate of fuel consumption $\dot{m}_f$ is equal to the mass of fuel m consumed during a time period τ while producing power bhp. The amount of fuel used in relation to the amount of *clean air* required is of course meager (about 15 parts of air per part fuel). We could discuss a brake specific air consumption. However, *volumetric efficiency*, η_v, defined as the ratio of the mass of air taken into the engine cylinder, $\dot{m}_v$, to the mass which theoretically could have been taken in at atmospheric conditions, $\dot{m}_t$, or

$$\eta_v = \frac{\dot{m}_a}{\dot{m}_t} \tag{10–20}$$

is a common parameter used to describe engine performance. Notice that volumetric efficiency is not a volume efficiency, but rather a mass efficiency.

The chemical reaction between the air and fuel produces heat, which in turn increases the temperature and pressure of the compressed gases and ultimately results in work *out*. The heat released during the air-fuel reaction is called the *heating value* of the fuel. When this reaction occurs, water is formed (either in a vapor or a liquid), and if we allow the water to condense to a liquid, the heating value is called the *higher heating value*, HHV, and the *lower heating value*, LHV, results if we retain the water in a vapor state. We can write

$$\text{HHV} - \text{LHV} = \text{Heat of vaporization of water formed during reaction} \tag{10–21}$$

and note that the heating values are tabulated in table B.7.

Example 10.7

A four-stroke engine having the *p-V* diagram shown in figure 10–14 has four cylinders and is operating at 4800 rpm. Using the polytropic Otto cycle and gas tables, determine the indicated horsepower and mechanical efficiency if bhp is found to be 183 horsepower. Also determine the bmep, the fuel heating value, and bsfc if 80 lbm of fuel are consumed per hour.

Solution For the engine considered, the ihp can be determined by the relation

$$\text{ihp} = Wk_{\text{cycle}} \times 4 \text{ cylinders} \times \frac{4800}{2} \text{ cycles/min}$$

and the work is determined from the enclosed area of the curve of figure 10–14. From the diagram we get the following data:

$$\begin{aligned} p_1 &= 14.7 \text{ psia} & V_1 &= 0.06 \text{ ft}^3 = V_4 \quad \text{(BDC)} \\ p_2 &= 125 \text{ psia} & V_2 &= 0.01 \text{ ft}^3 = V_3 \quad \text{(TDC)} \\ p_3 &= 550 \text{ psia} & p_4 &= 50 \text{ psia} \end{aligned}$$

We will also assume the air at state (1) has a temperature of 65°F. To determine the temperatures at the remaining states, we need the polytropic exponents of processes (1–2) and (3–4). For process (1–2)

$$\frac{p_1}{p_2} = \left(\frac{V_2}{V_1}\right)^n \quad \text{or} \quad \ln \frac{p_1}{p_2} = n \ln \frac{V_2}{V_1}$$

from which

$$\ln \frac{14.7}{125} = n \ln \frac{0.01}{0.06} = -\ln \frac{125}{14.7} = -n \ln 6$$

and

$$-\ln 8.5 = -n \ln 6$$

Therefore

$$n = \frac{\ln 8.5}{\ln 6} = \frac{2.14}{1.79} = 1.196 = n_{21}$$

For process (3–4)

$$\frac{p_3}{p_4} = \left(\frac{V_4}{V_3}\right)^n$$

$$\ln \frac{550}{50} = n \ln 6$$

yielding

$$n = \frac{\ln 11}{\ln 6} = 1.338 = n_{43}$$

From this we calculate the temperatures:

$$\frac{T_2}{T_1} = \left(\frac{V_1}{V_2}\right)^{n_{21}-1} = 6^{0.196} = 1.412$$

so

$$T_2 = T_1(1.42) = (65 + 460)(1.417) = 746\text{°R}$$

and, since process (1–4) is isometric

$$\frac{T_4}{T_1} = \frac{p_4}{p_1} = \frac{50}{14.7} = 3.4$$

giving us

$$T_4 = (525°\text{R})(3.4) = 1785°\text{R}$$

Also

$$\frac{T_3}{T_4} = \left(\frac{V_4}{V_3}\right)^{n_{43}-1} = 6^{0.338} = 1.832$$

yielding

$$T_3 = (1785)(1.832) = 3270°\text{R}$$

The work can then be easily obtained:

$$Wk_{\text{cycle}} = \frac{mR}{1-n}(T_4 - T_3 + T_2 - T_1)$$

$$= mR\left(\frac{T_4 - T_3}{1 - n_{43}} + \frac{T_2 - T_1}{1 - n_{21}}\right)$$

where

$$m = \frac{p_1 V_1}{RT_1} = \frac{14.7 \times 144 \times 0.06}{53.3 \times 525} = 0.00454 \text{ lbm}$$

Then

$$Wk_{\text{cycle}} = (0.00454 \text{ lbm})(53.3 \text{ ft-lbf/lbm} \cdot °\text{R})\left(\frac{1785 - 3270}{1 - 1.338} + \frac{746 - 525}{1 - 1.196}\right)$$

$$= 790 \text{ ft-lbf/cycle} = 1.016 \frac{\text{Btu}}{\text{cycles/cylinder}}$$

The ihp then is found from

$$\text{ihp} = Wk_{\text{cycle}} \times 4 \times 2400 = 1.016 \times 4 \times 2400$$
$$= 9754 \text{ Btu/min} = 9754 \text{ Btu/min} \times 1.41 \text{ hp} \cdot \text{s/Btu} \times 1 \text{ min/60 s}$$
$$= 229 \text{ hp}$$

Answer

Since the bhp is 183 hp, we easily calculate the mechanical efficiency from equation (10–16):

$$\eta_{\text{mech}} = \frac{\text{bhp}}{\text{ihp}} \times 100 = \frac{183}{229} \times 100 = 79.9\%$$

Answer

The bmep can be determined from equation (10–17):

$$\text{bmep} = \frac{\text{bhp}}{\text{stroke} \times \pi \times (\text{piston dia})^2/4 \times N \times n}$$

We know

$$\text{Cylinder displacement} = V_1 - V_2 = 0.05 \text{ ft}^3$$

$$\text{Cylinder displacement} = \text{stroke} \times \pi \times \frac{(\text{piston dia})^2}{4}$$

and then

$$\text{bmep} = \frac{183 \text{ hp}}{0.05 \text{ ft}^3 \times 2400 \text{ strokes/min} \times 4}$$

Since 33000 ft-lbf/min = 1 Btu we can readily give bmep the units of

$$\text{bmep} = \frac{183 \text{ hp} \times 33000 \text{ ft-lbf/min} \cdot \text{hp}}{0.05 \text{ ft}^3 \times 2400 \text{ strokes/min} \times 4} = 12581 \text{ lbf/ft}^2$$

or

$$\text{bmep} = 12581 \times \frac{1}{144} \text{ psi} = 87.4 \text{ psi} \qquad \textit{Answer}$$

The brake specific fuel consumption is obtained from equation (10–19):

$$\text{bsfc} = \frac{m}{(\text{bhp})\tau}$$

yielding

$$\text{bsfc} = \frac{80 \text{ lb/hr}}{183 \text{ hp}} = 0.437 \text{ lbm/bhp} \cdot \text{hr} \qquad \textit{Answer}$$

To obtain the fuel heating value, we notice that the energy increase (manifested by the pressure and the temperature increase) of process (2–3) represents approximately the energy released by the fuel during the combustion. From table B.6 we find, by interpolating

$$U_2 = 127.56 \text{ Btu/lbm @ } T_2 = 746°\text{R}$$
$$U_3 = 646.30 \text{ Btu/lbm @ } T_3 = 3270°\text{R}$$

and

$$U_3 - U_2 = 518.74 \times 0.00454 \text{ Btu/cycle} = 2.36 \text{ Btu/cycle}$$

The heating value of the fuel is the energy released by the fuel per unit mass of fuel. The energy released per minute is given by

$$2.36 \text{ Btu/cycle} \times 2400 \text{ cycles/min}$$

while the fuel consumed is 80/60 lbm/min. We therefore calculate the heating value HV from

$$\text{HV} = \frac{2.36 \text{ Btu/cycle} \times 2400 \text{ cycles/min}}{80/60 \text{ lbm/min}}$$

$$= 4248 \text{ Btu/lbm} \qquad \textit{Answer}$$

This value is significantly below the heating value of octane (or heat of combustion) listed in table B.7. The reason would no doubt be attributed to the incomplete combustion occurring in the test engine. This could be an area of improvement in the engine design.

10.6 Modifications

Let us discuss briefly the major design or technological changes or modifications made to Otto engines in attempting to increase efficiency, power, or some other performance parameter.

The designs of Otto engines have been altered to increase thermodynamic efficiency by increasing the piston stroke or decreasing the cylinder clearance volume. Equation (10–8) is invariably cited in justifying the increase of compression ratios in Otto engines, but while this relation is approximately correct for the actual cycles, the combustion of fuels is made less effective in some cases; and frequently a phenomenon known as *preignition* or *knock* is observed in high compression engines. Knock occurs because all fuel-air mixtures have a temperature at which they spontaneously combust or "explode." This temperature is reached in some cylinder chambers before the spark plug can ignite the mixture and consequently a sharp explosion ensues, producing a noise (knock) and a sharp reduction in power. This may be noted in figure 10–14 by visualizing the effects of the burning of fuel before the ignition point x (to the right on curve xb).

To combat preignition, the shape of the combustion chamber is frequently revised, as shown in figure 10–16. Designs are generally submitted to provide more continuous, expanding combustion of the fuel-air mixture with an elimination of potential corners or "hot spots" which could cause knock. The most common method of preventing preignition has been to add "ignition depressants," such as tetraethyl-lead, to fuel. It is also worth mentioning that the fuel required in high compression engines ($r_v \geq 6$, approximately) is a more refined extraction from crude oil than that required for low compression engines or steady state combustion processes. Water injected into the combustion chamber during compression has an effect similar to the additives (tetraethyl-lead) in reducing preignition.

It has been mentioned that the combustion process is that point of the Otto cycle during which energy is extracted from fuel and added to the

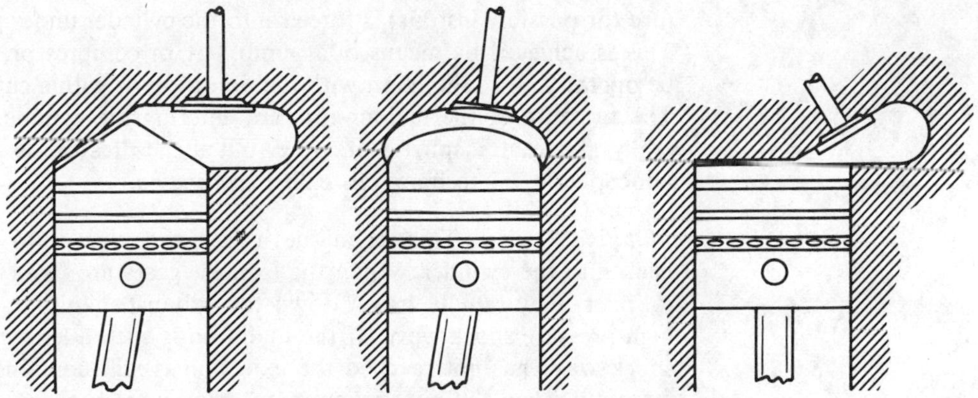

Figure 10–16 Possible configurations of pistons and cylinder chambers for improving the chemical-thermal-mechanical energy conversion in the fuel combustion process of the Otto engine.

cycle in the form of heat. We hope to extract the HHV or at least the LHV from the fuel, but this does not happen in the reciprocating IC engine. Combustion is a grossly irreversible process and in the Otto cycle, it is first suppressed from burning, then is ignited by a spark plug and finally extinguished — all this once every power revolution. The best that can be said for this combustion is that much higher temperatures are feasible than in a steady state combustion, because the materials making up the Otto engine will be subjected to the high temperatures for only small portions of the cycle. As a consequence, higher theoretical thermodynamic efficiencies [see equation (8–8)] are possible for the reciprocating IC engine than for engines which operate at steady state conditions.

Increasing the power or work of an engine is achieved by two general methods: (1) increasing the amount of fuel and air in the cylinder chamber; and (2) increasing the engine size, or displacement.

The second method is strictly a geometric problem. Increasing the piston and cylinder diameters is the only actual way to increase engine displacement, unless additional cylinders are added. The first method of increasing power is carried out in numerous ways: (1) fuel injection (2) supercharging, and (3) exhaust manifold tuning. We will discuss each briefly.

Fuel injection In this technique, fuel is forced into the cylinder by a pump. The devices are generally quite accurate in controlling the precise air/fuel ratio and they normally replace the carburetor. The carburetor, a device which allows air to mix with fuel by natural convection, is normally used to furnish the fuel-air mixture to the engine.

Supercharging This is a procedure in which the fuel and air mixture (or possibly just air) is forced into the cylinder under pressure. This is achieved by means of a pump, fan, or compressor, and can be operated in conjunction with fuel injection or with a carburetor. The pressure of the fuel-air mixture entering the cylinder is generally above atmospheric pressure, but if the pressure is less than atmospheric, the technique is called *scavenging.*

Manifold Tuning This technique uses a shock wave to force air back into the cylinder. When the exhaust gases are freed from the cylinder, they rapidly travel down the exhaust manifold, led by a high pressure shock wave. If the manifold is such a length that the shock wave has not reached the exit, and is still contained in the manifold when the piston begins to draw a fresh charge into the cylinder chamber, the shock wave may reverse and force the exhaust gases back into the cylinder, thus increasing the mass in the cylinder. This technique is not common in commercial engine design. There are, of course, many other methods of modifying the Otto engine to improve its performance. A brief superficial sketch is given; for more comprehensive information you are referred to appendix C.

The great use of the Otto engine is attributed to the fact that it is an internal combustion device delivering a positive, forceful output by means of a piston-cylinder. That is, it is expected to start easily and reach quickly its normal operating thermal condition, and it is expected to deliver high accelerations and decelerations when demanded. These characteristics are exhibited by the piston-cylinder Otto engine, but the configuration which provides those advantages is also the engine's greatest weakness. The combustion is unsteady (as we have said before); and the complicated mechanical devices required to keep the machine running (such as cams, valves, rocker arms, timing belts, distributors, oil and fuel pumps, crank shafts, connecting rods, and various assorted gadgets) are numerous, and as a result, it is noisy. It is a device which has caused a revolution in social patterns, but its continued popularity has stemmed more from vested financial interests than from its technical superiority over other power devices. Perhaps the most recent scientific improvement to come from this type of engine was the Diesel cycle concept introduced in 1892.

10.7 The Diesel Cycle

The ideal Diesel cycle is defined by the following four reversible processes:

(1–2) Adiabatic compression.
(2–3) Constant pressure heat addition.
(3–4) Adiabatic expansion.
(4–1) Constant volume heat rejection.

In figure 10–17 are shown the diagrams illustrating the paths of these processes on p-V and T-S coordinates, analogous to the Otto cycle. The enclosed area on the p-V plane is the work of the cycle, and the enclosed area on the T-S plane is the net heat added; the areas must be equal. The mechanism normally used to carry out these processes is the piston-cylinder device in figure 10–18. It has the same basic configuration and operation as that depicted in figure 10–2 except that no spark plug is present in the Diesel cycle and the fuel-air mixture is introduced by a fuel injector rather than by a carburetor and intake valve. Process (1–2) is a compression of a fresh charge of air and fuel (or just air); reaching state (2), the gases combust spontaneously due to the high pressure and the process continues at constant pressure to state (3). Most commonly, only air is introduced and compressed in process (1–2) while fuel is injected at a pressure near that of state (2). Process (2–3) classifies the Diesel engine as an internal combustion, compression-ignited engine (IC-CI engine).

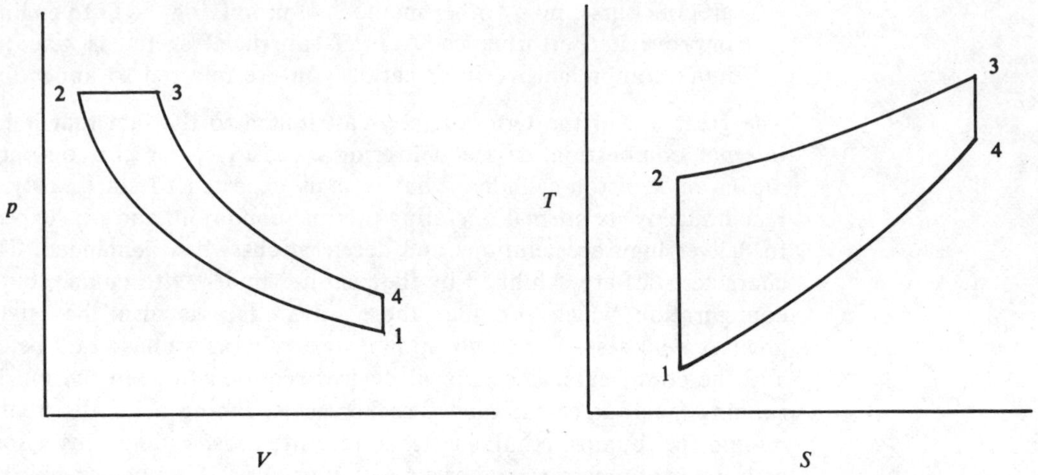

Figure 10–17 Property diagrams for ideal Diesel cycle.

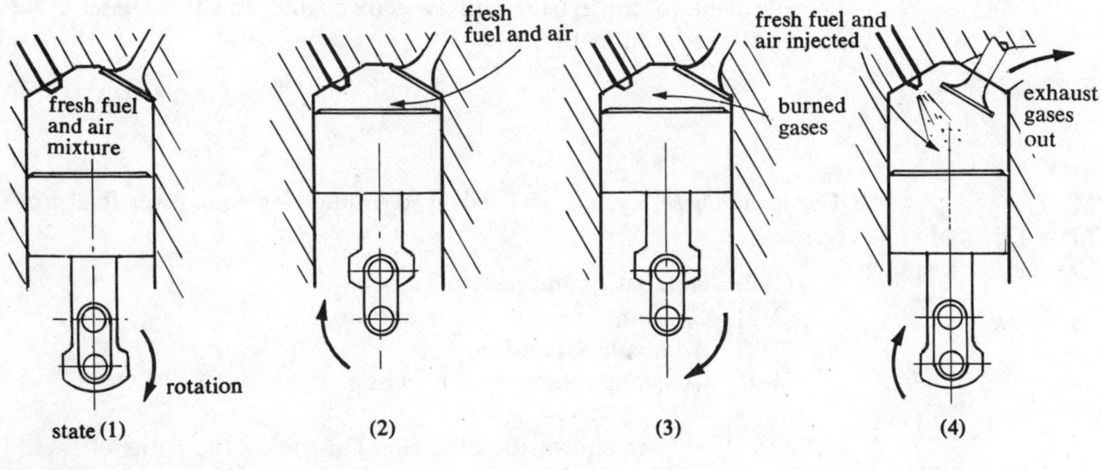

Figure 10–18 Physical operation of the Diesel engine.

Process (3–4) is an expansion of the exhaust gases until the piston reaches BDC (bottom dead center) at state (4). Then the exhaust is rejected into the atmosphere in process (4–1). Simultaneously air and fuel (or just fresh air) is injected into the chamber to be ready for a new cycle. This constitutes the two-stroke Diesel cycle with the exhaust and intake strokes of the piston together taking one revolution of the engine.

An actual Diesel engine is shown in figure 10–19. This illustration serves to give the reader a better picture of the major components of a Diesel engine. Notice the counter-weighted crankshaft located at the bottom of the engine. This is connected to the piston by means of a connecting rod. Four valves situated at the top of the cylinder are used for exhausting the spent gases. Midway down the cylinder, a row of openings around the

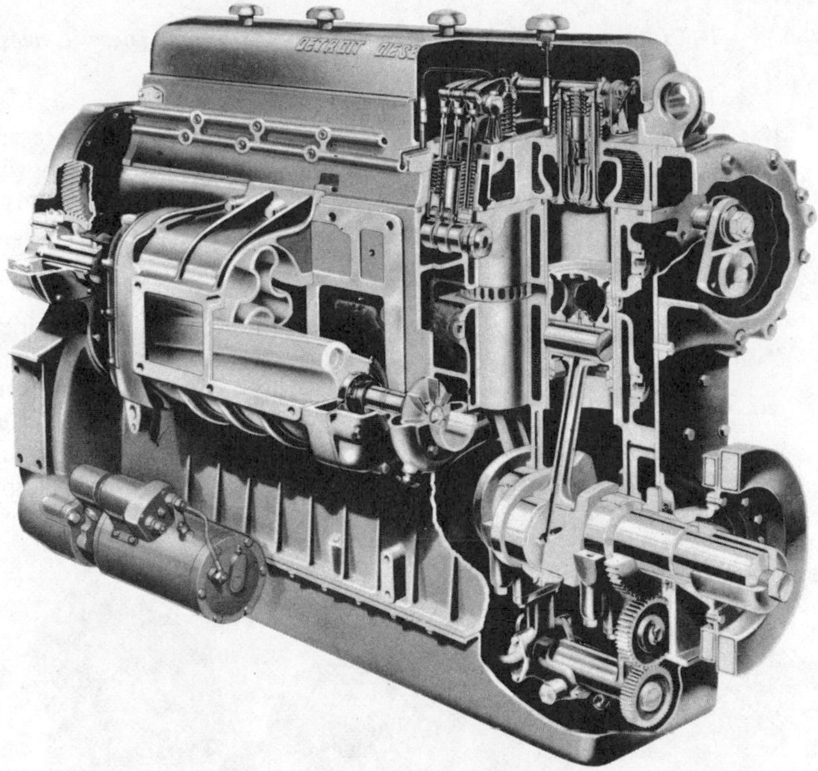

Figure 10–19 Cutaway view of typical Diesel engine. Reproduced with permission of Detroit Diesel Allison, Division of General Motors Corporation.

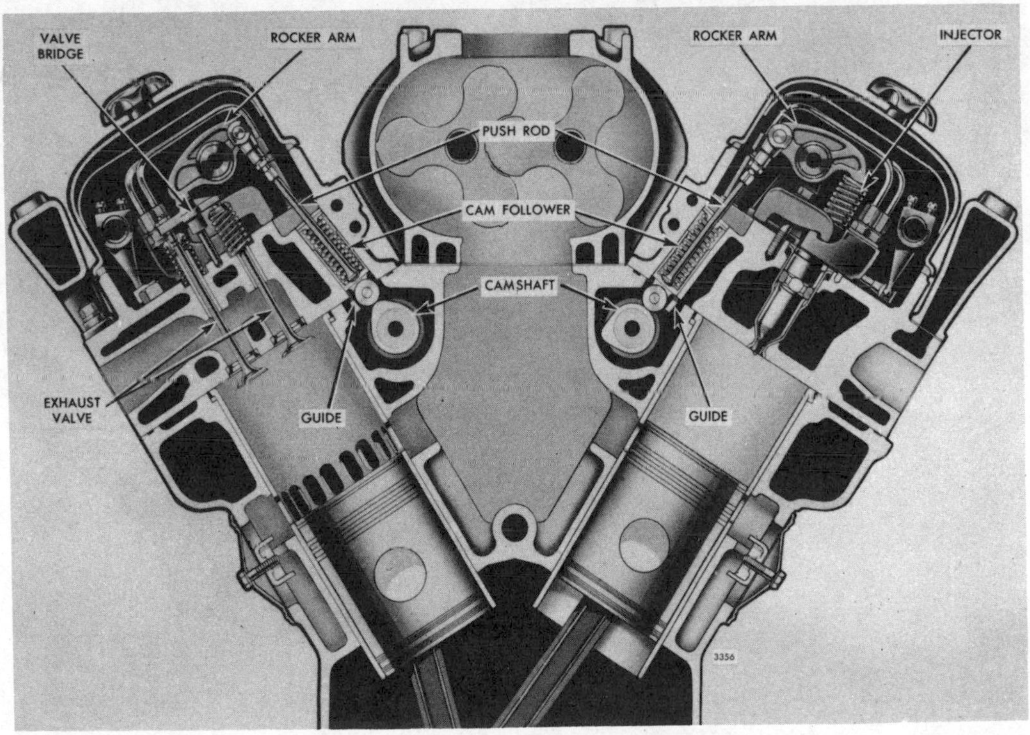

Figure 10–20 Cross-sectional view of typical Diesel engine. Reproduced with permission of Detroit Diesel Allison, Division of General Motors Corporation.

periphery of the cylinder can be seen. These are the intake openings where fresh air is introduced into the cylinder when the piston is at the bottom of its stroke. Fuel is injected from above after the compression stroke. The fuel injector used for this can be seen directly between the two exposed valves.

In figure 10–20 are shown the details of the critical parts which convert thermal energy to work through the piston-cylinder device. This view illustrates the manner of operation of the intake of air and fuel and the exhausting of the spent gases. At the top center of the figure are shown two three-lobed rotors which pump air into the center cavity where it is introduced into the cylinder through the peripheral openings in the cylinder walls when the piston is at the bottom of its travel. The exhaust

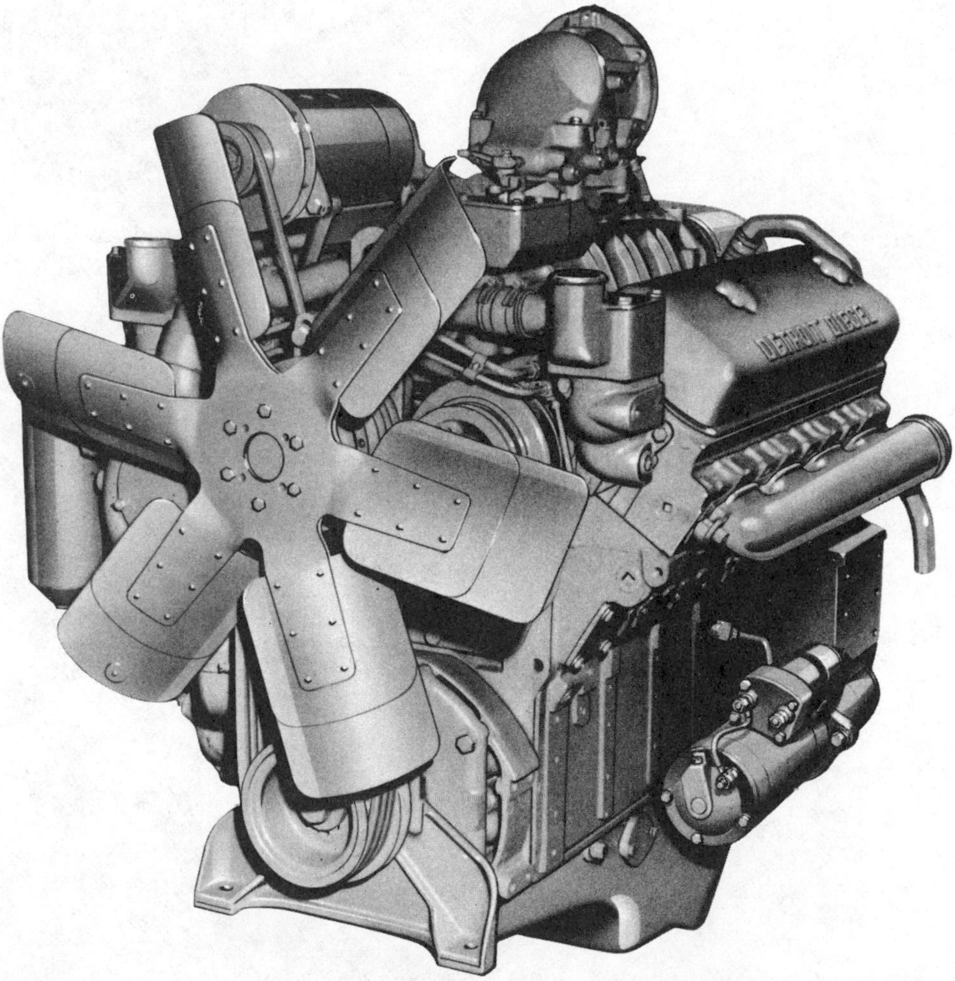

Figure 10–21 Diesel engine. Reproduced with permission of Detroit Diesel Allison, Courtesy of General Motors Corporation.

valves are actuated through a linkage system by the camshaft to allow the exhaust to escape from the cylinder at an appropriate time during the cycle. Notice that fuel is introduced into the cylinder through the injector while fresh air is introduced separately through openings in the cylinder walls. Finally, an external view of a typical Diesel engine is shown in figure 10–21. This view shows the external components of a typical Diesel engine. The large six-bladed fan is needed to draw air past a radiator to cool the engine when running. The radiator is not shown in this figure.

While the Diesel engine is commonly used in heavy applications such as trucks, tractors, and stationary power plants, a lighter engine has been used for some time in automotive applications. The Diesel, however, seems to be gaining in popularity here due to its economy and durability.

10.8 Air-Standard Analysis – Diesel Engine

To simplify calculations, the Diesel engine can be analyzed with the assumption that air is the only working medium and is a perfect gas. This is the air standard analysis and it ignores the fuel, except as a heat source. This was also done with the Otto engine. The work of a cycle, as was mentioned, is the enclosed area of the p-V plane which gives us the relation

$$Wk_{\text{cycle}} = Wk_{12} + W_{23} - Wk_{34} \tag{10–22}$$

For an air standard analysis we have

$$Wk_{12} = \frac{1}{1-k}(p_2V_2 - p_1V_1) \tag{10–23}$$

or

$$Wk_{12} = \frac{mR}{1-k}(T_2 - T_1) \tag{10–24}$$

Also,

$$Wk_{23} = p_2(V_3 - V_2) \tag{10–25}$$

and

$$Wk_{34} = \frac{mR}{1-k}(T_4 - T_3) \tag{10–26}$$

These results can quickly give us the net work of an ideal cycle, provided we know the properties at the four cycle corners. If we wish to fit the Diesel cycle to actual engines and find that processes (1–2) and (3–4) are not adiabatic, polytropic Diesel processes can easily be substituted. Then the work can be obtained from equations (10–22), (10–23), (10–24), (10–25), and (10–26), but k will be replaced by the polytropic exponent

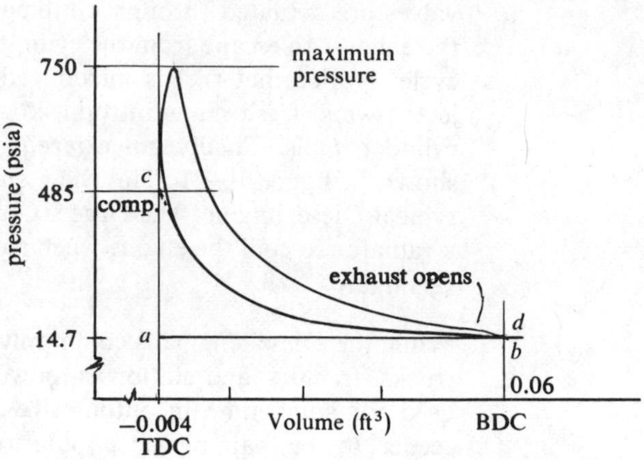

(*a*) slow speed, mechanical injection

Figure 10–22 Typical *p-V* diagram for compression-ignition engines operating at full load. Reproduced from E. F. Obert, *Internal Combustion Engines*, 2nd ed. (Scranton, 1959); with permission of International Textbook Company.

n. An analysis of the actual Diesel closely parallels the Otto cycle analysis. The terms bhp, ihp, mep, bmep, bsfc, stroke, bore, and *p-V* diagrams are all determined in a similar manner for the Diesel as for the Otto engine. A typical *p-V* diagram from an actual engine is shown in figure 10–22 from which we will develop an example problem to show the thermodynamic analysis of a Diesel engine.

Example 10.12 An ideal Diesel cycle engine operates with a compression ratio of 20 to 1, taking in air at 1.01 bars and 27°C. There is 0.05 kilogram of fuel added to each kilogram of air and the mixture behaves like air. During combustion the air-fuel mixture releases 2100 kJ/kg. Determine the optimum cutoff ratio r_c using gas table B.6.

Solution The cutoff ratio is defined by the relationship

$$r_c = \frac{V_3}{V_2}$$

and since process (2–3) is isobaric or constant pressure, we may write

$$\frac{V_3}{V_2} = \frac{T_3}{T_2}$$

We now determine these two temperatures with the aid of the data from table B.6. First, noting that the compression process (1–2) is reversible adiabatic, we have, from equation (10–15)

$$\frac{V_2}{V_1} = \frac{v_{r_2}}{v_{r_1}} = \frac{1}{20}$$

From table B.6 we find $v_{r_1} = 144.3$ at $T_1 = 300$ K. Then

$$v_{r_2} = \frac{144.3}{20} = 7.215$$

and from table B.6 we find that T_2 must be 1561 K. Also, the enthalpy at state (2) from this table is 969.5 kJ/kg. The heat added during process (2–3) is derived from the energy equation

$$q_{add} = \Delta u + wk_{cs}$$
$$= \Delta u + p\Delta v$$
$$= \Delta h = h_3 - h_2$$

from which we can find the enthalpy at state (3):

$$h_3 = q_{add} + h_2$$
$$= 2100 \text{ kJ/kg} + 969.5 \text{ kJ/kg} = 3069.5 \text{ kJ/kg}$$

From table B.6 we then find the temperature at (3) to be

$$T_3 = 2645 \text{ K}$$

The cutoff ratio may then be found from the equation

$$r_c = \frac{T_3}{T_2} = \frac{2645 \text{ K}}{1561 \text{ K}} = 1.69 \qquad \textit{Answer}$$

Example 10.13

From a two-stroke, six-cylinder Diesel engine operating at 1200 rpm is developed the p-V diagram shown in figure 10–22. If we use an air standard analysis and have the following additional information:

$$\text{bhp} = 115 \text{ hp @ } 1200 \text{ rpm}$$
$$= 58 \text{ hp @ } 600 \text{ rpm}$$

and 75.0 lbm/hr fuel usage at 1200 rpm, determine the Wk_{cycle}, ihp, mechanical efficiency, mep, and bsfc.

Solution

The following property values we extract from the curves of the p-V diagram in figure 10–22:

$p_1 = 14.7$ psia	$V_1 = 0.06 \text{ ft}^3 = V_4$
$p_2 = 750$ psia	$V_2 = 0.004 \text{ ft}^3$
$p_3 = 750$ psia	$V_3 = 0.008 \text{ ft}^3$ (estimated)
$p_4 = 20$ psia (estimated)	$T_1 = 70°\text{F}$ (assumed)

We now recognize that the polytropic exponents must be determined. This we do with the following calculations:

$$\frac{p_1}{p_2} = \left(\frac{V_2}{V_1}\right)^n$$

$$n_{21} = \frac{\ln (p_1/p_2)}{\ln (V_2/V_1)}$$

$$= \frac{\ln(14.7/750)}{\ln(0.004/0.06)} = \frac{\ln(750/14.7)}{\ln(0.06/0.004)} = \frac{3.94}{2.71} = 1.45$$

Similarly

$$\frac{p_3}{p_4} = \left(\frac{V_4}{V_3}\right)^n$$

and

$$n_{43} = \frac{\ln(p_3/p_4)}{\ln(V_4/V_3)} = \frac{\ln(750/20)}{\ln(0.06/0.008)} = \frac{3.62}{2.01} = 1.80$$

The compression ratio for this engine is 15 to 1 and the cutoff ratio, defined as V_3/V_2, can be computed as

$$r_c = \frac{V_3}{V_2} = \frac{0.008}{0.004} = 2$$

The amount of air introduced into the cylinder for each power stroke is obtained from the perfect gas equation

$$p_1V_1 = mRT_1$$

or

$$m = \frac{p_1V_1}{RT_1} = \frac{14.7 \text{ lbf/ln}^2 \times 144 \text{ in}^2/\text{ft}^2 \times 0.06 \text{ ft}^3}{53.3 \text{ ft-lbf/lbm°R} \times (460 + 70)\text{°R}}$$

$$= 0.0045 \text{ cycles/cylinder}$$

Since the engine is a two-stroke cycle arrangement, the number of cycles equals the number of revolutions. Then

$$m = 0.0045 \text{ revolutions/cylinder}$$

The work per cycle or revolution is obtained after we determine the temperature at states (2), (3), and (4):

$$T_2 = T_1 \left(\frac{V_1}{V_2}\right)^{n_{21}-1} = 530\text{°R} \left(\frac{0.06}{0.004}\right)^{0.45} = 1792\text{°R}$$

$$T_4 = T_1 \left(\frac{p_4}{p_1}\right) = 530\text{°R} \left(\frac{20.0}{14.7}\right) = 721\text{°R}$$

and

$$T_3 = T_2 \left(\frac{V_3}{V_2}\right) = 1792\text{°R} \left(\frac{0.008}{0.004}\right) = 3584\text{°R}$$

Check each process to make sure these equations are applicable.

Now the work is calculated:

$$Wk_{\text{cycle}} = \frac{mR}{1 - n_{21}}(T_2 - T_1) + p_2(V_3 - V_2) + \frac{mR}{1 - n_{43}}(T_4 - T_3)$$

$$= \left(\frac{0.0045 \text{ lbm} \times 53.3 \text{ ft-lbf/lbm} \cdot {}^\circ\text{R}}{1.00 - 1.45}\right)(1792 - 530)\ {}^\circ\text{R}$$

$$+ (750 \text{ lbf/in}^2 \times 144 \text{ in}^2/\text{ft}^2)(0.008 - 0.004)\text{ft}^3$$

$$+ \frac{0.0045 \text{ lbm} \times 53.3 \text{ ft-lbf/lbm} \cdot {}^\circ\text{R}}{1.0 - 1.8}(721 - 3584)\ {}^\circ\text{R}$$

$$= (-673 + 432 + 861) \text{ ft-lbf/cycle}$$

$$= 620 \text{ ft-lbf/cycle} = 0.797 \text{ Btu/cycle}$$ *Answer*

The total work for all six cylinders is

$$Wk_{\text{cycle}} = 6 \times 0.797 = 4.782 \text{ Btu/cycle}$$ *Answer*

and the indicated horsepower at 1200 rpm is obtained from

$$\text{ihp} = Wk_{\text{cycle}} \times N = 4.782 \times 1200 \text{ Btu/min} = 5738 \text{ Btu/min}$$

$$= 5738 \text{ Btu/min} \times \frac{60 \text{ min/hr}}{2545 \text{ Btu/hp} \cdot \text{hr}} = 135.3 \text{ hp}$$ *Answer*

Since the brake horsepower is 115 horsepower at 1200 rpm, the mechanical efficiency may be obtained from equation (10–16):

$$\eta_{mech} = \frac{\text{bhp}}{\text{ihp}} \times 100 = \frac{115}{135.3} \times 100 = 84.9\%$$ *Answer*

The mean effective pressure is obtained from equation (10–9):

$$\text{mep} = \frac{Wk_{\text{cycle/cylinder}}}{\text{cylinder displacement}} = \frac{620 \text{ ft-lbf/cycle}}{(0.06 - 0.004) \text{ ft}^3}$$

$$= 11{,}071 \text{ lbf/ft}^2 = 76.8 \text{ psi}$$ *Answer*

The bmep we calculate from equation (10–17):

$$\text{bmep} = \frac{\text{bhp}}{(\text{displacement})(N)(n)}$$

$$= \frac{115 \text{ hp} \times 33000 \text{ ft-lbf/min} \cdot \text{hp}}{(0.054 \text{ ft}^3)(6 \text{ cylinders})(1200 \text{ rpm})}$$

$$= 9750 \text{ lbf/ft}^2 = 67.7 \text{ psi}$$ *Answer*

The fuel consumption bsfc is calculated from equation (10–19):

$$\text{bsfc} = \frac{m}{(\text{bhp})\tau} = \frac{75 \text{ lbm/hr}}{115 \text{ bhp}}$$

$$= 0.652 \text{ lbm/bhp} \cdot \text{hr}$$ *Answer*

10.9 Diesel-Otto Comparison

The most common parameter for comparing various engines is the thermodynamic efficiency. Let us derive relationships for the efficiency of the Diesel engine and see how they compare with the Otto cycle. We first use the standard definition of thermodynamic efficiency

$$\eta_T = \frac{Wk_{\text{cycle}}}{Q_{\text{add}}} \times 100$$

and notice that for the Diesel cycle with a perfect gas having constant specific heats

$$Q_{\text{add}} = mc_p(T_3 - T_2) \tag{10–29}$$

and

$$\begin{aligned} Wk_{\text{cycle}} &= Q_{\text{add}} + Q_{\text{rej}} \\ &= mc_p(T_3 - T_2) + mc_v(T_1 - T_4) \end{aligned}$$

The efficiency can then be written

$$\begin{aligned} \eta_T &= \left[\frac{mc_p\,(T_3 - T_2) + mc_v\,(T_1 - T_4)}{mc_p\,(T_3 - T_2)}\right] \times 100 \\ &= \left[1 + \frac{c_v\,(T_1 - T_4)}{c_p\,(T_3 - T_2)}\right] \times 100 = \left[1 - \frac{1}{k}\left(\frac{T_4 - T_1}{T_3 - T_2}\right)\right] \times 100 \end{aligned} \tag{10–30}$$

This can be put in other terms (see problem 10.28) by a little more manipulation to give us

$$\eta_T = \left[1 - \frac{1}{(r_v)^{k-1}}\left(\frac{r_c^k - 1}{k(r_c - 1)}\right)\right] \times 100 \tag{10–31}$$

For the polytropic processes in the Diesel cycle, equation (10–31) does not hold, but equation (10–30) does apply. A comparison now between the Diesel and Otto cycles shows that at a given compression ratio, the Otto engine is theoretically more efficient [compare equation (10–31) to equation (10–8)] than the Diesel. This is further exemplified by the superimposed Otto cycle on the Diesel cycle in figure 10–23. It is shown

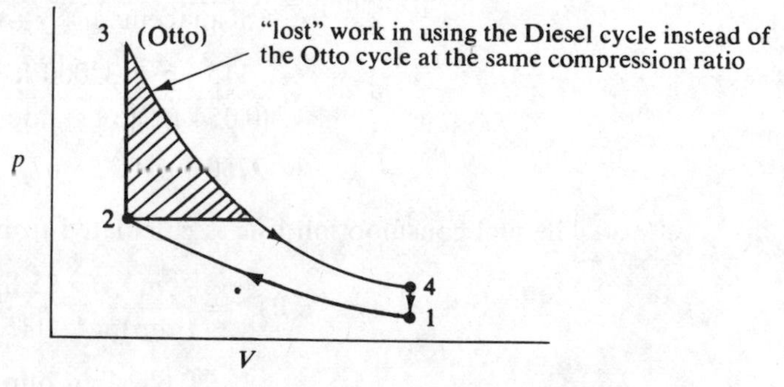

Figure 10–23 p-V diagram comparison of Otto and Diesel cycles at the same compression ratio.

that the shaded area represents work realized by the Otto engine and not by the Diesel. The primary reason for developing the Diesel, however, has been that higher compression ratios could be achieved. The fuel in an Otto engine exhibits preignition at elevated compression ratios, but in the Diesel it is just this preignition that caused the fuel to burn. Compression ratios can then be increased to values only limited by material strength and dynamics of the piston-cylinder. Most Diesel engines have thermodynamic efficiencies that are higher than those of Otto engines, because of the increased compression. As a side benefit, the Diesel can generally utilize a less refined fuel than the Otto. In fact, the inventor and first developer of the Diesel, Rudolph Diesel, visualized coal dust as the primary fuel, but this has since been replaced by liquid fuels. At any rate, the Diesel can generally use a less expensive fuel and use it more efficiently than the Otto engine.

As a review, we list the equations of the Otto and Diesel cycles in table 10–2, which can be used for elementary thermodynamic analyses.

10.10 Engine Design Considerations

In previous sections we have discussed the detailed calculations for predicting the energy transfers of the internal combustion engines, both Otto and Diesel. In addition, the Otto cycle has been discussed from the standpoint of modifications to make its performance more desirable. Here let us consider very briefly some general aspects of IC engine design, presented only to give the reader an overview of the inherent problems and advantages of these engines. There are a number of books, some of which are listed in appendix C, which can furnish finer detail. Here we will treat (1) engine balancing; (2) speed limitation; (3) gas flows and manifolding; (4) engine timing; (5) cooling systems; and (6) advantages of the IC engine.

Engine balancing The piston-cylinder device which reciprocates invariably transfers its mechanical work to a rotating shaft by means of a crank. This mechanism is shown in figure 10–24 with a counterweight superimposed to show its position on the crank. The counterweight is used to prevent extreme imbalance in the crank and rotating shaft when the power stroke of the piston produces a force on the crank. This method of balancing is complimented by reducing the mass of pistons, values, connecting rods, and other moving parts. However, a perfectly balanced reciprocating engine is impossible to achieve.

Speed Limitations A device which has an imbalance (even a small amount) will be subject to dynamic effects that become extremely intolerable at high speeds. Therefore, reciprocating engines have definite inherent speed limitations due to their characteristics of the piston-cylinder crank mechanism. In addition, the Otto and Diesel

Table 10–2

Equations for IC Engines

Term	*Otto*	*Diesel*
Q_{add}	$mc_v(T_3 - T_2)$	$mc_p(T_3 - T_2)$
Q_{rej}	$mc_v(T_1 - T_4)$	$mc_v(T_1 - T_4)$
Wk_{cycle} (ideal)	$\frac{mR}{1-k}(T_2 - T_1 + T_4 - T_3)$	$\frac{mR}{1-k}(T_2 - T_1 + T_4 - T_3) + p_2(V_3 - V_2)$
Wk_{cycle} (polytropic)	$\frac{mR}{1-n_{21}}(T_2 - T_1) + \frac{mR}{1-n_{43}}(T_4 - T_3)$	$\frac{mR}{1-n_{21}}(T_2 - T_1) + \frac{mR}{1-n_{43}}(T_4 - T_3) + p_2(V_3 - V_2)$
r_v	V_1/V_2	V_1/V_2
r_c	N A*	V_3/V_2
η_T (ideal)	$1 - \frac{1}{(r_v)^{k-1}}$	$1 - \frac{1}{(r_v)^{k-1}} \frac{r_c^k - 1}{k(r_c - 1)}$
mep or imep	$\frac{Wk_{cycle}}{\text{displacement}} = \frac{\text{ihp}}{(\text{displacement} \times N)}$	
bmep	$\frac{\text{bhp}}{(\text{displacement} \times N)}$	
η_{mech}	$\frac{\text{bhp}}{\text{ihp}} \times 100$	
ihp	$Wk_{cycle} \times N$	
N	engine speed in rpm or cycles/min	

* Not Applicable

engines utilizc intake and exhaust valves which have response limitations, either in mechanical or hydraulic actuation. That is, valves can physically be opened in a finite time, which will produce an upper limit to engine speeds. Of course, we have not mentioned irreversible effects, which increase with speed and which can become overwhelming in the expansion process where the gases are expected to produce the work of the engine.

The manner in which engine speed affects engine performance is seen in figure 10–25. Here is a typical performance chart which is

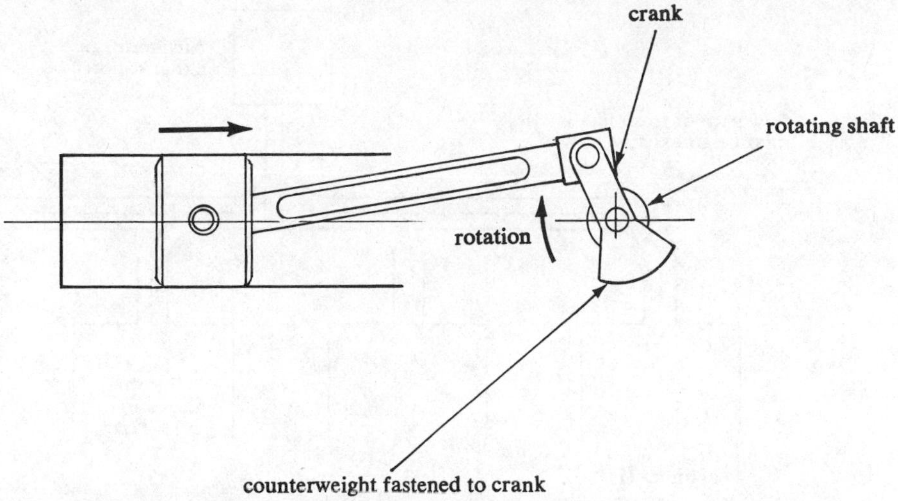

Figure 10–24 Piston-cylinder-crank mechanism of the reciprocating engine of the Otto and/or Diesel cycles.

developed from engine tests. Thermodynamics can give insights into why the parameters behave as they do, but cannot now accurately predict the full shapes of the curves. Notice that the individual curves do not have maximum or minimum values at the same speeds; that is, maximum power is developed at a higher speed than efficiency in the engine of figure 10–25. The optimization process, or selection of the speed which gives the all around best performance, is a problem which must be considered in all engine design or selection methods. We will not pursue this further except to say that the example problems we have considered in this chapter are essentially

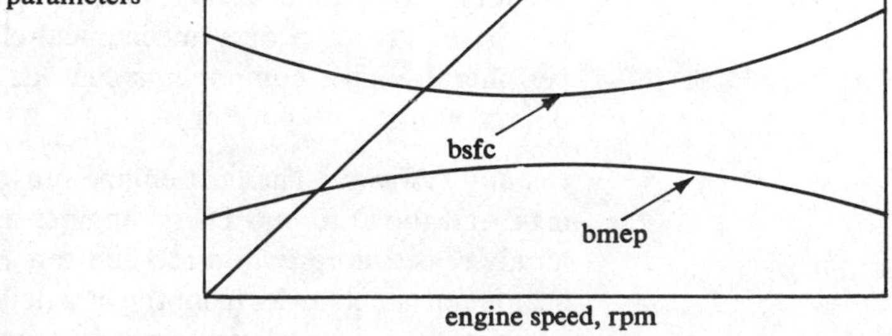

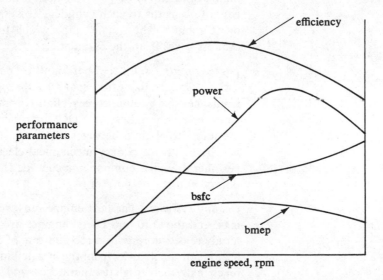

Figure 10–25 Typical performance chart of a reciprocating internal combustion engine.

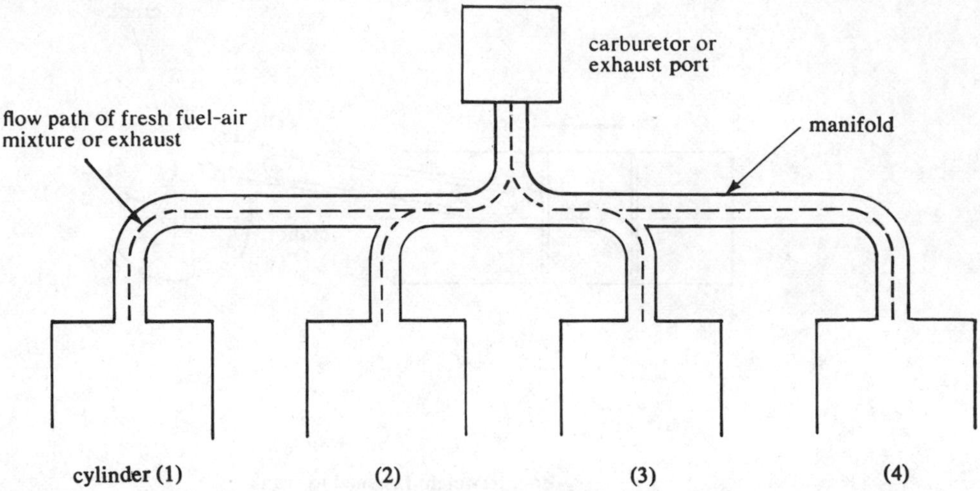

Figure 10–26 Typical configuration of a manifold and the flow pattern through it.

single speed conditions and the full analysis of engines must include the consideration of a performance chart as illustrated in figure 10–25.

Gas flows and manifolding Fuel injection is designed to replace carburetors and valves in supplying fuel to cylinders; however, the latter are more common. With the use of carburetors, however, there exists a problem of equal fuel-air mixture to the cylinders as illustrated in figure 10–26. Due to fluid or gas friction and the different distances, cylinders (2) and (3) will obviously get better service than cylinders (1) or (4). The design challenge here is to provide equal flow paths to each cylinder. This applies not only to the carburetor but to the exhaust pipe which tends to retard the flow of exhaust gases from the cylinder.

Engine timing With the spark ignition engine there exists the problem of synchronizing the spark to the piston position. The spark from the spark plug comes slightly before the piston has fully compressed the gases (at TDC) in general. As speeds are varied, the amount of time between the spark and TDC must also vary, which requires a proper mechanical-electrical coordination of the peripheral engine components such as the distributor, the ignition points, and the carburetor.

Cooling systems The heat engine must reject heat to the surroundings, and the Otto and Diesel engines are no exceptions. They commonly reject more than three-quarters of the heat added, and if this heat is not taken away from the engine block quickly, the engine will reach extremely high temperatures and permanently damage the materials. Air forced around the engine is the most obvious method

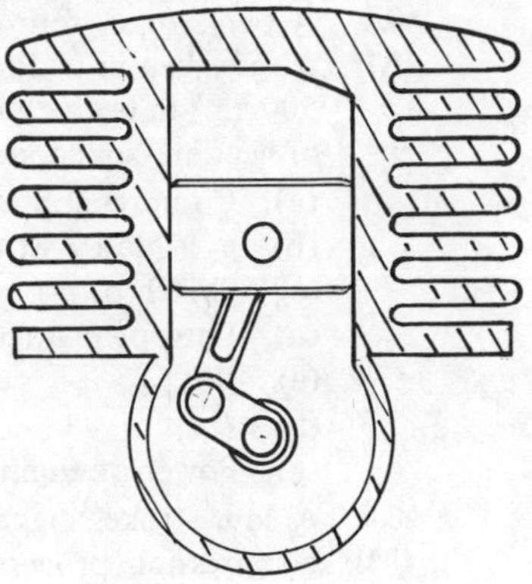

Figure 10–27 Typical cross section of air-cooled engine block.

of cooling, but this requires a large surface area to be effective. It is generally achieved by putting cooling fins or vanes in the engine block (see figure 10–27), an effective, but expensive process.

Water cooling is a common method of removing the rejected heat from the engine block. This is an effective method for large engines as well as small, but it requires additional equipment and thus reduces the reliability of the engine. Radiators, water pumps, and hoses are some of the items readily identified with a water cooling system.

Advantages of the IC reciprocating Engine We have pointed out a few of the problems of the IC engine, but let us reiterate its advantages. The reciprocating IC engine is an engine which is easily started, can produce its nominal power almost immediately upon starting, can be stopped easily and quickly, and can provide fast changes in speed (acceleration and deceleration).

The Diesel engine is well suited for providing large amounts of power at slow and medium speeds. It is frequently designed with large components to better withstand high compression ratios and thus provide an engine with extended lifetime.

Practice Problems

Those problems which use the SI units are indicated with the notation (M) under the problem number, and those using the English units are indicated with (E).

Section 10.2

10.1. **(M)** Determine the cylinder displacement and engine displacement of an engine having six cylinders with a bore of 102 millimeters and a stroke of 120 millimeters.

10.2. **(M)** Determine the compression ratio of an engine with a bore of 98 millimeters, a stroke of 100 millimeters, and a clearance volume of 75 cm^3.

10.3. **(M)** A two-stroke, four-cylinder, ideal Otto engine operates at 4800 rpm. Air is taken in at 1.01 bars and 20°C. The volume before compression is 500 cm^3 in each cylinder and 90 cm^3 after compression. If 250 kJ/kg of heat are added, determine
(a) Compression ratio.
(b) p, V, and T at the four corners.
(c) Sketch of p-V and T-s diagrams.
(d) Entropy change during compression.
(e) Wk_{cycle}.
(f) Q_{rej}.
(g) Power of engine.

10.4. **(M)** A four-stroke, eight-cylinder Otto engine operates at 2500 rpm. The compression process is reversible and polytropic, as is the expansion process. For both processes, $n = 1.3$ and the compression ratio is 6 to 1. The cylinder volume is 2000 cm^3 before compression and the intake air is at 1.01 bars and 30°C. Determine, if the heat addition is 900 kJ/kg
(a) Wk_{cycle}, or power.
(b) Thermodynamic efficiency.
(c) Heat rejected per cycle.

10.5. **(E)** Determine the displacement of an eight-cylinder engine having a bore of 3.25 inches and a stroke of 3 inches.

10.6. **(E)** Determine the stroke of an engine with a compression ratio of 8 to 1, a bore of 2.5 inches, and a clearance volume of 2 in^3.

10.7. **(E)** An ideal Otto engine with six cylinders operates on a four-stroke cycle with a compression ratio of 7.5 to 1. Given that $p_1 = 14.7$ psia, $T_1 =$ 100°F, $V_1 = 104$ in^3/cylinder, and the heat added is 400 Btu/lbm, determine for an air-standard analysis
(a) p and V at the four corners of the cycle.
(b) T at the four corners.
(c) Wk_{cycle}.
(d) Q_{rej}.
(e) Plot the p-V and T-s diagrams, labeling the coordinate points at (1), (2), (3), and (4).

10.8. **(E)** A four-stroke, two-cylinder Otto engine has a bore of 24 inches, a stroke of 20 inches, and a compression ratio of 7.2 to 1. The air taken in is at 14.7 psia and 40°F while the compression and expansion processes are reversible and polytropic ($n = 1.49$). If 1400 Btu/lbm air are added and the engine speed is 320 rpm, determine
(a) $\dot{W}k_{cycle}$.
(b) Thermodynamic efficiency.
(c) $\dot{Q}_{rej}$.

Section 10.3

10.9. **(M)** For the engine of problem 10.3, determine the thermodynamic efficiency.

10.10. **(M)** For the engine of problem 10.4, determine the thermodynamic efficiency and mep.

10.11. **(M)** Determine the mep of a six-cylinder engine which has a displacement of 630 cm^3 and which delivers 2600 J/cycle.

10.12. **(E)** For an eight-cylinder engine which has a stroke of 4 inches, a bore of 3.7. inches, and which delivers 1520 ft-lbf of work per cycle, determine the mep.

10.13. (E) For the engine of problem 10.7, determine the mep.

10.14. An Otto engine has a compression ratio of 6.8 to 1. What is its thermodynamic efficiency?

Sections 10.4 and 10.5

10.15. (M) If the fuel consumption is 5 kg/hr and the Bhp is 25 horsepower for the engine of problem 10.4, determine the mechanical efficiency, the bmep, and the bsfc.

10.16. (M) A four-stroke, four-cylinder engine, operating at 6000 rpm produces the *p-V* diagram shown in figure 10–28. Determine as best as possible

(a) Power.
(b) $\dot{Q}_{add}$.
(c) Thermodynamic efficiency.
(d) $\dot{Q}_{rej}$.

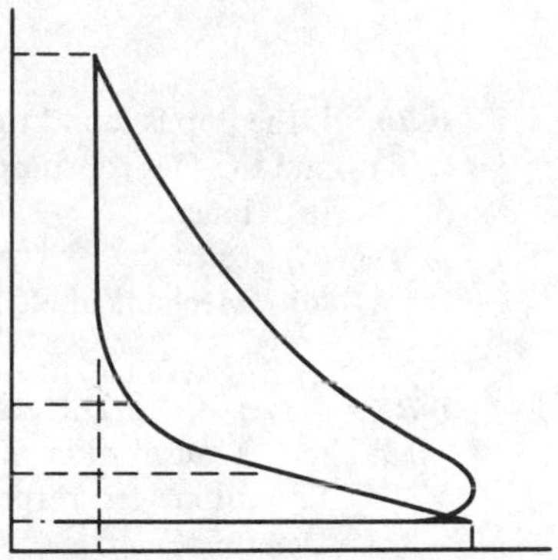

Figure 10–28

10.17. (M) For the engine of problem 10.16, assume the bhp is 180 horsepower and the fuel consumption is 600 g/min. Determine

(a) Mechanical efficiency.
(b) bmep.
(c) bsfc.

10.18. (E) An Otto engine operating on an ideal, four-stroke, air-standard cycle is required to develop 100 horsepower at 5000 rpm. It must take air in at 14.7 psia and 90°F, and is limited to four cylinders and a compression ratio of 6.5 to 1 and a maximum bore of 2.5 inches. If the proposed fuel to be used can deliver 850 Btu/lbm air, calculate

(a) p, v, and T at the four corners. Use gas table B.6.
(b) wk_{cycle}.
(c) Stroke. Use gas table B.6.
(d) $\dot{Q}_{rej}$. Use gas table B.6.

10.19. (E) An eight-cylinder four-stroke engine is operating at 6000 rpm when the *p-V* diagram is gotten (figure 10–29). Determine

(a) ihp.
(b) Thermodynamic efficiency.
(c) mep (or imep).

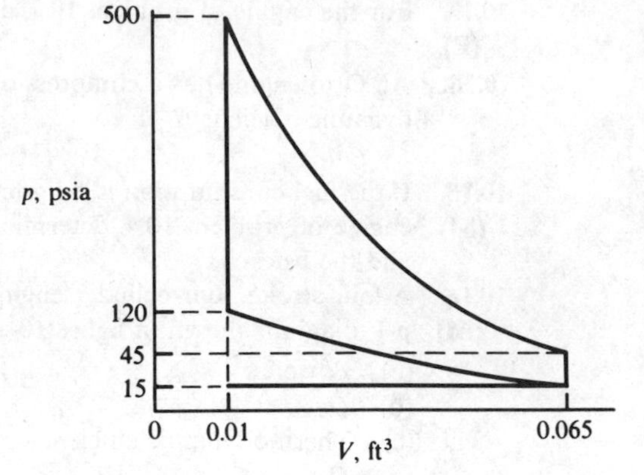

Figure 10–29

10.20. (E) If the bhp is found to be 450 horsepower for the engine of problem 10.19 and the fuel consumption is 400 lbm/hr, determine
(a) bmep.
(b) bsfc.
(c) Mechanical efficiency.

Sections 10.7, 10.8 and 10.9

10.21. (M) A two-stroke Diesel engine has six cylinders, a compression ratio of 15 to 1, a cutoff ratio of 3 to 1, and a bore and stroke of 14 centimeters and 12 centimeters respectively. The intake air is at 27°C and 1.01 bars. Determine
(a) p, V, T, at the cycle corners.
(b) Entropy change during process (1–2).
(c) Entropy change during process (2–3).
(d) Wk_{cycle}.
(e) Thermodynamic efficiency.

10.22. (M) If the engine of problem 10.21 consumes 0.9 kg/min of fuel when running at 3000 rpm, and delivers 800 hp. determine
(a) bsfc.
(b) bmep.
(c) Mechanical efficiency.

10.23. (M) There are 240 kJ/cycle of heat supplied to an ideal four-stroke, two-cylinder Diesel engine when running at 300 rpm. The following data are given for each cylinder:

$$p_1 = 1.01 \text{ bars}$$
$$T_1 = 27°\text{C}$$
$$V_1 = 0.04 \text{ m}^3$$
$$p_2 = 65 \text{ bars}$$

Using the air table B.6, determine
(a) Wk_{cycle}.
(b) mep.
(c) $\dot{W}k$.
(d) Thermodynamic efficiency.

10.24. **(M)** An ideal four-stroke, three-cylinder Diesel engine uses 0.4 kilogram of air at 1.01 bars and 20°C on each cycle. The compression ratio is 15 to 1 and the cutoff ratio is 2.6 to 1. Using the air table (B.6), determine

(a) Q_{add}.
(b) Q_{rej}.
(c) Wk_{cycle}.
(d) Thermodynamic efficiency.
(e) mep.

10.25. **(E)** A Diesel engine produces 1500 bhp at 280 rpm. It has eight cylinders, 18-inch bore, 18-inch stroke, and a two-stroke cycle. If 12 lbm/min of fuel with a LHV of 18,400 Btu/lbm are used and the imep is 82 psi, determine

(a) Mechanical efficiency.
(b) Thermodynamic efficiency.
(c) bmep.
(d) bsfc.

10.26. **(E)** There are 0.05 of fuel with an LVH of 18,000 Btu/lbm, and 0.9 lbm of air supplied to a Diesel engine. Given that

$$p_1 = 14.7 \text{ psi}$$
$$T_1 = 135°\text{F}$$
$$r_v = 14$$

determine r_c using the air tables.

10.27. **(E)** There are 91 Btu/cycle supplied to an ideal Diesel engine and 0.5 lbm of air at 14.7 psia and 130°F are supplied every cycle. At the end of compression, the pressure is 560 psia. If the engine is running at 900 rpm, determine (using the air table B.6), assuming a two-stroke cycle:

(a) r_v.
(b) r_c.
(c) Wk_{cycle}.
(d) $\dot{W}k_{cycle}$.
(e) Thermodynamic efficiency.
(f) mep.

10.28. From the efficiency equation

$$\eta_T = \frac{Wk_{cycle}}{Q_{add}} \times 100$$

derive the relation

$$\eta_T = \left[1 - \frac{1}{(r_v)^{k-1}}\left(\frac{r_c^k - 1}{k(r_c - 1)}\right)\right] \times 100$$

for an ideal Diesel cycle.

10.29. Derive a relation for the thermodynamic efficiency of a Diesel engine having polytropic processes for compression and expansion. Assume the polytropic exponents equal; i.e., $n_{21} = n_{43}$.

10.30. Compare the thermodynamic efficiencies of an Otto engine having a compression ratio of 10.5 to 1, and a Diesel engine having a compression ratio of 15 to 1, and a cutoff ratio of 2.5 to 1. Which is higher?

11

Gas Turbines and the Brayton Cycle

The *jet engine*, which extracts energy from a *gas turbine*, has become a popular power-producing device, replacing the reciprocating IC engine wherever higher power per unit engine mass, smoother operation, or increased maintainability is demanded. We will study the thermodynamics of the jet engine and other adaptations of the gas turbine in this chapter.

The *ideal Brayton cycle*, a thermodynamic heat engine cycle, will be defined and compared to the actual operation of the jet engine. To give the reader a better appreciation of the actual engines, the mechanics of the *gas turbine*, the *compressor*, the *combustor*, and *nozzles* and *diffusers* will be presented. These devices are the essential components of the common jet engine and a knowledge of their individual functions helps to make the Brayton cycle more believable.

The perfect gases and their processes, with which we are now familiar, will be used to analyze the Brayton cycle in its ideal configuration. We will see how power is mechanically extracted in a typical jet engine and in stationary power supplies.

By using gas tables, we will remove the perfect gas restriction from the Brayton cycle analysis. Here we will adhere closely to the ideal cycle and its constituting processes, but the generalization of using data from gas tables will provide a truer setting for Brayton cycle calculations.

Regenerative heating, the most common attempt to increase the efficiency of the Brayton cycle, will be considered in the context of a Brayton cycle application. Other variations of the gas turbine power plant, namely the *fan jet*, *ram jet*, and *pulse jet*, will be presented and briefly analyzed.

Finally, the *rocket engine* will be discussed, even though it is neither a gas turbine nor a complete heat engine. An elementary thermodynamic analysis of the solid and liquid rockets will then be made.

11.1 The Ideal Brayton Cycle and the Jet Engine

The ideal Brayton cycle is defined by the following four reversible processes:

(1–2) Adiabatic compression from state (1) to state (2).
(2–3) Constant pressure heat addition.
(3–4) Adiabatic expansion.
(4–1) Constant pressure heat rejection to state (1).

We can easily plot these four processes on property diagrams, as was done in figure 11–1 in *p-V* and *T-S* diagrams. Traditionally in describing a real operational engine, the Brayton cycle has been applied to the gas turbine engine. This engine has normally been utilized in powering vehicles (aircraft, trucks, and some experimental configurations) in which it is commonly called a *jet engine;* it has also been used in stationary, small electric power-generating units. In figure 11–2 is shown a typical arrangement of the critical components of the aircraft gas turbine, or jet engine. The major components of the typical gas turbine are listed in a cutaway view in figure 11–2: the fans; the compressor; the combustion chamber or combustor; the turbine and turbine nozzles; and the exhaust section. In the following two sections we will consider in detail the major components: the turbine, the compressor, and the combustor. In section 11.5, we will consider the complete engine and its cycle. Keep in mind that this is only an example and that there are many other engines which have physical characteristics different from those of the gas turbine, but which could be approximated by the Brayton cycle.

In figure 11–3 is shown the schematic diagram of the gas turbine operating on a Brayton cycle. If we apply the steady flow energy equation to the individual components and assume reversible processes and no significant kinetic or potential energy changes, we get

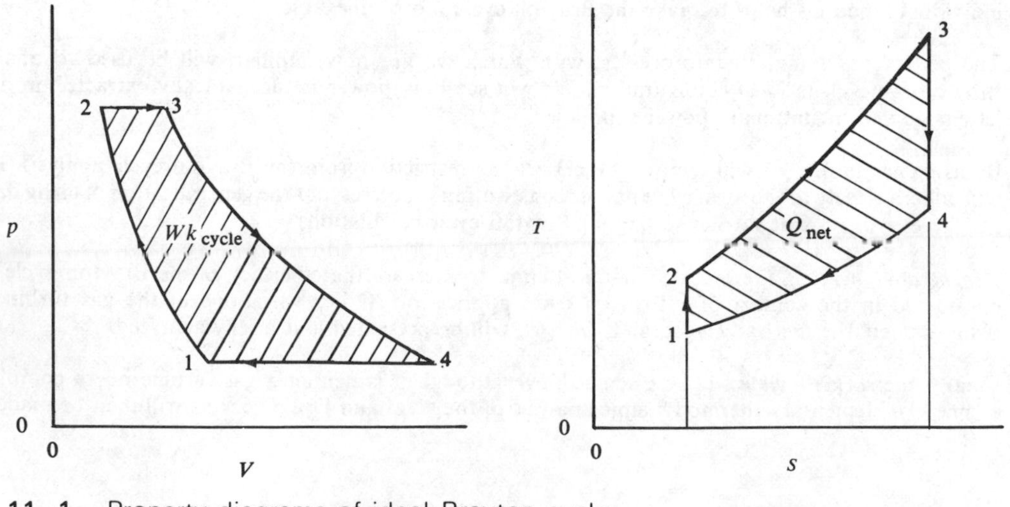

Figure 11–1 Property diagrams of ideal Brayton cycle.

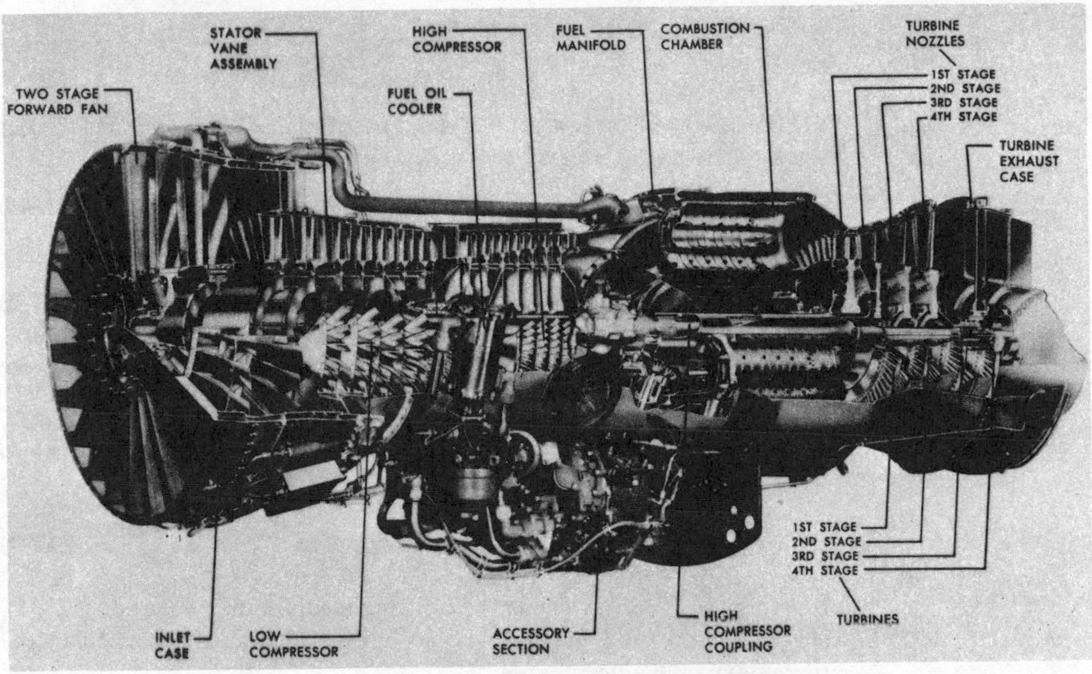

Figure 11–2 A typical aircraft gas turbine engine. Reproduced from "The Aircraft Gas Turbine Engine and Its Operation," August, 1970, ed.; with permission of Pratt & Whitney Aircraft Division.

1. for the compression (1–2):

$$h_2 - h_1 = -wk_{\text{comp}} \tag{11–1}$$

2. for the combustion (2–3):

$$h_3 - h_2 = q_{\text{add}} \tag{11–2}$$

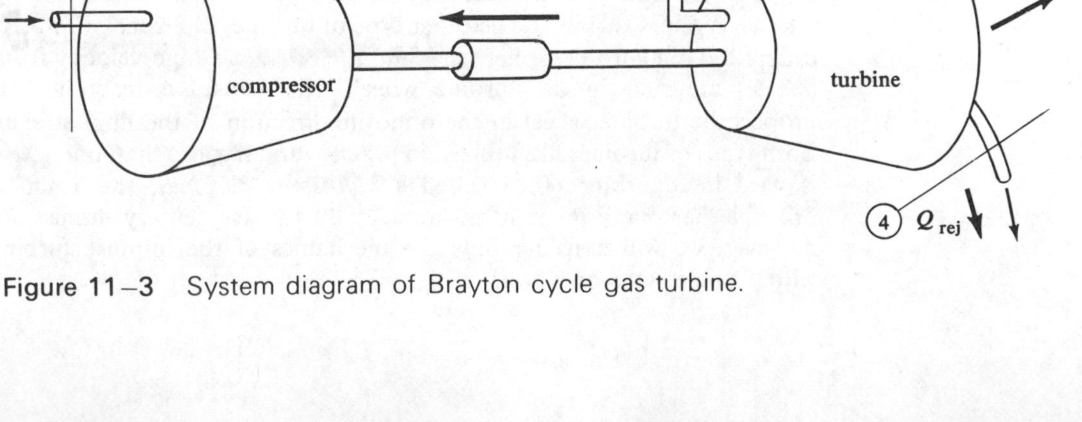

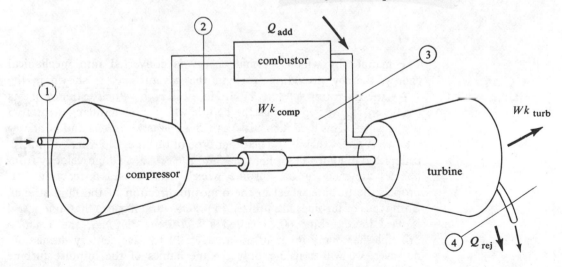

Figure 11–3 System diagram of Brayton cycle gas turbine.

3. for the turbine expansion (3–4):

$$h_4 - h_3 = -wk_{\text{turb}} \tag{11-3}$$

For the enclosed area in the *p-V* diagram of figure 11–1, which is the net work of the cycle, we get

$$wk_{\text{cycle}} = wk_{\text{turb}} + wk_{\text{comp}} \tag{11-4}$$

or

$$wk_{\text{cycle}} = h_3 - h_4 + h_1 - h_2 \tag{11-5}$$

The enclosed area in the *T-s* diagram must be the net heat added to the cycle and

$$q_{\text{net}} = q_{\text{add}} + q_{\text{rej}} \tag{11-6}$$

This also can be identified as

$$q_{\text{net}} = wk_{\text{cycle}} \tag{11-7}$$

If we assume a perfect gas as the working medium then equations (11–1), (11–2), (11–3), and (11–5) can be further altered. In particular the thermodynamic efficiency

$$\eta_T = \frac{Wk_{\text{cycle}}}{Q_{\text{add}}} \times 100 \tag{11-8}$$

reduces to

$$\eta_T = \left[1 - \frac{1}{(r_p)^{(k-1)/k}}\right] \times 100 \tag{11-9}$$

for perfect gases and reversible conditions (see problem 11.10) where r_p is the pressure ratio given by p_2/p_1 or p_3/p_4.

11.2 The Gas Turbine

The manner by which thermal energy is converted into mechanical energy, thereby providing work in the gas turbine, is shown in the schematic diagram of figure 11–4. Here the high velocity stream of gas (or liquid) is directed against the paddle wheel, thus inducing a rotation of the wheel. This arrangement is called an *impulse turbine* and is one of the two types of turbines. The other type of turbine, the *reaction turbine*, is depicted in figure 11–5; here the fluid is ejected at a high velocity from nozzles attached to the turbine wheel. This causes a reaction that propels the turbine wheel in the opposite direction of the fluid stream. Both types of turbines are utilized in practice and if more than one wheel is used in a turbine (then called a *multistage turbine*), the reaction principle has many inherent advantages. In this elementary discussion, however, we will consider only the mechanics of the impulse turbine with a single stage or wheel.

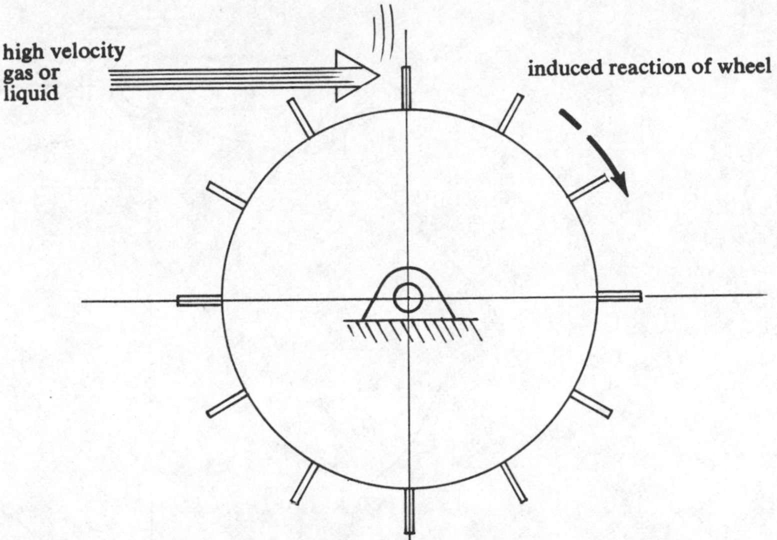

Figure 11–4 Basic principle of impulse turbine.

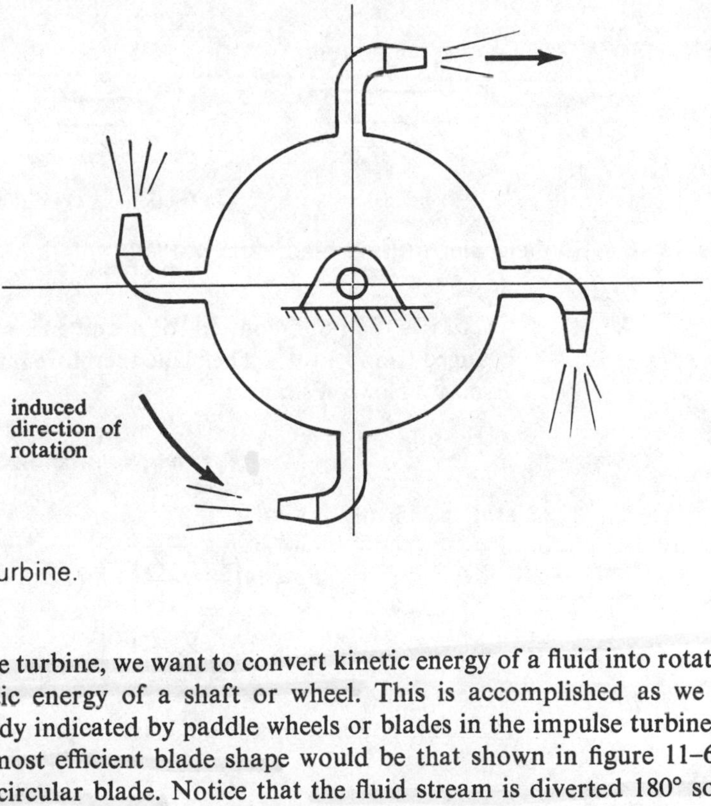

Figure 11–5 Reaction turbine.

In the turbine, we want to convert kinetic energy of a fluid into rotational kinetic energy of a shaft or wheel. This is accomplished as we have already indicated by paddle wheels or blades in the impulse turbine, and the most efficient blade shape would be that shown in figure 11–6, the semicircular blade. Notice that the fluid stream is diverted 180° so that its acceleration a_s is given by

$$a_s = \frac{d\bar{V}}{d\tau} = \frac{\bar{V}_2 - \bar{V}_1}{\tau} \qquad \textbf{(11–10)}$$

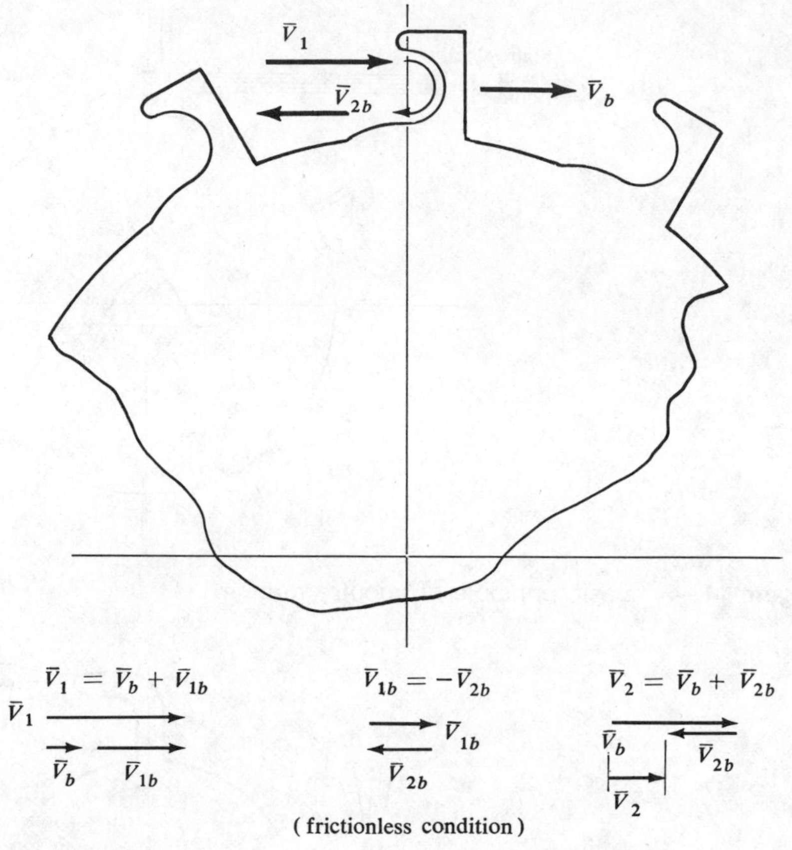

Figure 11–6 Semicircular turbine blade and velocity relationships.

where τ is the time required for a unit mass of fluid to have its velocity changed from $\bar{V}_1$ to $\bar{V}_2$. The blade therefore impresses, on the fluid, a force given by Newton's law.

$$F_s = ma_s = \text{force on fluid} \tag{11–11}$$

and specifically

$$\left.\begin{aligned} F_s &= m\left(\frac{\bar{V}_2 - \bar{V}_1}{\tau}\right) \qquad (F_s \text{ in N}) \\ &= \frac{1}{g_c} m\left(\frac{\bar{V}_2 - \bar{V}_1}{\tau}\right) \qquad (F_s \text{ in lbf}) \end{aligned}\right\} \tag{11–12}$$

The work gotten from the fluid stream per unit time τ is then, in SI units

$$Wk = -F_s x = mx\left(\frac{\bar{V}_1 - \bar{V}_2}{\tau}\right) \tag{11–13}$$

where x is the distance traveled by the blade during the unit time τ. The negative sign is shown to indicate that the work done by the fluid is determined by the force the fluid exerts on the blade ($-F_s$), and not by the blade force F_s. But

$$\frac{x}{\tau} = \bar{V}_b = \text{blade velocity} \tag{11–14}$$

so

$$Wk = m(\bar{V}_1 - \bar{V}_2)\bar{V}_b \tag{11–15}$$

and per unit of fluid mass

$$wk = (\bar{V}_1 - \bar{V}_2)\bar{V}_b \tag{11–16}$$

This, of course, is only true for a semicircular blade as shown in figure 11–6. Also, we have that the velocity of incoming fluid relative to the blade, $\bar{V}_{1b}$, is related to the actual velocity $\bar{V}_1$ by the equation

$$\bar{V}_1 = \bar{V}_b + \bar{V}_{1b} \tag{11–17}$$

and shown diagrammatically in figure 11–6. If the fluid is frictionless, then

$$\bar{V}_{1b} = -\bar{V}_{2b} \tag{11–18}$$

where $\bar{V}_{2b}$ is the outgoing fluid velocity relative to the blade and related to the actual velocity $\bar{V}_2$ by the equation

$$\bar{V}_2 = \bar{V}_b + \bar{V}_{2b} \tag{11–19}$$

It can then easily be shown that for the semicircular blade, the magnitude of the outgoing velocity $\bar{V}_2$ is given by the result

$$\bar{V}_2 = 2\bar{V}_b - \bar{V}_1$$

which then can be substituted into equation (11–16) to obtain

$$wk = (\bar{V}_1 - 2\bar{V}_b + \bar{V}_1)\bar{V}_b \tag{11–20}$$

or

$$wk = 2(\bar{V}_1\bar{V}_b - \bar{V}_b^2) \tag{11–21}$$

We see that the work is dependent on the blade and the fluid velocities. If we consider a fixed incoming velocity, the maximum work we can get will be determined from a well-known calculus theorem that states, in effect, that the work is greatest when its derivative with respect to its variable $\bar{V}_b$ is zero. Then, using equation (11–21) we get

$$\frac{dwk}{dV_b} = 2(V_1 - 2V_b) = 0$$

But this implies that, for maximum work,

$$\bar{V}_1 - 2\bar{V}_b = 0$$

or

$$\bar{V}_b = \frac{1}{2}\bar{V}_1 \tag{11–22}$$

That is, the work is greatest when the blade velocity is exactly one-half the magnitude of the incoming fluid velocity. The blade efficiency we define as the work extracted divided by the incoming fluid kinetic energy. This can be written

$$\eta_b = \left(\frac{(\bar{V}_1 - \bar{V}_2)\bar{V}_b}{\frac{1}{2}\bar{V}_1^2}\right) \times 100 \tag{11–23}$$

and, from equations (11–21) and (11–22) we get for the maximum efficiency

$$\eta_{b\,\max} = \left(4\frac{(\bar{V}_1\bar{V}_b - \bar{V}_b^2)}{\bar{V}_1^2}\right) \times 100 = \left(4\frac{\bar{V}_1^2/2 - \bar{V}_1^2/4}{\bar{V}_1^2}\right) \times 100$$
$$= 100\%$$

If the blade velocity drops below or goes above the value $\frac{1}{2}\bar{V}_1$ the efficiency will be less than 100% and ultimately reach zero efficiency as shown in figure 11–7. Obviously, this blade is ideal at a velocity of one-half the impinging fluid's velocity. There is, however, a difficulty associated with the semicircular blade; the fluid exiting the blade remains in an area which will be occupied by a following blade. This slow-moving fluid then acts as a resistance or barrier to the upcoming blade and produces a drop in the efficiency. In order to have a more practical arrangement for directing fluid past turbine blades, the configuration shown in figure 11–8(a) is used. Incoming fluid is directed at an angle θ into curved blades and subsequently ejected at an angle β. The side view in figure 11–8(b) more clearly shows the blade and fluid interaction, comparable to the simpler semicircular blade in figure 11–6. In figure 11–8(b) we are changing the fluid velocity into directions other than

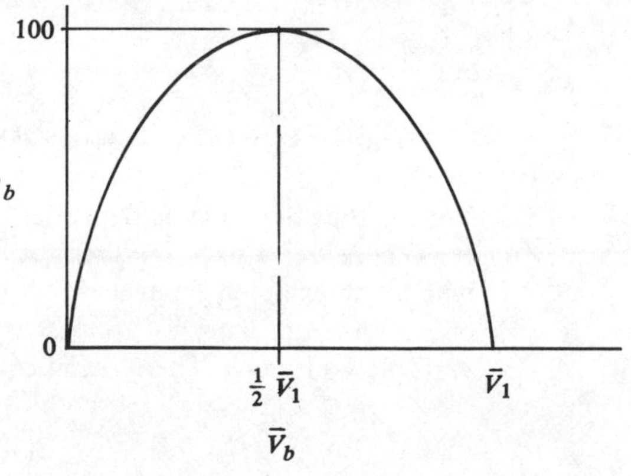

Figure 11–7 Blade efficiency of semicircular blade.

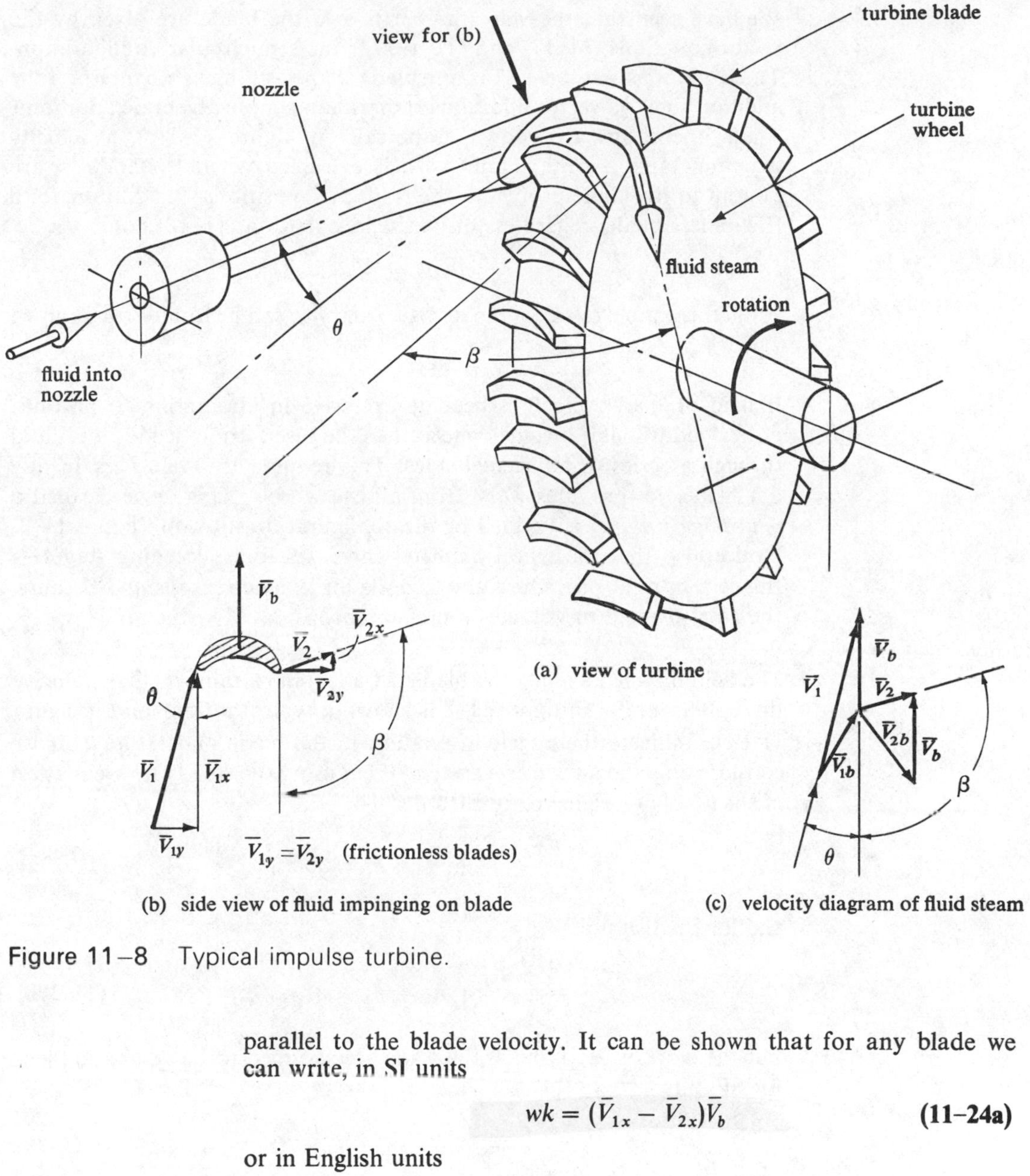

Figure 11–8 Typical impulse turbine.

parallel to the blade velocity. It can be shown that for any blade we can write, in SI units

$$wk = (\bar{V}_{1x} - \bar{V}_{2x})\bar{V}_b \tag{11–24a}$$

or in English units

$$wk = \frac{1}{g_c}(\bar{V}_{1x} - \bar{V}_{2x})\bar{V}_b \tag{11–24b}$$

where the $1x$ and $2x$ terms are just components of the velocity vectors:

$$\bar{V}_{1x} = \bar{V}_1 \cos\theta \tag{11–25}$$

$$\bar{V}_{2x} = \bar{V}_2 \cos\beta \tag{11–26}$$

We have seen that the velocities relative to the blade are given by the vector equations (11–17) and (11–19) for the semicircular configuration. These relations are correct for any blade shape as long as we interpret the addition signs as vector additions rather than simple algebraic additions. These operations are shown graphically in figure 11–8(c), a velocity diagram. Here the maximum work is extracted when $\bar{V}_{1b}$ and $\bar{V}_{2b}$ are tangent to the blade profiles at the respective positions. In addition, for a frictionless blade, regardless of its shape, equation (11–18) holds

$$|\bar{V}_{1b}| = |\bar{V}_{2b}|$$

From this simple relation, the terms $\bar{V}_2$ and $\bar{V}_{2x}$ can be found from known values for $\bar{V}_1$ and $\bar{V}_b$.

If fluid or blade velocities become excessive in attempting to produce power, additional turbine wheels can be used to cascade the fluid through a series of adjacent blades. This reduces the velocities in any one blade, yet provides work from all the wheels. This device, called a *multistage turbine*, is typified by the configuration shown in figure 11–2. Notice also that the wheel diameter increases for succeeding stages — this is a normal procedure and is done for kinetic reasons and because the gases expand in volume as pressure drops.

The fluid directed against the blades of a turbine attains its high velocity through a nozzle. In figure 11–8 is shown a typical nozzle, and in figure 11–9 we indicate the nozzle in relation to the blade and exhaust. If we consider only the nozzle as a system, then, using the steady flow equation of the first law, we have, for SI units

$$\frac{\bar{V}_1^2 - \bar{V}_0^2}{2} + h_1 - h_0 = q - wk_{os} \qquad \textbf{(11–27a)}$$

and for English units

$$\frac{\bar{V}_1^2 - \bar{V}_0^2}{2} + h_1 - h_0 = q - wk_{os} \qquad \textbf{(11–27b)}$$

but the work is zero and if the walls are assumed to be adiabatic then, for SI units

$$\frac{\bar{V}_1^2 - \bar{V}_0^2}{2} + h_1 - h_0 = 0 \qquad \textbf{(11–28)}$$

for the nozzle. For the whole turbine stage of nozzle, blade, and exhaust we get

$$\frac{\bar{V}_3^2 - \bar{V}_0^2}{2} + h_3 - h_0 = q - wk_{os}$$

and then the work wk_{os} is that done by the fluid in rotating the turbine wheel. If everything is adiabatic then

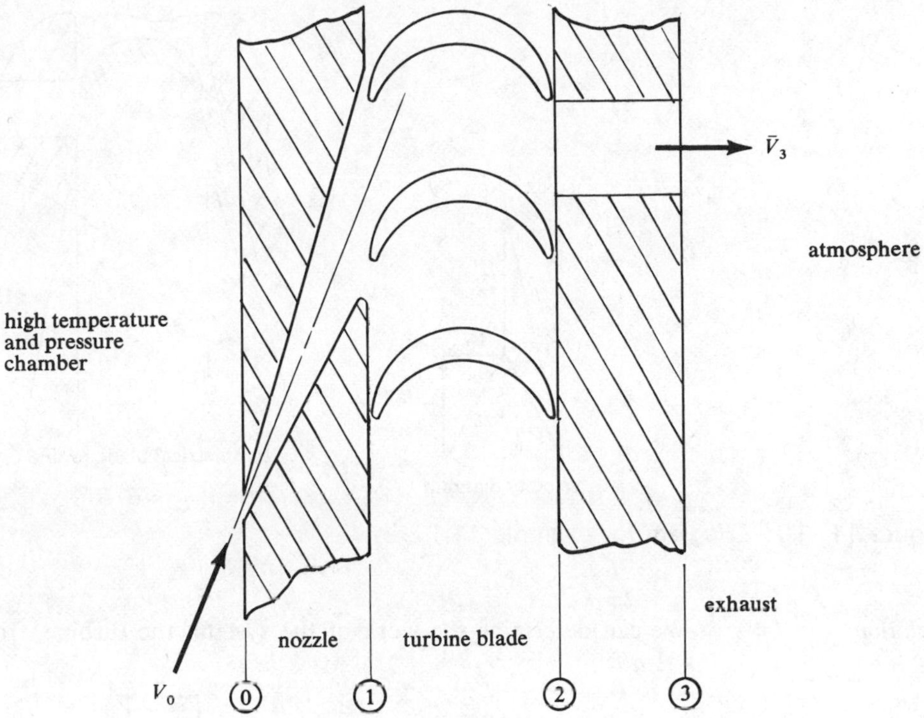

Figure 11–9 Typical turbine stage.

$$\frac{\bar{V}_3^2 - \bar{V}_0^2}{2} + h_3 - h_0 = -wk_{os} \qquad \textbf{(11–29)}$$

and we can simplify this by noting that $\bar{V}_0$ and $\bar{V}_3$ are negligible. Note that $\bar{V}_1$ is not negligible and is the whole reason for extracting work. Then

$$h_2 - h_1 = wk_{os} \qquad \textbf{(11–30)}$$

which is the work produced in a turbine if the process is frictionless and adiabatic, and if kinetic energy changes are neglected. The power is then found from

$$\dot{m}(h_2 - h_1) = -\dot{w}k_{os} \qquad \textbf{(11–31)}$$

which is a result similar to other processes. This equation is good with either SI or English units.

Example 11.1

A reversible adiabatic turbine receives 3.0 kg/s of air at 600°C and 20.0 bars. The air exhausts to a pressure of 1.01 bars. If the blade and nozzle profiles are those shown in figure 11–10 and the exhaust air temperature is 200°C, determine the turbine work per unit mass of air, the power extracted, the velocity of the air impinging on the blades, and the blade velocity.

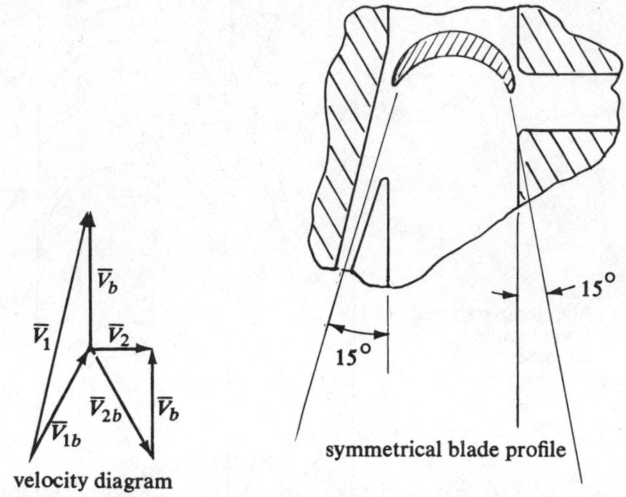

Figure 11–10 Diagram for example 11.1.

Solution We can determine the work of the system, the turbine, from equation (11–29)

$$-wk_{os} = h_3 - h_0 + \frac{\bar{V}_3^2 - \bar{V}_0^2}{2}$$

If we neglect the velocity of the air entering and leaving the turbine and assume constant specific heat values, we then have

$$-wk_{os} = c_p(T_3 - T_0)$$

Then

$$-wk_{os} = (1.007\ \text{kJ/kg} \cdot \text{K})(473\ \text{K} - 873\ \text{K})$$

or

$$wk_{os} = 402.8\ \text{kJ/kg}$$ *Answer*

The power is obtained from

$$\dot{W}k_{os} = \dot{m}wk_{os} \qquad \textbf{(11–32)}$$

which yields

$$\dot{W}k_{os} = (3\ \text{kg/s})(402.8\ \text{kJ/kg}) = 1208\ \text{kJ/s}$$

or

$$\dot{W}k_{os} = 1208\ \text{kW}$$ *Answer*

The blade velocity can be determined by first recalling equation (11–24a):

$$wk = (\bar{V}_{1x} - \bar{V}_{2x})V_b$$

and noting that, from equation (11–26)

$$\bar{V}_{1x} = \bar{V}_1 \cos 15°$$
$$V_{2x} = V_2 \cos 90° = 0$$

From the velocity diagram of figure 11–10 we see that

$$|\bar{V}_{1b}| = |\bar{V}_{2b}|$$

so

$$\bar{V}_2 = \bar{V}_{1y} = \bar{V}_1 \sin 15°$$

Also, from this same diagram

$$2\,\bar{V}_b = \bar{V}_1 \cos 15°$$

or

$$\bar{V}_b = \frac{V_1 \cos 15°}{2} \qquad \textbf{(11–33)}$$

Using these trigonometric results then in equation (11–24a) we get

$$wk = (\bar{V}_1 \cos 15°)\left(\frac{\bar{V}_1 \cos 15°}{2}\right)$$

but *wk* is equal to 402.8 kJ/kg, so

$$402.8 \text{ kJ/kg} = (\tfrac{1}{2})(\bar{V}_1 \cos 15°)^2$$

yielding, for V_1

$$\bar{V}_1 = \frac{\sqrt{(402.8 \text{ kJ/kg})(2)}}{\cos 15°}$$
$$= \frac{\sqrt{805.6 \times 10^3 \text{ N} \cdot \text{m/kg}}}{0.96593}$$
$$= \frac{\sqrt{805.6 \times 10^3 \text{ N} \cdot \text{m/N} \cdot \text{s}^2\text{/m}}}{0.96593}$$
$$= 929 \text{ m/s}$$ *Answer*

Since this represents a velocity greater than sound,* there would be phenomena occurring in the turbine which we have not accounted for. Let us merely note here that we would probably recommend a series of turbine stages in order that nozzle velocities may be reduced. At any rate the blade velocity expected from the above situation can be determined from equation (11–33)

$$\bar{V}_b = \frac{\bar{V}_1 \cos 15°}{2} = \frac{(929 \text{ m/s})(0.96593)}{2}$$
$$= 449 \text{ m/s}$$ *Answer*

* Velocity of sound is near 360 m/s or 1200 ft/s for common earth atmospheric conditions.

11.3 Combustors and Compressors

The high temperature gases furnished to the gas turbine nozzle are produced in the combustor. A combustor is merely a chamber at a constant pressure which allows for the burning or combustion of fuel and air. It can be visualized as a pipe with fuel injectors situated as shown in figure 11–11. The flame wall is a continuous, steady state chemical reaction of fuel and air, producing the hot gases for the turbine. We can consider then the flame wall as a heat addition process and the whole combustor can be represented by a system diagram shown in figure 11–12.

Notice in this figure that to conserve mass we must have that

$$\dot{m}_2 = \dot{m}_1 + \dot{m}_f \quad \textbf{(11–34)}$$

The fuel/air ratio, $\dot{m}_f/\dot{m}_1$, is generally about 1 to 30, but can vary greatly from this because of the fuel, air conditions, or operating requirements. The first law written for the combustor is

$$\dot{m}_2 h_2 + \dot{m}_f h_f - \dot{m}_1 h_1 = \dot{Q}_{\text{added}} \quad \textbf{(11–35)}$$

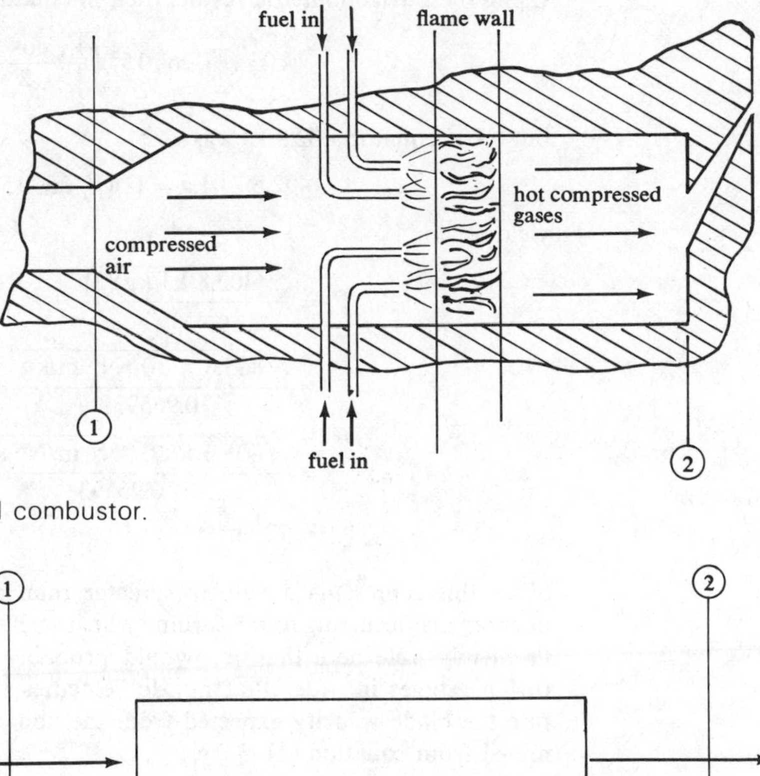

Figure 11–11 Typical combustor.

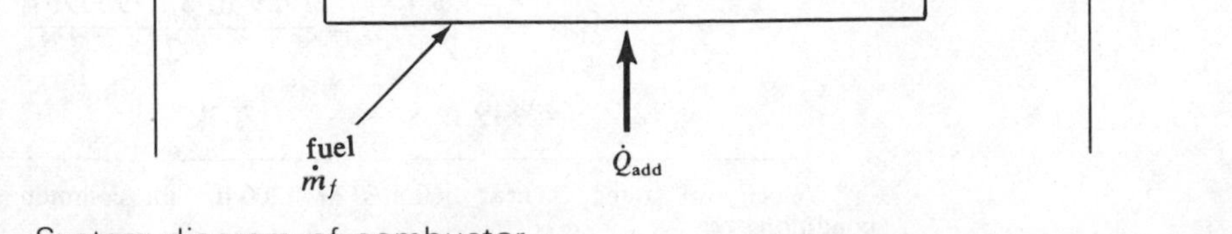

Figure 11–12 System diagram of combustor.

assuming that no kinetic or potential energy changes occur. The work or power is obviously zero in the combustor and if we further neglect the fuel mass, equation (11–35) can be written

$$\dot{m}_2(h_2 - h_1) = \dot{Q}_{\text{added}} \tag{11–36}$$

or

$$h_2 - h_1 = q_{\text{added}} \tag{11–37}$$

where q is based on unit mass of air flowing through the combustor.

The gases flowing through the combustor are at an essentially constant high pressure. To get the air or gases up to a high pressure as they enter the combustor, a pump or compressor must be used. We will see in the ram jet that other methods exist to compress the gas, but characteristically in the pump or compressor, gases are received at a low velocity, are accelerated through blades or fans to a high velocity, and finally restricted to increase the gas density and pressure while reducing the velocity to a low value again. There are many varied configurations to produce compressed gases, but here we will consider only the typical one used with a gas turbine. It is essentially a reversed turbine which has been considered in some detail already. Gases are taken in at low pressure, as indicated in the system diagram of figure 11–13, and are compressed through a series of turbine-type blades and nozzles to get a high pressure gas. The work to drive the compressor is normally supplied by a shaft from the gas turbine itself. The first law of thermodynamics can take the form

$$\frac{V_2^2 - V_1^2}{2} + h_2 - h_1 = q - wk_{\text{comp}} \tag{11–38}$$

for the compressor and if the walls are assumed to be adiabatic, we can eliminate q. The velocity terms are also neglected in some cases, although this can represent a significant error for high speed jet engine aircraft.

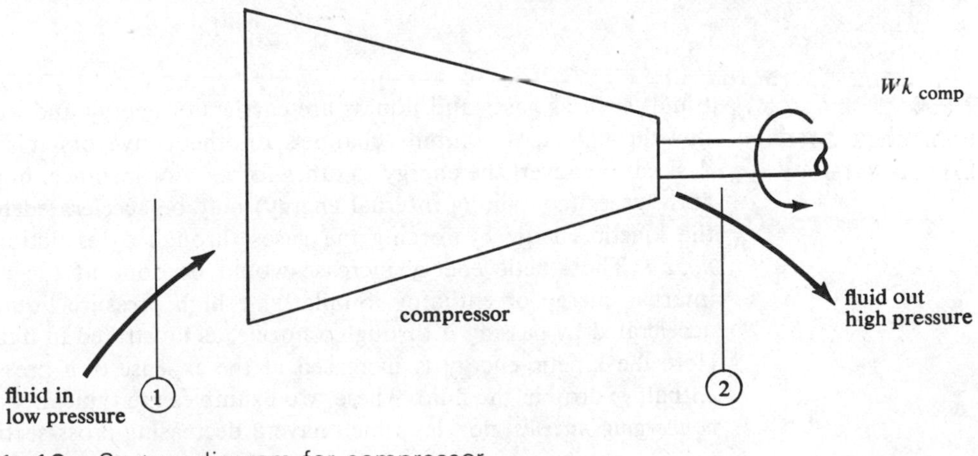

Figure 11–13 System diagram for compressor.

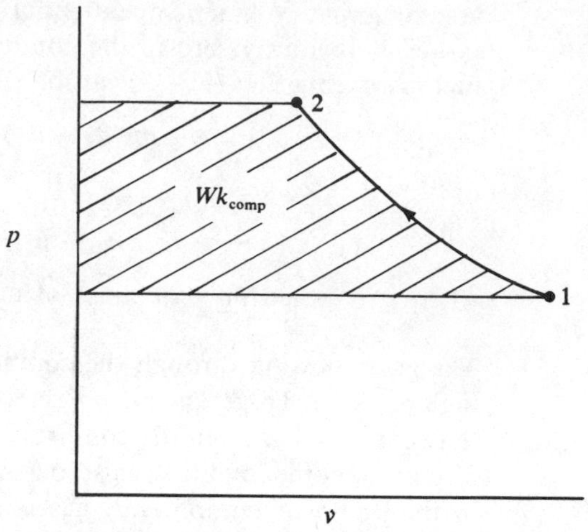

Figure 11–14 Diagram of compressor.

At any rate, if we do make these simplifying assumptions, then

$$h_2 - h_1 = -wk_{comp} \tag{11–39}$$

and we can then also determine work, if we have a reversible compressor, from the area to the left of a curve in a p-v diagram. This is seen in figure 11–14 where the area is found from equation (6–49)

$$-wk_{comp} = \frac{k}{1-k}(p_2v_2 - p_1v_1)$$

By using the polytropic relation for work

$$-wk_{comp} = \frac{n}{1-n}(p_2v_2 - p_1v_1)$$

we can consider *nonadiabatic* compressors as well, but then we must use equation (11–38) instead of (11–39) for any energy balances.

11.4 Nozzles and Diffusers

Fluids such as gases and liquids are carriers of energy and during their flow through pipes, conduits, channels, or other conveyors, it is frequently desired to convert the energy to other forms. For instance, high temperature gases (containing internal energy) may be accelerated to increase the kinetic energy by forcing the gases through a restriction called a *nozzle*. The kinetic energy increase would be done at the expense of internal energy or enthalpy. Similarly, a high pressure liquid may be accelerated by passing it through a nozzle, as illustrated in figure 11–15. Here the kinetic energy is increased at the expense of a pressure (and enthalpy) drop in the fluid. These two examples are typical of the use of *converging nozzles*, nozzles which have a decreasing cross-sectional area

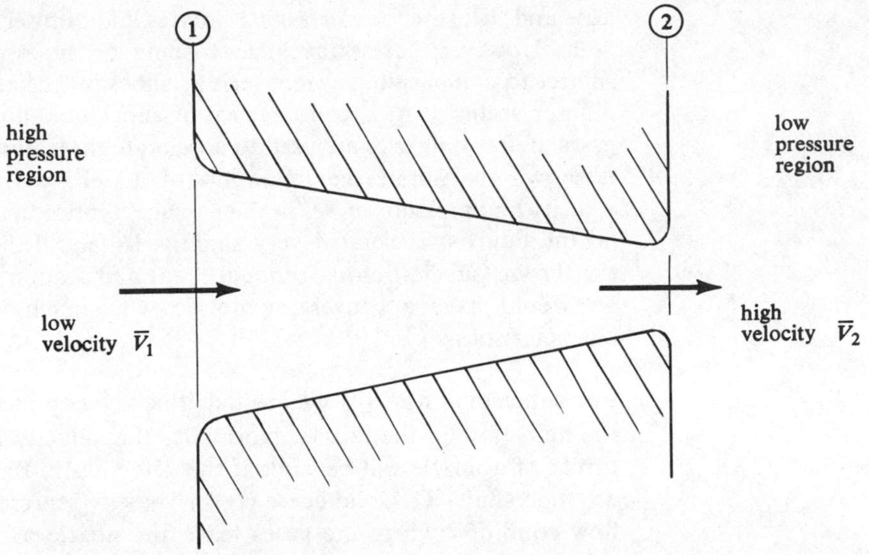

Figure 11–15 Typical converging nozzle.

downstream in the flow. There are, however, other types of nozzles: *diverging* nozzles or *diffusers*, and *converging-diverging* nozzles or *deLaval* nozzles. Diffusers are commonly used to decelerate high velocity gases or liquids from regions of low pressure to those at a higher pressure. That is, diffusers compress fluids at the expense of kinetic energy. A typical one is shown in figure 11–16.

Fluids which travel through nozzles at velocities less than sonic (less than the speed of sound) behave rather predictably. No uncommon situations

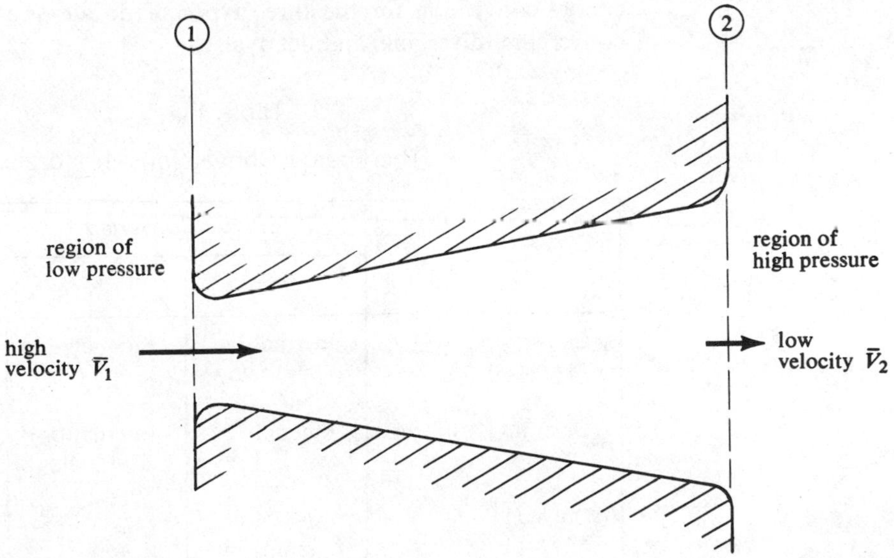

Figure 11–16 Typical diffuser.

arise and their use for converging nozzles and diffusers can be intuitively seen. However, gases traveling at sonic or supersonic velocities are subject to complicating events such as shock waves and accelerations in diffuser sections. As a consequence of supersonic flow of compressible gases, deLaval nozzles are used to accelerate these fluids. In figure 11–17 is shown the normal configuration of a deLaval nozzle and typical velocity and pressure curves in the passage. Notice that for subsonic flow (a) the fluid is accelerated very slightly. In fact, if the section between the throat (smallest cross-sectional area) and station (2) were removed, we would have a converging nozzle with much better accelerating characteristics.

For supersonic flow (b) we see that this velocity increases throughout the fluid flow in the nozzle. Normally, the velocity of the fluid at the throat of a nozzle will be sonic if the velocities in the divergent section are supersonic. The solid curve (I) for velocity represents the supersonic flow condition where the gases leave the nozzle at a high supersonic velocity. At some place external to the nozzle, a shock wave will be present which quickly decelerates the gases to subsonic speed. The discontinuous curve (II) represents a condition where a shock wave occurs in the diverging section of the nozzle. The shock wave can be seen to be a jump discontinuity in pressure and velocity and is normally described as an irreversible adiabatic process. The velocity downstream of the shock wave is subsonic and decreases to the exit in the normal fashion of subsonic diffusers. A more complete description and analysis of nozzle flows of compressible or incompressible fluids can be found in many good fluid mechanics or gas dynamics texts.

Let us now review the meaning of nozzles. In table 11–1 we see listed the energy conversion for the three types of nozzles we have considered: converging, diverging, and deLaval.

Table 11–1

The Energy Conversions in Nozzles

Nozzle Type	*Energy Converted* from	*Energy Converted* to	*Type of Flow*
Converging	Internal or enthalpy	Kinetic	Subsonic or supersonic
Diverging, diffuser	Kinetic	Internal or enthalpy	Subsonic or supersonic
DeLaval	Internal or enthalpy	Kinetic	Supersonic

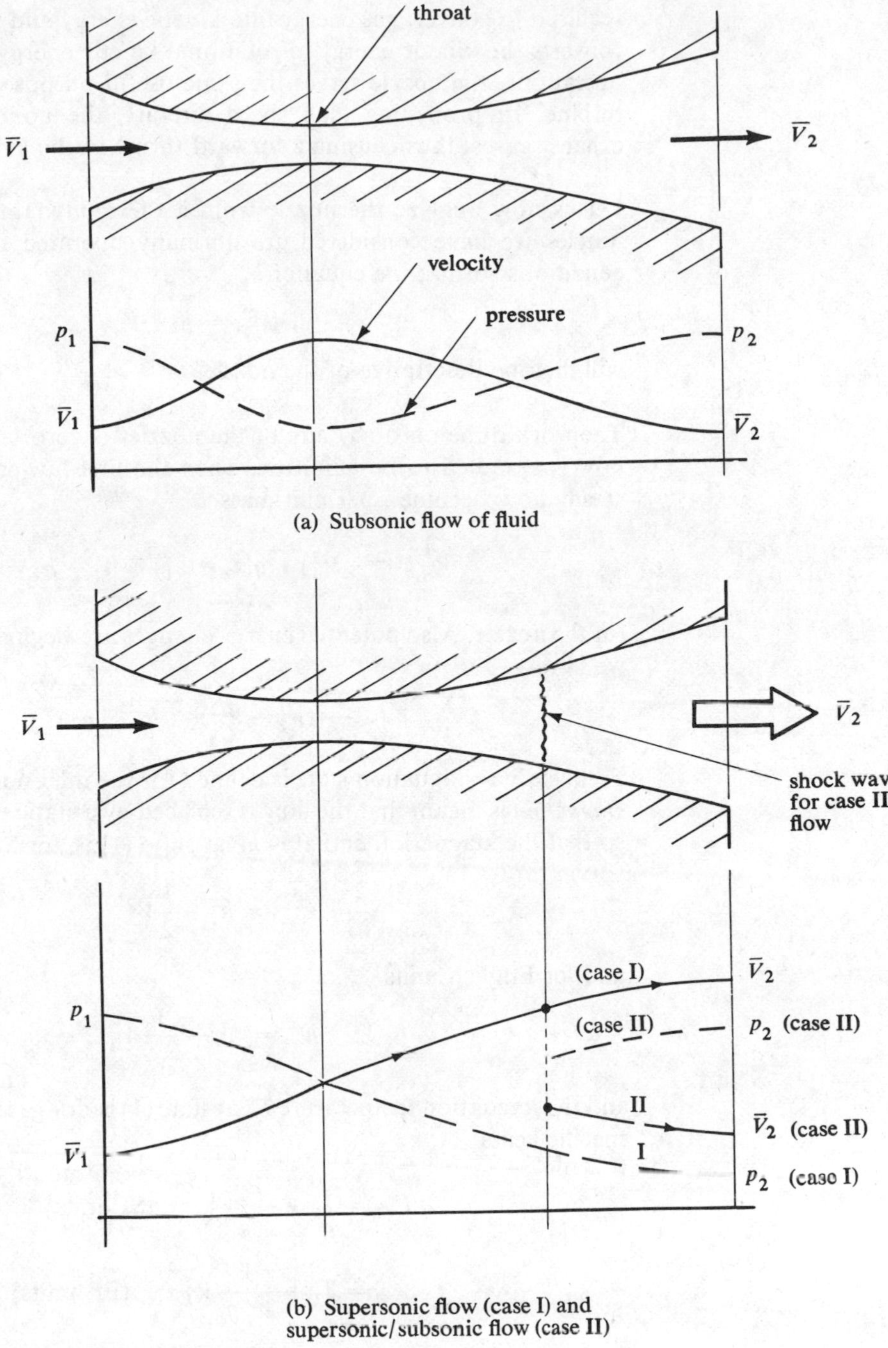

Figure 11–17 Flow through a deLaval nozzle.

We have seen before that gas turbines are visualized as devices converting the internal energy of a high temperature/pressure gas into rotational kinetic energy by means of turbine blades. The nozzle, however, is required to convert gas energy into kinetic energy and the turbine merely converts the kinetic energy to rotational kinetic energy. In steam turbine operations, the nozzle serves the same useful purpose as it does for gas turbine. In propelling high speed aircraft, the nozzle accelerates the exhaust gases, thus causing a forward thrust in the aircraft.

Let us now analyze the nozzle from a thermodynamic viewpoint. The nozzles we have considered are normally operated under steady flow conditions so that the equation

$$\rho_1 A_1 \bar{V}_1 = \rho_2 A_2 \bar{V}_2 \tag{11–40}$$

will then be descriptive of the nozzle.

The work done on or by any of the nozzles is zero, and commonly the flow is assumed to be adiabatic. Then the first law energy equation for steady flow becomes, per unit mass

$$\frac{1}{2}(\bar{V}_2^2 - \bar{V}_1^2) + g(z_2 - z_1) + h_2 - h_1 = 0$$

for the nozzle. Also potential energy changes are negligible ($z_2 \approx z_1$) and we have

$$\frac{1}{2}(\bar{V}_2^2 - \bar{V}_1^2) = h_1 - h_2 \tag{11–41}$$

Frequently a stagnation state is defined for the inlet nozzle condition. By *stagnation* is meant that the flow is replaced by a static or stagnating state so that the stagnation enthalpy h^* at state (1) is, for SI units

$$h^*_1 = h_1 + \frac{1}{2}\bar{V}_1^2 \tag{11–42a}$$

and for English units

$$h^*_1 = h_1 + \frac{1}{2g_c}\bar{V}_1^2 \tag{11–42b}$$

and the stagnation temperature T^* at state (1) is, for gases having constant specific heats

$$T^*_1 = T_1 + \frac{1}{2c_p}\bar{V}_1^2 \qquad \text{(SI units)} \tag{11–43a}$$

$$= T_1 + \frac{1}{2c_p g_c}\bar{V}_1^2 \qquad \text{(EE units)} \tag{11–43b}$$

so

$$h^*_1 = c_p T^*_1 \tag{11–44}$$

Then equation (11–41) can be written as

$$\frac{1}{2}\bar{V}_2^2 = h_1 + \frac{1}{2g_c}\bar{V}_1^2 - h_2 = h^*_1 - h_2$$

or

$$\bar{V}_2 = \sqrt{2(h^*_1 - h_2)} \tag{11-45}$$

This can be reduced to

$$\bar{V}_2 = \sqrt{2g_c c_p(T^*_1 - T_2)} \tag{11-46a}$$

or, in English units

$$\bar{V}_2 = \sqrt{2g_c c_p(T^*_1 - T_2)} \tag{11-46b}$$

for gases having constant specific heats.

Since nozzles are assumed to be adiabatic and if they are also reversible, the flow of fluid would be conducted at constant entropy. On a *T-s* diagram, shown in figure 11–18, the reversible adiabatic flow through a converging nozzle would be represented by path (1–2). We say this is 100% efficient and note that an expansion to the same pressure p_2 conducted in an irreversible manner, as in path (1–2′) would reduce the nozzle efficiency η_N. We define nozzle efficiency by the equation

$$\eta_n = \left(\frac{h^*_1 - h_{2'}}{h^*_1 - h_2}\right) \times 100 \tag{11-47}$$

where $h_{2'}$ is the actual enthalpy of the exiting fluid. For gases having constant specific heats, equation (11–47) becomes

$$\eta_n = \left(\frac{T^*_1 - T_{2'}}{T^*_1 - T_2}\right) \times 100 \tag{11-48}$$

Example 11.2

Air at a stagnation state of 277°C and 10.0 bars enters a converging nozzle. Determine the exit velocity if the nozzle is reversible and the pressure on the exit side is 1.01 bars.

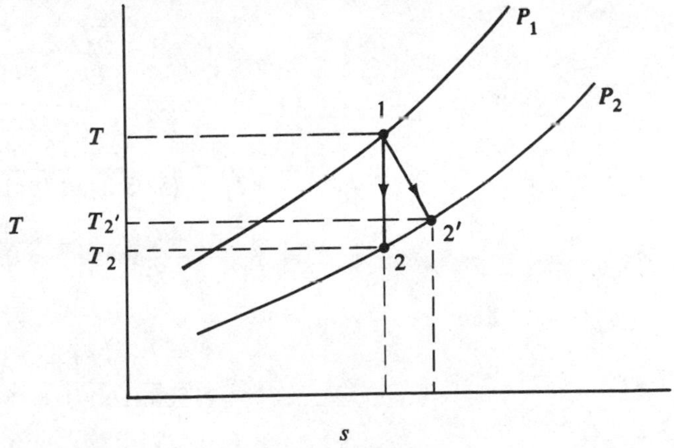

Figure 11–18 *T-s* diagram of flow through a converging nozzle.

Solution We may use equation (11–46a) since the system, a converging nozzle, is conducting air which we assume has constant specific heats. The exit temperature can be computed from the perfect gas relationship and the reversible adiabatic process equation $pv^k = C$. From the equation

$$\frac{p_1}{p_2} = \left(\frac{T_1}{T_2}\right)^{k/k-1}$$

we solve for T_2:

$$T_2 = T_1\left(\frac{p_2}{p_1}\right)^{(k-1)/k} = T_1{}^*\left(\frac{p_2}{p_1{}^*}\right)^{(k-1)/k}$$

$$T_2 = (277°\text{C} + 273\ \text{K})\left(\frac{1.01\ \text{bars}}{10.0\ \text{bars}}\right)^{0.4/1.4}$$
$$= 286\ \text{K}$$

Then, from equation (11–46a) we obtain

$$\bar{V}_2 = \sqrt{(2)(1.007\ \text{kJ/kg}\cdot\text{K})(550\ \text{K} - 286\ \text{K})(10^3\ \text{N}\cdot\text{m/kJ})}$$
$$= 729\ \text{m/s}$$

Answer

Example 11.3 Air enters a diffuser at 1500 ft/s, 600°R, and 15 psia. If 3 lbm/s of air flows through the diffuser, determine the entrance area of the diffuser, the pressure at the exit, and the mach number of the air at the entrance. Assume the velocity at the exit is negligible and the diffuser efficiency is 90%.

Solution The entrance area can be calculated from the continuity equation (11–40)

$$\rho_1 A_1 \bar{V}_1 = \dot{m}_1$$

or

$$A_1 = \frac{\dot{m}_1}{\bar{V}_1 \rho_1}$$

We have that

$$v_1 = \frac{RT_1}{p_1} = \frac{1}{\rho_1}$$

or

$$\rho_1 = \frac{p_1}{RT_1} = \frac{15\ \text{lbf/in}^2}{(53.3\ \text{ft-lbf/lbm}\cdot°\text{R})(600°\text{R})}$$
$$= \frac{15 \times 144}{53.3 \times 600}\ \text{lbm/ft}^3 = 0.0675\ \text{lbm/ft}^3$$

Then

$$A_1 = \frac{(3\ \text{lbm/s})}{(1500\ \text{ft/s})(0.0676\ \text{lbm/ft}^3)}$$
$$= 0.0296\ \text{ft}^2 = 4.26\ \text{in}^2$$

Answer

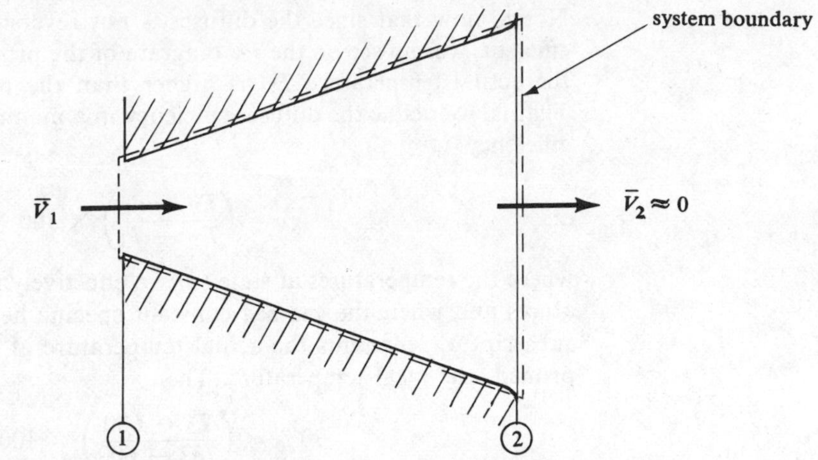

Figure 11–19 Adiabatic diffuser.

If we assume that the diffuser, shown in figure 11–19, is adiabatic then we may apply equation (11–35) in the form

$$\frac{1}{2g_c}\bar{V}_1^2 = h_2 - h_1 = c_p(T_{2'} - T_1)$$

Then

$$T_{2'} = T_1 + \frac{1}{2c_p g_c}\bar{V}_1^2$$

$$= 600°\text{R} + \frac{1}{2(0.24\ \text{Btu/lbm}\cdot°\text{R})(32.2\ \text{ft-lbm/s}^2\cdot\text{lbf})}\ \frac{(1500)^2\ \text{ft}^2/\text{s}^2}{(778\ \text{ft-lbf/Btu})}$$

$$= 787°\text{R}$$

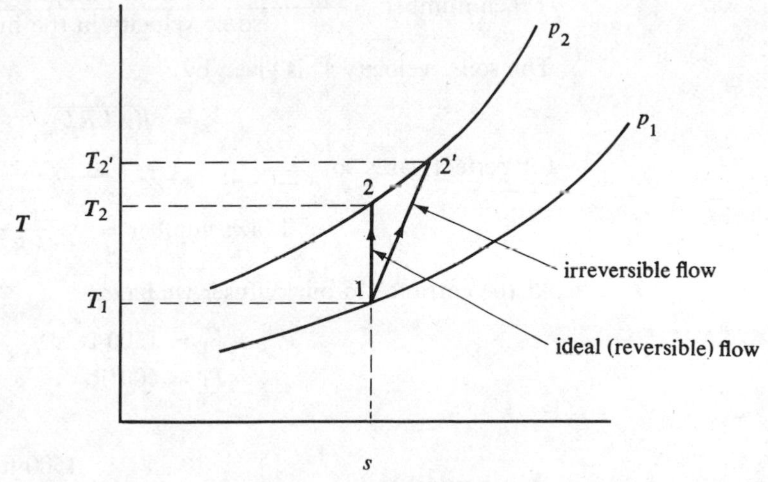

Figure 11–20 *T-s* diagram for diffuser flow in example 11.3.

Notice now that since the diffuser is not reversible, but rather is 90% efficient, we can see on the T-s diagram of the process (figure 11–20) that the actual temperature $T_{2'}$ is higher than the reversible adiabatic T_2. That is, we define the diffuser efficiency in a manner similar to the nozzle efficiency

$$\eta_D = \left(\frac{T_2 - T_1}{T_{2'} - T_1}\right) \times 100 \tag{11–49}$$

where the temperatures at state (2) are effectively the stagnation temperatures and where the gas has constant specific heat values. The primed subscript 2′ represents the actual temperature at state (2′) and the unprimed, the ideal temperature. Then

$$90\% = \left(\frac{T_2 - 600}{787 - 600}\right) \times 100$$

or

$$T_2 = 768°\text{R}$$

and using the reversible isentropic relation

$$\frac{p_1}{p_2} = \left(\frac{T_1}{T_2}\right)^{k/(k-1)}$$

we may obtain the exit pressure. Calculating

$$p_2 = p_1\left(\frac{T_2}{T_1}\right)^{k/(k-1)} = (15\text{ psia})\left(\frac{768}{600}\right)^{1.4/0.4}$$

$$= 35.6\text{ psia} \qquad \textit{Answer}$$

The *mach number* is a term used frequently in gas dynamics or fluid mechanics analysis. It is defined by the relationship

$$\text{Mach number} = \frac{\text{velocity of fluid with respect to the system boundary}}{\text{sonic velocity in the fluid at that point}}$$

The sonic velocity $\bar{V}_s$ is given by

$$\bar{V}_s = \sqrt{(g_c kRT)}$$

for perfect gases so

$$\text{Mach number} = \frac{\bar{V}}{\sqrt{g_c kRT}} \tag{11–50}$$

At the entrance to our diffuser we have

$$\bar{V}_1 = 1500\text{ ft/s}$$
$$T_1 = 600°\text{R}$$

so

$$\text{Mach number} = \frac{1500\text{ ft/s}}{\sqrt{(32.2\text{ ft-lbm/s}^2\cdot\text{lbf})(1.4)(53.3\text{ ft-lbf/lbm}\cdot°\text{R})(600°\text{R})}}$$

$$= 1.25 \qquad \textit{Answer}$$

Notice that a mach number equal to unity (1) would describe sonic velocity conditions. In our problem, the velocity is greater than sonic so it is called *supersonic*. We have, in general, that

$$\begin{aligned} & \text{(implies)} \\ M &= 1 \longrightarrow \text{sonic velocity} \\ M &> 1 \longrightarrow \text{supersonic} \\ M &< 1 \longrightarrow \text{subsonic} \end{aligned}$$

Also, observe that mach number is dependent upon the air or fluid temperature as well as upon the velocity, and the condition of flow (whether subsonic, sonic, or supersonic) is not only determined by velocity but by the fluid properties of temperature, pressure, or volume as well.

11.5 Perfect Gas Analysis of Brayton Cycle

Two applications of the Brayton cycle will be treated here. In each case the gas turbine will be used and the medium will be considered a perfect gas.

Example 11.4

An aircraft is propelled by a reversible gas turbine, or turbo-jet engine, shown in figure 11–21. The aircraft is flying at 40,000 feet altitude under standard day atmospheric conditions at 500 mph. Assume the compressor has a pressure ratio of 10 to 1, the fuel/air ratio is 1 to 35, and the fuel is kerosene. Assume also that the kerosene has the same LHV as octane (C_8H_{18}) and burns completely, releasing the full LHV in the combustor. If the engine operates on an ideal Brayton cycle and the gases are ideal, determine the values of p, v, T, and s for the Brayton cycle corners. Then determine the thermodynamic efficiency and the mass flow rate of air required to power the aircraft if the wind resistance (drag) is 600 lbf and the flight is level.

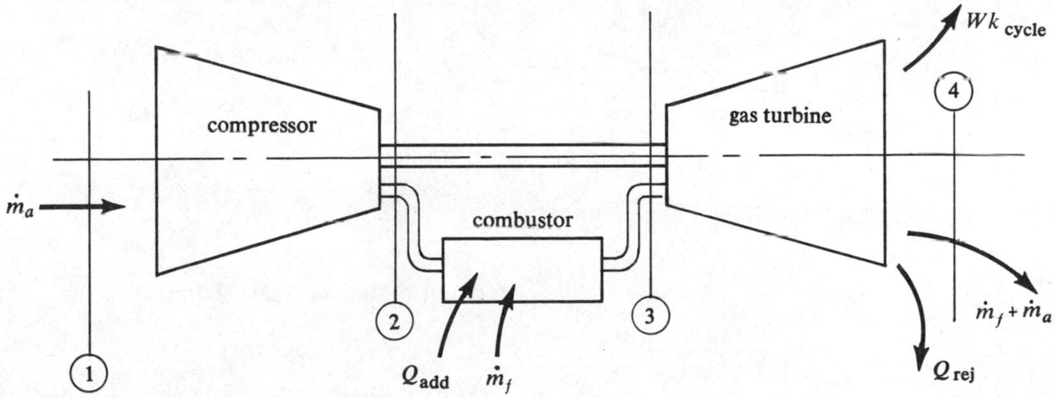

Figure 11–21 Turbo-jet engine.

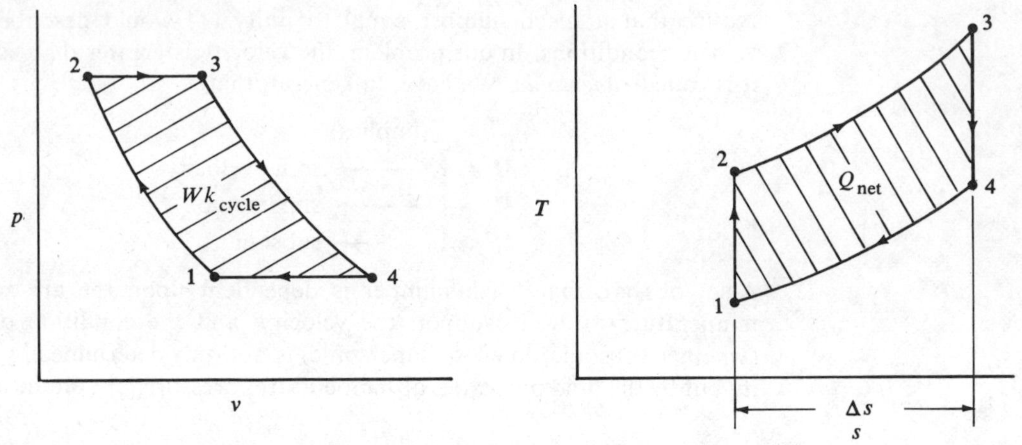

Figure 11–22 Property diagrams for reversible jet engine.

Solution In figure 11–22 are shown the *p-v* and *T-s* diagrams for the jet engine. The standard day atmospheric condition fixes the properties at state (1). From table B.16 they are $p_1 = 5.56$ inches Hg; $\rho_1 = 0.0189$ lbm/ft³; and $T_1 = 389.97°R$. The pressure at state (2) is obtained from the compressor pressure ratio $p_2/p_1 = 10$, which gives us

$$p_2 = 10\,p_1 = 55.6 \text{ in Hg}$$

Also

$$p_3 = p_2 = 55.6 \text{ in Hg} \times 0.491 \text{ psi/in Hg} = 27.2 \text{ psia}$$

and

$$p_4 = p_1 = 2.72 \text{ psia}$$

Since the engine is operating in accordance with an ideal Brayton cycle, we can write

$$p_1 v_1^k = p_2 v_2^k$$

and

$$\frac{p_1}{p_2} = \left(\frac{v_2}{v_1}\right)^k$$

or

$$v_2 = \left(\frac{p_1}{p_2}\right)^{1/k} v_1$$

But

$$v_1 = \frac{1}{\rho_1} = 1 \text{ ft}^3/0.0189 \text{ lbm} = 52.9 \text{ ft}^3/\text{lbm}$$

so

$$v_2 = 52.9 \text{ ft}^3/\text{lbm} \left(\frac{2.73}{27.3}\right)^{1/1.4}$$

$$= 10.2 \text{ ft}^3/\text{lbm}$$ *Answer*

Using the perfect gas relation

$$T_2 = \frac{p_2 v_2}{R}$$

we have

$$T_2 = \frac{(27.2 \text{ lbf/in}^2)(10.2 \text{ ft}^3\text{/lbm})(144 \text{ in}^2\text{/ft}^2)}{(53.3 \text{ ft-lbf/lbm} \cdot {}^\circ\text{R})}$$

$$= 750^\circ\text{R}$$ *Answer*

To determine the properties at state (3), we now observe that process (2–3) is a constant pressure heat addition in the combustor. Using equation (11–35)

$$\dot{m}_3 h_3 + \dot{m}_f h_f - \dot{m}_2 h_2 = \dot{Q}_{\text{added}}$$

we observe that the heat added is just the LHV of the fuel. For kerosene, as for octane, this has a value of 19,256 Btu/lbm fuel. The enthalpy of the liquid fuel h_f can be calculated from an equation from *Gas Tables* by Keenan and Kaye*

$$h_f = (0.5T - 287) \text{ Btu/lbm fuel} \tag{11–51}$$

where T is in degrees Rankine. For SI units, with T measured on the Kelvin scale, equation (11–51) is

$$h_f = (2.1T - 668) \text{ kJ/kg fuel}$$

This gives us, assuming in our problem a fuel temperature of 40°F or 500°R, that

$$h_f = (0.5)(500) - 287 = -37 \text{ Btu/lbm fuel}$$

Then

$$\dot{m}_3 h_3 + \dot{m}_f(-37) - \dot{m}_2 h_2 = \dot{m}_f(19{,}256)$$

or

$$\left(\frac{\dot{m}_3}{\dot{m}_f}\right) h_3 - 37 - \left(\frac{\dot{m}_2}{\dot{m}_f}\right) h_2 = 19{,}256$$

But

$$\frac{\dot{m}_2}{\dot{m}_f} = 35$$

$$\frac{\dot{m}_3}{\dot{m}_f} = \frac{(\dot{m}_2 + \dot{m}_f)}{\dot{m}_f} = \left(\frac{\dot{m}_2}{\dot{m}_f} + 1\right) = 36$$

yielding

$$\left(\frac{\dot{m}_2}{\dot{m}_f} + 1\right) h_3 - 37 - \frac{\dot{m}_2}{\dot{m}_f} h_2 = 19{,}256$$

---

* From J. H. Kennan and J. Kaye, *Gas Tables* (New York, 1948), John Wiley & Sons, Inc., with permission of authors and publisher.

or

$$36\,h_3 - 37\text{ Btu/lbm} - 35\,h_2 = 19{,}256\text{ Btu/lbm}$$

If we assume the same values of specific heat for both air and hot gases, then we can write

$$36c_pT_3 - 37 - 35c_pT_2 = 19{,}256$$

$$36(0.24\text{ Btu/lbm}\cdot{}^\circ\text{R})T_3 - 37 - 35(0.24\text{ Btu/lbm}\cdot{}^\circ\text{R})(750^\circ\text{R}) = 19{,}256\text{ Btu/lbm}$$

giving us

$$T_3 = (19{,}256 + 35 \times 0.24 \times 750 + 37)\frac{1}{36 \times 0.24}$$

$$= 2962^\circ\text{R}$$ *Answer*

From the perfect gas relationship

$$v_3 = \frac{RT_3}{p_3} = \frac{53.3 \times 2962}{144 \times 27.2} = 40.3\text{ ft}^3\text{/lbm}$$ *Answer*

Using isentropic expansion equations for process (3–4), we have

$$\frac{p_4}{p_3} = \left(\frac{v_3}{v_4}\right)^k$$

or

$$v_4 = \left(\frac{p_3}{p_4}\right)^{1/k} v_3 = (10)^{1/1.4}(40.3)$$

$$= 209\text{ ft}^3\text{/lbm}$$ *Answer*

The temperature at state (4) we can determine from the perfect gas relation

$$T_4 = \frac{p_4v_4}{R} = \frac{2.72 \times 144 \times 209}{53.3} = 1536^\circ\text{R}$$ *Answer*

The entropy change is obtained from process (2–3) or process (4–1). Both of these processes are isobaric and we can write, from equation (7–21)

$$\Delta s = s_4 - s_1 = s_3 - s_2 = c_p \ln\frac{T_4}{T_1} = c_p \ln\frac{T_3}{T_2}$$

$$= 0.24\text{ Btu/lbm}\cdot{}^\circ\text{R} \ln\frac{1536}{389.97} = 0.329\text{ Btu/lbm}\cdot{}^\circ\text{R}$$ *Answer*

The thermodynamic efficiency we can calculate from either equation (11–8) or (11–9). From the latter

$$\eta_T = \left(1 - \frac{1}{10^{0.4/1.4}}\right) \times 100 = 48.2\%$$ *Answer*

The power required to propel the aircraft can be determined from

$$\dot{W}k = F\bar{V} \tag{11–52}$$

where F is the force (in this case the drag), and $\bar{V}$ is the velocity. We get then

$$\dot{W}k = (600 \text{ lbf})\left(500 \times \frac{5280}{3600} \text{ ft/s}\right)$$
$$= 440{,}000 \text{ ft-lbf/s}$$
$$= 440{,}000 \times \frac{1}{550} \text{ hp} = 800 \text{ hp}$$

The mass flow rate of air required can be obtained from the relationship

$$\dot{W}k = \dot{m}wk_{\text{cycle}} = 800 \text{ hp}$$

The work of the Brayton cycle can be obtained from equations (11–6) and (11–7)

$$wk_{\text{cycle}} = q_{\text{net}} = q_{\text{add}} + q_{\text{rej}}$$

We have

$$q_{\text{add}} = 19{,}256 \text{ Btu/lbm fuel}$$
$$= \frac{1}{35} \times 19{,}256 \text{ Btu/lbm air} = 550 \text{ Btu/lbm air}$$

The heat rejected is just process (4–1), occurring in the surroundings

$$q_{\text{rej}} = h_1 - h_4 = c_p(T_1 - T_4)$$
$$= 0.24 \text{ Btu/lbm} \cdot {}^\circ\text{R } (389.97^\circ\text{R} - 1528^\circ\text{R})$$
$$= -273 \text{ Btu/lbm air}$$

Then

$$\dot{m}_{\text{air}} = \frac{800 \text{ hp}}{273 \text{ Btu/lbm air}} = \frac{440{,}000 \text{ ft-lbf/s}}{277 \times 778 \text{ ft-lbf/lbm air}}$$
$$= 2.07 \text{ lbm/s}$$

Answer

An interesting mechanics problem is embodied in the problem just considered. The gas turbine drives the compressor through a shaft, thus providing the high pressure gases. But the question of how the aircraft is propelled by such a device is not answered. The turbine can rotate but cannot exert a direct force on the aircraft. In fact, the manner of driving the vehicle through air is the mechanics problem referred to. The answer is found in an application of Newton's law; namely, as shown in figure 11–23, the air is increased in velocity from $\bar{V}_1$ (which would be the negative aircraft velocity) to $\bar{V}_4$. This is an acceleration, $(\bar{V}_4 - \bar{V}_1)/\tau$, and from the relation $F = m\bar{a}$ we get

$$F = m\left(\frac{\bar{V}_4 - \bar{V}_1}{\tau}\right) = \dot{m}(\bar{V}_4 - \bar{V}_1)$$

But every force has an opposing force and we call this force F_I the impluse which actually drives the aircraft. In addition; the difference in

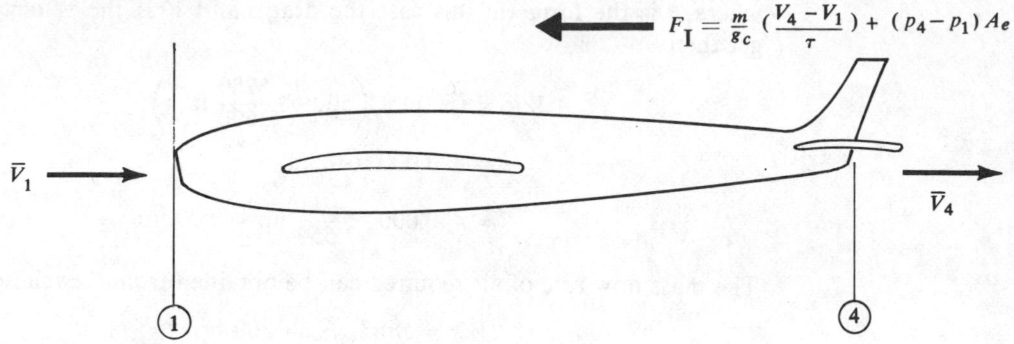

Figure 11–23 Reaction principle of jet aircraft propulsion.

pressure between the entrance at atmospheric pressure and the exit can cause a force given by $(p_4 - p_1)A_e$ where A_e is the area of the exhaust nozzle. Adding this to the impulse force of the above equation, we then get

$$F_I = m\left(\frac{\overline{V}_4 - \overline{V}_1}{\tau}\right) + (p_4 - p_1)A_e \tag{11–53}$$

Frequently the specific impulse, defined by

$$I_{sp} = \frac{F_I}{\dot{m}} = \overline{V}_4 - \overline{V}_1 + \frac{(p_4 - p_1)A_e}{m} \tag{11–54}$$

is referred to in the literature. This quantity provides a quick measure of the propulsive capability of a particular design.

Let us now consider a gas turbine with some irreversibilities included.

Example 11.5

A closed turbine engine operates with 10 kilograms/s of air as the working medium operates on an air-standard Brayton cycle. The air-standard Brayton cycle is defined as a Brayton cycle in which the properties of all the gases are assumed to be like those of air. In addition, the mass of the fuel added in the combustor is generally neglected in the air-standard cycle. For this example problem, the engine is subject to the following restrictions: the compressor is 90% efficient (assuming reversible adiabatic to be ideal); the turbine is 88% efficient; and the combustor utilizes 70% of the LHV of the fuel, kerosene. The combustor allows the kerosene to burn and the heat is then transferred by conduction to the confined constant pressure air in the closed cycle. Assume 1 kilogram of kerosene is used for 40 kilograms of air flowing through the combustor. The engine is shown in figure 11–24 where it can be seen that the power produced is used to provide electric power through a generator. The following data apply: $p_1 = 1.0$ bar, $t_1 = 27°C$, and $p_2 = 7.0$ bars. Determine the critical property values on *p-v* and *T-s* diagrams and determine the power produced. Also calculate the cycle efficiency and the heat rejected.

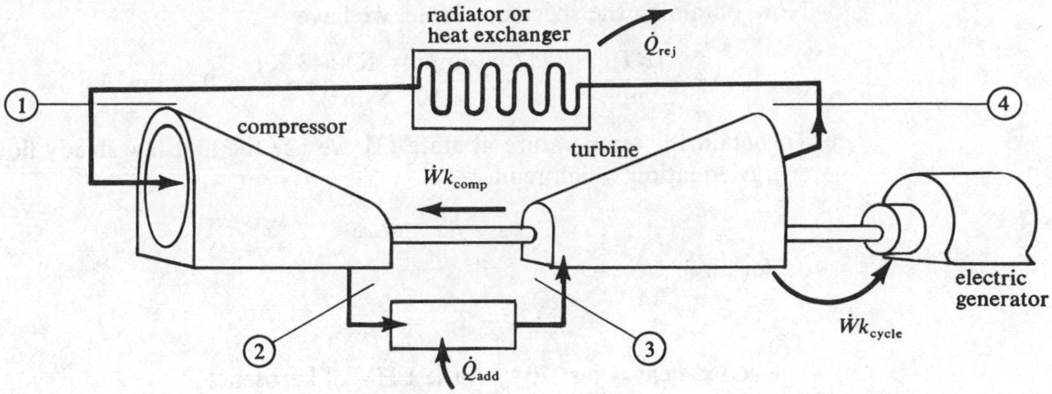

Figure 11–24 Closed air-standard Brayton cycle.

Solution

Since we are faced with an irreversible process, the process equations developed in chapter 6 will not be of full assistance. First we note that $p_1 = p_4$ and $p_3 = p_2$. Also, from the perfect gas relation

$$v_1 = \frac{RT_1}{p_1} = \frac{287 \text{ N} \cdot \text{m/kg} \cdot \text{K} \times 300 \text{ K}}{1 \times 10^5 \text{ N/m}^2} = 0.861 \text{ m}^3/\text{kg}$$

and we have determined state (1). The adiabatic compressor efficiency we write as

$$\eta_c = \frac{\text{ideal } Wk}{\text{actual } Wk} \times 100 = \frac{h_2 - h_1}{h_{2'} - h_1} \times 100 \qquad \textbf{(11–55)}$$

where $h_{2'}$ is the enthalpy value of the gas if it were compressed reversibly and adiabatically to the given pressure p_2. We can then write, for this engine, assuming constant specific heat values, that

$$90\% = \frac{c_p(T_2 - T_1)}{c_p(T_{2'} - T_1)} \times 100$$

Using the reversible adiabatic process as the idealized compression

$$\frac{T_2}{T_1} = \left(\frac{p_2}{p_1}\right)^{(k-1)/k}$$

or

$$T_2 = (300 \text{ K})\left(\frac{7.0 \text{ bars}}{1.0 \text{ bar}}\right)^{0.4/1.4} = 523 \text{ K}$$

We then compute the true temperature at state (2) from equation (11–55):

$$0.90 = \frac{T_2 - T_1}{T_{2'} - T_1} = \frac{523 \text{ K} - 300 \text{ K}}{T_{2'} - 300 \text{ K}}$$

or

$$T_{2'} = 548 \text{ K}$$

Now, obtaining the specific volume, we have

$$v_{2'} = \frac{RT_{2'}}{p_{2'}} = \frac{(287 \text{ N} \cdot \text{m/kg} \cdot \text{K})(548 \text{ K})}{(7 \times 10^5 \text{ N/m}^2)} = 0.225 \text{ m}^3/\text{kg}$$

To obtain the temperature at state (3), we use the first law steady flow energy equation which reduces to

$$h_3 - h_{2'} = q_{\text{add}}$$

or, for air

$$c_p(T_3 - T_{2'}) = q_{\text{add}}$$

The added heat is just 70% of the LHV of kerosene:

$$(44{,}900 \text{ kJ/kg})(0.70) = 31{,}430 \text{ kJ/kg}$$

and per unit of air this is

$$q_{\text{add}} = (31{,}430 \text{ kJ/kg})\left(\frac{1 \text{ kg fuel}}{40 \text{ kg air}}\right)$$

$$= 786 \text{ kJ/kg air}$$

Then, using the above equation

$$(1.007 \text{ kJ/kg} \cdot \text{K})(T_3 - 548) \text{ K} = 786 \text{ kJ/kg}$$

and

$$T_3 = \frac{786 \text{ kJ/kg}}{1.007 \text{ kJ/kg} \cdot \text{K}} + 548 \text{ K}$$

$$= 1329 \text{ K}$$

Also, from the perfect gas equation

$$v_3 = \frac{RT_3}{p_3} = \frac{(287 \text{ N} \cdot \text{m/kg} \cdot \text{K})(1329 \text{ K})}{(7 \times 10^5 \text{ N/m}^2)} = 0.545 \text{ m}^3/\text{kg}$$

The turbine efficiency we can write as

$$\eta_{\text{turb}} = \left(\frac{\text{actual work}}{\text{ideal work}}\right) \times 100 = \left(\frac{h_3 - h_{4'}}{h_3 - h_4}\right) \times 100 \quad \textbf{(11–56)}$$

where h_4 is the enthalpy if the gases were expanded reversibly and adiabatically through the turbine. The actual turbine exhaust enthalpy is given by $h_{4'}$. Then

$$88\% = \left(\frac{h_3 - h_{4'}}{h_3 - h_4}\right) \times 100 \quad \text{and} \quad 0.88 = \frac{c_p(T_3 - T_{4'})}{c_p(T_3 - T_4)} = \frac{T_3 - T_{4'}}{T_3 - T_4}$$

We obtain T_4 from the ideal cycle relation

$$\frac{T_4}{T_3} = \frac{T_1}{T_2}$$

so

$$T_4 = (1329\ \text{K})\left(\frac{300\ \text{K}}{523\ \text{K}}\right) = 762\ \text{K}$$

and

$$0.88 = \frac{1329\ \text{K} - T_{4'}}{1329\ \text{K} - 762\ \text{K}}$$

or, the temperature is

$$T_{4'} = 830\ \text{K}$$

Further, we can calculate the specific volume at state (4) which becomes $v_{4'} = 2.39\ \text{m}^3/\text{kg}$ and then plot the *p-v* and *T-s* diagrams as shown in figure 11-25. To calculate the entropy changes for the various processes we may use the constant pressure reversible process equation:

$$\Delta s = c_p \ln \frac{T_f}{T_i}$$

and, specifically, for process (1–2)

$$\Delta s = s_{2'} - s_1 = s_{2'} - s_2 = c_p \ln \frac{T_{2'}}{T_2} = (1.007\ \text{kJ/kg} \cdot \text{K}) \ln \frac{548\ \text{K}}{523\ \text{K}}$$

$$= 0.047\ \text{kJ/kg} \cdot \text{K}$$

For process (2–3) we have

$$s_3 - s_{2'} = c_p \ln \frac{T_3}{T_{2'}} = (1.007\ \text{kJ/kg} \cdot \text{K}) \ln \frac{1329\ \text{K}}{548\ \text{K}}$$

$$= 0.892\ \text{kJ/kg} \cdot \text{K}$$

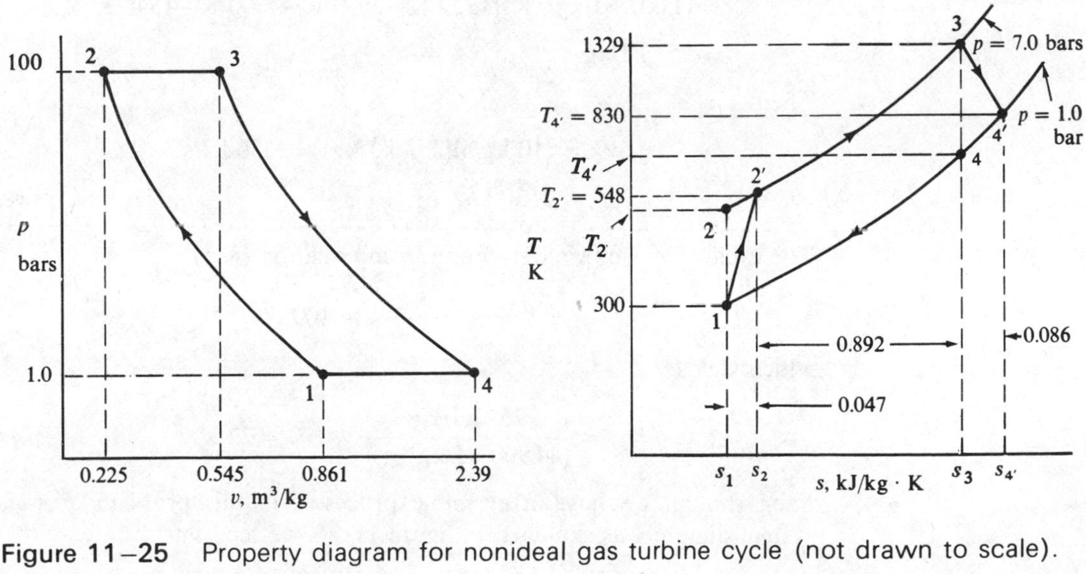

Figure 11–25 Property diagram for nonideal gas turbine cycle (not drawn to scale).

For process (3–4)

$$s_3 - s_{4'} = s_4 - s_{4'} = (1.007 \text{ kJ/kg} \cdot \text{K})\ln \frac{762 \text{ K}}{830 \text{ K}}$$

$$= -0.086 \text{ kJ/kg} \cdot \text{K}$$

And for process (4–1)

$$s_4 - s_1 = c_p \ln \frac{T_4}{T_1} = (1.007 \text{ kJ/kg} \cdot \text{K})\ln \frac{762 \text{ K}}{300 \text{ K}}$$

$$= 0.939 \text{ kJ/kg} \cdot \text{K}$$

Notice in the *T-s* diagram that the irreversible cycle has a higher heat rejection than the ideal Brayton cycle, both operating between the same pressures. At the same time the net heat added is approximately the same.

The power produced by the cycle and supplied to the electric generator is obtained from

$$\dot{W}k = \dot{m}wk_{\text{cycle}}$$

where

$$\dot{m} = 10 \text{ kg/s}$$

and

$$\begin{aligned} wk_{\text{cycle}} &= wk_{\text{turb}} + wk_{\text{comp}} \\ &= h_3 - h_{4'} + h_1 - h_{2'} \\ &= c_p(T_3 - T_{4'} + T_1 - T_{2'}) \\ &= (1.007 \text{ kJ/kg} \cdot \text{K})(1329 \text{ K} - 830 \text{ K} + 300 \text{ K} - 548 \text{ K}) \\ &= 253 \text{ kJ/kg} \end{aligned}$$

Then

$$\dot{W}k = (10 \text{ kg/s})(253 \text{ kJ/kg}) = 2530 \text{ kJ/s}$$

$$= 2530 \text{ kW} \qquad \textit{Answer}$$

The cycle efficiency we determine from equation (8–5)

$$\eta_T = \frac{wk_{\text{cycle}}}{q_{\text{add}}} \times 100$$

which yields

$$\eta_T = \frac{253 \text{ kJ/kg}}{(44{,}900 \text{ kJ/kg})(\frac{1}{40})} \times 100 = 22.5\% \qquad \textit{Answer}$$

Even though we have irreversible processes in this problem, from a system diagram, as sketched in figure 11–26, we see that

$$\dot{Q}_{\text{rej}} = \dot{Q}_{\text{add}} - \dot{W}k_{\text{cycle}}$$

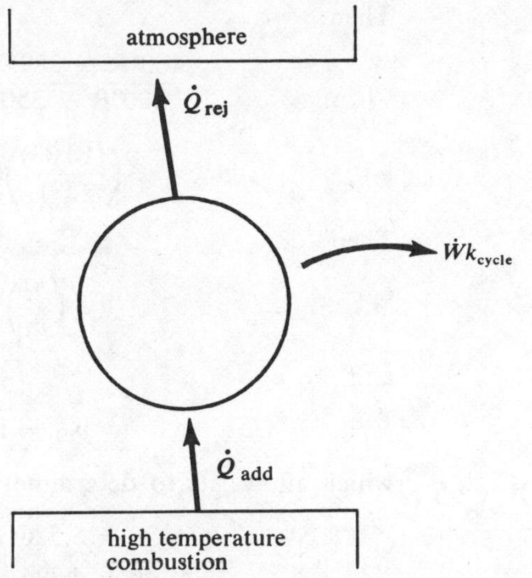

Figure 11–26 System diagram of gas turbine.

which gives us

$$\dot{Q}_{rej} = (10 \text{ kg/s})(-44{,}900/40 \text{ kJ/kg}) + 2530 \text{ kJ/s}$$

$$= -8695 \text{ kJ/s} \qquad \textit{Answer}$$

11.6 Gas Tables and Brayton Cycle Calculations

The gas turbine can be analyzed with the aid of gas table data; this will increase the accuracy of predicted results or property values over the perfect gas analysis as given previously. The following example problem should aid in clarifying this method. A reading of section 10.4 should help in clarifying the use of gas tables.

Example 11.6

For the turbo-jet engine of example 11.4, determine the property values at the four corners of the ideal Brayton cycle. (Use the air table B.6.) Using these results, obtain the required air flow for level flight and for a 10° climb angle assuming a drag of 600 lbf and an aircraft mass of 40,000 lbm.

Solution

The initial conditions are still the same as for example 11.4: $p_1 = 2.73$ psia; $\rho_1 = 0.0189$ lbm/ft³; and $T_1 = 389.97$°R. From the following calculations we determine p_{r1}.

From table B.6 we have

$$p_r = 0.4858 \quad \text{at} \quad T = 400°\text{R}$$
$$p_r = 0.3048 \quad \text{at} \quad T = 350°\text{R}$$

Then

$$\frac{400°\text{R} - 389.97°\text{R}}{400°\text{R} - 350°\text{R}} = \frac{0.4858 - p_{r1}}{0.4858 - 0.3048}$$

$$p_{r1} = -\left(\frac{(10.03)}{(50)}\right)(0.1810) + 0.4858 = 0.4495$$

Then

$$\left(\frac{p_2}{p_1}\right) = 10 = \left(\frac{p_{r2}}{p_{r1}}\right)$$

and

$$p_{r2} = 10 \times p_{r1} = 4.495$$

which allows us to determine T_2. We again interpolate, here between

$$p_r = 5.526 \text{ at } T = 800°\text{R}$$
$$p_r = 4.396 \text{ at } T = 750°\text{R}$$

Then

$$\frac{750 - T_2}{750 - 800} = \frac{4.396 - 4.495}{4.396 - 5.526}$$

so

$$-T_2 = \left(\frac{0.099}{1.130}\right)(50) - 750$$

or

$$T_2 = 746°\text{R}$$

The pressure at state (2) is 27.3 psia since $p_2/p_1 = 10$ and $p_2 = 10\,p_1$. For the combustor we use equation (11–35)

$$\dot{m}_3 h_3 + \dot{m}_f h_f - \dot{m}_2 h_2 = \dot{Q}_{\text{add}}$$

$$h_f = -37 \text{ Btu/lbm fuel} \qquad \text{See equation (11–51).}$$

Here

$$\frac{\dot{m}_f}{\dot{m}_2} = 35 q_{\text{add}} = 19{,}256 \text{ Btu/lbm fuel}$$

and

$$h_2 = 178.7 \text{ Btu/lbm air} \qquad \text{(from table B.6)}$$

We then write the energy equation (11–35) for the combustor

$$\left(\frac{\dot{m}_2 + \dot{m}_f}{\dot{m}_f}\right) h_3 + h_f - \left(\frac{\dot{m}_2}{\dot{m}_f}\right) h_2 = q_{\text{add}}$$

or

$$36 h_3 - 37 - 35(178.7) = 19{,}256$$

and then

$$h_3 = \left(\frac{1}{36}\right)(19256 + 6300 + 37)$$
$$= 710.0 \text{ Btu/lbm}$$

Now we can find T_3 from table B.6. Interpolating between 2600°R and 3000°R we obtain

$$\frac{3000 - T_3}{3000 - 2600} = \frac{790.68 - 710.0}{790.68 - 674.49}$$

$$T_3 = 3000 - \left(\frac{80.68}{116.19}\right)(400) = 2722.2°\text{R}$$

At state (3) we calculate the relative pressure by using the interpolation

$$p_{r3} = p_{r3000°\text{R}} - \left(\frac{80.68}{116.19}\right)(p_{r3000°\text{R}} - p_{r2600°\text{R}})$$

Substituting values into this from table B.6 we obtain

$$p_{r3} = 941.4 - (0.694)(427.9) = 644.4$$

which gives us

$$p_{r4} = \left(\frac{1}{10}\right)(p_{r3}) = 64.44$$

At this relative pressure we get, from table B.6

$$T_4 = 1556.2°\text{R}$$

In figure 11–27 are plotted the property diagrams for the cycle. The specific volumes are obtained from the following calculations:

$$v_1 = \frac{1}{\rho_1} = \frac{1}{0.0189} = 53 \text{ ft}^3/\text{lbm}$$

$$v_2 = v_1\left(\frac{v_{r2}}{v_{r1}}\right) = (53)\left(\frac{64.1}{331.8}\right) = 10.24 \text{ ft}^3/\text{lbm}$$

$$v_3 = v_2\left(\frac{T_3}{T_2}\right) = 10.24\left(\frac{2722.2}{746}\right) = 37.36 \text{ ft}^3/\text{lbm}$$

and

$$v_4 = v_3 \frac{v_{r4}}{v_{r3}} = 37.36\left(\frac{8.958}{1.664}\right) = 201 \text{ ft}^3/\text{lbm}$$

The entropy change is obtained from

$$\Delta s = s_4 - s_1 = \phi_4 - \phi_1 = 0.863 - 0.523$$
$$= 0.340 \text{ Btu/lbm} \cdot °\text{R}$$

To determine the air flow required, we must calculate the cyclic work from the equation

$$wk_{\text{cycle}} = wk_{\text{turb}} + wk_{\text{comp}}$$
$$= h_3 - h_4 + h_1 - h_2$$

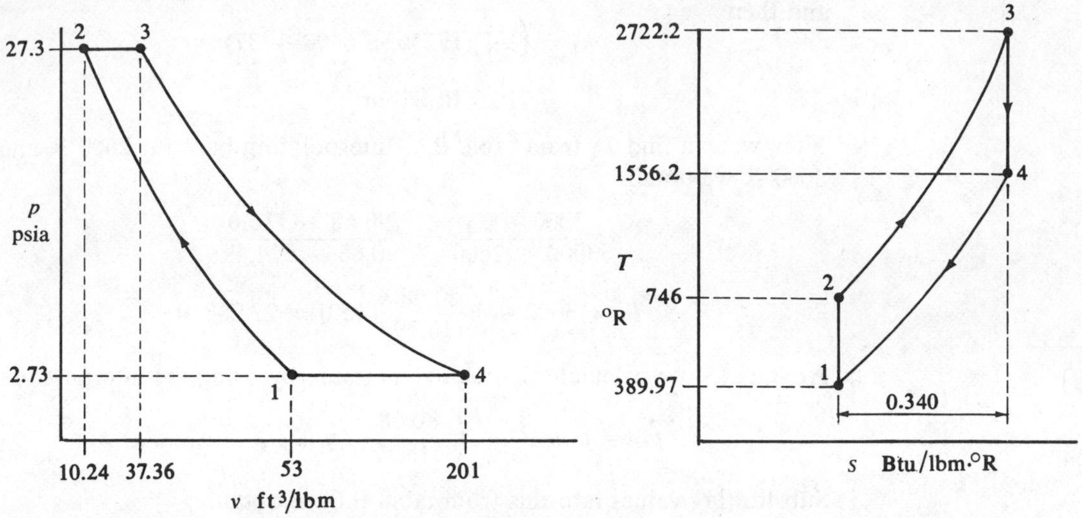

Figure 11–27 Property diagram of Brayton cycle (not drawn to scale).

The enthalpy values we obtain from table B.6:

$$h_1 = 93.2 \text{ Btu/lbm}$$
$$h_2 = 178.7 \text{ Btu/lbm}$$
$$h_3 = 711.0 \text{ Btu/lbm}$$
$$h_4 = 384.1 \text{ Btu/lbm}$$

and then

$$wk_{\text{cycle}} = 711.0 - 384.1 + 93.2 - 178.7$$
$$= 241.4 \text{ Btu/lbm air}$$

The power required for level flight at 500 mph we found to be 800 horsepower, so

$$800 \text{ hp} = \dot{m}wk_{\text{cycle}} = \dot{m}(241.4) \text{ Btu/lbm}$$

This gives us

$$\dot{m} = \frac{800 \text{ hp}}{241.4 \text{ Btu/lbm} \times 1.41} \text{ hp} \cdot \text{s/Btu}$$
$$= 2.35 \text{ lbm/s}$$ *Answer*

This is slightly more than 2.04 lbm/s of air required when we assumed perfect gas conditions in example 11.4. The trend is altogether to be expected.

For the aircraft climbing at 10° the power required is determined by adding the drag and the component of the weight to be overcome. From figure 11–28 we see that this gives a required thrust force of

$$F = 600 \text{ lbf} + W_x = 600 + 40{,}000 \sin 10°$$
$$= 7550 \text{ lbf}$$

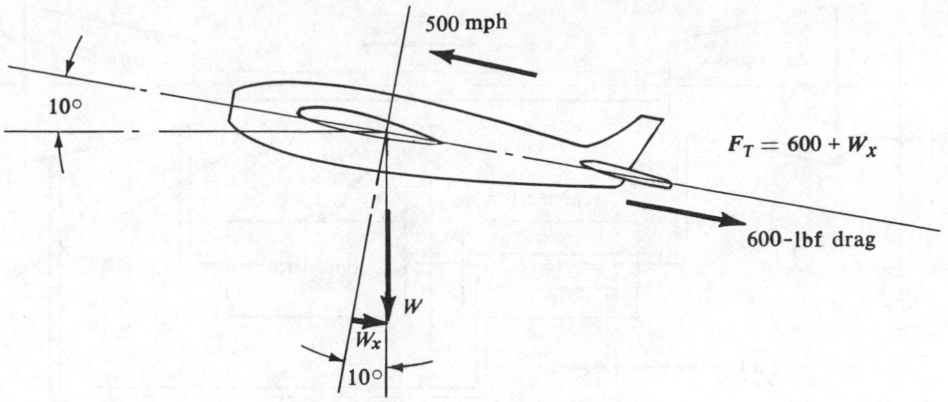

Figure 11–28 Forces on aircraft in climb.

The power required at 500 mph is then

$$\dot{W}k = 7550 \text{ lbf} \times 500 \times \frac{5280}{3600} \text{ ft/s}$$
$$= 5{,}536{,}666 \text{ ft-lbf/s}$$

and the air flow is quickly determined;

$$\dot{m} = \frac{5{,}536{,}666 \text{ ft-lbf/s}}{241.4 \text{ Btu/lbm} \times 778 \text{ ft-lbf/Btu}} = 29.5 \text{ lbm/s} \quad \textit{Answer}$$

We thus see that an appreciable increase in air is required to maneuver the aircraft up. The engine we have considered here in driving an aircraft, the turbo-jet, is an open cycle type of engine and is an "air breather." That is, it requires a large atmosphere of air to function properly. For this reason, the jet aircraft cannot operate at high altitudes or in space. In addition, it is not reusing the gases from the exhaust as was done in the closed cycle of exercise 11.5. We will see a scheme to reduce this wastefulness in the next section.

11.7 Regenerative Heating and Engine Design Considerations

The open cycle Brayton engine exhausts a high temperature gas which must also have a high velocity if it is to propel the aircraft. In applications where the power may be extracted from the turbine shaft, such as was done in exercise 11.5 and could be done in a ground vehicle, the hot exhaust gases are a total loss. To reduce this waste of energy and increase efficiency, a regenerative heater can be attached to the normal gas turbine cycle, as diagrammed in figure 11–29. The technique merely transfers some of the energy from the exhaust to the fresh gases and the ideal Brayton cycle property diagrams then take the form shown in figure 11–30. In each diagram, the shaded area represents the work or heat added by virtue of the regeneration process. From an application of

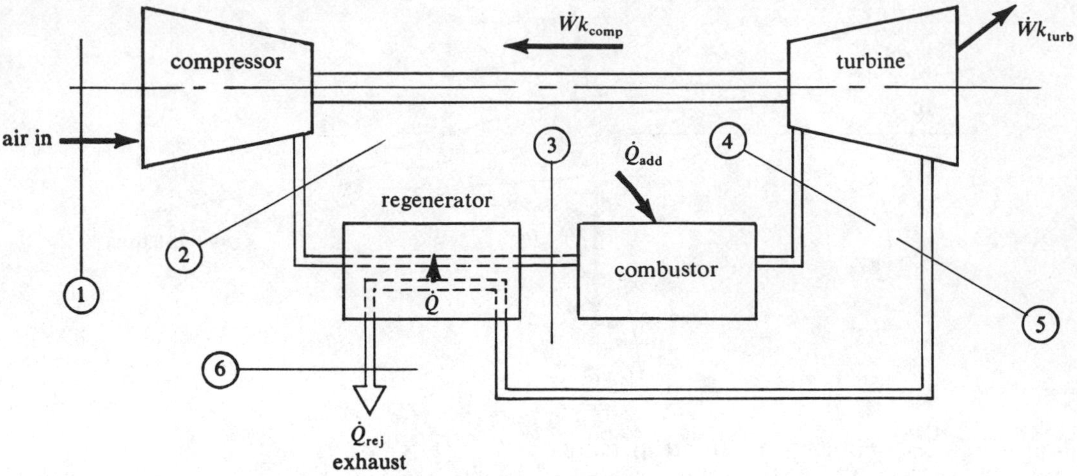

Figure 11–29 Regenerative heating Brayton cycle.

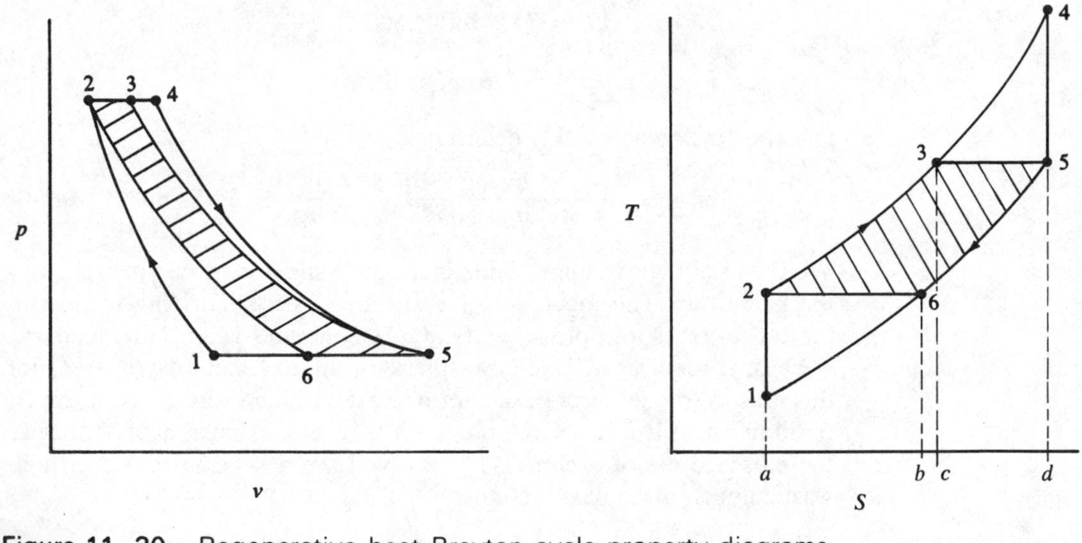

Figure 11–30 Regenerative heat Brayton cycle property diagrams.

the steady flow energy to the components we get

1. for the compressor:

$$h_2 - h_1 = -wk_{comp} \tag{11–57}$$

2. for the regenerator:

$$h_5 - h_6 = h_3 - h_2 \tag{11–58}$$

3. for the combustor:

$$h_4 - h_3 = q_{add} \tag{11–59}$$

4. for the turbine:

$$h_4 - h_5 = -wk_{turb} \tag{11-60}$$

The thermodynamic efficiency can then be written

$$\eta_T = \frac{wk_{net}}{q_{add}} = \left(\frac{h_4 - h_5 + h_1 - h_2}{h_4 - h_3}\right) \times 100 \tag{11-61}$$

Now, if we assume a perfect gas medium with constant specific heats, we can make the substitution $h = c_pT$ and get

$$\eta_T = \left(\frac{T_4 - T_5 + T_1 - T_2}{T_4 - T_3}\right) \times 100$$

or if we have $T_3 = T_5$

$$\eta_T = \left(1 - \frac{T_2 - T_1}{T_4 - T_5}\right) \times 100 \tag{11-62}$$

This can be revised further by some algebra:

$$\eta_T = \left(1 - \frac{T_1\,(T_2/T_1 - 1)}{T_4\,(1 - T_5/T_4)}\right) \times 100 = \left[\left(1 - \frac{T_1}{T_4}\right)(r_p{}^{k-1/k})\right] \times 100 \tag{11-63}$$

where

$$r_p = \frac{p_2}{p_1} = \frac{p_4}{p_5} \quad \text{and} \quad \frac{T_2}{T_1} = \frac{T_4}{T_5}$$

This interesting result shows that an increase in compression or pressure ratio will decrease the efficiency of a regenerative Brayton cycle engine. This is directly opposite to the result of the simple Brayton cycle, as given by equation (11–9), where increases in pressure ratios increase the cycle efficiency. This should provide enough indication that any given engine needs to be analyzed from a fresh outlook. Keep the basic thermodynamic tools as foundations, and from this base make assumptions and restrictive developments in context with the actual situation.

Example 11.7

A reversible, regenerative gas turbine uses 1.0 kg/s of air. Its operating pressure ratio is 15 to 1 and the compressor inlet conditions are 1.01 bars pressure and 17°C. If the turbine exhaust is at 700 K determine the following from an air-standard analysis assuming perfect gas behavior: (1) thermodynamic efficiency, (2) power developed, and (3) heat added and rejected.

Solution

1. We calculate the efficiency from equation (11–63):

First, however, we need to determine T_4:

$$\frac{T_4}{T_5} = \left(\frac{p_4}{p_5}\right)^{(k-1)/k} = (r_p)^{(k-1)/k}$$

and we can then substitute this into the efficiency equations to obtain

$$\eta_T = \left(1 - \frac{T_1}{T_5}\right) \times 100$$

which yields

$$\eta_T = \left(1 - \frac{290\text{ K}}{700\text{ K}}\right) \times 100 = 58.5\% \qquad \textit{Answer}$$

Notice that the efficiency of a simple Brayton cycle gas turbine operating with the same pressure ratio is

$$\eta_T = \left(1 - \frac{1}{(r_p)^{(k-1)/k}}\right) \times 100 = \left(1 - \frac{1}{15^{0.286}}\right) \times 100 = 53.9\%$$

We therefore gain 4.6% efficiency by using the reheater on this engine.

2. The power developed can be obtained from the relation $\dot{W}k = \dot{m}wk_{\text{cycle}}$. For air-standard analysis

$$wk_{\text{cycle}} = c_p(T_4 - T_5 + T_1 - T_2)$$

and

$$T_4 = T_5\left(\frac{p_4}{p_5}\right)^{(k-1)/k} = (700\text{ K})(15)^{0.286} = 1519\text{ K}$$

Similarly, for the temperature T_2

$$T_2 = T_1\left(\frac{p_2}{p_1}\right)^{(k-1)/k} = (290\text{ K})(15)^{0.286} = 629\text{ K}$$

and then

$$wk_{\text{cycle}} = (1.007\text{ kJ/kg}\cdot\text{K})(1519\text{ K} - 700\text{ K} + 290\text{ K} - 629\text{ K})$$
$$= 483\text{ kJ/kg}$$

The delivered power is then computed from

$$\dot{W}k_{\text{cycle}} = (1.0\text{ kg/s})(483\text{ kJ/kg})$$
$$= 483\text{ kW} \qquad \textit{Answer}$$

3. We can calculate the heat added in at least two ways:

$$q_{\text{add}} = \left(\frac{wk_{\text{cycle}}}{\eta_T}\right) \times 100$$

or

$$q_{\text{add}} = c_p(T_4 - T_5)$$

From the first relationship

$$q_{\text{add}} = \frac{483\text{ kJ/kg}}{58.5\%} \times 100 = 826\text{ kJ/kg} \qquad \textit{Answer}$$

From the second method we have

$$q_{add} = (1.007 \text{ kJ/kg} \cdot \text{K})(1519 \text{ K} - 700 \text{ K})$$
$$= 825 \text{ kJ/kg} \qquad \textit{Answer}$$

Thus, there is close agreement between the two methods.

The heat rejected can be computed from the balance

$$q_{rej} = wk_{cycle} - q_{add}$$
$$= 483 \text{ kJ/kg} - 825 \text{ kJ/kg}$$
$$= -342 \text{ kJ/kg} \qquad \textit{Answer}$$

The gas turbine, operating in accordance with the Brayton cycle, is a device which is capable of high performance. Operating efficiences of 40% to 50% have been achieved, and it is capable of high bursts of power (with an accompanied decrease in efficiency). Mechanically it is quite simple to operate and maintain. It is inherently stable, and by rotating symmetrically about a single axis, it can be easily balanced. Its main detriment is its limitation on a high temperature and consequent efficiency limitation; that is, T_3 in the simple cycle or T_4 in the regenerative cycle cannot exceed maximum temperatures that materials can constantly withstand. The cycle is in steady state and the upper temperature is retained for long periods of time — not like in the Otto cycle. (See chapter 10.) The developing of materials which can withstand ever higher temperature, of course, helps the push for higher gas turbine efficiency, but a limit is still there. Additionally, a gas turbine can represent a relatively high initial investment in construction, which can be considered a drawback, but more constant criticism is toward its slow acceleration characteristics for ground vehicles. While the gas turbine is a device which functions best at one constant, optimum speed (as is true of most all engines), the lack of acceptable acceleration in propelling surface vehicles has not been proven and in fact, the exact opposite has frequently been demonstrated.

Two techniques which provide the turbo-jet engine with quick bursts of power and increased performance for short durations are *after-burning* and *water-injection.* The former involves an additional combustion of fuel immediately behind the turbine, which increases the temperature of the already hot exhaust gases and induces an increased exit velocity. As we have seen from equation (11–53), the force on the aircraft is increased and thus drives it faster.

Water-injection is a scheme whereby water is sprayed into the exhaust from the turbine and, while this does not increase the gas temperature, it increases the mass of the expelled gases. This requires equation (11–53)

$$F_I = m\left(\frac{\bar{V}_4 - \bar{V}_1}{\tau}\right) + (p_4 - p_1)A_e$$

to be read as

$$F_I = \left(\frac{m_4\bar{V}_4 - m_1\bar{V}_1}{\tau}\right) + (p_4 - p_1)A_e$$

which then can be seen as an increase in the thrust force F_I due to m_4 being greater than m_1.

11.8 The Ram Jet

We have considered the gas turbine as the prime engine described by the Brayton cycle. As specific examples we looked at the turbo-jet engine and the closed cycle gas turbine. Now let us look at the case of a vehicle traveling at such a high speed in air that the turbo-jet compressor is not needed. That is, at high velocities the entering air will be already traveling faster than a turbine driven compressor can induce. Then, we can neglect the compressor and just restrict the air to increase its pressure. This high pressure air is then allowed to combust with fuel and is subsequently expelled as hot exhaust gases. Since the turbine merely drove the compressor in the turbo-jet engine, we can also neglect this component and direct the exhaust through a nozzle to increase the velocity for impulse. This whole series of processes describes the workings of the ram jet engine. It derives its name from the fact that air is "rammed" into the engine from the high velocity of the air relative to the engine. In figure 11–31 is shown the basic schematic of the engine and the mechanics of it are depicted in figure 11–32. The shock waves shown as lines in this figure serve to decrease the relative velocity of the air entering the engine and consequently increase the pressure. Shock waves are a normal occurrence in supersonic speeds, which are the operating regime of the ram jet.

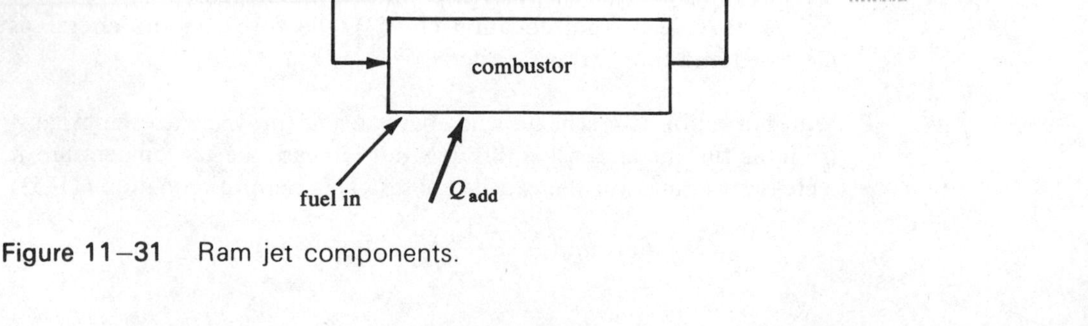

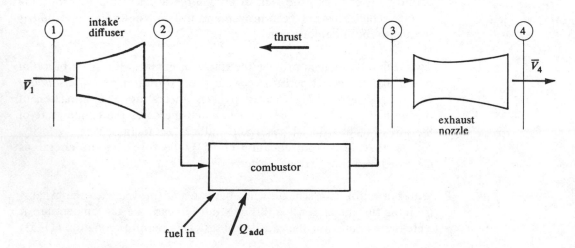

Figure 11–31 Ram jet components.

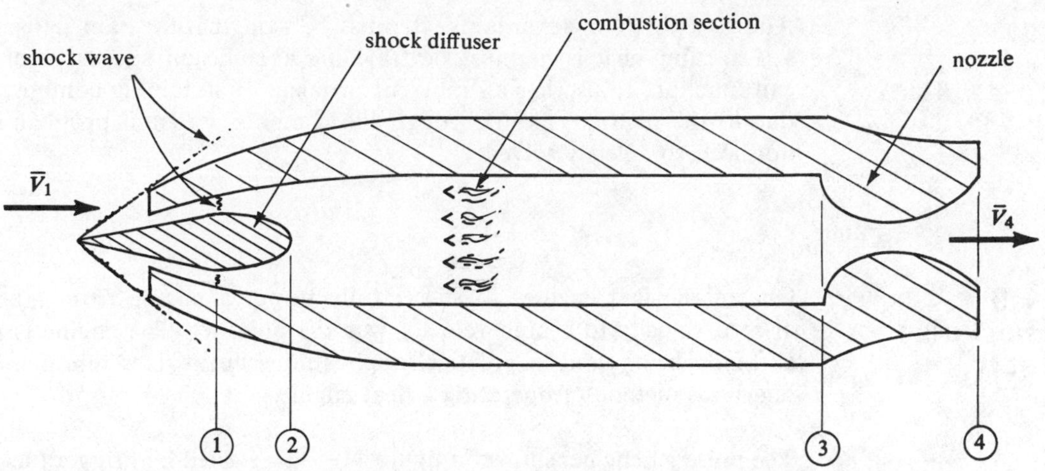

Figure 11–32 Typical ram jet engine.

The thermodynamic analysis of the engine is identical to that of the turbo-jet; that is, for an air-standard analysis we have

$$q_{\text{add}} = c_p(T_3 - T_2) \tag{11-64}$$

$$q_{\text{rej}} = c_p(T_1 - T_4) \tag{11-65}$$

and

$$wk_{\text{cycle}} = q_{\text{add}} + q_{\text{rej}} = c_p(T_3 - T_2 + T_1 - T_4) \tag{11-66}$$

The efficiency is gotten from

$$\eta_T = \left(1 - \frac{1}{(r_p)^{(k-1)/k}}\right) \times 100 \tag{11-67}$$

In equations (11–64), (11–65), and (11–66) above, all the temperatures should properly be replaced by total temperature T^* defined by equation (11–43):

$$T^* = T + \frac{\overline{V}^2}{2c_p}$$

Total temperature obviously includes the kinetic energy of the gases. In the ram jet in particular the kinetic energy effects cannot generally be ignored.

The ram jet has the following advantages over a turbo-jet engine:

1. Higher operating temperature due to a simplified construction. This implies an increased efficiency.
2. More power per unit engine weight.
3. Higher operating speeds.
4. Less maintenance.

Along with these advantages, it must be said that the ram jet is not self-starting; that is, it must be traveling at sufficient speeds before it can function. It also has an inherent problem of sustaining combustion due to the high air speed through the chamber, and this problem has not been completely solved.

11.9 The Pulse Jet

One of the first engines to successfully power a missile from takeoff at zero velocity to a high velocity was the pulse jet. The engine is not currently being used in practical applications, but it does represent an ingenious method of operating a heat engine.

The pulse jet engine, shown in figure 11–33, is started by firing a fuel air mixture with an electrical spark. This increases the pressure and temperature in the combustion chamber (process (2–3)) isometrically. Then the hot gases are expelled through the exhaust nozzle, thus imparting an impulse force to the vehicle. Immediately following this process, the air is drawn into the inlet diffuser and then through the flapper valves into the combustion chamber. The flapper valves allow air to flow into the chamber, but they close if air tends to flow back. In figure 11–33 are shown the two normal positions of the flapper valves; the closed position during combustion (process (2–3)) and expansion (process (3–4)), and the open position during compression (process (1–2)). The engine has a distinct cycle during which the flapper valves open and close to allow for air intake and combustion respectively. Depending on the engine's size,

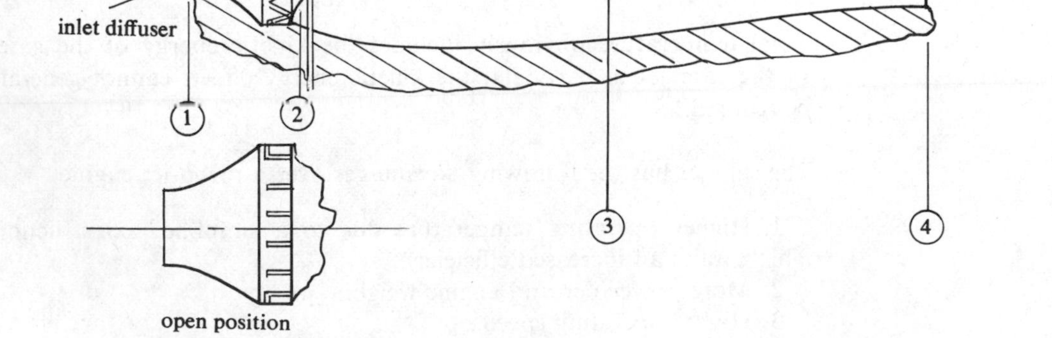

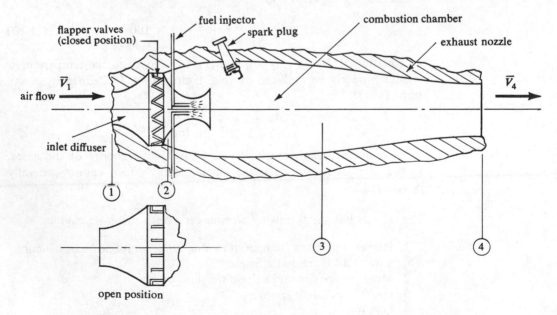

Figure 11–33 Cross section of pulse jet.

the cycle will proceed at a speed inherent to the design. That is, no external controls are required to actuate the flapper valves as they are moved by differential pressures between the diffuser and the combustion chamber. One of the earliest and the most famous applications of the pulse jet was the German V-1 rocket, a pilotless warcraft, used in World War II, which operated at approximately 40 cycles/s. Because of this particular frequency, it produced a loud buzzing sound when in flight, and for this reason, the V-1 was referred to as the "buzz bomb." Other engines of smaller size than the V-1 power plant have been built with operating speeds of 250 to 300 cycles/s.

The thermodynamic cycle of the ideal pulse jet is described by the *p-V* diagram of figure 11–34; however, the deviation from this is great in actual engines. As can be seen, the cycle is a hybrid of the Brayton and Otto cycles. Today, however, there are no serious efforts at adapting the pulse jet to produce actual engines, even though it is relatively inexpensive, can produce power at zero velocity, and increases its effectiveness with increased vehicle velocity.

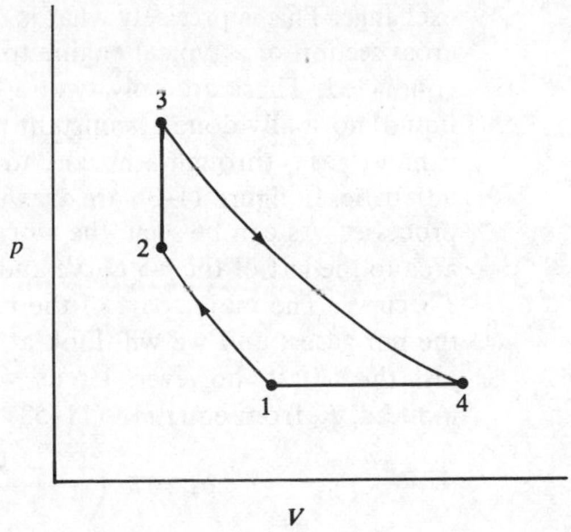

Figure 11–34 *p-V* diagram of ideal pulse jet.

11.10 Rockets

The rocket engine is included here because it is frequently compared to the jet engine as a means of producing power. It has some distinct advantages over the gas turbine engine:

1. It can develop maximum power at zero velocity.
2. It is not "air-breathing" as are all other gas turbine or jet engines.
3. It can develop much higher thrust-to-weight ratios.

These characteristics have made the rocket engine the primary means of power for travel beyond the earth's atmosphere. The rocket engine has a distinct advantage when compared to other heat engines, however, since the rocket is not a true cyclic device, that is, it is not a heat engine,

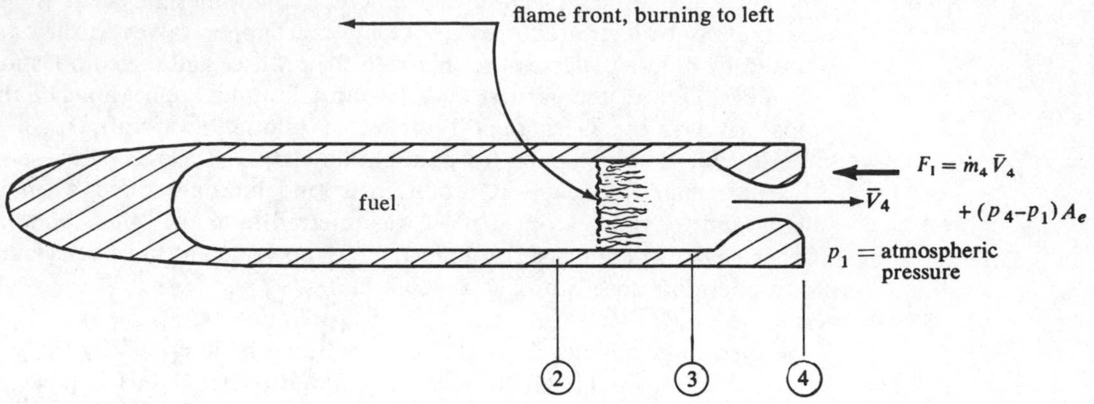

Figure 11–35 Cross section of typical rocket.

but is rather a device executing processes which *never* return the engine to its initial state. This means the rocket need not be limited to the second law of thermodynamics — it may produce work with only one heat exchange. This is precisely what is done, and in figure 11–35 is shown the cross section of a typical engine to indicate the physical processes being conducted. There are only two: a combustion of a fuel (either solid or liquid) normally done at constant pressure; and an expansion of the hot exhaust gases through a nozzle. Ideally, this expansion is reversible and adiabatic. In figure 11–36 are drawn the process curves of these two ideal processes. As can be seen, the work available to propel the rocket is the area to the left of the *p-v* curve and the heat added is the area under the *T-s* curve. The major part of the rocket is the nozzle, which accelerates the hot gases; and we will look at an example problem concerned with just this. First, however, let us write the general equation for thrust induced, F_I from equation (11–53)

$$F_I = m\left(\frac{\bar{V}_4 - \bar{V}_1}{\tau}\right) + (p_4 - p_1)A_e$$

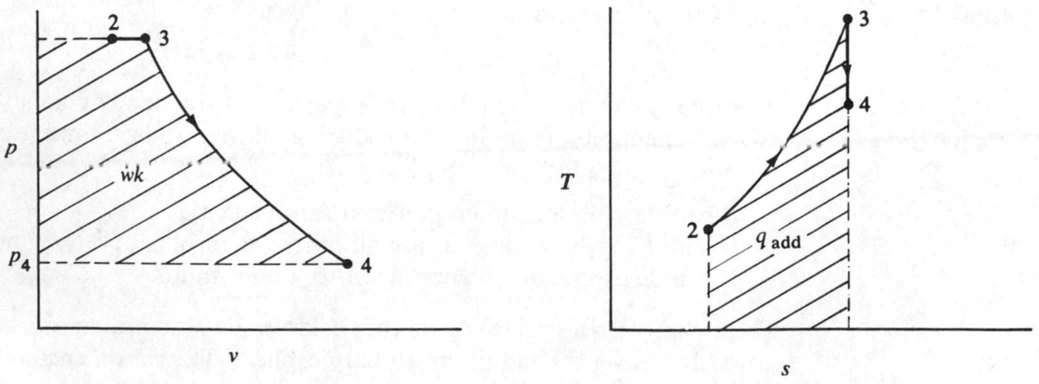

Figure 11–36 Property diagrams for rocket engine.

In the rocket, however, no mass is entering the system so $\bar{V}_1 = 0$, yielding

$$F_I = m\,(\bar{V}_4) + (p_4 - p_1)A_e \tag{11–68}$$

The specific impulse of the turbine is then written

$$I_{sp} = \frac{F_I}{\dot{m}} = \bar{V}_4 + \left(\frac{p_4 - p_3}{\dot{m}}\right)A_e \tag{11–69}$$

Example 11.8 A rocket burns 200 lbm/s of fuel and oxidizer at 2700°F and 300 psia. The nozzle exhaust area is 2 ft², and the pressure is assumed to be atmospheric (14.7 psia) at the exit. Determine the exhaust velocity, the specific impulse, impulse, and power produced by the engine if the nozzle is reversible and adiabatic, and we assume air-standard conditions.

Solution For a reversible adiabatic nozzle we write the steady flow energy equation

$$h_4 - h_3 + \frac{\bar{V}_4^2 - \bar{V}_3^2}{2g_c} = 0$$

Since $\bar{V}_3 = 0$ we then have

$$\bar{V}_4 = \sqrt{2g_c(h_3 - h_4)}$$

Invoking air-standard conditions

$$h_3 - h_4 = c_p(T_3 - T_4)$$

Then, for reversible adiabatic processes

$$T_4 = T_3\left(\frac{p_4}{p_3}\right)^{(k-1)/k} = (3160°\text{R})\left(\frac{14.7}{300}\right)^{0.286}$$
$$= 1334°\text{R}$$

and the velocity can then be calculated:

$$\bar{V}_4 = \sqrt{2 \times 32.2 \text{ ft-lbm/s}^2 \cdot \text{lbf} \times 0.24 \text{ Btu/lbm} \cdot °\text{R} \times 778 \text{ ft-lbf/lbm } (3160 - 1334)\ °\text{R}}$$

$$= 4685 \text{ ft/s}$$ *Answer*

The specific impulse is then obtained from equation (11–69), assuming $p_4 = p_3$

$$I_{sp} = \frac{4685 \text{ ft/s}}{32.2 \text{ ft-lbm/lbf} \cdot \text{s}^2} + 0 = 145 \text{ lbf} \cdot \text{s/lbm}$$ *Answer*

The impulse F_I can be quickly calculated from equation (11–68)

$$F_I = \frac{\dot{m}}{g_c}(V_4) = \dot{m}I_{sp} = 200 \text{ lbm/s} \times 145 \text{ lbf} \cdot \text{s/lbm}$$
$$= 29{,}000 \text{ lbf}$$ *Answer*

Using our common equations, we obtain the power from the equation

$$\dot{W}k = \dot{m}wk$$

where

$$wk = c_p(T_3 - T_4)$$

Then

$$\dot{W}k = 200 \text{ lbm/s} \times 0.24 \text{ Btu/lbm} \cdot {}^\circ\text{R } (3160 - 1334)\ {}^\circ\text{R}$$

or

$$\dot{W}k = 87648 \text{ Btu/s} = 123{,}584 \text{ hp}$$ *Answer*

Practice Problems

Problems that use the SI units are indicated with the notation (M) under the problem number, and those problems using English units are indicated with (E).

Problems designated with an asterisk * preceding the number should only be attempted by those having a background which includes the knowledge of integral calculus.

Section 11.1

11.1. Determine the thermodynamic efficiency of an ideal gas turbine operating with a pressure ratio of 20.

11.2. Determine the thermodynamic efficiency of an ideal Brayton cycle engine if the pressure increase across the compressor is 18 times the inlet pressure.

11.3. If the pressure ratio across the compressor of an ideal Brayton cycle engine is 22 to 1 and the working medium is air, determine the temperature ratio across the compressor.

11.4. (M) If the inlet air temperature to the compressor of problem 11.3 is 20°C, determine the work required of the compressor.

11.5. (M) An ideal Brayton cycle engine operates with a pressure ratio of 8, the working media is air, and the compressor inlet temperature is 30°C. If the entropy change across the combustor is 0.7 kJ/kg · K, determine the pressure, specific volume, and temperature at the four corners of the cycle. Assume the inlet pressure to the compressor is 1.01 bars.

11.6. (E) If the inlet air temperature to the compressor of problem 11.2 (assuming air to be the working medium) is 100°F, determine the compressor work.

11.7. Sketch the *p-V* and *T-S* diagrams for an ideal Brayton cycle engine. Then, if the heat added is 300 Btu/lbm air, $r_p = 10$, $p_1 = 14.7$ psia, $T_1 = 100$°F, and the working medium is air, determine the properties p, ρ, and T at the four corners of the cycle.

11.8. (E) A 4000-hp ideal gas turbine engine operates between 15 psia and 160 psia pressures. Determine the rate of heat addition to the cycle.

11.9. Starting from the steady flow energy equation, state the necessary assumptions for arriving at equations (11–1), (11–2), and (11–3).

11.10. Prove that the thermodynamic efficiency of an ideal Brayton cycle is given by

$$\eta_T = 1 - \frac{1}{(r_p)^{(k-1)/k}} \times 100$$

for a perfect gas working medium.

Sections 11.2 and 11.3

11.11. (M) A turbine is designed with semicircular blades. If the blade velocity is expected to be 30 m/s, determine the most desireable velocity of fluid impinging on the blades.

11.12. (M) A turbine receives 60 kg/min of air at 20 m/s at 1200 K. The exhaust air is at 600 K and 10 m/s. Determine the work produced by the turbine per unit mass of air and the power produced.

11.13. (M) A gas turbine is designed to produce 3000 horsepower when using air at a maximum inlet temperature of 650 K and assuming a heat loss of 70 kJ/s. If the exhaust temperature is desired to be no greater than 200°C, determine the mass flow rate of air required. Also determine the velocity of air leaving the nozzle and the blade velocity. If the blade is not to exceed sonic speeds, would you recommend using additional turbine stages?

11.14. (M) A turbine expands air from a region where the pressure and temperature are 20.0 bars and 1000°C to a region at 1.01 bars. Neglecting kinetic and potential energy changes of the air and assuming the process through the turbine is polytropic with an exponent $n = 1.5$, determine

(a) wk_{turb}.

(b) q.

(c) $\dot{W}k_{turb}$ if mass flow of air is 6 kg/s.

11.15. (M) If 60,000 kg/h of air are supplied to a combustor at 5.0 bars and 160°C and the leaving gases are to be at 800°C, determine the rate of heat addition required by the combustion of fuel and air. Assume the kinetic energy changes and the mass of fuel to be negligible.

11.16. (M) A compressor receives air at 1.01 bars, 40°C, and 6 m/s. If the gases are compressed reversibly adiabatically to 9.0 bars and they leave the compressor at 90 m/s velocity, determine the amount of work per kilogram of air that is required by the compressor. If the process is reversible polytropic with $n = 1.34$ and kinetic energy changes are neglected, determine the pressure of the air leaving the compressor and the work required by the compressor per kilogram of air. Assume that the temperature T_2 is 540 K.

11.17. (E) An ideal turbine receives 30 lbm/min of air at 2100°R and exhausts it at 700°F. Determine the power produced by the turbine if kinetic energy changes are neglected.

11.18. (E) A turbine nozzle shown in figure 11–37 receives air at 1600°F and 200 psia. Assuming the air impinging on the blades has a velocity of 800

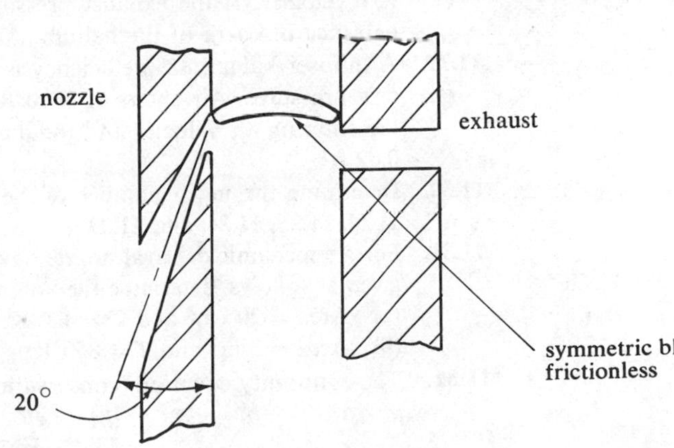

Figure 11–37

ft/s, determine the blade velocity and the exhaust temperature if heat losses as well as kinetic energy changes are neglected in the turbine.

11.19. (E) A turbine reversible adiabatically expands air from 280 psia to 15 psia and 800°F. Determine the inlet air temperature and the work gained from the turbine per pound-mass of air if kinetic and potential energy changes are neglected in the turbine.

11.20. (E) Air enters a combustor at 100 psia, 400°F, and 200 ft/s. Fuel having a heating value of 17,000 Btu/lbm is burned and the gases leaving the combustor (assuming they have similar properties to air) are at 1700°F and 400 ft/s. Determine the amount of fuel required per pound-mass of air, $\dot{m}_f/\dot{m}_a$.

11.21. (E) Air having a density of 0.06 lbm/ft³ and at a pressure of 14.7 psia is compressed to 0.10 lbm/ft³. Determine the pressure of the air leaving the compressor and the work per pound-mass of air required if the process is reversible polytropic with $n = 1.34$ and the velocity changes of the air flow through the compressor are neglected.

Section 11.4

11.22. (M) Air enters a reversible adiabatic converging nozzle at 5 m/s, 1000 K, and 12.0 bars. Determine the velocity of air leaving the nozzle if the pressure there is 1.1 bars.

11.23. (M) Determine the entrance and exit areas of the nozzle in problem 11.22 if the mass flow rate is to be 120 kg/min.

11.24. (M) Air at 500 m/s velocity enters a diffuser at 0.8 bar. If the diffuser is reversible and adiabatic and the exit pressure is 2.4 bars, determine the inlet air temperature. The air is quiescent when it leaves the diffuser.

11.25. (M) An adiabatic diffuser is used to direct oxygen gas into a compressor. Oxygen is furnished at 650 K and 760 mm Hg through an area of 10 cm². The diffuser is 88 percent efficient when the oxygen leaves at 1000 K and 3000 mm Hg. Determine the entrance velocity and mass flow of oxygen.

11.26. (E) A nozzle has 85% efficiency and operates between 180 psia and 12 inches of Hg pressure. Air flows in at 1600°R and leaves at 3200 ft/s. Determine the mass flow of air through the nozzle if the entrance area is 0.5 in².

11.27. (E) For the nozzle of problem 11.26 determine the exit area of the nozzle.

11.28. (E) A reversible adiabatic diffuser receives helium gas at 800°R and 1950 ft/s velocity. If the exhaust pressure must be 50 psia, determine the entrance pressure of the helium. Assume the exhaust velocity is zero.

11.29. (E) A diffuser having 82% efficiency is operating between 15 psia and 35 psia pressures. Air enters at 1650°R and leaves at 2200°R. Determine the entering air velocity and the mass flow rate if the entrance area is 0.02 ft³.

11.30. Determine the mach number of the flow at the entrance in problems 11.24, 11.25, 11.26, and 11.29.

11.31. For a supersonic deLaval nozzle having a throat area and temperature given as follows, determine the volume flow through the nozzle:
(a) Area = 20 cm² and $T = 420$ K.
(b) Area = 3 in² and $T = 850$°R.

***11.32.** The continuity equation (conservation of mass) in differential form is

$$\frac{dA}{A} + \frac{dV}{V} - \frac{dv}{v} = 0$$

and the steady flow energy equation can be written

$$\frac{VdV}{g} = -vdp$$

Using these two and the relation $pv^k = C$ prove that

$$\frac{dA}{A} = (M^2 - 1)\frac{dV}{V}$$

for reversible adiabatic nozzle flow. (M is the mach number.)

Section 11.5

11.33. **(M)** An ideal Brayton cycle gas turbine operates with a pressure ratio of 7.6. The entering air is at a pressure of 1.01 bars and the temperature is 27°C. The air/fuel ratio is 35 to 1 and the fuel has a LHV of 45,000 kJ/kg. Determine

(a) p-v and T-s diagrams.
(b) q_{add}.
(c) wk_{turb}.
(d) wk_{comp}.
(e) wk_{cycle}.
(f) Thermodynamic efficiency.

11.34. **(M)** An aircraft flying at 8000 meters in a standard day atmosphere is traveling with an airspeed of 200 m/s. The craft is propelled by an ideal gas turbine which operates on a pressure ratio of 9 to 1 and which burns kerosene. If the air/fuel ratio is 40 to 1 calculate the following:

(a) wk_{turb}.
(b) wk_{comp}.
(c) wk_{cycle}.
(d) q_{rej}.

11.35. **(E)** An ideal closed gas turbine engine operates with air at 14.7 psia and 50°F. After compression the air is at 860°R. Then, entering the turbine the gases are at 1400°F. If the combustor is burning octane fuel, determine

(a) Air/fuel ratio.
(b) wk_{turb}.
(c) wk_{comp}.
(d) Thermodynamic efficiency.

11.36. **(E)** For the gas turbine cycle shown in figure 11–38 the following data apply:

$\eta_{comp} = 90\%$	$p_2 = 100$ psia
$\eta_{turb} = 90\%$	$T_3 = 2500°R$
$\eta_{combust} = 65\%$	fuel LHV = 15,000 Btu/lbm
$p_1 = 14$ psia	$\dot{W}k_{turb} = 7000$ hp
$t_1 = 65°F$	

Determine

(a) $\dot{m}_1$ *and* $\dot{m}_f$.
(b) Thermodynamic efficiency.

Section 11.6

11.37. **(M)** Air enters an ideal stationary gas turbine engine at 20°C and 10^6 N/m^2. The pressure ratio is 8 and the air leaves the combustor at 1100 K. Assume the mass of fuel is negligible and the mass flow of air is 100 kg/s. Determine

(a) Cycle thermodynamic efficiency.

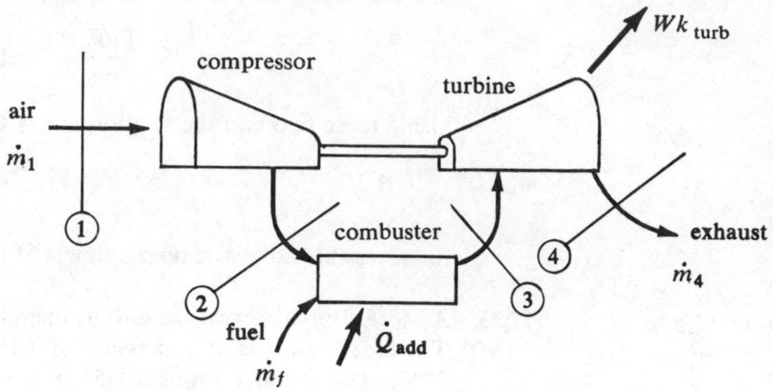

Figure 11–38

(b) Heat added per unit time.
(c) Heat rejected per unit time.
(d) Power developed.

11.38. (M) Using the gas tables, solve problem 11.34.

11.39. (E) Using the gas tables, solve problem 11.35.

11.40. (E) An ideal Brayton cycle gas turbine using air operates with a pressure ratio of 7.5 to 1. The entrance conditions are 14.7 psia and 60°F. If 500 Btu/lbm of heat are added in the combustor, using the gas tables, determine

(a) p, v, T at the four cycle corners.
(b) Δs for compression and expansion.
(c) wk_{cycle}.
(d) Thermodynamic efficiency.

11.41. (E) Using the gas tables, solve problem 11.36.

Section 11.7

11.42. (M) The effectiveness of a regenerator is defined by the ratio r_g given by

$$r_g = \left[\frac{(T_3 - T_2)\,\dot{m}_2}{(T_5 - T_6)\,(\dot{m}_2 + \dot{m}_{fuel})}\right] \times 100\%$$

where the subscripts refer to figure 11–29 and the specific heats are constant and equal for air and the exhaust. Determine the effectiveness to the combustor is at 620°C and the gases leaving the regenerator are at 370°C and the gases leaving the turbine are at 800°C. Assume the air/fuel ratio is 30 to 1. Plot the T-s diagram of a cycle which could be using this regenerator, indicating the significant temperature data.

11.43. (E) A regenerative gas turbine uses 200,000 lbm/hr of air and 4000 lbm/hr of fuel having a LHV of 17,500 Btu/lbm. The air is taken in at 14.7 psia, 65°F, and compressed reversible adiabatically to 120 psia. A regenerator with an effectiveness of 60% (see problem 11.42) heats the air to 1100°R. Assume the turbine is reversible and adiabatic. Then determine

(a) $\dot{Q}_{add}$
(b) $\dot{W}k_{cycle}$

(c) Exhaust temperatures
(d) $\dot{Q}_{rej}$
(e) Cycle thermodynamic efficiency

Section 11.8

11.44. Determine the thermodynamic efficiency of a ram jet engine operating between pressures of
(a) 0.9 bar and 5.4 bars.
(b) 12.7 psia and 80 psia.

11.45. An aircraft is powered by a ram jet engine. For the two cases given determine: q_{add}, q_{rej}, $T^*{}_1$, $T^*{}_4$, and efficiency.
(a) Aircraft velocity of 1000 m/s, $t_1 = 5°C$, $t_2 = 150°C$, $t_3 = 1000°C$, $t_4 = 350°C$.
(b) Aircraft velocity of 4000 ft/s, $t_1 = 50°F$, $t_2 = 300°F$, $t_3 = 1800°F$, $t_4 = 600°F$.
Assume $\bar{V}_1 = \bar{V}_2 = \bar{V}_3 = \bar{V}_4$ for both cases.

Section 11.10

11.46. (M) A rocket engine burns 700 kg/s of fuel and oxidizer at 1300°C and 20 atmospheric pressure. The nozzle is frictionless and adiabatic and has a throat area equal to 0.6 m^3. Assume exhaust gases have a density of 0.004 g/cm^3 and $c_p = 0.24$ cal/g · K. If the rocket is operating in an atmosphere at 760 millimeters Hg, determine
(a) Exhaust temperature.
(b) Impulse.
(c) Power produced by the rocket.

11.47. (E) Determine the specific impulse of a rocket which burns 3600 lbm fuel and air per second. The rocket nozzle has a cross-sectional area of 10 ft^2 allowing the exhaust gases to expand from the combustion chamber at 600 psia to a 10 psia atmospheric pressure surrounding. Assume the density of gases leaving the nozzle is 0.05 lbm/ft^3.

12

The Closed Steam Power Cycle and the Rankine Cycle

Electrical power represents the form in which the greatest amount of energy is directly used by society. However, the overwhelming majority of this power is generated by steam and the *steam turbine*. Fossil fuels (coal, oil, and gas), or nuclear fuels provide the chemical availability to produce the steam which in turn drives the steam turbine and electric generator. This cycle has demonstrated the highest thermodynamic efficiency for vast power production in the myriad of technical devices and it appears destined to be the major source of mechanical power for quite some time. A typical steam turbine power plant is shown in figure 12–1, where the raw material is coal, transported by river barge or railroads. When burned, the coal produces thermal energy which boils water to drive steam turbines and, in turn, electrical generators. The workings of this system constitute the typical *Rankine* or *steam turbine cycle*, which we will examine in this chapter.

The Rankine cycle, an idealization, is defined and compared to the operation of the actual steam turbine cycle. The major components of the closed steam cycle — boiler or steam generator, condenser, feedwater pumps, and steam turbine — are discussed to show the physical processes corresponding to the Rankine cycle and to assist in the cycle analysis.

The *phase changes*, so important in steam power applications, are discussed along with the concept of *superheated steam*. Property diagrams (T-s and p-v) are used to help expand the details of the phase change beyond the brief introduction from section 8.10.

Use of the *steam tables* and the *Mollier diagram* are explained (the perfect gas assumption is never used with steam except at very high temperatures) to give the reader a storehouse of ready and useful data for problem solving. Some traditional steam turbines operating on the Rankine cycle are then analyzed with the thermodynamic tools.

The modifications of the Rankine steam turbine cycle to increase power output and/or efficiency are introduced with treatments of the *regenerative* and *reheat cycles*. Finally, some general considerations are made regarding the design of future steam turbine cycles.

Figure 12–1 The modern steam turbine, electric power generating station. Reproduced with permission of Dayton Power and Light Company; Dayton, Ohio.

12.1 The Rankine Cycle

The thermodynamic cycle which most properly describes the workings of the ideal steam turbine is the Rankine cycle. This is defined by the four reversible processes:

(1–2) Adiabatic compression of liquid (water).
(2–3) Isobaric heat addition to convert liquid to a vapor (steam).
(3–4) Adiabatic expansion of vapor to low pressure.
(4–1) Isobaric heat rejection to condense vapor to liquid.

These four processes are the same combination which describes the gas turbine or Brayton cycle (see chapter 11), but here we are involved with phase changes in the working media which lend unique characteristics to the Rankine cycle not present in the Brayton cycle. In figure 12–2 are depicted the typical property diagrams of the Rankine cycle, and in figure 12–3 the schematic of the major components of the steam turbine cycle is described by these property diagrams. Notice that a saturation line is drawn in the property diagrams to give the reader an idea of the approximate relation of the process to the phase change of H_2O which occurs "under" the saturation line.

For the ideal steam turbine cycle of figures 12–2 and 12–3, we can apply the steady flow energy equation to the individual components to arrive

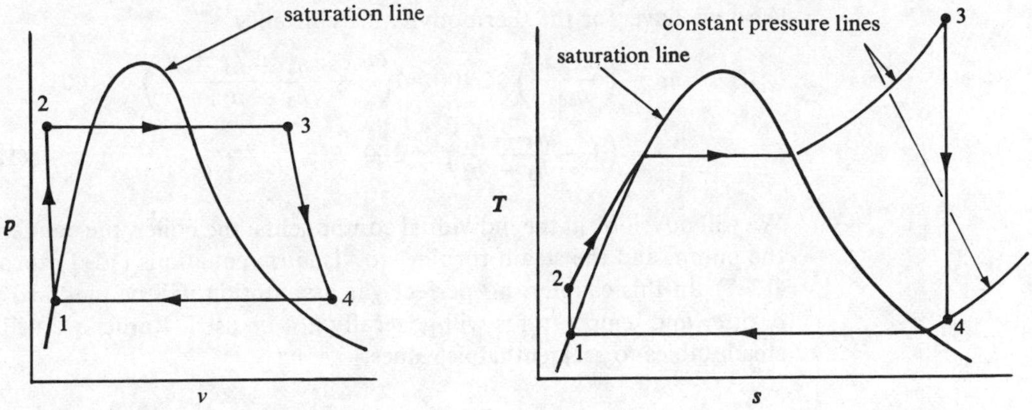

Figure 12–2 Property diagrams of the Rankine cycle.

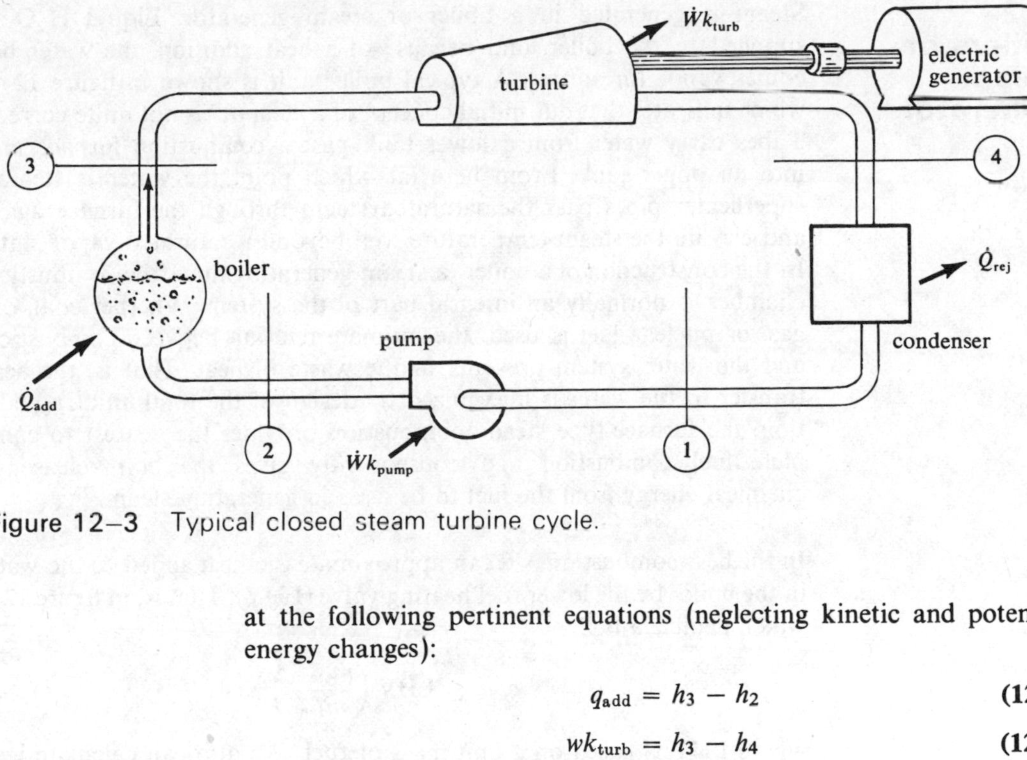

Figure 12–3 Typical closed steam turbine cycle.

at the following pertinent equations (neglecting kinetic and potential energy changes):

$$q_{add} = h_3 - h_2 \tag{12–1}$$

$$wk_{turb} = h_3 - h_4 \tag{12–2}$$

$$q_{rej} = h_1 - h_4 \tag{12–3}$$

$$wk_{pump} = h_1 - h_2 \tag{12–4}$$

For the full cycle then

$$wk_{cycle} = wk_{turb} + wk_{pump} \tag{12–5}$$

and by using the results we get

$$wk_{cycle} = h_3 - h_4 + h_1 - h_2 \tag{12–6}$$

Also we have, for the thermodynamic efficiency

$$\eta_T = \left(\frac{wk_{\text{cycle}}}{q_{\text{add}}}\right) \times 100 = \left(\frac{h_3 - h_4 + h_1 - h_2}{h_3 - h_2}\right) \times 100$$

$$= \left(1 - \frac{h_4 - h_1}{h_3 - h_2}\right) \times 100 \qquad \textbf{(12–7)}$$

We will now look at the individual components: the boiler, the condenser, the pump, and the steam turbine, to visualize equations (12–1) through (12–7). In this chapter, no perfect gas assumption will be made so that c_v or c_p and temperatures will generally not be used. Rather we will use steam tables to seek enthalpy values.

12.2 Boilers or Steam Generators

Steam is generated in a boiler or steam generator. Liquid H_2O is supplied to the boiler, and because of a heat addition, the water becomes vapor (or steam). A typical boiler unit is shown in figure 12–4, which indicates that our initial concept of a "teapot" is not quite correct. Tubes carry water from a lower tank past a combustion furnace and into an upper tank. From here (at which point the water is steam), superheater pipes pass the saturated steam through the furnace again and elevate the steam temperature well beyond a saturated vapor state. In the construction of a boiler (a steam generating unit), the combustion chamber is normally an integral part of the system. Whether coal, oil, gas, or nuclear fuel is used, the intimate relation between combustion and the water system prevents undue waste of heat. That is, the heat transfer to the water is maximized by design of the total unit. In addition, the furnace-type steady combustion provides the nearest to complete fuel combustion and consequently gives the best release of chemical energy from the fuel to be used in generating steam.

In the best combustion, we can approximate the heat added to the water in the boiler by the lower fuel heating valve (LHV). That is, in figure 12–3 which depicts a boiler

$$q_{\text{add}} \leq \text{LHV}\left(\frac{\dot{m}_{\text{steam}}}{\dot{m}_{\text{fuel}}}\right) \qquad \textbf{(12–8)}$$

where LHV is based on a unit mass of fuel. We also can calculate heat added to the water from equation (12–1)

$$q_{\text{add}} = h_3 - h_2$$

With these two results we can define the boiler efficiency as

$$\eta_{\text{boiler}} = \left(\frac{h_3 - h_2}{\text{LHV}}\right)\left(\frac{\dot{m}_{\text{steam}}}{\dot{m}_{\text{fuel}}}\right) \qquad \textbf{(12–9)}$$

The boiler is an important component in the steam cycle and with the search for new energy sources, it has been the center of attention.

Figure 12–4 Typical steam generating boiler. Field erected industrial boiler. Reproduced with permission of Combustion Engineering, Inc., Windsor, Conn.

12.3 Steam Turbines

Steam at a high pressure and temperature has a large amount of internal energy, but a device is needed to convert this energy into mechanical work or power. The steam turbine is the machine which does this converting and which operates on exactly the same principles as the historic waterwheel, or the gas turbine. The reader is advised to read section 11.2 at this time if he has not already done so, as the description of the gas turbine and its operation is completely analogous to that of the steam turbine. While the steam turbine is occasionally constructed as a reaction turbine, the majority of the machines are impulse-reaction turbines. In this device, steam is supplied from a boiler, and its internal energy is then converted into kinetic energy in a nozzle, which directs the steam against buckets attached to a turbine wheel. This steam is diverted by the buckets into an exhaust passage or another nozzle, and turbine wheel. The diverting of the steam causes the turbine wheel to rotate and thus produce a shaft work which can easily be utilized by mechanically connecting an electric generator or other device to the turbine shaft. For a frictionless turbine allowing reversible adiabatic expansion of the steam, we have found that by writing the steady flow energy equation, neglecting kinetic and potential energy changes, equation (12–2) resulted

$$wk_{\text{turb}} = h_3 - h_4$$

The work can also be obtained by using steam velocity values in equation (11–24a):

$$wk = (\bar{V}_{1x} - \bar{V}_{2x})\bar{V}_b$$

The power is found by multiplying the work by the mass flow of steam through the turbine

$$\dot{W}k_{\text{turb}} = \dot{m}(h_3 - h_4) \qquad \textbf{(12–10)}$$

The expansion of steam through a turbine is most commonly conducted in a manner to achieve approximately saturated vapor at the turbine exhaust. That is, at state (4) we wish to have extracted as much energy as possible from the steam in producing turbine work, but if any moisture or liquid particles are in the steam while it is anywhere in the turbine, these particles can act as abrasives and thus seriously damage parts of the turbine. Additionally, moisture in a turbine can induce corrosion of vital parts. For this reason we desire a steam which is a vapor that is ready to condense (saturated vapor) at the turbine exit.

For turbines which have irreversible effects (this includes *all* turbines), the work produced will be less than in the ideal case. Perhaps the best way to see what happens here is through a *T-s* diagram. In figure 12–5 is shown the ideal expansion from pressure p_3 to p_4. As is indicated, the steam has just achieved a saturated vapor state at the exhaust, at state (4). In the irreversible process, fluid friction expansion of steam between the same pressures p_3 and p_4 the entropy *increased* and thus the exhaust steam is still superheated at a pressure p_4, but at a temperature T_4. The work for the irreversible process is then gotten from

$$wk_{\text{turb}} = h_3 - h_{4'} \qquad \textbf{(12–11)}$$

with an entropy increase of $s_{4'} - s_3$ in the steam.

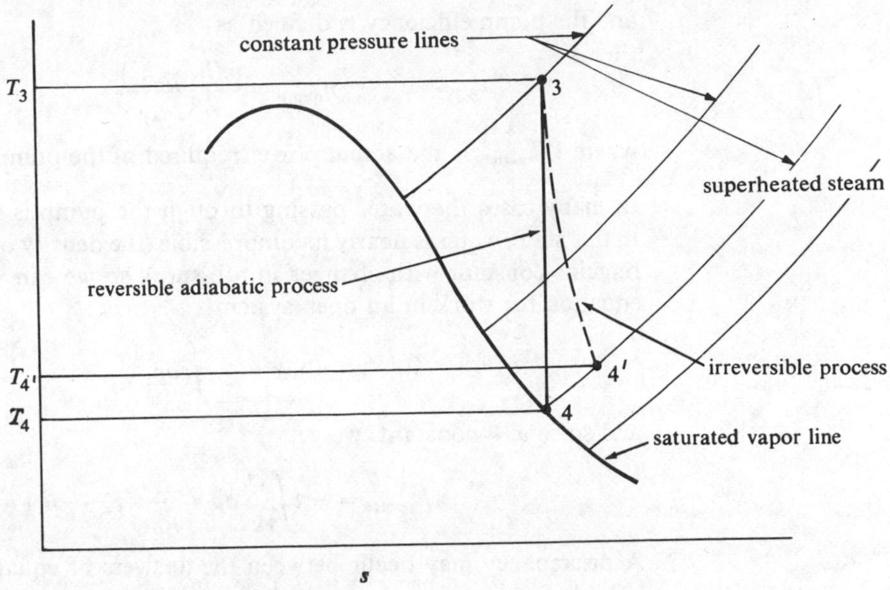

Figure 12–5 Expansion processes through a turbine.

We use the reversible adiabatic expansion as the ideal process and define the turbine efficiency by

$$\eta_{\text{turb}} = \left(\frac{wk_{\text{actual}}}{wk_{\text{ideal}}}\right) \times 100 = \left(\frac{h_3 - h_{4'}}{h_3 - h_4}\right) \times 100 \qquad \textbf{(12–12)}$$

12.4 Pumps

In the Rankine steam power cycle the steam is exhausted from the turbine to a low pressure (perhaps at atmospheric pressure or lower) and the boiler is furnished liquid water at a high pressure. To raise the steam or water from a low to a high pressure and then force it into the boiler is the task of the pump. In some cycles we may call the device a *compressor*, but here we are handling water, and *pump* is a more common term. Pumps are continuous flow mechanisms which are essentially reversed turbines, or they are piston-cylinder devices which provide intermittent flow. In either case, we may consider the pump to be a steady flow open system when observed over a sufficient time span. By applying the steady flow energy equation, neglecting kinetic and potential energy changes, and assuming an adiabatic compression in the pump, we obtain equation (12–4):

$$wk_{\text{pump}} = h_1 - h_2$$

The power required to convey $\dot{m}$ flow rate of water through the pump is determined from the equation

$$\dot{W}k_{\text{pump}} = \dot{m}(h_1 - h_2) \qquad \textbf{(12–13)}$$

and the pump efficiency is defined as

$$\eta_{\text{pump}} = \frac{\dot{m}(h_1 - h_2)}{\dot{W}k_{\text{actual}}}$$

where $\dot{W}k_{\text{actual}}$ is the actual power required of the pump.

In many cases the water passing through the pump is saturated liquid. In this state, water is nearly incompressible (the density or specific volume remains constant with changes in pressure), so we can write the general equation for work in an open system

$$wk = -\int v dp$$

and since v = constant, we get

$$wk_{\text{pump}} = -v\int_{p1}^{p2} dp = -v(p_2 - p_1) \qquad \textbf{(12–14)}$$

A discrepancy may occur between the answers of equations (12–4) and (12–14) for pump work in which case the answer obtained from (12–4) should be used. The reason for this discrepancy is that in equation (12–14), the specific volume is assumed constant which is incorrect for all real fluids or gases. If we identified v as the average (v_{av}), then we would be correct to use equation (12–14) and obtain agreement with equation (12–4), but the average specific volume is difficult to predict in many problems.

Pumps are normally driven by electric motors or small steam turbines. In either case the pump work must be accountable in determining the net cycle work. Also, if some of the H_2O is lost due to leakage (usually through turbine seals) in the cycle, fresh feed water can be added to the cycle as shown in figure 12–6.

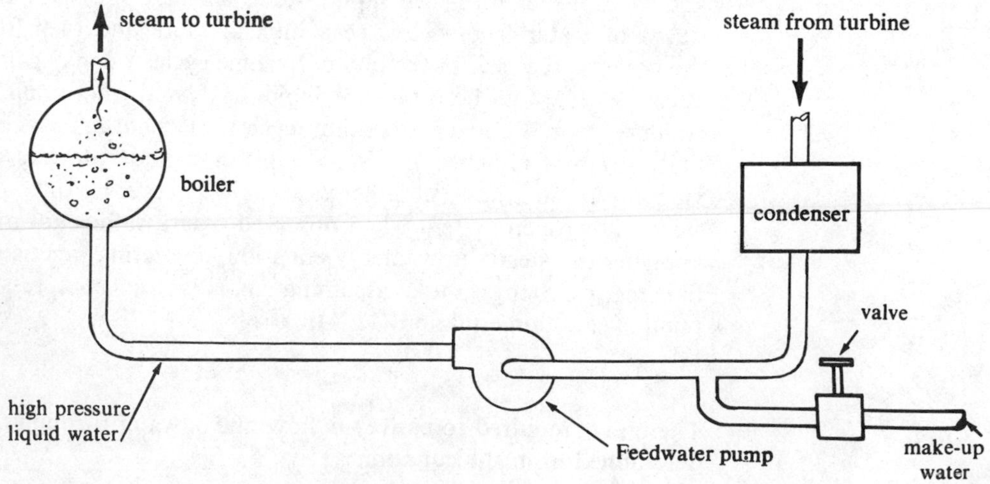

Figure 12–6 Pumps in a steam turbine cycle.

Occasionally, steam injectors are used instead of pumps to convey water to the boiler. A discussion of these devices is included in chapter 13.

12.5 Condensers

Steam exhausted from a turbine in a Rankine cycle is pumped back into the boiler to recycle the water. If, however, the steam exhausted by the turbine is directly furnished to the feedwater pump, the ideal work required by the pump to deliver high pressure steam to the boiler would be exactly the work gotten out of the reversible adiabatic expansion in the turbine operating between the same pressures. In effect then, one steam turbine would be using all its output to drive the pump, and the cycle work would be zero. If there were irreversibilities in the cycle, the steam turbine work would not be enough to drive the pump and we would then need to add work from the surroundings. To prevent these undesirable results, a condenser receives the exhaust steam, condenses this steam to a liquid, and then feeds this liquid water to the pump. This device is shown in the cycle of figure 12–3 where it is indicated that heat is rejected. This is the precise role of the condenser — it transfers heat to the surroundings so that the heat engine can operate. We have seen that the second law of thermodynamics demands a transfer of heat between two regions for any cyclic heat engine. Of course, a mechanical device does not obey a law; a law rather obeys the operation of a mechanical device, but without the condenser, the steam turbine cycle is involved (ideally) in only one heat transfer, that at the boiler. In any case, if the condenser is used as an open steady flow system and if kinetic and potential energy changes in the system are neglected, the first law of thermodynamics gives us

$$q_{rej} = h_1 - h_4 \quad \textbf{(12–3)}$$

The rate of heat transfer $\dot{Q}_{rej}$ can be obtained from

$$\dot{Q}_{rej} = \dot{m}(h_1 - h_4) \quad \textbf{(12–15)}$$

where $\dot{m}$ is the mass flow through the condenser. The mechanical reason that the condenser reduces the pump work, and thus makes the steam turbine practical, is that the steam vapor is converted (condensed) into a liquid. Consequently, the state of the medium leaving the condenser should ideally be a saturated liquid, and then enthalpy h_1 so determined.

The condenser can be built in a number of different physical configurations. It may be a sealed condenser, as shown in figure 12–7(a), which transfers heat from the steam to a coolant across an intermediate boundary. The coolant may be air, river water, or some other fluid, but it is not in direct contact with the working medium (H_2O). Most condensers are constructed in this general manner. A variation of this device is the cooling tower, which relies on heat transfer to the surrounding air.

A second method of manifesting the phase change of steam vapor to liquid is the aerator or open condenser, illustrated in figure 12–7(b). This

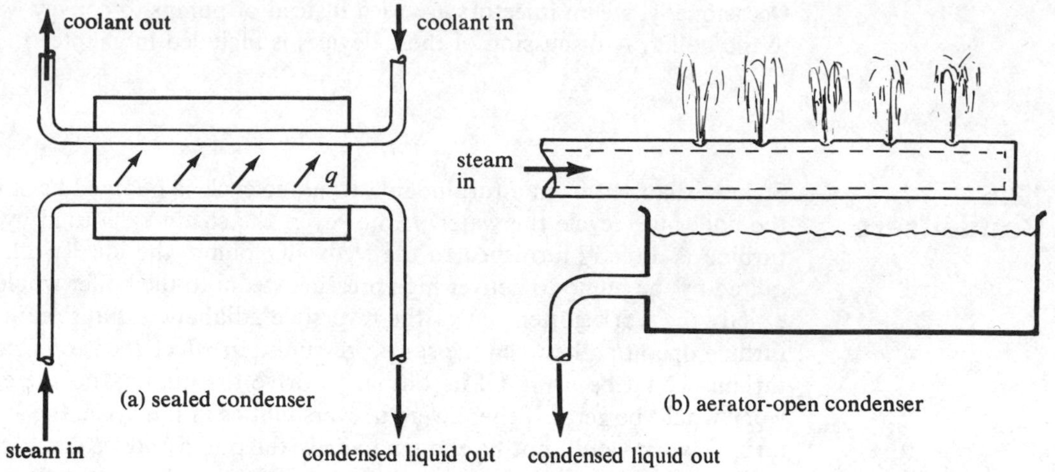

Figure 12–7 Types of condensers in their elemental form.

device literally exhausts the steam into the surroundings; due to this intimate mixing, the heat transfer is rapid and the condensing of the steam to liquid is equally rapid. The drawback to this type of condenser, while it is less expensive than the closed condenser, is that the exhaust pressure of the turbine cannot be less than the atmospheric pressure. The closed condenser can operate at pressures less than atmospheric and thus allow for exhaust vacuum pressures in the turbine and increase cycle efficiencies.

12.6 The Phase Change and Superheating

The working medium used in the steam turbine cycle is water. Chemically, we symbolize it as H_2O and then call it a pure substance if no impurities are present. Water, of course, can take the three common phases with which we are familiar — solid (or ice), liquid, and vapor (or steam). In figure 12–8 is shown the phase diagram of water. Notice that at the triple point, all phases converge and can all three exist in equilibrium and beyond the critical point (to the right of it) the liquid and vapor phases cannot be differentiated. For our purposes, the liquid-vapor phase change and the vapor or steam phase itself will be sufficient. Suppose, for instance, that we have a constant pressure container of water. If we add energy to the contents, monitoring the temperatures as we add energy, we notice that the water begins to have an increased temperature. In figure 12–9, we see the result of this plotted on a *T-E* diagram, indicated by the curve *A-B*. At point *B*, the water begins to boil; that is, it changes from a liquid to a vapor. If the boiler pressure is 1.01 bars, then the temperature at point *B* will be 373 K; but if the pressure is higher (or lower) than 1.01 bars, then the boiling temperature

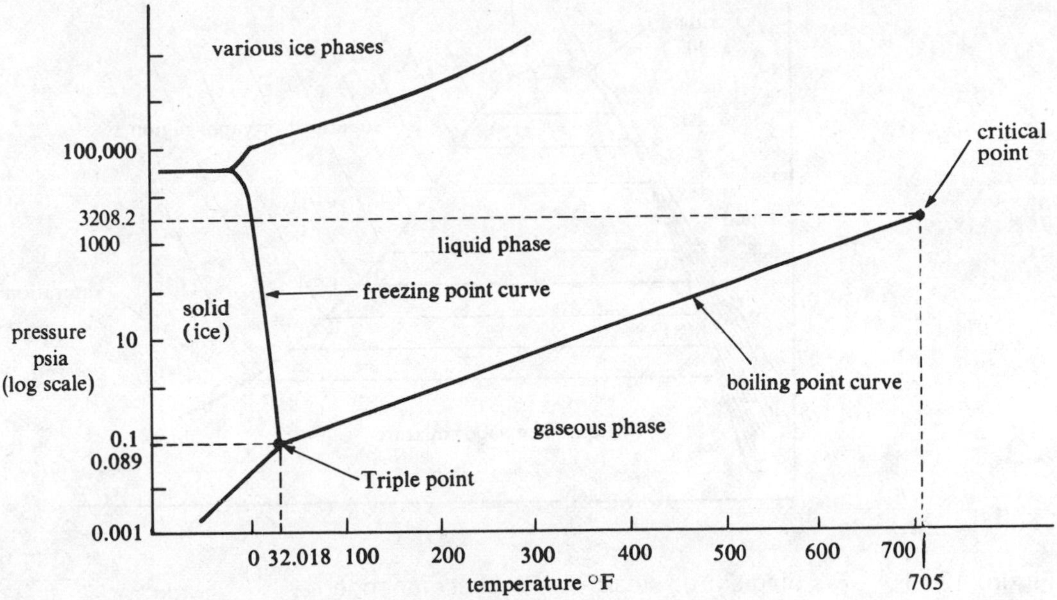

Figure 12–8 Phase diagram for water (H_2O) (not drawn to scale).

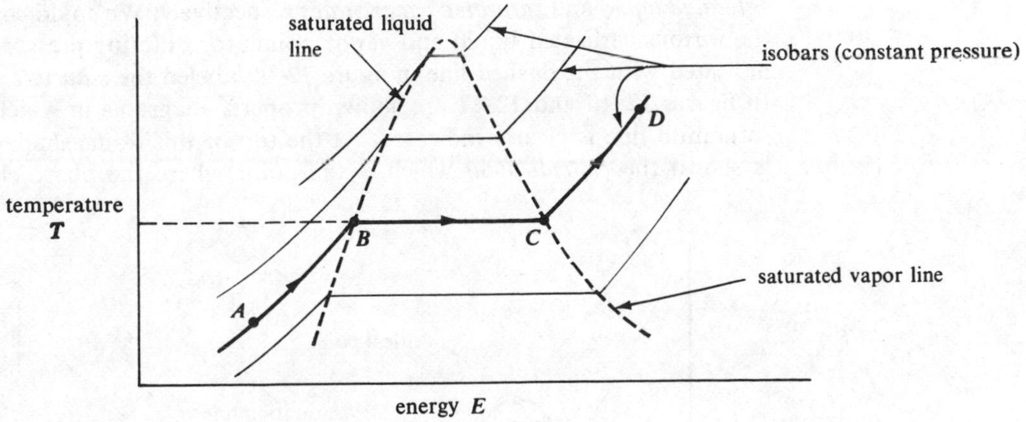

Figure 12–9 Temperature-energy relation of material going through liquid-gas phase change at various pressures.

will be greater than (or less than) the 373 K. Various isobaric lines are shown in the graph of figure 12–9 to show this effect.

Once the water begins to boil, we may keep adding energy without changing the temperature until we reach a point when all the water is now converted to vapor or steam. This is the state C on our isobaric line, and if we continue adding energy, the temperature increases again for the steam. Any particular state beyond the point C is called a *superheated*

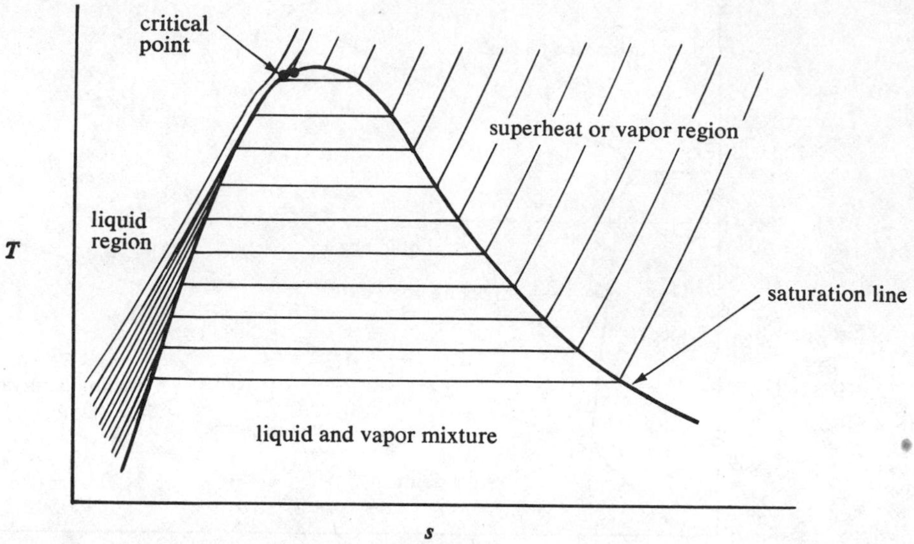

Figure 12–10 *T-s* diagram of steam-water phase change.

steam (such as point *D*), and the unique points *B* and *C* are called the *saturated liquid* and *saturated vapor states* respectively. We could connect the various saturated liquid and vapor points for differing pressures, as indicated with the dashed line in figure 12–9, labeled the *saturation line*. In figures 12–10 and 12–11 are shown property diagrams in which this saturation line is clearly indicated. At the top of this dome-shaped line is shown the *critical point* which is the point where the phase change

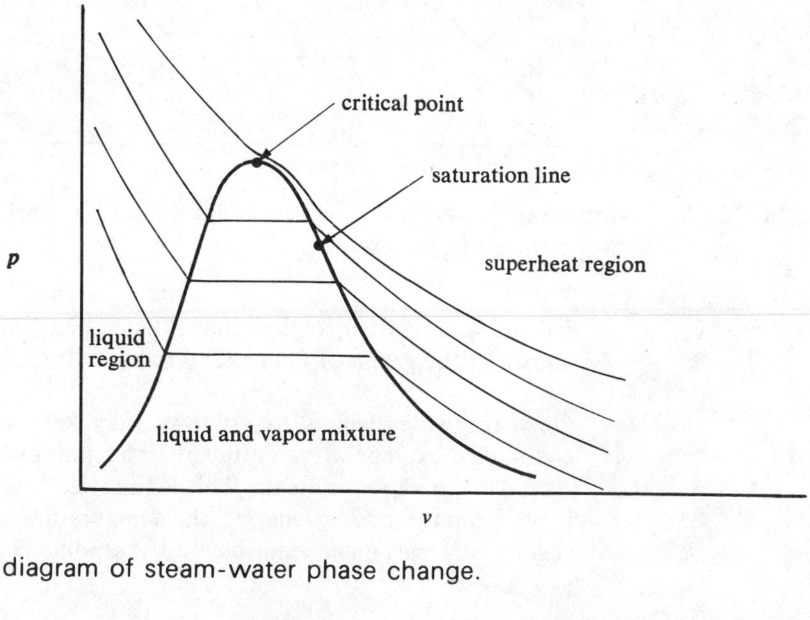

Figure 12–11 *p · v* diagram of steam-water phase change.

becomes undefined. The *T-s* diagram of figure 12–10 depicts isobaric lines as in figure 12–9. Also, the area under the *T-s* diagram curves should be visualized as "heat transfers," as we have always done.

12.7 Use of Steam Tables

The appendix tables B.1, B.2, and B.3 will furnish the data for the properties of water at saturated conditions, superheated steam conditions, and high temperature liquid conditions. In table B.1 are listed the saturation properties as functions of temperature as well as of pressure. The properties specific volume, enthalpy, and entropy are tabulated. Notice that the subscript f denotes saturated liquid and g denotes saturated vapor. For enthalpy and entropy third and fourth terms are tabulated, h_{fg} and s_{fg}. These are the differences of the saturated states and are defined as

$$h_{fg} = h_g - h_f \tag{12–16}$$

and

$$s_{fg} = s_g - s_f \tag{12–17}$$

The term h_{fg} is obviously the measure of the energy added or lost during the vapor-liquid phase change. It is referred to as the *heat of vaporization, heat of condensation, latent heat, latent heat of vaporization*, or *latent heat of condensation.*

Through the above property data, we may determine the precise values of the properties of a state intermediate between saturated liquid and saturated vapor, provided we know the quality x of the state. We defined quality previously (see section 8.10) as the ratio of the vapor mass to the mixture mass. The state we wish to define, for example, state a, is depicted in figure 12–12 by use of the entropy property. Here we can write

$$100 \times \left(\frac{s_a - s_f}{s_g - s_f}\right) = x\% = \text{quality at state } a \text{ in percent}$$

and then

$$s_a = \frac{x}{100}(s_g - s_f) + s_f$$

or

$$s_a = \frac{x}{100} s_{fg} + s_f \tag{12–18}$$

Similarly, we could write

$$h_a = \frac{x}{100} h_{fg} + h_f \tag{12–19}$$

$$v_a = \frac{x}{100}(v_g - v_f) + v_f \tag{12–20}$$

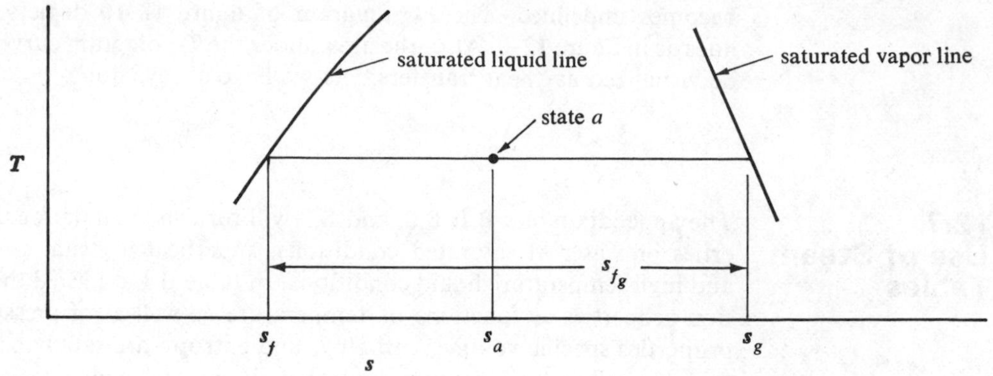

Figure 12–12 Relation between the entropy of a state intermediate between a saturated liquid and vapor and the entropy of the saturation points.

and

$$u_a = \frac{x}{100}(u_g - u_f) + u_f \tag{12–21}$$

Internal energy can be easily calculated from the data in tables B.1 and B.2 by recalling that $u = h - pv$. The superheated steam properties are given in table B.2 where the pressure and temperature together determine the particular state. Specific volume, enthalpy, and entropy are tabulated at each of these points.

Subcooled liquid properties are listed in an abbreviated table, B.3 which furnishes the data of differences between saturated liquid and the specified subcooled or compressed state properties.

The following example problems should help in seeing how data are extracted from these tables.

Example 12.1

A steam generating unit supplies 1000 lbm/min of superheated steam at 400 psia and 800°F. Determine the entropy, enthalpy, and specific heat of the steam.

Solution

From table B.2 we obtain the answers

$$s = 1.6850 \text{ Btu/lbm} \cdot {}^\circ\text{R}$$

$$h = 1417.0 \text{ Btu/lbm}$$

$$v = 1.8151 \text{ ft}^3\text{/lbm}$$

Answer

Example 12.2

Steam is condensing at a pressure of 0.8 bar. If the quality of the steam is 30% at a particular instant, determine the temperature, enthalpy, entropy, and internal energy of the steam.

Solution The steam is in the phase change, so we obtain the required data from the saturated steam table B.1. At a pressure of 0.8 bar the temperature is 93.5°C. We then use equations (12–18), (12–19), and (12–21):

$$h_a = \frac{x}{100} h_{fg} + h_f$$

$$s_a = \frac{x}{100} s_{fg} + s_f$$

$$u_a = \frac{x}{100} (u_g - u_f) + u_f$$

and from table B.1 find

$$h_{fg} = 2273 \text{ kJ/kg} \qquad h_f = 392 \text{ kJ/kg} \qquad h_g = 2665 \text{ kJ/kg}$$

$$s_{fg} = 6.201 \text{ kJ/kg} \cdot \text{K} \qquad s_f = 1.233 \text{ kJ/kg} \cdot \text{K} \qquad v_g = 2.09 \text{ m}^3\text{/kg}$$

$$u_f = h_f - v_f p \qquad v_f = 0.001039 \text{ m}^3\text{/kg}$$

$$= 392 \text{ kJ/kg} - (0.001039 \text{ m}^3\text{/kg})(0.8 \times 10^5 \text{ N/m}^2)$$

$$\simeq 302 \text{ kJ/kg}$$

$$u_g = h_g - v_g p$$

$$= 2665 \text{ kJ/kg} - (2.09 \text{ m}^3\text{/kg})(0.8 \times 10^5 \text{ N/m}^2)$$

$$= 2558 \text{ kJ/kg}$$

and from these results we compute

$$h_a = (0.30)(2273 \text{ kJ/kg}) + 392 \text{ kJ/kg}$$

$$= 1074 \text{ kJ/kg}$$ *Answer*

$$s_a = (0.30)(6.201 \text{ kJ/kg} \cdot \text{K}) + 1.233 \text{ kJ/kg} \cdot \text{K}$$

$$= 3.094 \text{ kJ/kg} \cdot \text{K}$$ *Answer*

$$u_a = (0.30)(2558 - 302) \text{ kJ/kg} + 302 \text{ kJ/kg}$$

$$= 979 \text{ kJ/kg}$$ *Answer*

Example 12.3 A feedwater pump compresses saturated liquid water at 200°F to 400 psia and 200°F. Determine the enthalpy of the water leaving the pump.

Solution The enthalpy of the entering fluid is easily found in table B.1:

$$h_f = 168.1$$

Note that the pressure is 11.53 psia at this point. The difference between the enthalpy of the exit water and that at the inlet is obtained from table B.3:

$$h - h_f = +0.88 \text{ Btu/lbm}$$

Then

$$h = +0.88 + h_f = 0.88 + 168.1$$
$$= 168.98 \text{ Btu/lbm} \qquad \textit{Answer}$$

Example 12.4 Determine the work done on each kilogram of steam for a pump isentropically compressing water from a saturated liquid at 90°C to 100.0 bars.

Solution From equation (12–4)

$$wk_{\text{pump}} = h_1 - h_2$$

we may use the values for enthalpy listed in the tables B.1 and B.3. From table B.1 we find, at 90°C

$$h_f = h_1 = 377 \text{ kJ/kg}$$

and then, from table B.3

$$h - h_f = 10.2 \text{ kJ/kg @ 90°C and 100.0 bars}$$

or

$$h = h_2 = 377 \text{ kJ/kg} + 10.2 \text{ kJ/kg} = 387.2 \text{ kJ/kg}$$

Thus, the pump work is simply

$$wk_{\text{pump}} = 377 \text{ kJ/kg} - 387.2 \text{ kJ/kg}$$
$$= -10.2 \text{ kJ/kg} \qquad \textit{Answer}$$

This is the value listed in table B.3 as the pump work. An alternate solution to obtaining the pump work is to assume a constant volume compression and write

$$wk_{\text{pump}} = (v_{\text{av}})(p_1 - p_2)$$

where v_{av} is the average value of the specific volume of the steam. For this case, $p_1 = 0.701$ bar, $p_2 = 100.0$ bars, and v_{av} is obtained approximately as follows:

$$v_{\text{av}} = \frac{v_1 + v_2}{2}$$

$$= \frac{(0.00104 + 0.00145 \text{ m}^3/\text{kg})}{2}$$

$$= 0.00125 \text{ m}^3/\text{kg}$$

Then the pump work is

$$wk_{\text{pump}} = (0.00125 \text{ m}^3/\text{kg})(99.3 \times 10^5 \text{ N/m}^2)$$
$$= -12.4 \text{ kJ/kg} \qquad \textit{Answer}$$

Notice that there is a discrepancy between the two methods. The first answer should be accepted as the more accurate but either method will give results that are acceptable for most engineering work.

Example 12.5 A boiler supplies 1000 kg/m steam at 30.0 bars and 720°C. It burns coal having a lower heating value of 25,000 kJ/kg coal. If the effectiveness of the boiler is 90% and the water supplied to the boiler has an enthalpy of 600 kJ/kg, determine the rate of coal consumption.

Solution Let us first define the generating unit and the fluxes involved in it (see figure 12–13). We find the heat added from

$$\dot{Q}_{add} = \dot{m}_{steam}(h_3 - h_2)$$

and from table B.2 we find

$$h_3 = 3957 \text{ kJ/kg}$$

Then

$$\dot{Q}_{add} = (1000 \text{ kg/m})(3957 \text{ kJ/kg} - 600 \text{ kJ/kg})$$
$$= 3.36 \times 10^6 \text{ kJ/m}$$

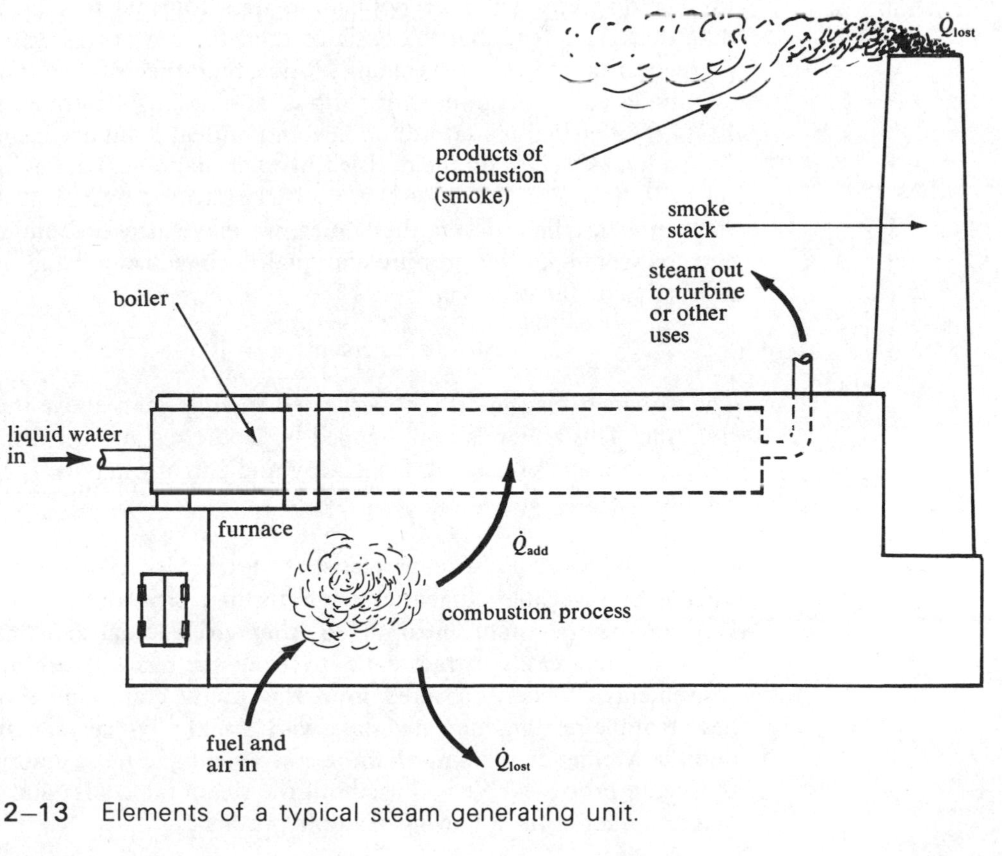

Figure 12–13 Elements of a typical steam generating unit.

The boiler effectiveness is defined as

$$\eta_{\text{boiler}} = \frac{\dot{Q}_{\text{add}}}{\dot{m}_{\text{fuel}}\ (\text{LHV})} \times 100 \tag{12–22}$$

For this problem we have

$$90\% = \frac{3.36 \times 10^6 \text{ kJ/m}}{\dot{m}_{\text{fuel}}\ (25{,}000 \text{ kJ/kg})} \times 100$$

and the mass of fuel used is then

$$\dot{m}_{\text{fuel}} = 149 \text{ kg/m}$$ *Answer*

12.8 The Mollier Diagram

Many people prefer to extract data from graphs or charts rather than from tables. While the steam tables contain the required data to properly analyze the behavior of steam in the vapor-liquid, superheated, and subcooled regions, the Mollier diagram duplicates this information in chart form. This diagram is included in appendix B in a reduced form. Mollier diagrams are quite popular in sizes fourfold to tenfold larger than that given here, but the basic method for use of this chart can be grasped from this small version. Notice that the Mollier diagram is essentially an *h-s* diagram with various iso-lines, as illustrated in figure 12–14. Notice that the saturation line and critical point are identified on the sketch as well as on the detailed Mollier diagram. The region below the saturation line, the phase change, is crisscrossed by isobars and constant-moisture lines. Using these lines, we may easily find enthalpy and entropy from a known pressure and quality since the moisture is related to quality by the equation

$$\text{Moisture percentage} = 100 - x \tag{12–23}$$

The isobars extend into the superheated steam region above the saturation line. This region is also mapped by isotherms and between a given temperature and pressure, the enthalpy and entropy may be found from the Mollier diagram.

One of the most useful characteristics of the Mollier diagram is the tracing of reversible adiabatic steam turbine processes. These follow vertical lines (constant entropy), so from given initial and final pressures, we may easily extract enthalpy from the chart. There are other conveniences to be discovered from this useful graph and those who have trouble relating tabulated data with actual physical processes may find the Mollier diagram much more convenient and more descriptive of the steam processes. We will use both the steam table data and Mollier diagram data in the following example problems.

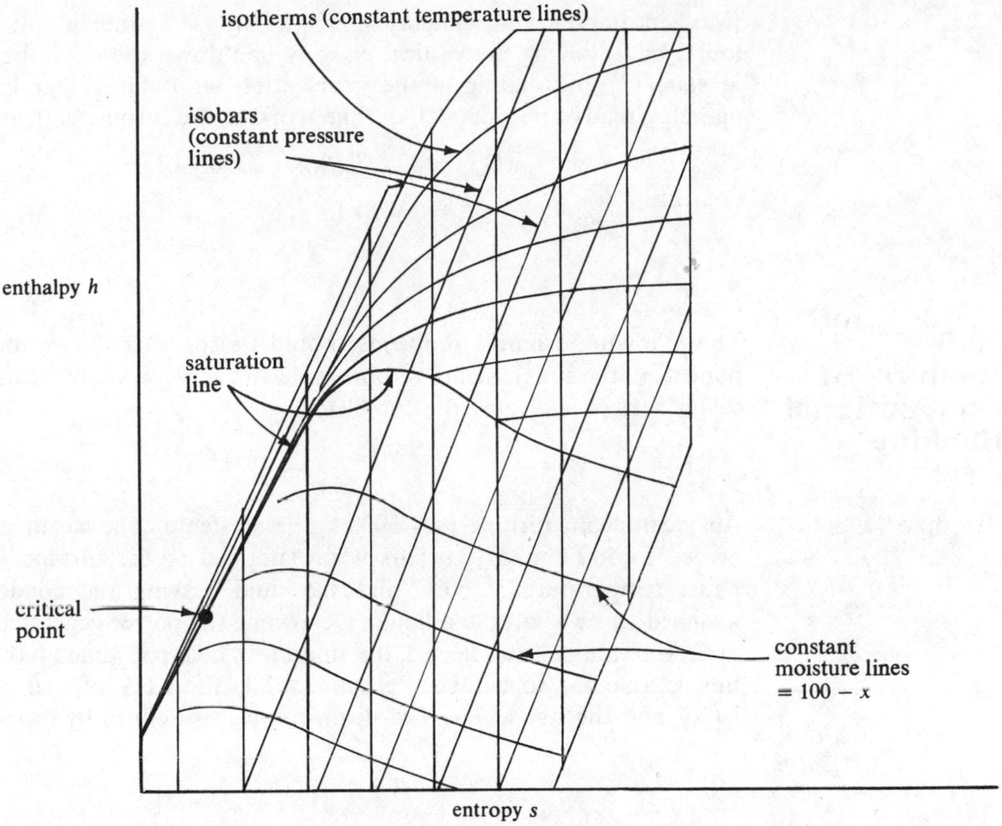

Figure 12–14 Sketch of Mollier diagram (not drawn to scale).

Example 12.6 Determine from the Mollier diagram the percent moisture in steam expanded isentropically from 800°F and 1000 psia to 2 psia.

Solution Referring to chart B.1 we find the entropy of steam at 800°F and 1000 psia to be 1.56 Btu/lbm · °R. Following this entropy line vertically down the chart until it intersects the 2.0 psia line, we then read 78.8% vapor or a moisture percentage of 100% − 78.8% = 21.2% *Answer*

Example 12.7 A steam turbine expands steam at 50.0 bars and 640°C isentropically to 0.2 bar. Determine the work done by the turbine through this expansion.

Solution The work of the turbine is, referring to figure 12–3

$$wk_{turb} = h_3 - h_4$$

We will obtain the enthalpy values from the Mollier diagram although we could find them from table B.1 and/or B.2 as well. From chart B.1

(Mollier diagram) we find $h_3 = 3750$ kJ/kg. The state at (4) can be found by following the vertical entropy line down the chart, beginning at state (3) and ending at the intersection with the 0.2-bar line. The enthalpy reads $h_4 = 2390$ kJ/kg. The work of the turbine is then

$$wk_{\text{turb}} = 3750 \text{ kJ/kg} - 2390 \text{ kJ/kg}$$

$$= 1360 \text{ kJ/kg} \qquad \textit{Answer}$$

12.9 Analysis of Conventional Rankine Cycle

The following example problems should tie together the various components in the simple steam turbine cycle and serve to show the use of the steam data in analyzing the complete cycle.

Example 12.8

An ideal steam turbine uses 5000 kg/hr of steam. The steam is superheated to 560°C and 20.0 bars when supplied to the turbine. The exhaust temperature is 60°C and the fluid leaving the condenser is assumed to be a saturated liquid. Determine the power generated by the cycle, the rate of heat rejected, the amount of coal consumed if the boiler unit is assumed to be 100% effective while the LHV of coal is 30,000 kJ/kg, and the steam rate. The steam rate $\dot{m}_{sr}$ is defined by the equation

$$\dot{m}_{sr} = \frac{\dot{m}_{\text{steam}}}{\dot{W}k_{\text{cycle}}} \qquad \textbf{(12–24)}$$

Solution

This heat engine is sketched in figure 12–15 and the property diagrams are depicted in figures 12–16 and 12–17. The property data labeled in the

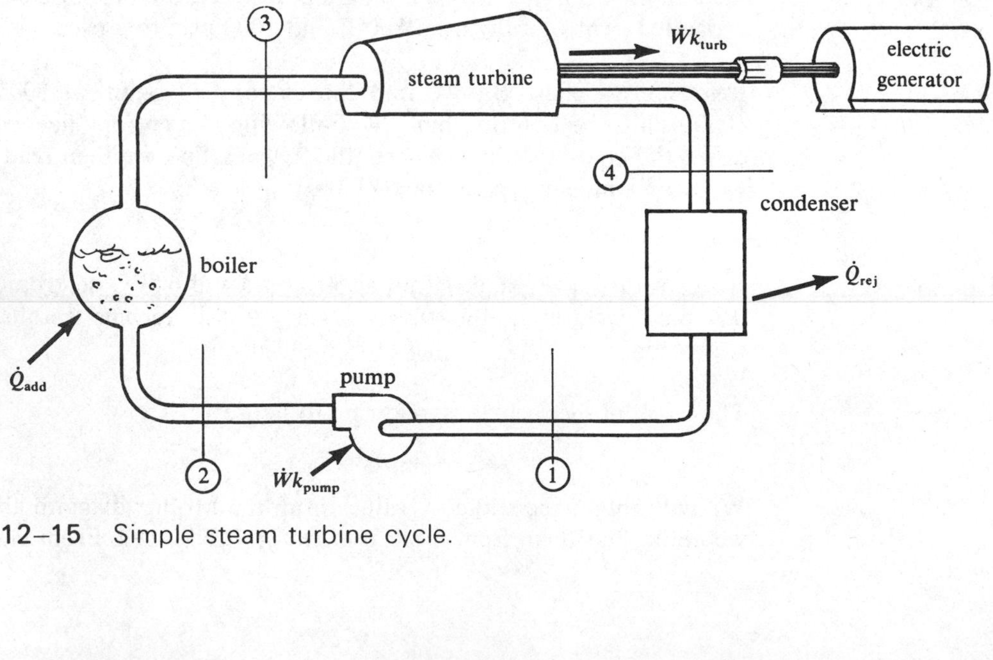

Figure 12–15 Simple steam turbine cycle.

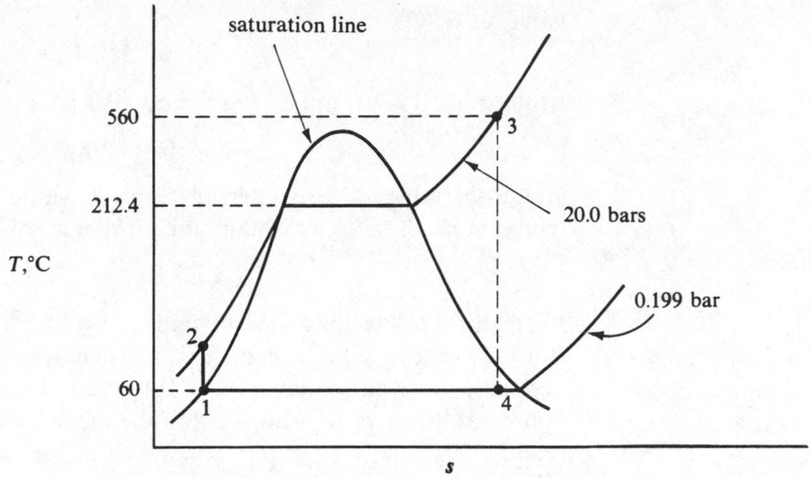

Figure 12–16 *T-s* diagram of steam turbine cycle of example 12.8.

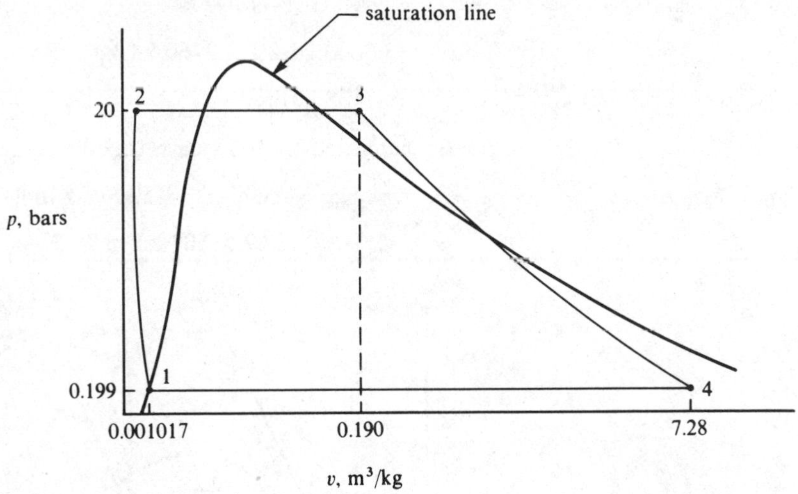

Figure 12–17 *p-V* diagram of steam turbine cycle of example 12.8.

graphs are either given in the problem statement or easily obtained from the steam tables. The power generated by the cycle is obtained from

$$\dot{W}k_{\text{cycle}} = \dot{m}wk_{\text{cycle}}$$

where

$$wk_{\text{cycle}} = wk_{\text{turb}} + wk_{\text{pump}} \tag{12–25}$$

The ideal cycle involves adiabatic processes in the pump and turbine, so we use equations (12–2) and (12–4). We obtain from table B.1 the value, at 60°C

$$h_1 = h_{f_1} = 251 \text{ kJ/kg}$$

From table B.2 we find at 560°C and 20.0 bars pressure

$$h_3 = 3600 \text{ kJ/kg}$$

and from table B.1 we interpolate for a value of $h - h_f$. Thus, at a temperature of 60°C we obtain, for a pressure of 20.0 bars, that

$$h_2 = h_1 + 2.0 \text{ kJ/kg} = 253 \text{ kJ/kg}$$

From the Mollier diagram, we find h_4 by first locating the position of state (3) in the superheated region. Then we follow the entropy line vertically down to the pressure of 0.199 bar. This technique is shown in figure 12–18 where we also see that the moisture can be determined. We read

$$h_4 = 2460 \text{ kJ/kg}$$

and the moisture is 7.5%, so $x = 92.5\%$. Then

$$\begin{aligned} wk_{\text{cycle}} &= h_3 - h_4 + h_1 - h_2 \\ &= 3600 \text{ kJ/kg} - 2460 \text{ kJ/kg} + 251 \text{ kJ/kg} - 253 \text{ kJ/kg} \\ &= 1138 \text{ kJ/kg} \end{aligned}$$

The power generated by the steam turbine is then

$$\begin{aligned} \dot{W}k_{\text{cycle}} &= \dot{m}wk_{\text{cycle}} = (5000 \text{ kg/hr})(1138 \text{ kJ/kg}) \\ &= 5.69 \times 10^6 \text{ kJ/hr} \end{aligned}$$

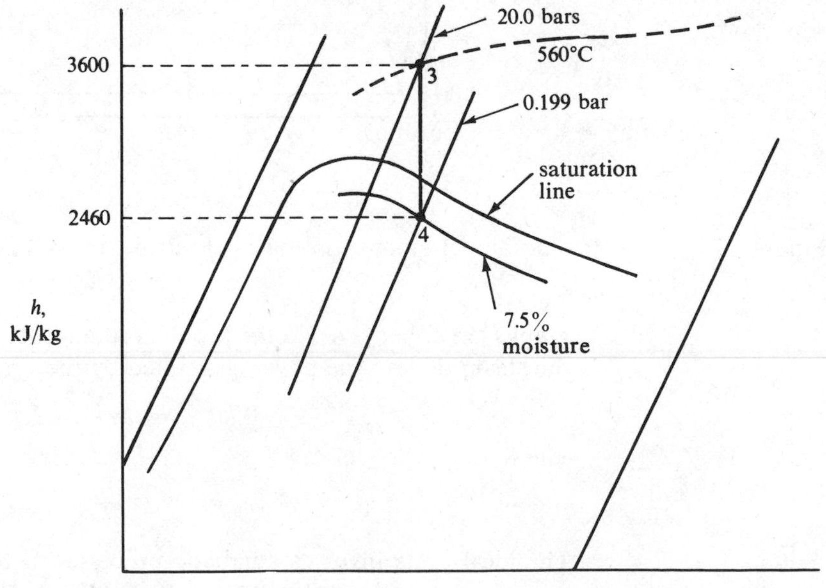

Figure 12–18 Mollier diagram of reversible adiabatic expansion in example 12.8.

or

$$\dot{W}k_{\text{cycle}} = 1580 \text{ kW}$$ *Answer*

From equation (12–3) we obtain

$$q_{\text{rej}} = h_1 - h_4 = 251 \text{ kJ/kg} - 2460 \text{ kJ/kg}$$
$$= -2209 \text{ kJ/kg}$$

and then

$$\dot{Q}_{\text{rej}} = \dot{m}q_{\text{rej}} = 11.0 \times 10^6 \text{ kJ/hr}$$ *Answer*

The rate of heat added we get from

$$\dot{Q}_{\text{add}} = \dot{m}q_{\text{add}} = \dot{m}(h_3 - h_2)$$
$$= (5000 \text{ kg/hr})(3600 \text{ kJ/kg} - 253 \text{ kJ/kg})$$
$$= 16.7 \times 10^6 \text{ kJ/hr}$$

Since the boiler has a 100% effectiveness we may write

$$\dot{m}_{\text{fuel}} \text{ (LHV)} = \dot{Q}_{\text{add}} = 16.7 \times 10^6 \text{ kJ/hr}$$

Hence

$$\dot{m}_{\text{fuel}} = 558 \text{ kg/hr}$$ *Answer*

The thermodynamic efficiency can be obtained from

$$\eta_T = \frac{\dot{W}k_{\text{cycle}}}{\dot{Q}_{\text{add}}} \times 100 = 34\%$$

The steam rate may be computed from equation (12–24):

$$\dot{m}_{sr} = \frac{\dot{m}_{\text{steam}}}{\dot{W}k_{\text{cycle}}} = \frac{5000 \text{ kg/hr}}{1580 \text{ kW}} = 3.17 \text{ kg/kW} \cdot \text{hr}$$ *Answer*

Let us now consider the cycle with inefficiencies or irreversibilities.

Example 12.9

A steam turbine uses 5000 kg/hr of steam. The steam generator furnishes superheated steam to the turbine at 560°C and 20.0 bars. The exhaust pressure is 0.199 bar and the quality of the steam here is 99%. Assume the boiler has an effectiveness of 91%. Determine the power produced, the turbine efficiency, and the steam rate.

Solution

The system is defined in figure 12–15; however, our property diagrams given in figures 12–19 and 12–20 deviate slightly at state (4) from those of example 12.8. The enthalpy values correspond to the data of example 12.8 except at state (4):

$$h_1 = 251 \text{ kJ/kg}$$
$$h_2 = 253 \text{ kJ/kg}$$
$$h_3 = 3600 \text{ kJ/kg}$$

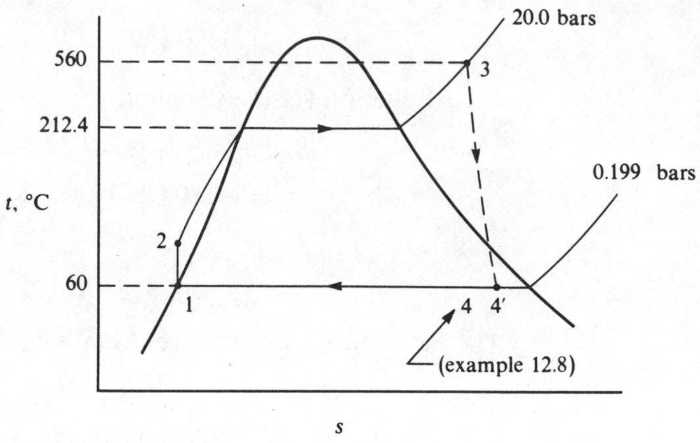

Figure 12–19 Example 12.9.

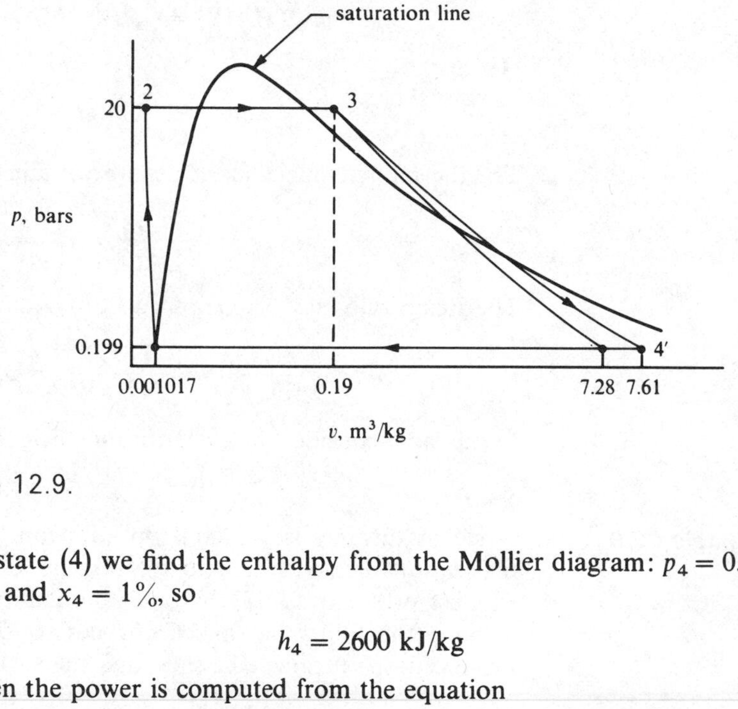

Figure 12–20 Example 12.9.

At state (4) we find the enthalpy from the Mollier diagram: $p_4 = 0.199$ bar and $x_4 = 1\%$, so

$$h_4 = 2600 \text{ kJ/kg}$$

Then the power is computed from the equation

$$\dot{W}k_{\text{cycle}} = \dot{m}wk_{\text{cycle}} = \dot{m}(h_3 - h_4 + h_1 - h_2)$$

$$= (5000 \text{ kg/hr})(3600 \text{ kJ/kg} - 2600 \text{ kJ/kg}$$

$$+ 251 \text{ kJ/kg} - 253 \text{ kJ/kg})$$

$$= 4.99 \times 10^6 \text{ kJ/hr} \qquad \textit{Answer}$$

or

$$\dot{W}k_{\text{cycle}} = 1386 \text{ kW} \qquad \textit{Answer}$$

During the irreversible expansion (3–4) there is an increase in the entropy. This increase is

$$s_4 - s_3$$

The value for s_3 is found to be 7.596 kJ/kg · K from table B.2 and s_4 is 7.78 kJ/kg · K from the Mollier diagram. Thus, the entropy increases 0.184 kJ/kg · K.

Various theories have been postulated, and research continues in methods to predict the increase in entropy during irreversible processes, yet there exists at this time no way to proceed from a predicted entropy change to the defined state. That is, if we had a rational, analytical method for predicting the change in entropy for an irreversible process, we could easily predict the outcome of processes and the efficiency of unbuilt engines. But we cannot.

The efficiency of the turbine we calculate from equation (12–12)

$$\eta_{\text{turb}} = \left(\frac{h_3 - h_{4'}}{h_3 - h_4}\right) \times 100$$

where

$$h_3 - h_4 = 3600 \text{ kJ/kg} - 2460 \text{ kJ/kg} = 1140 \text{ kJ/kg}$$

and

$$h_3 - h_{4'} = 3600 \text{ kJ/kg} - 2600 \text{ kJ/kg} = 1000 \text{ kJ/kg}$$

Then

$$\eta_{\text{turb}} = \frac{1000}{1140} \times 100 = 87.7\% \qquad \textit{Answer}$$

The rate of heat rejected can be found from the equation

$$\begin{aligned}\dot{Q}_{\text{rej}} &= \dot{m}(h_{4'} - h_1) \\ &= (5000 \text{ kg/hr})(2600 \text{ kJ/kg} - 251 \text{ kJ/kg}) \\ &= 11.7 \times 10^6 \text{ kJ/hr}\end{aligned} \qquad \textit{Answer}$$

The thermodynamic efficiency is then

$$\begin{aligned}\eta_T &= \frac{4.99 \times 10^6 \text{ kJ/hr}}{4.99 \times 10^6 \text{ kJ/hr} + 11.7 \times 10^6 \text{ kJ/hr}} \times 100 \\ &= 27.1\%\end{aligned} \qquad \textit{Answer}$$

The steam rate is

$$\dot{m}_{sr} = \frac{\dot{m}_{\text{steam}}}{\dot{W}k_{\text{cycle}}} = \frac{5000 \text{ kg/hr}}{1386 \text{ kW}} = 3.6 \text{ kg/kW} \cdot \text{h}$$

Clearly the efficiency of the steam cycle decreases with inefficiencies in the turbine. The methods of increasing the efficiencies will now be considered.

12.10 The Reheat Cycle

The most obvious method to increase the efficiency of the steam turbine cycle, apparent from the *T-s* diagram of figure 12–19, is to increase the steam pressure and temperature — thereby increasing the turbine work the same amount as the additional heat added (ideally). There are, unfortunately, upper limits to operating pressures and temperatures, generally determined by the boiler, turbine, pump, pipe, and condenser materials. Consequently, to achieve higher thermodynamic efficiencies than the simple Rankine cycle, the reheat device is used. This involves the addition of energy to steam after it has expanded through a high pressure turbine and is subsequently expanding through a low pressure turbine. The typical cycle is sketched in figure 12–21, and the property diagrams, in figure 12–22. The cycle is still closed, and the system is all the water contained in the various components. Notice in the property diagrams that the shaded areas represent the additional heat and work of the reheat cycle over the simple Rankine cycle. Of course, these diagrams are descriptive of the reversible processes only, and with this restriction we can determine the following relationships for the ideal reheat cycle:

$$q_{add} = h_3 - h_2 + h_5 - h_4 \tag{12–26}$$

$$q_{rej} = h_1 - h_6 \tag{12–27}$$

$$wk_{pump} = h_1 - h_2 \tag{12–28}$$

$$wk_{turb} = h_3 - h_4 + h_5 - h_6 \tag{12–29}$$

The reheat cycle is best adapted to systems where very high pressure and temperature superheated steam are used so that the initial expansion

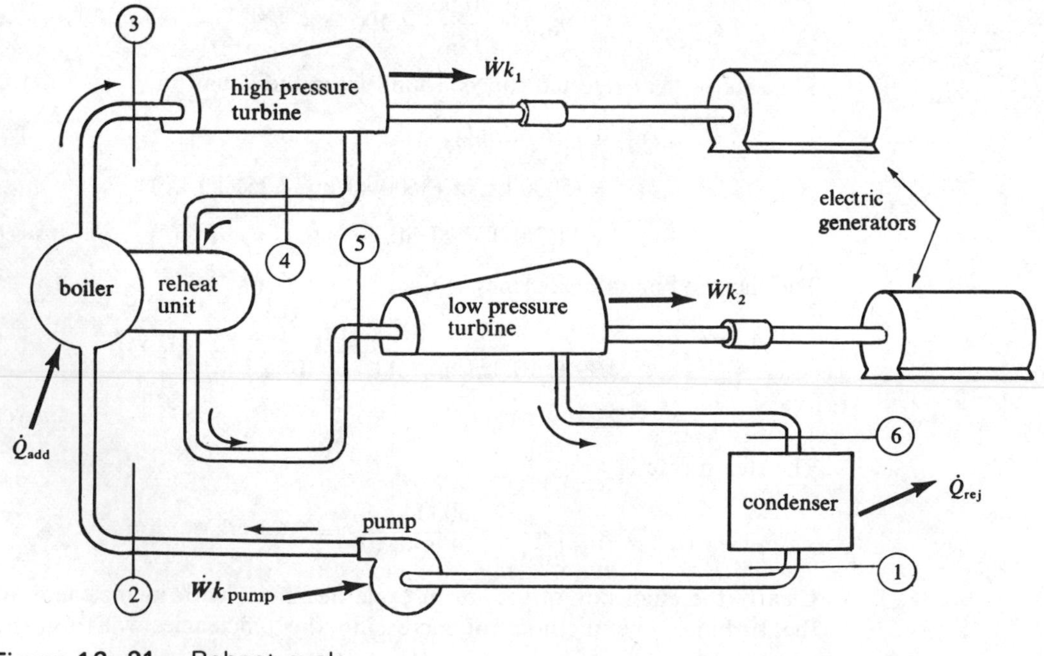

Figure 12–21 Reheat cycle.

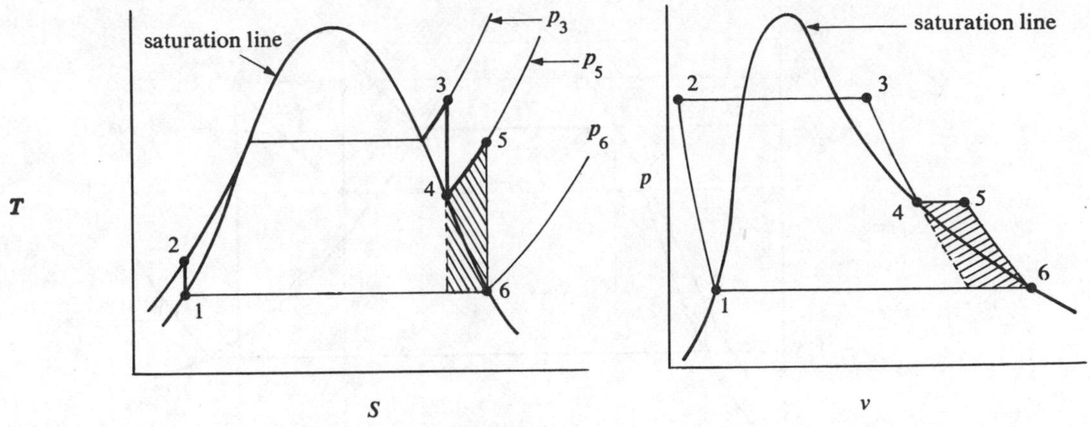

Figure 12–22 Reheat cycle property diagram.

through the high pressure turbine will leave a sufficiently high steam pressure for reheating. Let us now look at an example of this engine.

Example 12.10

Steam at 2000 psia and 1000°F enters the high pressure turbine stages of a reheat cycle and expands reversibly and adiabatically to 200 psia. This steam is then directed through the reheater from which it emerges at a temperature of 950°F. The low pressure turbine expands the steam reversibly and adiabatically to 2 psia. If 7 lbm/s of steam are used, determine the power produced, the rate of heat added, the cycle efficiency, and the steam rate. Assume the circulating pump operates ideally and the fluid leaving the condenser is saturated liquid.

Solution

We will obtain the power from the relation $\dot{W}k_{\text{cycle}} = \dot{m}wk_{\text{cycle}}$ and where we have for the ideal cycle

$$wk_{\text{cycle}} = h_3 - h_4 + h_5 - h_6 + h_1 - h_2$$

The subscripts refer to the physical points of the reheat cycle of figure 12–21. We shall assume the engine here fits the same description. In figure 12–23 is shown the *T-s* diagram of the cycle with certain states defined by given temperatures. From table B.2 we obtain the following:

$$h_3 = 1474.1 \text{ Btu/lbm}$$

$$h_5 = 1503.2 \text{ Btu/lbm}$$

Also, since $s_4 = s_3$, we can see that at state (4) the steam is still superheated (from table B.2 or the Mollier diagram), we obtain

$$h_4 \approx 1211 \text{ @ } 200 \text{ psia}$$

We have that $s_5 = s_6$ and since $s_5 = 1.8236$ Btu/lbm· °R we get that

$$1.8236 = \frac{x}{100} s_{fg6} + s_{f6}$$

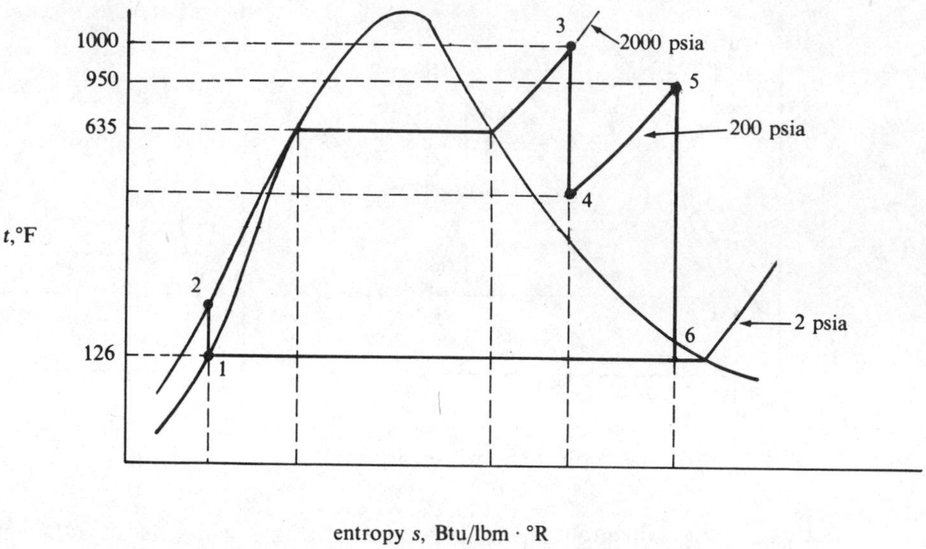

Figure 12–23 *T-s* diagram of reheat cycle (not drawn to scale).

where

$$s_{f_6} = 0.1749 \text{ Btu/lbm} \cdot {}^\circ\text{R @ 2 psia} \qquad \text{(table B.1)}$$

and

$$s_{fg6} = 1.7451 \text{ Btu/lbm} \cdot {}^\circ\text{R} \qquad \text{(table B.1)}$$

Then

$$\frac{x}{100} = \frac{1.8236 \text{ Btu/lbm} \cdot {}^\circ\text{R} - 0.1749 \text{ Btu/lbm} \cdot {}^\circ\text{R}}{1.7451 \text{ Btu/lbm} \cdot {}^\circ\text{R}}$$

$$= 0.945$$

and the enthalpy at state (6) is

$$h_6 = \frac{x}{100} h_{fg6} + h_{f6}$$

$$= (0.945)(1022.2 \text{ Btu/lbm}) + 93.99 \text{ Btu/lbm}$$

$$= 1060 \text{ Btu/lbm}$$

From table B.1 we find

$$h_1 = h_{f1} = 93.99 \text{ Btu/lbm} = h_{f6}$$

The pump work can be computed from equation (12–28):

$$wk_{\text{pump}} = h_1 - h_2$$

or from equation (12–14):

$$wk_{\text{pump}} = -\int_{p1}^{p2} v\,dp \approx -v_{\text{av}}(p_2 - p_1)$$

From equation (12–14) we first approximate the average specific volume by v_1. Then

$$v_1 \approx v_{f1} \approx 0.01623 \text{ ft}^3/\text{lbm}$$

and the pump work is then

$$wk_{\text{pump}} = -(0.01623 \text{ ft}^3/\text{lbm})(2000 - 2 \text{ lbf/in}^2)\left(\frac{144 \text{ in}^2/\text{ft}^2}{778 \text{ ft-lbf/Btu}}\right)$$
$$= 6.0 \text{ Btu/lbm}$$

We then have

$$h_2 = h_1 - wk_{\text{pump}}$$
$$= 93.99 - (-6.0) = 99.99 \text{ Btu/lbm}$$

and we can then calculate the net work of the cycle

$$wk_{\text{cycle}} = h_3 - h_4 + h_5 - h_6 + h_1 - h_2$$
$$= 1474.1 \text{ Btu/lbm} - 1211 \text{ Btu/lbm} + 1503.2 \text{ Btu/lbm}$$
$$- 1060 \text{ Btu/lbm} + 93.99 \text{ Btu/lbm} - 99.99 \text{ Btu/lbm}$$
$$= 700.3 \text{ Btu/lbm}$$

The power is computed from

$$\dot{W}k_{\text{cycle}} = \dot{m}wk_{\text{cycle}} = (7 \text{ lbm/s})(700 \text{ Btu/lbm})$$
$$= 4900 \text{ Btu/s} = 6909 \text{ hp}$$ *Answer*

The rate of heat addition to the cycle is obtained from

$$\dot{Q}_{\text{add}} = \dot{m}_{q\text{add}}$$

where we compute q_{add} from the equation

$$q_{\text{add}} = h_3 - h_2 + h_5 - h_4$$
$$= 1474.1 \text{ Btu/lbm} - 99.99 \text{ Btu/lbm}$$
$$+ 1503.2 \text{ Btu/lbm} - 1211 \text{ Btu/lbm}$$
$$= 1666.3 \text{ Btu/lbm}$$

Then

$$\dot{Q}_{\text{add}} = (7 \text{ lbm/s})(1666.3 \text{ Btu/lbm}) = 11664 \text{ Btu/s}$$ *Answer*

We compute the cycle efficiency from the equation

$$\eta_T = \frac{\dot{W}k_{\text{cycle}}}{\dot{Q}_{\text{add}}} \times 100 = \frac{4900 \text{ Btu/s}}{11664 \text{ Btu/lbm}} \times 100 = 42\%$$ *Answer*

The steam rate is obtained by the defining equation (12–24):

$$\dot{m}_{sr} = \frac{\dot{m}_{\text{steam}}}{\dot{W}k_{\text{cycle}}} = \frac{7 \text{ lbm/s}}{6909 \text{ hp}} \times 3600 \text{ s/hr}$$
$$= 3.65 \text{ lbm/hp} \cdot \text{h}$$ *Answer*

12.11 The Regenerative Cycle

A method which can increase steam cycle efficiency without increasing the superheated steam pressure and temperature is the regenerative heating process which is, essentially, a method of adding heat, like a reversible process at a higher temperature. Regenerative heating, in practice, is the process of expansion of steam in the turbine well into the phase change region. Moisture is withdrawn mechanically from the turbine to reduce the effects of wear and corrosion. In figure 12–24 are shown the physical features of the regenerative steam turbine cycle and the property diagrams corresponding to this cycle are depicted in figure 12–25. The saturated liquid from the condenser is fed to a mixing chamber by a pump (*c*). The chamber allows the liquid to be heated by mixing with steam bled from the turbine; then this mixture is fed to the

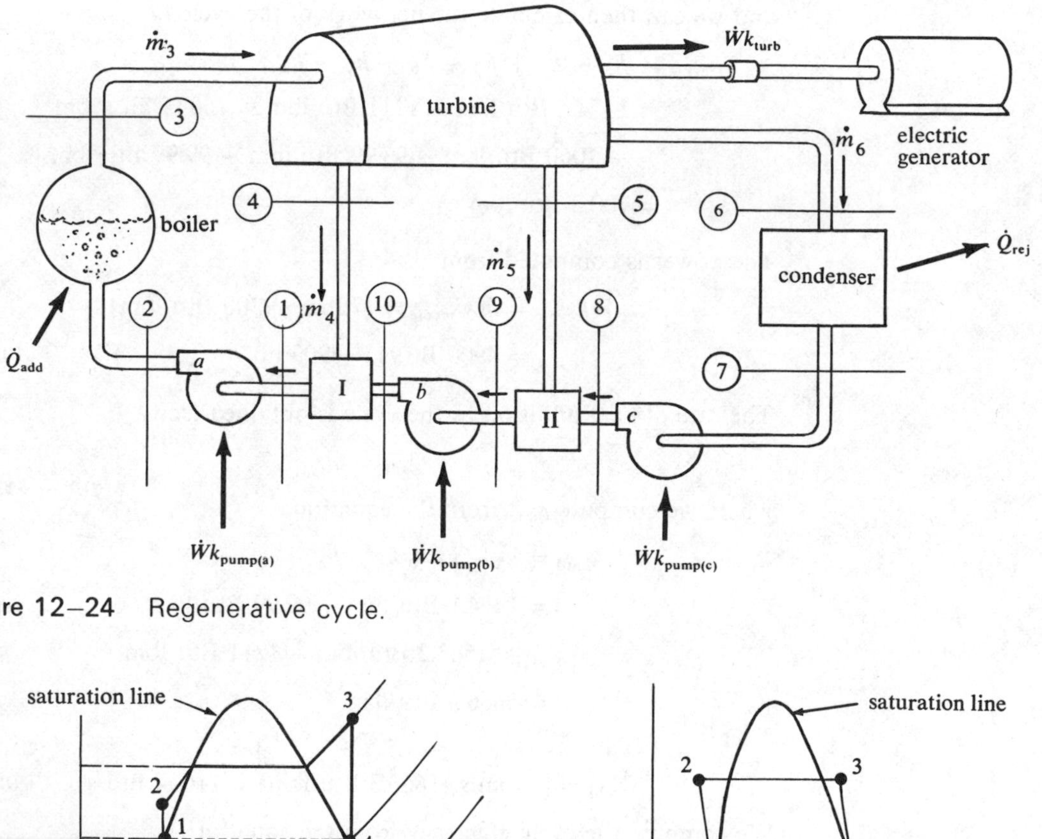

Figure 12–24 Regenerative cycle.

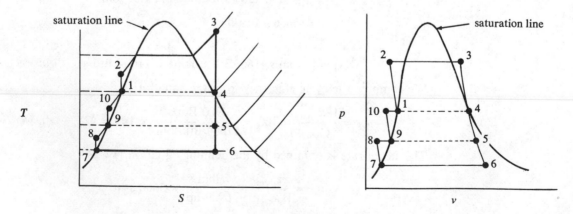

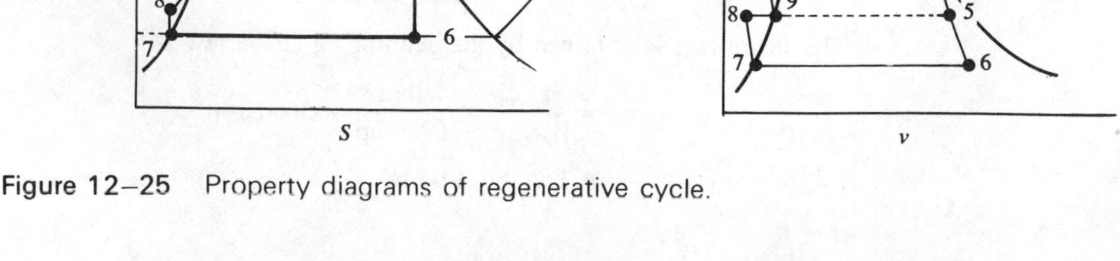

Figure 12–25 Property diagrams of regenerative cycle.

next chamber and ultimately back to the boiler for recirculation. In the cycle of figure 12–24, two mixing chambers are utilized, but more could easily be added. In general, the first regenerative withdraw, corresponding to state (4) in figures 12–24 and 12–25, is at a superheated state. The diagram in figure 12–25 indicates that this withdrawal is done at a saturated vapor condition.

The computational techniques for the regenerative cycle are identical to those of the simple Rankine cycle with the following major deviation: mass must be accounted for throughout the cycle in a more detailed manner than in the simple Rankine cycle analysis. If we look at the conservation of mass through the turbine we get

$$\dot{m}_3 = \dot{m}_4 + \dot{m}_5 + \dot{m}_6 \tag{12–30}$$

and from this relationship we can obtain the total power produced by the turbine from the equation

$$\dot{W}k_{\text{turb}} = \dot{m}_3(wk_{\text{turb}(3-4)}) + (\dot{m}_5 + \dot{m}_6)(wk_{\text{turb}(4-5)}) + (\dot{m}_6)(wk_{\text{turb}(5-6)}) \tag{12–31}$$

where the $wk_{\text{turb}(3-4)}$ term is the work produced per unit mass of steam expanding from state (3) to state (4). Disregarding kinetic and potential energy changes, we have

$$wk_{\text{turb}(3-4)} = h_3 - h_4 \tag{12–32}$$

and, in a like manner

$$wk_{\text{turb}(4-5)} = h_4 - h_5 \tag{12–33}$$

and

$$wk_{\text{turb}(5-6)} = h_5 - h_6 \tag{12–34}$$

Consequently we can write for an ideal turbine with regeneration

$$\dot{W}k_{\text{turb}} = \dot{m}_3(h_3 - h_4) + (\dot{m}_5 + \dot{m}_6)(h_4 - h_5) + \dot{m}_6(h_5 - h_6) \tag{12–35}$$

If we look at the mixing chambers we can write the steady flow energy equation for them. Neglecting kinetic or potential energy changes we have for chamber (I)

$$\dot{m}_1 h_1 = \dot{m}_4 h_4 + \dot{m}_{10} h_{10}$$

but

$$\dot{m}_{10} = \dot{m}_1 - \dot{m}_4$$

so

$$\dot{m}_1 h_1 = \dot{m}_4 h_4 + (\dot{m}_1 - \dot{m}_4) h_{10} \tag{12–36}$$

For mixing chamber (II) we get

$$\dot{m}_9 h_9 = \dot{m}_8 h_8 + \dot{m}_5 h_5$$

or, since $\dot{m}_9 = \dot{m}_{10}$ and $\dot{m}_8 = \dot{m}_6$

$$(\dot{m}_1 - \dot{m}_4)h_9 = \dot{m}_6 h_8 + \dot{m}_5 h_5$$

Also we can write from conservation of mass in the last turbine stages that

$$\dot{m}_6 = \dot{m}_3 - \dot{m}_4 - \dot{m}_5 = \dot{m}_1 - \dot{m}_4 - \dot{m}_5$$

so

$$(\dot{m}_1 - \dot{m}_4)h_9 = (\dot{m}_1 - \dot{m}_4 - \dot{m}_5)h_8 + \dot{m}_5 h_5 \qquad \textbf{(12–37)}$$

Using these relations we can analyze quite precisely the expected flow rates from a knowledge of the enthalpy values. The following example problem will show some of the numerical computation involved in a regenerative heating cycle.

Example 12.11 An ideal regenerative steam cycle operates with 200,000 kg/hr of steam. The steam is at 80.0 bars pressure and 560°C as it enters the turbine. The exhaust steam is at 0.2 bar. Assume that steam is extracted at two stages for regeneration. The first stage corresponds to saturated vapor and the second when the moisture content is 14% in the turbine steam. Determine the amounts of steam extracted for regeneration in the two states, the power produced, and the thermodynamic efficiency.

Solution The system is shown in figure 12–24 since the cycle includes two stages of regeneration. The *T*-*s* diagram is sketched in figure 12–26 to show the various states and the properties. From table B.2 we find $h_3 = 3544$ kJ/kg. The enthalpies at states (4), (5), and (6) can then be obtained

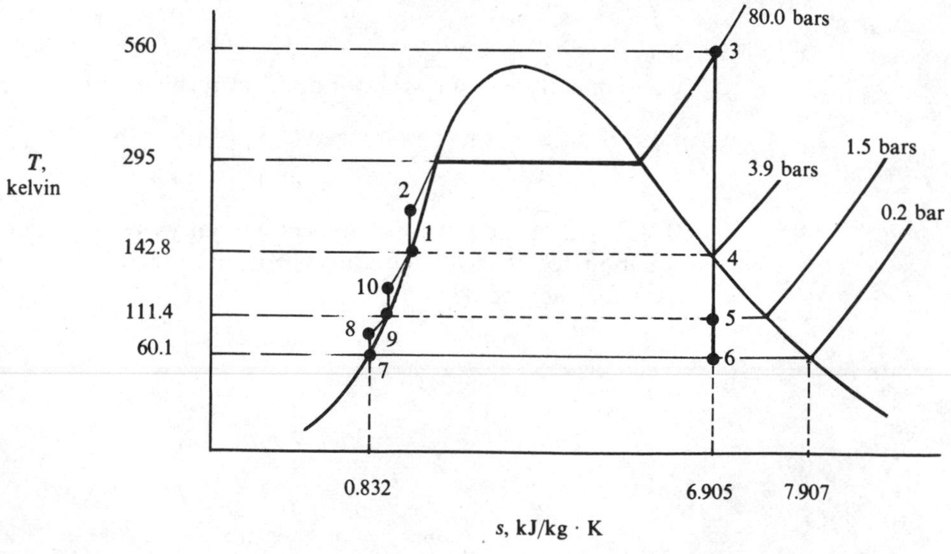

Figure 12–26 *T*-*s* diagram for two-stage regenerative cycle (not drawn to scale).

from table B.1 or the Mollier diagram. Using the Mollier diagram we find $h_4 = 2740$ kJ/kg, $h_5 = 2570$ kJ/kg, and $h_6 = 2280$ kJ/kg. From table B.1 we find $h_7 = 251$ kJ/kg, $h_9 = 467$ kJ/kg, and $h_1 = 601$ kJ/kg which are enthalpies of the saturated liquids at the corresponding pressures. Using equation (12–28)

$$\begin{aligned} h_2 &= h_1 - wk_{\text{pump }(a)} \\ h_{10} &= h_9 - wk_{\text{pump }(b)} \\ h_8 &= h_7 - wk_{\text{pump }(c)} \end{aligned}$$

and from equation (12–14) we get

$$\begin{aligned} wk_{\text{pump }(a)} &\cong -v_{f1}(p_2 - p_1) \\ &= -(0.001 \text{ m}^3/\text{kg})(80.0 \text{ bars} - 3.9 \text{ bars})(10^5 \text{ N/m}^2/\text{bars}) \\ &= -7.6 \text{ kJ/kg} \end{aligned}$$

and the enthalpy at state (2) is then

$$h_2 = 601 \text{ kJ/kg} - (-7.6 \text{ kJ/kg}) = 609 \text{ kJ/kg}$$

Similarly we find that

$$wk_{\text{pump }(b)} = -0.24 \text{ kJ/kg}$$

so

$$h_{10} = 467 \text{ kJ/kg} - (-0.24 \text{ kJ/kg}) = 467 \text{ kJ/kg}$$

and

$$wk_{\text{pump }(c)} = -0.13 \text{ kJ/kg}$$

which yields

$$h_8 = 251 \text{ kJ/kg} - (-0.13 \text{ kJ/kg}) = 251 \text{ kJ/kg}$$

From these data we can then proceed with our analysis. Using equation (12–36), we can obtain the flow rate of steam at state (4)

$$\dot{m}_1 h_1 = \dot{m}_4(h_4 - h_{10}) + m_1 h_{10}$$

or

$$\dot{m}_4 = \frac{\dot{m}_1(h_1 - h_{10})}{h_4 - h_{10}} = \frac{(2 \times 10^5 \text{ kg/hr})(601 \text{ kJ/kg} - 467 \text{ kJ/kg})}{2740 \text{ kJ/kg} - 467 \text{ kJ/kg}}$$

$$= 11790 \text{ kg/hr}$$ *Answer*

Then, from equation (12–37)

$$(\dot{m}_1 - \dot{m}_4)h_9 = (\dot{m}_1 - \dot{m}_4 - \dot{m}_5)h_8 + \dot{m}_5 h_5$$

we get

$$(\dot{m}_1 - \dot{m}_4)(h_9 - h_8) = \dot{m}_5(h_5 - h_8)$$

or

$$\dot{m}_5 = \frac{(\dot{m}_1 - \dot{m}_4)(h_9 - h_8)}{(h_5 - h_8)}$$

$$= \frac{(200{,}000 \text{ kg/hr} - 11{,}790 \text{ kg/hr})(467 \text{ kJ/kg} - 251 \text{ kJ/kg})}{2510 \text{ kJ/kg} - 251 \text{ kJ/kg}}$$

$$= 17{,}996 \text{ kg/hr}$$ *Answer*

We have that $\dot{m}_3 = \dot{m}_1 = 200{,}000$ kg/hr and

$$\dot{m}_6 = \dot{m}_1 - \dot{m}_4 - \dot{m}_5 = 170{,}214 \text{ kg/hr}$$

The power is computed from equation (12–35)

$$\dot{W}k_{\text{turb}} = \dot{m}_3(h_3 - h_4) + (\dot{m}_5 + \dot{m}_6)(h_4 - h_5) + \dot{m}_6(h_5 - h_6)$$

which gives us

$$\dot{W}k_{\text{turb}} = (200{,}000 \text{ kg/hr})(3544 - 2740)$$
$$+ (188210 \text{ kg/hr})(2740 - 2510)$$
$$+ (169580 \text{ kg/hr})(2510 - 2280) = 243 \times 10^6 \text{ kJ/hr}$$
$$= 67500 \text{ kW}$$ *Answer*

The net power produced we obtain by subtracting the power of the pumps; i.e.

$$\dot{W}k_{\text{cycle}} = \dot{W}k_{\text{turb}} + \dot{W}k_{\text{pump }(a)} + \dot{W}k_{\text{pump }(b)} + \dot{W}k_{\text{pump }(c)}$$

we have

$$\dot{W}k_{\text{pump }(a)} = \dot{m}\dot{W}k_{\text{pump }(a)} = (200{,}000 \text{ kg/hr})(-7.6 \text{ kJ/kg})$$
$$= -1.52 \times 10^6 \text{ kJ/hr}$$
$$= -422 \text{ kW}$$

Similarly we find

$$\dot{W}k_{\text{pump }(b)} = -12.5 \text{ kW}$$

and

$$\dot{W}k_{\text{pump }(c)} = -6.1 \text{ kW}$$

Then

$$\dot{W}k_{\text{cycle}} = 67{,}500 \text{ kW} - 422 \text{ kW} - 12.5 \text{ kW} - 6.1 \text{ kW}$$
$$= 67{,}059 \text{ kW}$$ *Answer*

The rate of heat addition is found from the equation

$$\dot{Q}_{\text{add}} = \dot{m}_1(h_3 - h_2)$$

so

$$\dot{Q}_{add} = (200{,}000 \text{ kg/hr})(3544 \text{ kJ/kg} - 609 \text{ kJ/kg})$$
$$= 587 \times 10^6 \text{ kJ/hr} = 163{,}000 \text{ kW}$$

The thermodynamic efficiency is then found from the equation

$$\eta_T = \left(\frac{\dot{W}k_{cycle}}{\dot{Q}_{add}}\right) \times 100$$

yielding

$$\eta_T = 41.1\%$$ *Answer*

12.12 The Reheat-Regenerative Cycle

The reheat-regenerative steam cycle operates with a higher efficiency than any of the steam power cycles operating between similar temperature and pressure regimes. For this reason most of the modern electric power generating systems using steam power utilize this cycle. The cycle components are generally a high pressure turbine, one or more reheaters, one or more low pressure turbines, condensers, regenerative

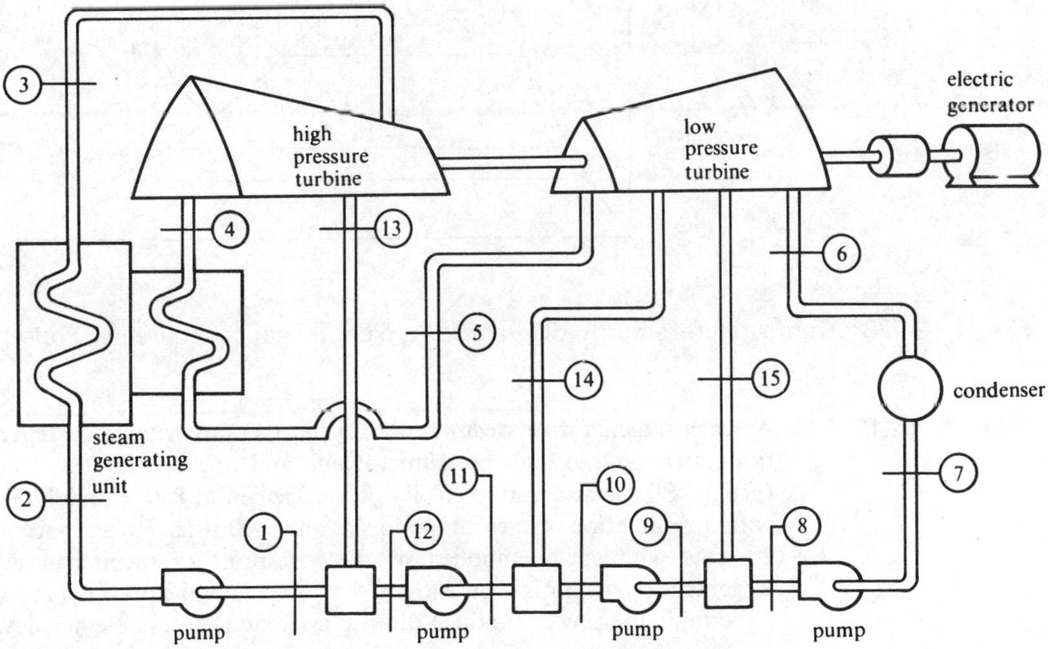

Figure 12–27 Typical elements of a reheat-regenerative cycle.

heaters, pumps, and a steam generator. The regenerative heaters are used with either or both the high and low pressure turbines. A typical arrangement for the reheat-regenerative cycle is shown in figure 12–27. The particular cycle shown has one reheat, one regeneration for the high pressure turbine, and two regenerations for the low pressure turbine. The temperature-entropy diagram in figure 12–28 indicates some of the typical states corresponding to the cycle shown in figure 12–27. Remember that regeneration can occur in the superheat region for states (13) and (14) and that the exhaust from the turbine at state (4) could be superheated steam although this would not be most effective in general. The analysis of a reheat-regenerative cycle follows along the same manner of the individual reheat cycles and the regenerative cycles. The enthalpy values at the individual states can be found by the techniques previously discussed and demonstrated. The mass flow rates through the regenerative stages are determined in the same manner as in the simple regenerative cycle.

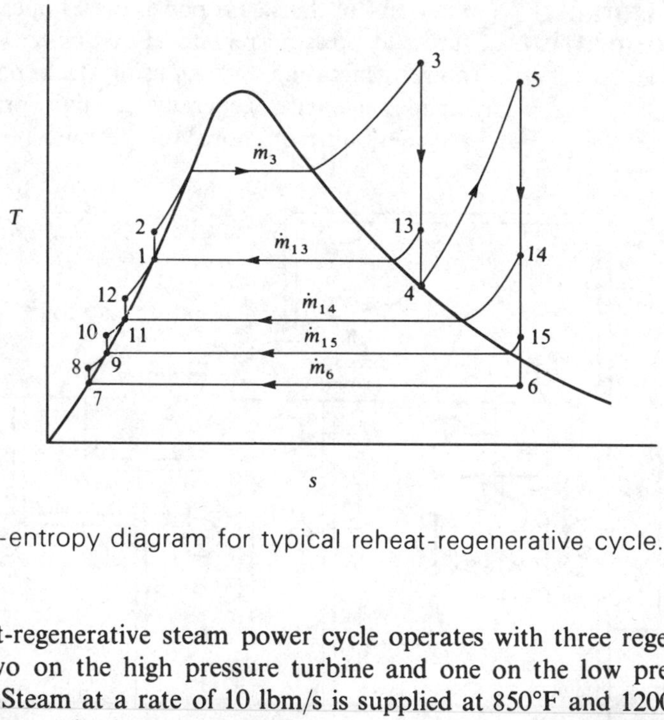

Figure 12–28 Temperature-entropy diagram for typical reheat-regenerative cycle.

Example 12.12 A reheat-regenerative steam power cycle operates with three regenerations, two on the high pressure turbine and one on the low pressure turbine. Steam at a rate of 10 lbm/s is supplied at 850°F and 1200 psia with regeneration stages at 600 psia and 400 psia. Reheat occurs at 200 psia with steam supplied to the low pressure turbine at 800°F. Regeneration steam is then extracted at 30 psia and exhaust is at 1 psia. Determine the power produced in the cycle and the cycle efficiency.

Solution Using the cycle shown in figure 12–29 as a representation of the cycle under consideration, we proceed to determine the enthalpy values at the various station points. Using the Mollier diagram, chart B.1, we find

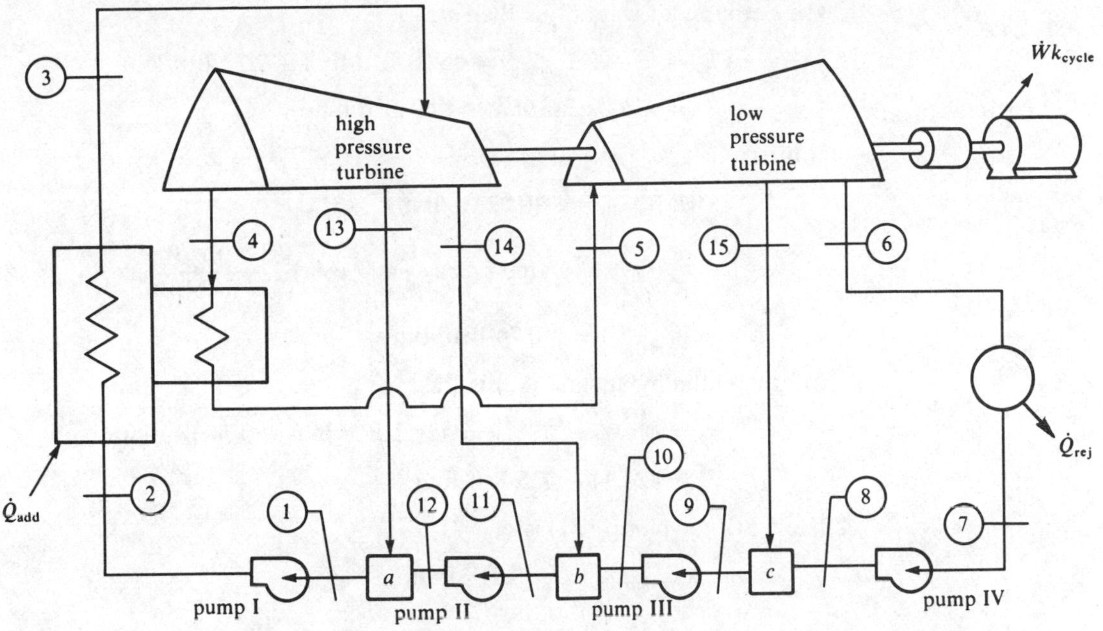

Figure 12–29 Reheat-regenerative cycle of example 12.12.

$h_3 = 1410$ Btu/lbm $\quad h_{13} = 1326$ Btu/lbm

$h_4 = 1216$ Btu/lbm $\quad h_{14} = 1272$ Btu/lbm

$h_5 = 1425$ Btu/lbm $\quad h_{15} = 1215$ Btu/lbm

$h_6 = 987$ Btu/lbm

From table B.1 we find

$$h_7 = h_f = 69.7 \text{ Btu/lbm}$$

$$h_9 = h_f \text{ (@ 30 psia)} = 218.9 \text{ Btu/lbm}$$

$$h_{11} = h_f \text{ (@ 400 psia)} = 424.2 \text{ Btu/lbm}$$

$$h_1 = h_f \text{ (@ 600 psia)} = 467.0 \text{ Btu/lbm}$$

The work required by the pumps may be approximated with the equation

$$wk_{\text{pump}} = -v_{\text{av}}(\Delta p)$$

Then, for pump (I) the average specific volume is near 0.02 ft³/lbm between 600 and 1200 psia. The work is then simply

$$wk_{\text{pump (I)}} = -(0.02 \text{ ft}^3/\text{lbm}) \times (1200 \text{ psia} - 600 \text{ psia}) \frac{144 \text{ in}^2/\text{ft}^2}{778 \text{ ft-lbf/Btu}}$$

$$= -2.22 \text{ Btu/lbm}$$

The enthalpy at state (2) is then

$$h_2 = h_1 - wk_{\text{pump (I)}} = 467.0 \text{ Btu/lbm} + 2.22 \text{ Btu/lbm}$$
$$= 469.22 \text{ Btu/lbm} \approx 469 \text{ Btu/lbm}$$

Similarly,

$$wk_{\text{pump (II)}} = -(0.02 \text{ ft}^3/\text{lbm})$$
$$\times (600 \text{ psia} - 400 \text{ psia}) \frac{144 \text{ in}^2/\text{ft}^2}{778 \text{ ft-lbf/Btu}}$$
$$= -0.74 \text{ Btu/lbm}$$

and the enthalpy at state (12) is then

$$h_{12} = h_{11} - wk_{\text{pump (II)}} = 424.2 \text{ Btu/lbm} + 0.74 \text{ Btu/lbm}$$
$$= 424.94 \approx 425 \text{ Btu/lbm}$$

The work for pump (III) is

$$wk_{\text{pump (III)}} = -(0.018 \text{ ft}^3/\text{lbm})$$
$$\times (400 \text{ psia} - 30 \text{ psia}) \frac{144 \text{ in}^2/\text{ft}^2}{778 \text{ ft-lbf/Btu}}$$
$$= -1.23 \text{ Btu/lbm}$$

and the enthalpy for state (10) is then

$$h_{10} = h_9 - wk_{\text{pump (III)}}$$
$$h_{10} = 218.9 \text{ Btu/lbm} + 1.23 \text{ Btu/lbm} = 220 \text{ Btu/lbm}$$

The work of pump (IV) is

$$wk_{\text{pump (IV)}} = -(0.0166 \text{ ft}^3/\text{lbm})$$
$$\times (30 \text{ psia} - 1.0 \text{ psia}) \frac{144 \text{ in}^2/\text{ft}^2}{778 \text{ ft-lbf/Btu}}$$
$$= -0.089 \text{ Btu/lbm}$$

and the enthalpy at state (8) is then

$$h_8 = 69.7 \text{ Btu/lbm} + 0.089 \text{ Btu/lbm} = 69.8 \text{ Btu/lbm}$$

The mass flow rates are obtained from the methods shown in section 12.11. Thus, for the regenerator (*a*) we have as an energy balance that

$$\dot{m}_1 h_1 = \dot{m}_{12} h_{12} + \dot{m}_{13} h_{13}$$

and as a mass balance that

$$\dot{m}_{12} = \dot{m}_{14} + \dot{m}_{10} = \dot{m}_{14} + \dot{m}_1 - \dot{m}_{13} - \dot{m}_{14}$$

or

$$\dot{m}_{12} = \dot{m}_1 - \dot{m}_{13}$$

If this equation is then substituted into the energy balance we find that

$$\dot{m}_1 h_1 = \dot{m}_{13} h_{13} + (\dot{m}_1 - \dot{m}_{13}) h_{12}$$

and then

$$\dot{m}_{13} = \frac{\dot{m}_1 (h_1 - h_{12})}{(h_{13} - h_{12})}$$

Since $\dot{m}_1 = 10$ lbm/s we can compute the mass flow at state (13):

$$\dot{m}_{13} = 10 \text{ lbm/s}\left(\frac{467 \text{ Btu/lbm} - 425 \text{ Btu/lbm}}{1326 \text{ Btu/lbm} - 425 \text{ Btu/lbm}}\right) = 0.47 \text{ lbm/s}$$

and the mass flow rate at (12) is

$$\dot{m}_{12} = 10 \text{ lbm/s} - 0.47 \text{ lbm/s} = 9.53 \text{ lbm/s}$$

From a similar analysis of regenerator (*b*) we find that

$$\dot{m}_{14} = 1.85 \text{ lbm/s}$$

so

$$\dot{m}_4 = 7.68 \text{ lbm/s}$$

and from regenerator (*c*) we obtain, where $\dot{m}_9 = \dot{m}_4$

$$\dot{m}_{15} = \dot{m}_9 \left(\frac{h_9 - h_8}{h_{15} - h_8}\right)$$
$$= 1.00 \text{ lbm/s}$$

and also

$$\dot{m}_6 = \dot{m}_4 - \dot{m}_{15} = 6.68 \text{ lbm/s}$$

The power produced from the cycle is computed from the equation

$$\begin{aligned}
\dot{W}k_{\text{cycle}} &= \dot{W}k_{\text{turb}} + \dot{W}k_{\text{pumps}} \\
&= \dot{m}_{13}(h_3 - h_{13}) + \dot{m}_{14}(h_3 - h_{14}) \\
&\quad + \dot{m}_4(h_3 - h_4) + \dot{m}_{15}(h_5 - h_{15}) \\
&\quad + \dot{m}_6(h_5 - h_6) + \dot{m}_7\, wk_{\text{pump (IV)}} \\
&\quad + \dot{m}_9\, wk_{\text{pump (III)}} + \dot{m}_{11}\, wk_{\text{pump (II)}} + \dot{m}_1\, wk_{\text{pump (I)}} \\
&= (0.47 \text{ lbm/s})(1410 - 1326) + (1.85) \\
&\quad \times (1410 - 1272) + (7.68)(1410 - 1216) \\
&\quad + 1.00(1425 - 1215) + (6.68)(1425 - 987) \\
&\quad + (6.68)(-0.089) + (7.68)(-1.23) \\
&\quad + (9.53)(-0.74) + (10)(-2.22) \\
&= 4898 \text{ Btu/s} = 6906 \text{ hp}
\end{aligned}$$

Answer

we may convert this answer to electrical power units to obtain

$$\dot{W}k_{\text{cycle}} = 5153 \text{ kW} \qquad \textit{Answer}$$

The heat rejected by the condenser is

$$\begin{aligned} \dot{Q}_{\text{rej}} &= \dot{m}_6(h_7 - h_6) \\ &= (6.68 \text{ lbm/s})(69.7 \text{ Btu/lbm} - 987 \text{ Btu/lbm}) \\ &= -6128 \text{ Btu/s} \end{aligned}$$

and the heat added is obtained from the balance

$$\dot{Q}_{\text{add}} = \dot{W}k_{\text{cycle}} - \dot{Q}_{\text{rej}} = 11{,}026 \text{ Btu/s}$$

The efficiency is computed in the usual manner to obtain

$$\eta_T = \frac{\dot{W}k_{\text{cycle}}}{\dot{Q}_{\text{add}}} = 44.4\% \qquad \textit{Answer}$$

12.13 Other Considerations of the Rankine Cycle

We have seen that the Rankine cycle in operation is a closed steam turbine cycle and that we can improve its efficiency by using reheat or regenerative heating devices. Steam turbines operating with both reheaters and regenerators are the most efficient power-producing devices available for large power demands. Increased improvements in thermodynamic efficiencies can also be realized with higher superheated steam pressures and temperatures; but, of course, an upper limit is set here by material properties. The upper limits for continuous operations seem to be around 80.0-bar (1200 psia) pressures with 650°C (1200°F) temperatures. Higher temperatures and pressures are always considered in newer designs but generally there are significant compromises of safety.

One of the most attractive aspects of the steam turbine cycle is its independence of the source of thermal energy. While internal combustion engines are critically dependent on their working media (requiring a precise refined fuel), the steam turbine merely requires energy to boil water. This energy can come from coal, petroleum, or natural gas combustion, from high temperature exhaust gases of some other engine, from solar energy, or from nuclear reactors. Beyond the steam generating unit there are few, if any, design alterations needed to accommodate these different types of heat sources. We can see then that the steam turbine cycle is well adapted to convert thermal energy to mechanical and electrical energy, and it allows us to hope that future power requirements may be met.

The combustion of fuel (if this type of heat source is utilized for generating steam) can be accomplished under well-controlled conditions and thus effect more efficient combustion and less obnoxious exhaust or products of combustion.

An attractive use of the steam turbine cycle is "behind" or tandem with another engine operating at a much higher temperature. This leading engine is commonly proposed to be a magneto-hydrodynamic (MHD) converter which operates on very high temperature; ionized gases and exhaust are also at a high temperature. The steam generator could then produce superheated steam from this exhaust and get additional power that might be lost. The MHD converter produces electrical power, and the two separate units then can operate as an efficient generating station.

The steam turbine has been characterized as a constant speed device, probably because its greatest development has come in large power units. In small units, however, the steam turbine may be applicable to variable speed applications although this utilization has not been fully attempted.

Finally, the promise of nuclear power as a practical source of power can, at this time, only be realized with the steam power cycles which we have considered. The basic elements of a typical cycle for this application are shown in figure 12-30. The nuclear reactor which generates the thermal energy during the reaction has heat carried away from it by a nuclear coolant. This coolant generally is separate and distinct from the steam power cycle. The coolant passes through a heat exchanger, in the typical system, and transfers heat to the steam. This coolant then returns to the reactor to receive more thermal energy. Some of the materials which have been used as coolants for this operation are water, liquid sodium, helium, and carbon dioxide.

Although the nuclear power plant is becoming a feasible and common mode of electrical power generation, its total practicality and safety are still in doubt. It is not clear that the total energy required to produce the processed nuclear fuel and the special handling of the fuel and waste

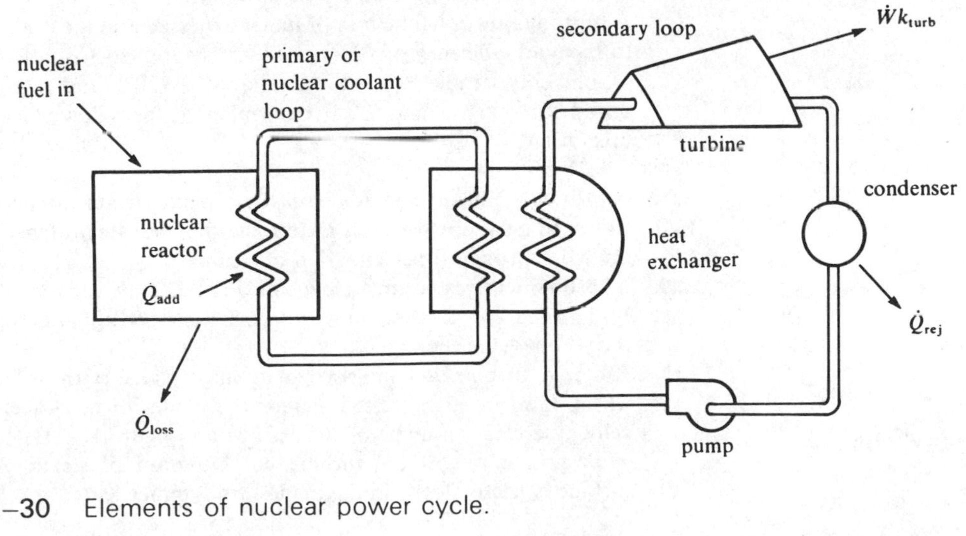

Figure 12–30 Elements of nuclear power cycle.

products are offset by the power actually produced. In many ways the customary fuels, coal, gas, and oil, are still subsidizing the operations of the nuclear power plants.

Practice Problems

Problems using SI units are indicated with the notation (M) under the problem number, and those problems using English units are indicated with (E). Those using mixed units are marked with (C) for combined under the problem number.

Section 12.1

12.1. A closed steam power plant delivers 10 megawatts (10 million watts) of electrical power. If 10,000 watts are required to drive water-recirculating pumps and 1.2×10^6 Btu/min of heat are added to the cycle, determine the amount of heat rejected and the cycle efficiency.

12.2. Sketch the p-v and T-s diagrams for a Rankine cycle and indicate which processes have work and heat present.

12.3. What are the source of power and the working medium in the common steam turbine cycle?

Sections 12.1 and 12.7

12.4. (M) There are 20 kg/s of water furnished to a boiler at 15.0 bars and as a saturated liquid. If the steam leaving the boiler is at 15.0 bars and 1400 kilojoules of heat are added per kilogram of water, determine
(a) Rate of heat addition.
(b) Steam temperature.

12.5. (M) A steam generator operates with an effectiveness of 85% when using coal as a fuel. If the incoming water is a saturated liquid at 10.0 bars and if there are 2000 kilograms of steam produced per hour at 400°C, determine the amount of coal needed per hour. Assume the LHV of coal is 28,000 kJ/kg.

12.6. (E) A boiler must supply steam at 250 psia and 650°F. If the water supplied to the boiler is saturated liquid at 3 psia, determine the amount of heat added per unit mass steam.

12.7. (E) A gas-fired steam generator uses a fuel with a LHV of 16,000 Btu/lbm. A burning rate of 6 lbm/min of fuel is expected and the boiler is assumed to have an efficiency of 90%. If water is supplied to the boiler at conditions comparable to a saturated liquid at 130°F and if the steam produced is at 400 psia and 600°F, determine the mass flow of water through this unit.

Sections 12.3 and 12.7

12.8. (M) A turbine reversibly and adiabatically expands steam from 40.0 bars and 640°C to a saturated vapor. Determine the final steam temperature and the work produced per kilogram of steam.

12.9. (M) A turbine receives steam at 40.0 bars and 800°C. It expands the steam to 1.0 bar and 240°C. Determine the work done per kilogram of steam and the turbine efficiency.

12.10. (M) A steam turbine is comprised of a single stage with a 1-m-diameter wheel. If steam is impressed against the blades of the wheel at 240 m/s velocity and at an angle of 10° as shown in figure 12–31, determine the work gotten out of the turbine per kilogram of steam. Assume the turbine is frictionless and the blades are symmetrical.

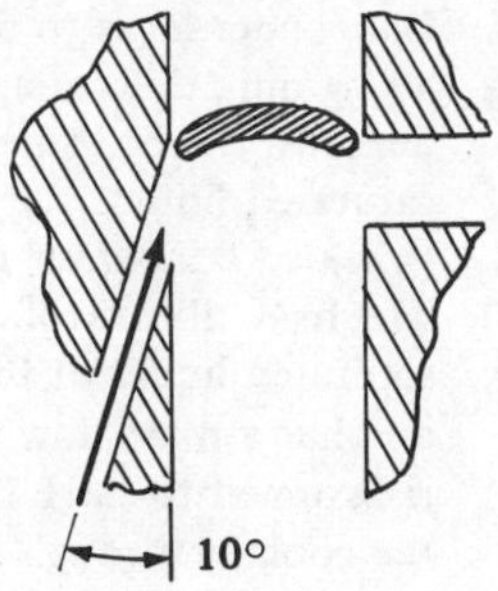

Figure 12–31

12.11. **(E)** In a turbine 700 lbm/min of steam are isentropically expanded from 350 psia and 700°F to 10 psia. Estimate the power produced by the turbine.

12.12. **(E)** A turbine has an efficiency of 90% in expanding steam from 300 psia and 650°F to 3 psia. Determine the work done and the final steam temperature.

12.13. **(E)** A turbine produces 20 Btu of work per pound-mass of steam. Determine the blade speed V_b if the blade is frictionless and symmetrical. The steam is directed against the blades at an angle of 18°.

Sections 12.4 and 12.7

12.14. **(C)** Determine the size of motor you would recommend to drive the following pumps:

(a) A feedwater pump rated at 30,000 kg/min of water at 90°C from 0.7 bar pressure to 100.0 bars.

(b) A feedwater pump rated at 50,000 lbm/min of water at 180°F from 2 psia to 380 psia.

12.15. **(C)** Saturated liquid water is pumped into a steam generator. Determine the enthalpy of the steam entering the generator, if the water has a constant density through the pump, for the following:

(a) Density of water is 970 kg/m^3, inlet pressure to pump is 5.0 bars, and outlet pressure is 40.0 bars.

(b) Density of water is 60 lbm/ft^3, inlet pressure to pump is 5 psia, and outlet pressure is 600 psia.

12.16. **(C)** For the two pumps given, determine the work done by the pumps per unit of mass of water in delivering saturated liquid water at a low pressure to a steam generator at a high pressure. Use the steam tables in appendix B and assume the pumps are reversible and adiabatic.

(a) Pump inlet pressure = 4.76 bars; pump discharge pressure = 60 bars.

(b) Pump inlet pressure = 10 psia; pump discharge pressure = 600 psia.

12.17. What is the general equation to calculate pump work, starting from the steady flow energy and assuming an adiabatic process as the only restriction?

Sections 12.5 and 12.7

12.18. **(M)** A condenser receives steam at 0.8 bar and 95% quality. Determine the heat rejected per unit mass of steam if the exit condition is a saturated liquid.

12.19. **(M)** If a condenser is to reject 2000 kJ/s of heat and the mass flow of water is 80 kg/min, determine the steam quality of that steam entering the condenser if the pressure is 0.2 bar and the water leaving the condenser is a saturated liquid.

12.20. **(E)** A sealed condenser is used to reject heat from a steam turbine cycle in which 10 lbm/s of steam are converted from a saturated vapor to a saturated liquid at 10 psia (figure 12–32). The coolant is water at 15 psia and has a mass flow rate of 200 lbm/s. If the specific heat of the coolant is assumed to be 1 Btu/lbm · °R, determine the temperature increase of the coolant through the condenser.

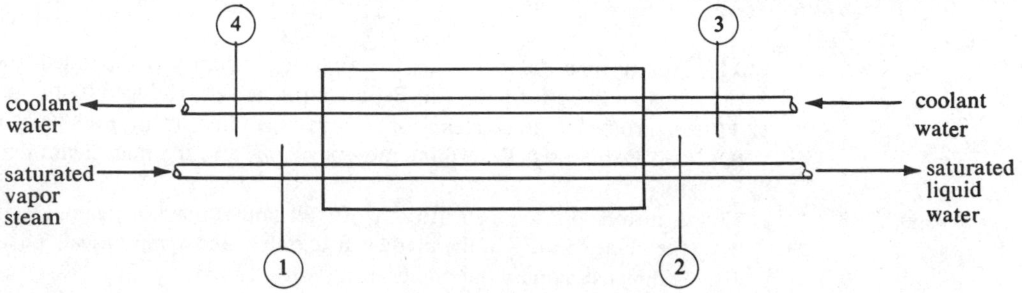

Figure 12–32

12.21. **(E)** A condenser rejects heat from a steam turbine cycle and thereby converts the working medium (steam) from a saturated vapor to a saturated liquid at 120°F. Determine the entropy decrease per pound-mass of steam in the condenser and the entropy increase of the atmosphere. What is the total increase in entropy of the universe for the condenser process?

Use the Mollier diagrams for problems 12.22 through 12.25.

Section 12.8

12.22. **(C)** Enthalpy of steam at
(a) 720°C and 10.0 bars.
(b) 1000°F and 150 psia.

12.23. **(C)** Entropy and enthalpy of steam at
(a) 2.0 bars and 6% moisture.
(b) 30 psia and 5% moisture.

12.24. **(C)** Temperature and enthalpy of steam at
(a) 0.1 bar and $s = 7.80$ kJ/kg · K.
(b) 1.5 psia and $s = 1.9$ Btu/lbm · °R.

12.25. **(C)** Enthalpy and pressure of steam with
(a) Temperature of 260°C and entropy of 7.0 kJ/kg·K.
(b) Temperature of 500°R and entropy of 1.65 Btu/lbm · °R.

Section 12.9

12.26. **(M)** A steam turbine produces 700 horsepower through a reversible adiabatic expansion from 20.0 bars and 480°C to 80°C. Neglecting pump work, determine the steam rate of the cycle.

12.27. **(M)** An ideal closed steam turbine cycle uses 5500 kg/hr of steam. The steam leaves the boiler at 40.0 bars pressure and is expanded reversible adiabatically to 0.6 bar and 95% quality in the turbine. The water leaving the condenser is a saturated liquid. Using the Mollier diagram, compute

(a) Turbine power produced.
(b) Heat rejected.
(c) Pump work.
(d) Heat added.
(e) Cycle efficiency and steam rate.

12.28. (M) A steam generating unit supplies 6000 kg/hr of steam at 70.0 bars and 800°C. Assume that the generator burns pulverized coal having an LHV of 30,000 kJ/kg and that the heat transfer effectiveness is 88%. A turbine expands this steam at 95% efficiency to 30°C and a condenser allows the steam to revert to a saturated liquid. If the pumps are assumed to be 100% efficient, determine

(a) Heat added to cycle.
(b) Heat added to water in boiler.
(c) Work of turbine.
(d) Heat rejected.
(e) Net cycle work.
(f) Overall cycle efficiency.
(g) Steam rate.

12.29. (E) A 200-megawatt power station operates on a closed steam turbine cycle, as shown in figure 12–33. The mechanical efficiency of the turbine-electric generator unit is 95% and the turbine efficiency is 90% in extracting energy from steam. The steam expands from 300 psia and 1050°F to 12 inches Hg in the turbine and is condensed to a saturated liquid in the condenser. The condenser is assumed to lose 5% of the steam to the atmosphere. A feedwater pump inserts 65°F water into the system after the circulating pump. Determine for the system

(a) Mass flow rate of steam through turbine.
(b) Heat added to water in boiler.
(c) Heat rejected in condenser.
(d) Work of both pumps. (Assume the water to be incompressible at all points.)
(e) Cycle efficiency.

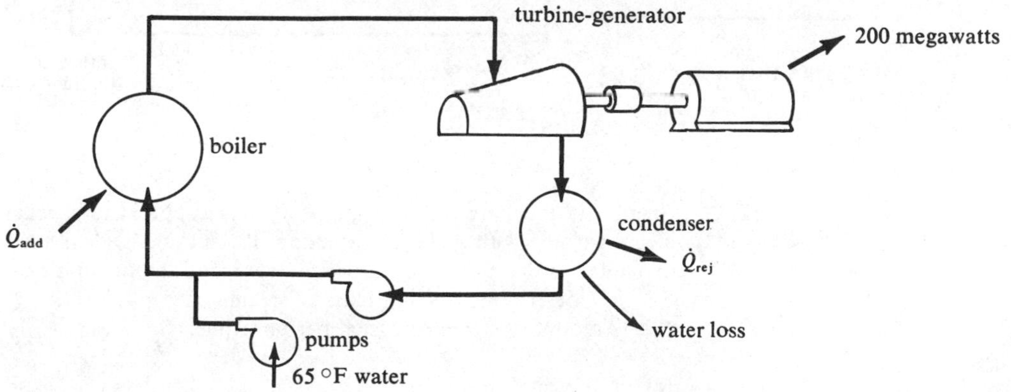

Figure 12–33

12.30. **(E)** A steam turbine cycle produces a net power of 10 megawatts. The turbine generator unit has a mechanical efficiency of 97% while steam is expanded reversibly and adiabatically through the turbine from 190 psia and 900°F to 5 psia. The water leaving the condenser is saturated liquid and is incompressible. If the boiler unit has a heat transfer efficiency of 90% and burns coal having an LHV of 11,000 Btu/lbm determine
(a) Pump work per unit mass of steam.
(b) Power produced by turbine.
(c) Rate of coal consumption (lbm/hr).
(d) Rate of heat rejection.

Section 12.10

12.31. **(M)** In an ideal steam turbine reheat cycle, the turbine expands reversible adiabatically 60 kg/s of steam from 30.0 bars and 880°C to 4.0 bars. The steam is then reheated to 880°C and expanded further to 0.10 bar. Determine
(a) Steam rate.
(b) Cycle efficiency.

12.32. **(E)** For the reheat cycle shown in figure 12–34, determine
(a) Turbine output (horsepower).
(b) Rate of heat added.
(c) Steam rate.
(d) Cycle efficiency.

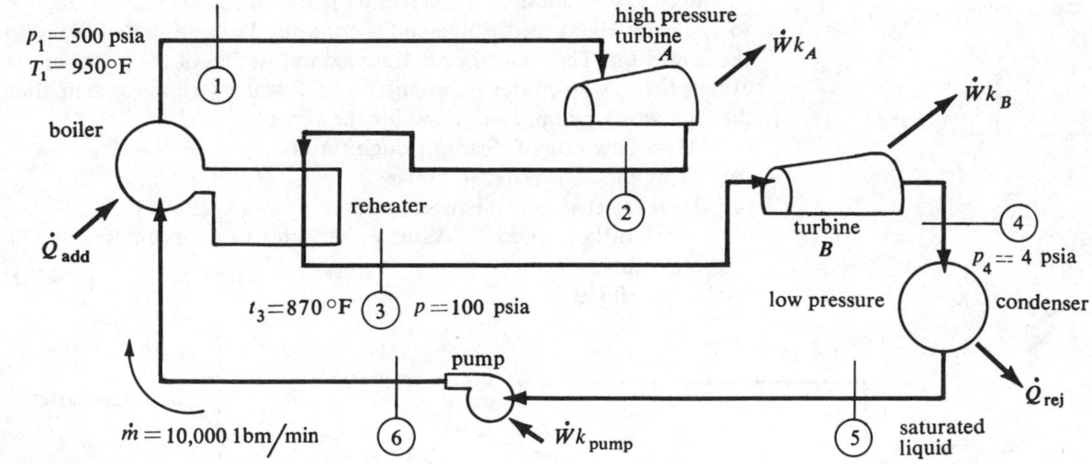

Figure 12–34

12.33. **(M)** Steam at 70.0 bars pressure and 960°C is supplied to a regenerative steam turbine with a single extraction. If the cycle is ideal, the steam is expanded in the turbine to 0.1 bar pressure, and the steam is extracted at 5.0 bars, determine, per unit mass of steam
(a) Amount of steam extracted per unit time.
(b) wk_{turb}
(c) Cycle efficiency.

12.34. For the ideal regenerative cycle shown in figure 12–35, determine
(E) (a) $\dot{m}_4$ and $\dot{m}_5$
(b) $\dot{W}k_c$, $\dot{W}k_b$, and $\dot{W}k_a$
(c) $\dot{W}k_{turb}$
(d) Cycle efficiency
(e) $\dot{Q}_{rej}$

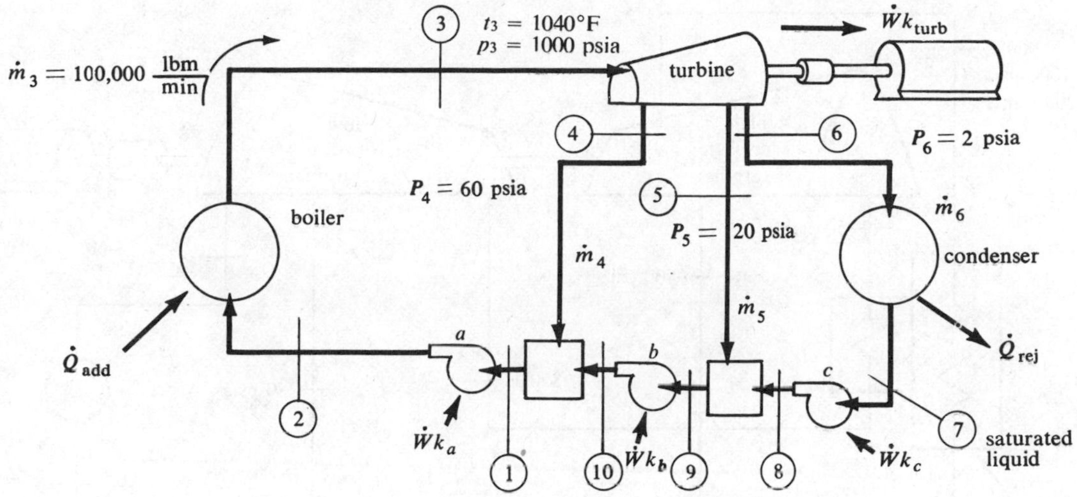

Figure 12–35

Section 12.12

12.35. For the reheat-regenerative cycle shown in figure 12–36, determine the
(M) (a) Cycle efficiency.
(b) Power produced.
(c) Steam rate.

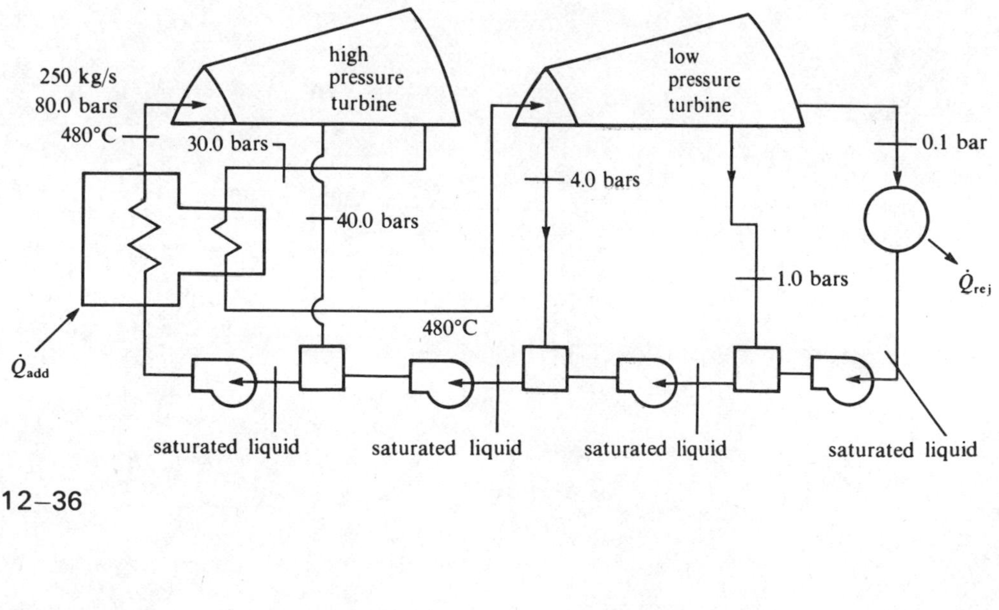

Figure 12–36

12.36. For the reheat-regenerative cycle shown in figure 12–37, determine
(E) (a) $\dot{W}k_{turb}$.
(b) $\dot{W}k_{cycle}$.
(c) $\dot{Q}_{rej}$.
(d) Cycle efficiency.
(e) Steam rate.

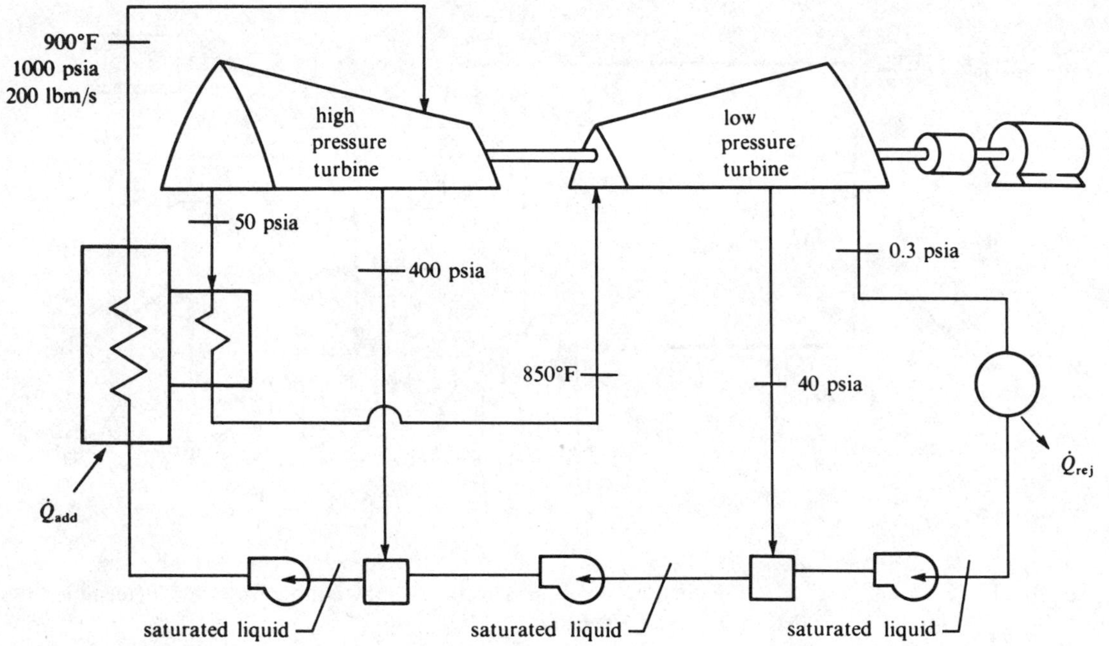

Figure 12–37

13 The Steam Engine

There is no other device that has contributed as much to the development of modern society as the *steam engine* has. Although it is now regarded almost as a museum piece, there are subtle features about the steam engine which could very easily return it to the wide use it enjoyed during the early years of the industrial revolution. From the early 1800s to the mid 1900s, the steam engine represented the major power-producing heat engine of society, and its popularity was the single most important reason for the development of thermodynamics. Early engineers and scientists, in seeking to explain and predict the steam engine's operation, refined and expanded the concepts and tools we now consider classical thermodynamics. We will discuss the steam engine here, not only to provide historical information, but also to see the promised capabilities of this most ingenious device.

The *ideal thermodynamic cycle of the steam expansion cycle* (which approximates the working of the actual steam engine) is defined. The work of the steam engine is then derived and calculated for typical engines, followed by a discussion of the efficiency. The mechanics which make up the typical steam engine are described in general terms to provide the reader with some idea of the components involved. Two of the most critical aspects of this engine are *valve arrangement* and *kinematics*. We will point out some of the important variations of these details and finally summarize the inherent advantages and problems associated with the steam engine.

13.1 The Steam Expansion Cycle

The steam engine is a heat engine utilizing the piston-cylinder device to provide mechanical work. It typically operates on a Rankine cycle which we define, ideally, by the following reversible processes:

(1–2) Adiabatic compression (liquid).
(2–3) Isobaric heat addition (vaporization).
(3–4) Adiabatic expansion (steam).
(4–1) Isobaric heat rejection (condensation).

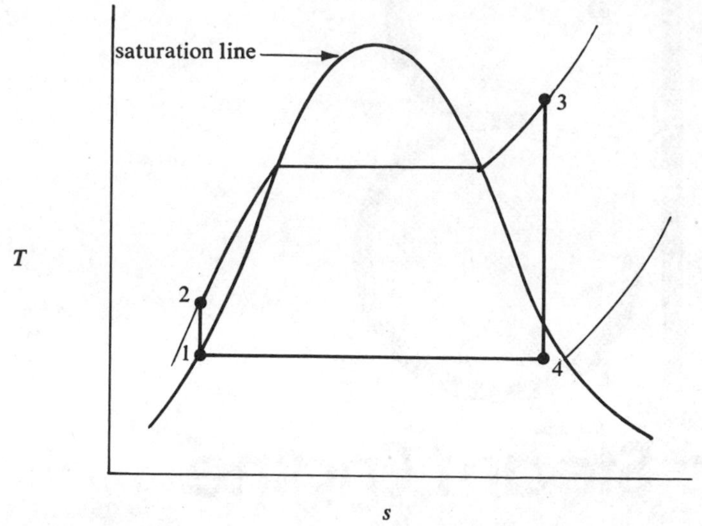

Figure 13–1 *T-s* diagram of steam engine.

These processes are sketched in the property diagrams in figures 13–1 and 13–2. Notice that these diagrams and the processes they describe are identical to the steam turbine cycle of section 12.1; the only difference is in the components which execute the various processes. The components of the closed cycle steam engine are shown in figure 13–3, The historically more common steam engine is the open cycle, shown in figure 13–4, which eliminates the condenser and circulating pump but which must continuously be supplied feedwater by an injector or a pump. The open cycle is generally simpler to operate in practice, but it is not as efficient

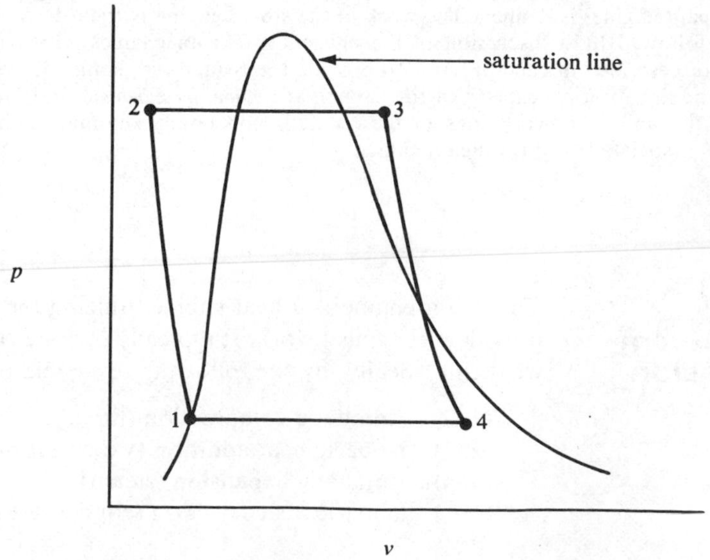

Figure 13–2 *p-v* diagram of steam engine.

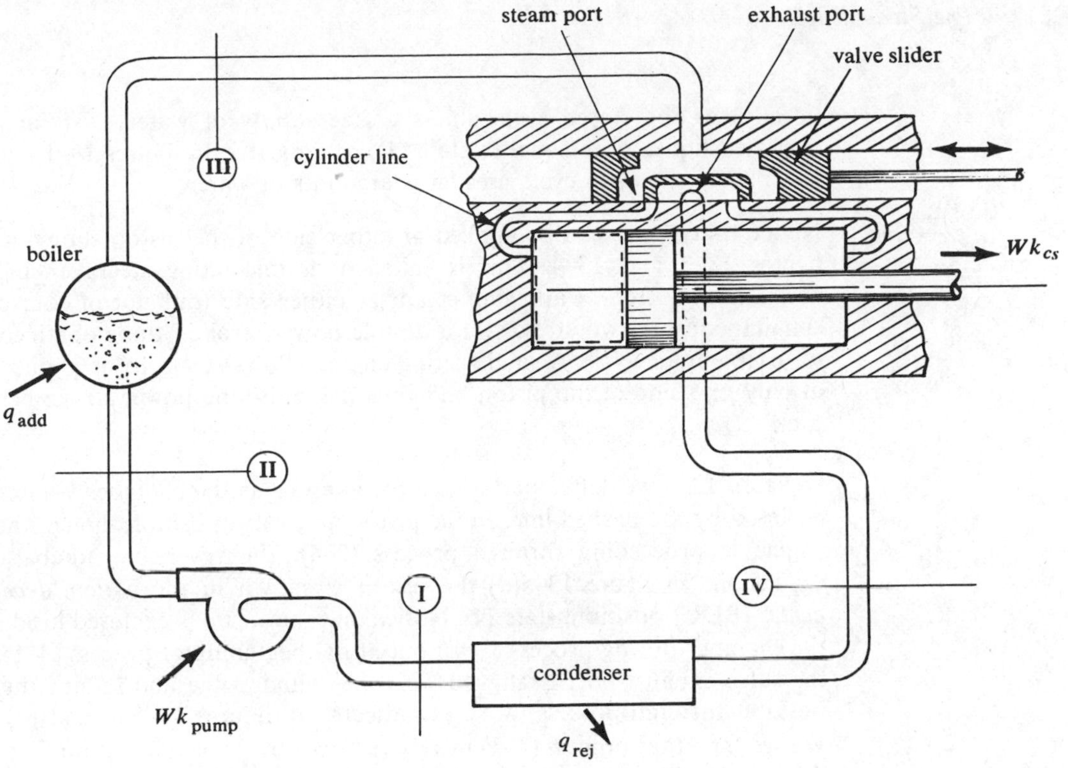

Figure 13–3 Sketch of typical closed cycle steam engine.

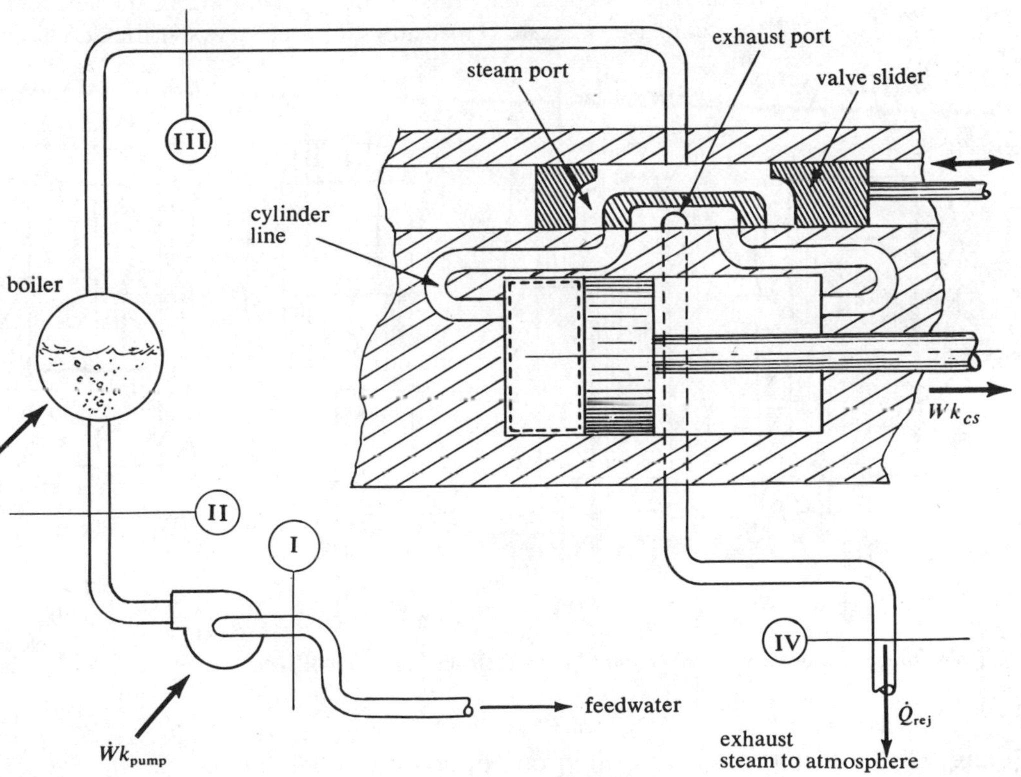

Figure 13–4 Sketch of typical open system steam engine.

(ideally) as the closed cycle unless a free supply of water exists at a saturated liquid state equal to state (1), as indicated in figures 13–1 and 13–2. Also, the open cycle uses large amounts of water.

Notice that steam can be applied at either side of the piston shown in figures 13–3 and 13–4. This is called a double-acting steam engine because, by virtue of applying steam at either side (but not of course simultaneously), two strokes or a double power stroke can be obtained for each engine cycle. A single-acting engine allows steam to be applied at only one side of the piston and thus has only one power stroke per cycle.

In figure 13–3 we define our system for analysis as the chamber volume enclosed by the dashed line. In the piston and valve position shown, the engine is proceeding through process (3–4), the reversible adiabatic expansion. In figure 13–5(d) the piston is shown in the *bottom dead center* (BDC) position, state (4). Notice that the steam is enclosed inside the chamber during process (3–4); but at the beginning of process (4–1), the valve exhaust port is aligned with the cylinder line and retains this position through process (4–1), as indicated in figures 13–5(a) and (d). We see then that process (4–1) merely removes the expanded steam from the cylinder. The properties of steam at state (4) should ideally correspond to those at state (IV) in figure 13–3. Pressure losses and heat transfer, however, will reduce the pressure and temperature of the steam entering the condenser at state (IV) below those at state (4) for actual engines.

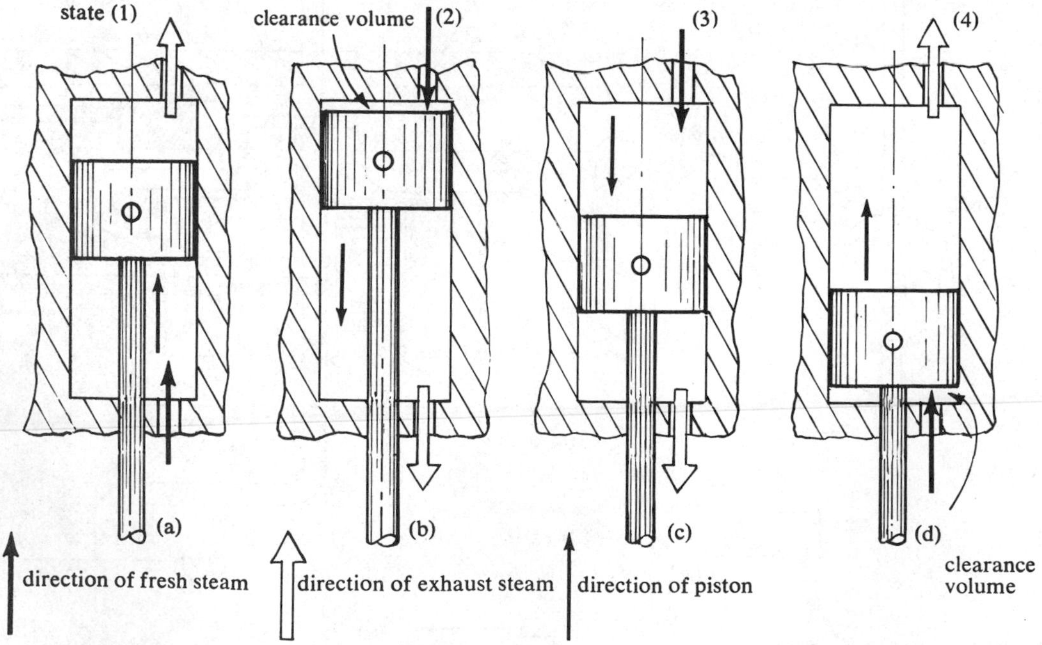

Figure 13–5 Piston positions during one cycle of a steam engine.

When the piston reaches a position such that the remaining steam in the cylinder condenses to state (1), the engine furnishes fresh, superheated steam at state (3) to the cylinders by repositioning the valve, as indicated in figure 13–5(b). As the engine then proceeds through process (1–2), the steam mixture (fresh and some expanded steam) is reversibly and adiabatically compressed to pressure p_2. State (2) represents the extreme innermost position of the piston inside the cylinder, called the *top dead center* (TDC), from which the piston is forced back out at constant steam pressure. At state (3), the steam port is closed to the cylinder and expansion process (3–4) then proceeds again.

To repeat, process (3–4) is ideally a reversible adiabatic expansion of steam carried out in a closed system. (The slide valve does not permit steam to enter or leave the cylinder.) Process (4–1) is an exhaust process which eliminates much of the expanded steam from the cylinder. Process (1–2) is a reversible adiabatic compression of the remaining steam in a closed system to a pressure and state equal to the state (2), indicated in the diagrams of figures 13–1 and 13–2. Process (2–3) is a constant pressure process, carried out with the steam port positioned over the cylinder line in the mechanism of figure 13–3. This is an open system process, since steam is added to the system or cylinder. The valve slide is then repositioned to that position shown in figure 13–3 and the reversible adiabatic process (3–4) proceeds again.

There are some apparent inconsistencies which should be explained. First of all, we have stated that ideally states (IV) and (4) are identical (in figure 13–3 and 13–1 respectively). Writing the steady flow energy equation, with kinetic and potential energy changes ignored, for the condenser as an open system we get

$$q_{\text{rej}} = h_{\text{I}} - h_{\text{IV}} \tag{13–1}$$

For ideal or reversible cycles the heat rejected by the piston cylinder is just

$$q_{\text{rej}} = h_1 - h_4 \tag{13–2}$$

and since we have already stated that, ideally, $h_4 = h_{\text{IV}}$ we can imply $h_1 = h_{\text{I}}$ and surmise that states (I) and (1) are identical although they are physically removed from each other in the engine. Also, the steam entering the engine, described as state (III), can be equated to state (3) if pressure and thermal losses are neglected in the lines and more importantly, if we neglect the loss of specific enthalpy and temperature due to the initial mixing of the steam at state (III) with the remaining steam from process (4–1). This allows us to write $h_3 = h_{\text{III}}$ and since

$$q_{\text{add}} = h_{\text{III}} - h_{\text{II}} = h_3 - h_2 \tag{13–3}$$

we can then say $h_{\text{II}} = h_2$ and infer that states (II) and (2) are identical.

Quite generally we can write

$$q_{\text{cycle}} = wk_{\text{cycle}}$$

from which we obtain

$$q_{\text{add}} - q_{\text{rej}} = wk_{\text{cycle}} \tag{13–4}$$

or

$$q_{\text{add}} - q_{\text{rej}} = wk_{\text{engine}} - wk_{\text{pump}} \tag{13–5}$$

Easily then, from equations (13–2), (13–3), and (13–4), we get

$$wk_{\text{cycle}} = h_3 - h_2 + h_1 - h_4 \tag{13–6}$$

and from a standard definition of thermodynamic efficiency

$$\eta_T = \frac{wk_{\text{cycle}}}{q_{\text{add}}} \times 100$$

we obtain

$$\eta_T = \frac{h_3 - h_2 + h_1 - h_4}{h_3 - h_2} \times 100 = \left(1 - \frac{h_4 - h_1}{h_3 - h_2}\right) \times 100 \tag{13–7}$$

13.2 Work of the Steam Engine

We may be tempted to further reduce the equation for calculating net work of a steam engine, equation (13–6), by recalling the caloric relation $h = c_pT$. This would be correct, but over the temperature ranges of the steam engine, the specific heat at constant pressure c_p is not a constant value and in fact has sharp changes during a phase change. For this reason we will use equation (13–6) for obtaining net work of the ideal steam engine. However, where there are small temperature changes and where no phase changes exist, $h = cT$ can be used. It might also be mentioned that net work should be equivalent to the area enclosed on the *p-v* diagram of figure 13–2. In this case, then, geometric means may be used to calculate the cyclic work.

The net work can also be equated to the average pressure of the steam cylinder during the full piston travel. If we call the average pressure the *mean effective pressure*, mep, the piston area A, and the piston travel L, then we have that

$$(\text{mep})(L)(A) = Wk_{\text{cycle}} \tag{13–8}$$

and the rate of work produced, the power, is obtained from

$$\dot{W}k_{\text{cycle}} = Wk_{\text{cycle}} \times N = (\text{mep})(LAN) \tag{13–9}$$

Equation (13–9) is frequently written in the mnemonic form

$$\dot{W}k_{\text{cycle}} = PLAN \tag{13–10}$$

where

$$P = \text{mep}$$

In the actual steam engine test procedure, an engine indicator is used to generate an engine diagram. The engine diagram is the actual *p-V* diagram of the steam piston-cylinder device. (A typical one is shown in figure 13–6.) When measured with a planimeter, the enclosed area will approximate the work produced by the piston. The actual work which can be gotten from the engine is, however, bound to be slightly less than the area, due to mechanical friction. Of course, the engine diagram is considerably altered for increasing or decreasing speeds, and for this reason the work of a cycle will be a function of the engine speed. In figure 13–7

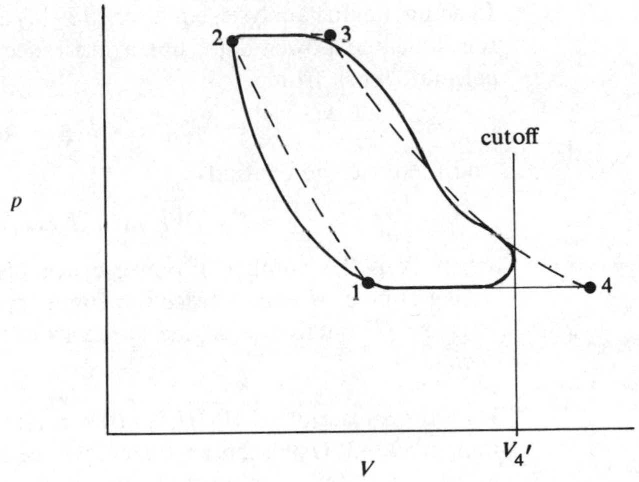

Figure 13–6 Typical engine diagram.

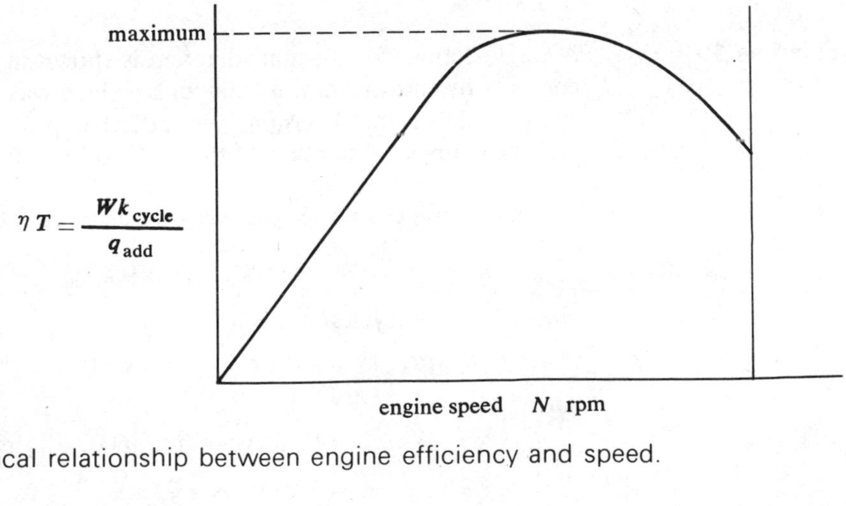

Figure 13–7 Typical relationship between engine efficiency and speed.

is shown a typical efficiency curve of a steam engine which reflects this variation in cyclic work due to engine speed.

In figure 13–6, the ideal cycle is superimposed to provide a visual comparison between the actual and ideal cases. Notice that the actual engine cycle is truncated at the right side; that is, the volume never expands to state (4), but rather is "cut off" at state (4′). This is a common procedure in actual steam engines, as the final portion of the expansion of steam is carried on at very slow speeds. In order to accommodate more acceptable engine speeds, therefore, the piston is reversed at a point corresponding to state (4′) rather than (4). Some potential work is lost due to this variation, but the increase in operating speeds is generally considered a good replacement for the loss.

In actual engine analysis, equation (13–10) is frequently utilized to obtain the indicated power data, but a more accurate approach would be to calculate work from

$$wk_{\text{cycle}} = h_{\text{III}} - h_{\text{IV}} + h_{\text{I}} - h_{\text{II}} \tag{13–11}$$

and then use the relation

$$\dot{W}k_{\text{cycle}} = NWk_{\text{cycle}} \tag{13–12}$$

where N is the number of power cycles per unit of time. For a double-acting engine, N equals twice the engine speed whereas for single-acting engines, N equals the engine speed in cycles per minute or cycles per second.

Here the subscripts I, II, III, and IV correspond to the points indicated in figure 13–3. Using this approach, we see that line energy losses may be more realistically included in the engine analysis, although we still assume adiabatic processes in the cylinder and circulating pump to arrive at equation (13–11).

Example 13.1

A single-acting steam engine diagram is shown in figure 13–8. The area enclosed by this diagram is 6000 joules which was recorded at an engine speed of 250 rpm. Determine the indicated power produced, the mep, and the cylinder dimensions if the bore and the stroke are to be equal.

Solution

We can obtain the power produced by using equation (13–12)

$$\dot{W}k_{\text{cycle}} = NWk_{\text{cycle}}$$

where N is 250 cycles/min and

$$Wk_{\text{cycle}} = 6000 \text{ J}$$

Then

$$\dot{W}k_{\text{cycle}} = (250 \text{ cycles/min})(6000 \text{ J})(1 \text{ min}/60 \text{ s})$$

$$= 25{,}000 \text{ J/s} = 25 \text{ kW}$$

Answer

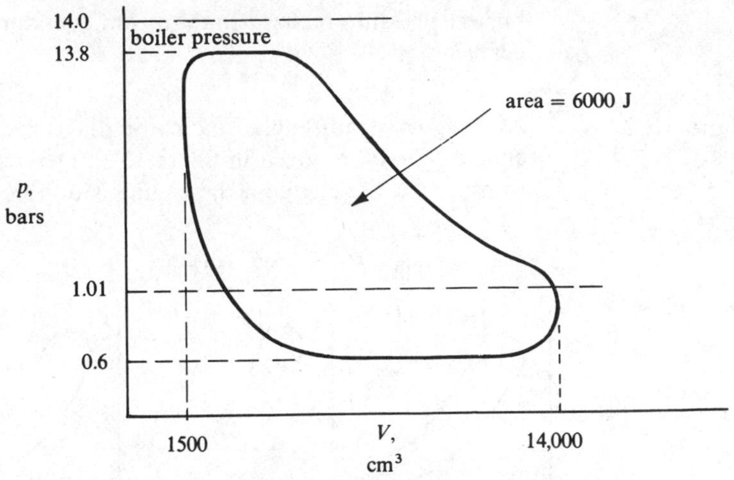

Figure 13–8 Engine diagram.

The mean effective pressure we can calculate from equation (13–9)

$$\text{mep} = \frac{\dot{W}k_{\text{cycle}}}{LAN}$$

Since here

$$LA = \text{volume swept by piston} = 14{,}000 \text{ cm}^3 - 1500 \text{ cm}^3$$
$$= 12{,}500 \text{ cm}^3$$

we get

$$\text{mep} = \frac{(6000 \text{ N} \cdot \text{m})(250 \text{ cycles/min})}{(12{,}500 \text{ cm}^3)(250 \text{ cycles/min})}$$
$$= 4.8 \times 10^5 \text{ N/m}^2$$
$$= 4.8 \text{ bars}$$ *Answer*

The dimensions of the cylinder can be obtained since we know L = stroke = piston diameter. Thus

$$LA = (L)(\pi)\left(\frac{L^2}{4}\right) = \pi\left(\frac{L^3}{4}\right) = 12{,}500 \text{ cm}^3$$

Then

$$L = \sqrt[3]{\frac{(4)(12{,}500 \text{ cm}^3)}{\pi}} = 25.2 \text{ cm}$$ *Answer*

The cylinder diameter and the piston stroke must both be 25.2 cm.

Notice in figure 13–8 that the boiler pressure is indicated as 14.0 bars. Can you justify why the boiler pressure does not correspond to the highest pressure realized in the steam cylinder as indicated by the closed curve?

Example 13.2

A series of steam engine indicator diagrams, each taken at a different engine speed, are given in figure 13–9. Determine the speed-power relationship for the engine if the engine is double acting.

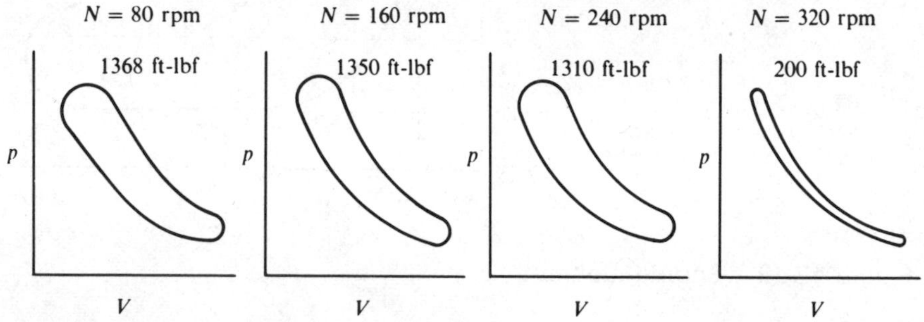

Figure 13–9 Engine diagrams for example 13.2.

Solution

For each case the power can be computed from the area enclosed by the curve. Thus

$$\dot{W}k = NWk_{cycle}$$

where N is twice the engine speed.

For state (a) we obtain

$$\begin{aligned}\dot{W}k &= (1368 \text{ ft-lbf})(80 \text{ cycles/min} \times 2)\\ &= 218{,}880 \text{ ft-lbf/min}\\ &= 218{,}880/33{,}000 \text{ hp} = 6.63 \text{ hp}\end{aligned}$$

The remaining results are tabulated in table 13–1 and a plot of the data is shown in figure 13–10. Notice that the power reaches a maximum near 240 rpm and drops sharply from that speed. Most engines, whether steam or other type, will exhibit a performance similar to this.

Table 13–1

Engine Speed (rpm)	*Power Produced* (hp)
80	6.63
160	13.09
240	19.05
320	3.88

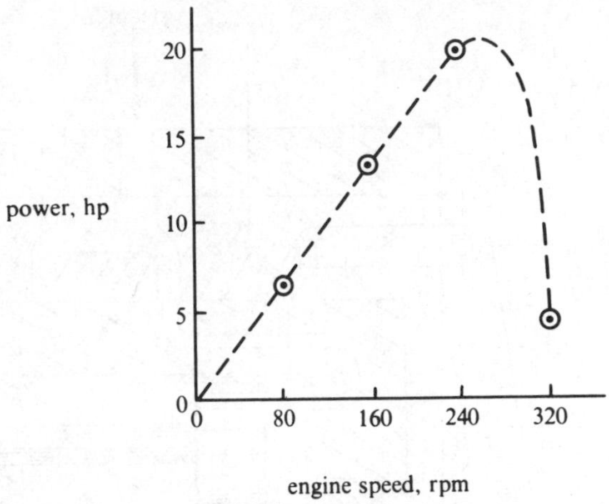

Figure 13–10 Speed-power diagram for engine of example 13.2.

13.3 Cycle Components

The major parts of a common steam engine have been indicated in figure 13–3: the boiler, the piston-cylinder, the condenser, and the circulating pump or injector.

The boiler we recognize as the component which generates steam and we could quite properly consider it as a steam generator. Physically, it may function as a simple pressure cooker, but in most applications, it is a series of tubes to allow for faster, more efficient heat transfer between the water and the heat source. The combustion of fuel such as coal, oil, wood, or gas has commonly been the heat source to generate steam; but nuclear or thermoelectric heat sources could also be used. The steam generator for turbines is identical to those used for steam engines.

The piston-cylinder which transforms thermal or internal energy of steam into mechanical work represents the heart of the cycle. We have already considered ways in which the work produced by the operating cylinder may be determined, so let us here look only at some configurations which have been used. The double-acting cylinder, shown in figure 11 3, operates with two power strokes per revolution. That is, there are two separate systems operating in opposition, with the result that for every revolution or cycle of the piston, both systems traverse a cycle and produce a power stroke. The advantage to the double-acting cylinder is, of course, its compactness.

Another piston-cylinder arrangement which has had frequent use is the double-expansion cylinder, or compound engine. This is sketched in figure 13–11 along with a typical *p-V* diagram, shown in figure 13–12, to illustrate that the double expansion is carried out in two separate cylinders, a high pressure cylinder and a low pressure one. The operation is,

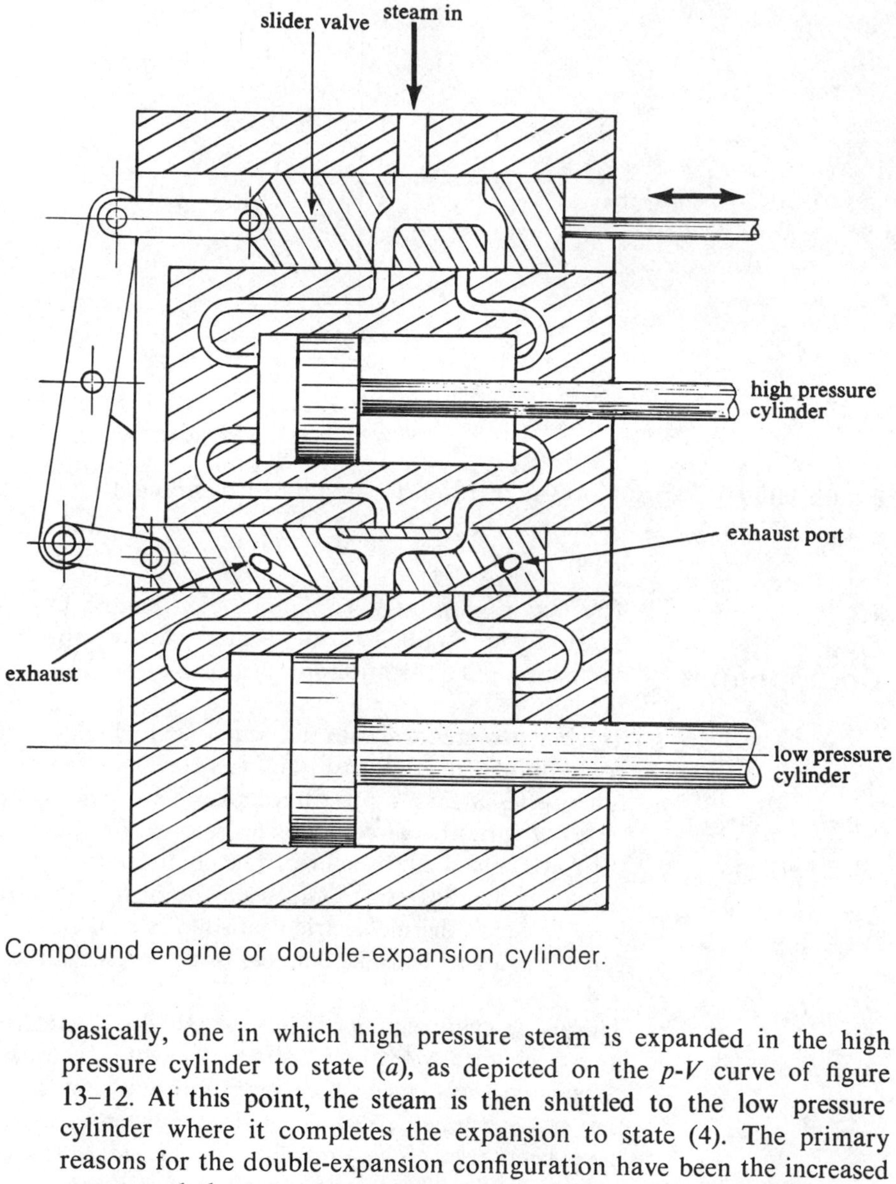

Figure 13–11 Compound engine or double-expansion cylinder.

basically, one in which high pressure steam is expanded in the high pressure cylinder to state (*a*), as depicted on the *p-V* curve of figure 13–12. At this point, the steam is then shuttled to the low pressure cylinder where it completes the expansion to state (4). The primary reasons for the double-expansion configuration have been the increased power and the smoother running characteristics. The practical reason for gaining power over the simple steam engine is that a more complete expansion can be achieved without a premature cutoff as is necessitated in the simple engine. (See figure 13–6.)

The condenser is required on all steam engines which recirculate the H_2O. This device is demanded from the implications of the second law of thermodynamics; that is, we must reject heat to a sink to have an operable heat engine. There are numerous configurations to the condenser, but here we will merely state that moist steam (mixture of liquid and

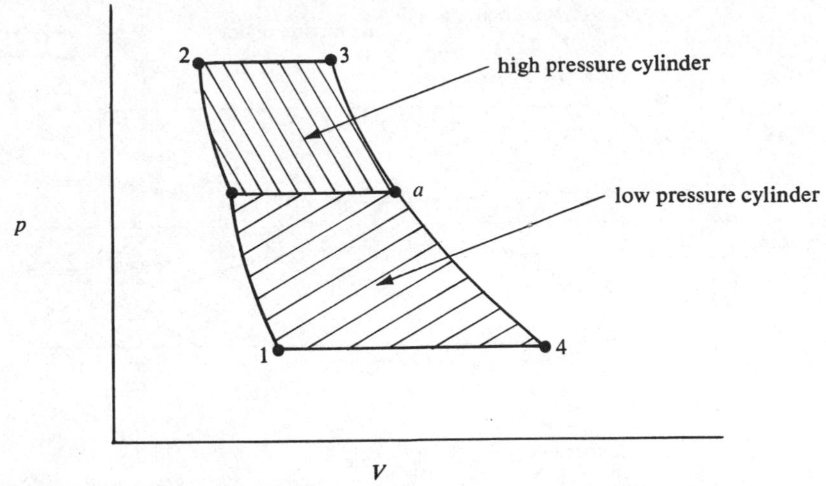

Figure 13–12 Double-expansion cylinder.

vapor) enters the condenser and will leave, ideally at a saturated liquid state — we say the steam has been condensed.

The water or circulating pump is a necessary part of any continuously operating steam engine. The function of this device is to recirculate the water in closed cycles and feed fresh water into the boiler in open cycles. In either case, the pump must, figuratively, push the water up a pressure gradient from a low pressure to a high pressure in the boiler or steam generator.

In the open cycle, illustrated in figure 13–4, the injector is commonly used in place of the pump. This device will be exposed in the following example problem to show the application of thermodynamics to an almost mystical process. This problem should also serve to show that the injector may be used in closed steam cycles, as well as in the open cycle.

Example 13.3 A nonlifting injector operates with superheated steam at 200 psia and 500°F. The injector receives water at 10 psia and 180°F and discharges it at 200 psia and 280°F into a boiler feed pipe. Determine the amount of steam required to supply 10 lbm of water to the boiler.

Solution A nonlifting injector, shown in figure 13–13, receives water and steam through separate ports. The water is free to flow into a chamber and steam is forced into this same chamber through a nozzle. Irreversible mixing of the steam and water takes place in the region designated as *a* in the chamber and this mixture then reaches an equilibrium at a sufficiently high dynamic pressure p_D $[p_D = p + (\rho V^2/2g_c)]$ that it can be introduced into the boiler or other high pressure region. The check value

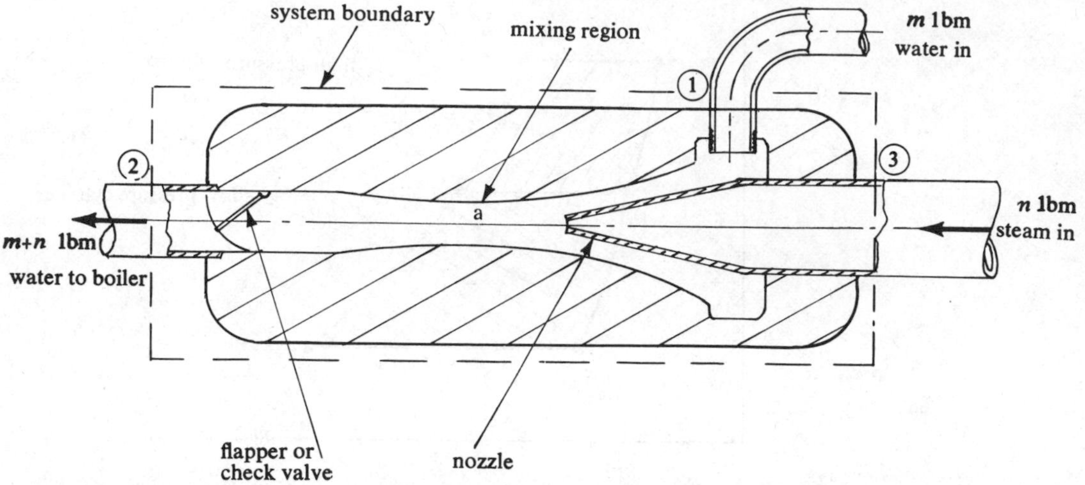

Figure 13–13 Nonlifting injector.

shown in the figure merely prevents backward flow when the injector is not operating — it does not retard forward fluid flow.

We now proceed to write the steady flow energy equation for the defined system enclosed by the boundary in figure 13–13. First we assume that the process is adiabatic ($Q = 0$), and that no changes in potential energy exist. Also, we observe that no work is being done. Then using subscripts which are compatible with the figure, we have

$$(m + n)h_2 - mh_1 - nh_3 + \frac{m + n}{2g_c} V_2^2 - \frac{m}{2g_c} V_1^2 - \frac{n}{2g_c} V_3^2 = 0$$

Further, it can be shown that the velocity terms are negligible with respect to the enthalpy terms so that we can then reduce our energy equation to

$$(m + n)h_2 - mh_1 - nh_3 = 0 \qquad \textbf{(13–13)}$$

If we desire, we may now write a general equation for the amount of steam required to transport 1 lbm of water through the injector. This term n/m is found by rearranging equation (13–13) to read

$$\frac{n}{m} = \frac{h_1 - h_2}{h_2 - h_3} \qquad \textbf{(13–14)}$$

For our problem, since the water flowing through the injector is well below the boiling point and since we do not know the value of h_1 nor can we extract it from the steam tables, we shall use the approximation $h = cT$ where c is near 1 Btu/lbm · °R. Using this we can write that

$$h_f - h_1 = c(T_f - T_1) = c(t_f - t_1) \qquad \textbf{(13–15)}$$

where h_f = 161.2 Btu/lbm corresponding to the saturation temperature t_f of 193.21°F as given in table B.1 for steam at 10 psia. Then we can directly calculate h_1; that is

$$h_1 = h_f - c(t_f - t_1)$$
$$= 161.2 - 1.0(193.21 - 180) = 147.99 \text{ Btu/lbm}$$

Using this result, we calculate h_2 from the equation

$$h_2 - h_1 = 1 \text{ Btu/lbm} \cdot {}^\circ\text{R } (t_2 - t_1)$$

since we assume that the water has not boiled. Then

$$h_2 = h_1 + 1(t_2 - t_1)$$

where

$$t_2 = 280^\circ\text{F} \quad \text{and} \quad t_1 = 180^\circ\text{F}$$

and

$$h_2 = 147.99 + 1(280 - 180)$$
$$= 247.99 \text{ Btu/lbm}$$

The steam supplied to the injector is at 500°F and 200 psia so h_3 = 1269.0 Btu/lbm from table B.2. Then, using equation (13–14)

$$\frac{n}{m} = \frac{h_1 - h_2}{h_2 - h_3}$$

we get

$$\frac{n}{m} = \frac{147.99 - 247.99}{247.99 - 1269.0} = 0.098 \text{ lbm steam/lbm water}$$

and for 10 lbm of water supplied to the boiler, we say $n + m = 10$ lbm. But we now can use the result from above, $m = n/0.098$, to get

$$n + \frac{n}{0.098} = 10$$

or

$$n = \frac{0.980}{1.098} = 0.892 \text{ lbm of steam} \qquad \textit{Answer}$$

for every 10 lbm of water supplied.

This device, the steam injector, is irreversible in its operation, yet the internal energy of the steam which is degraded to a lower temperature cannot be counted as lost — it merely recirculates through the boiler or steam generator to be used again. Also, in this example problem we see that the water received by the injector could very likely be coming from the condenser and this operation we would associate with a closed cycle. As we have earlier stated, injectors have commonly been used in open systems to supply feedwater to a boiler.

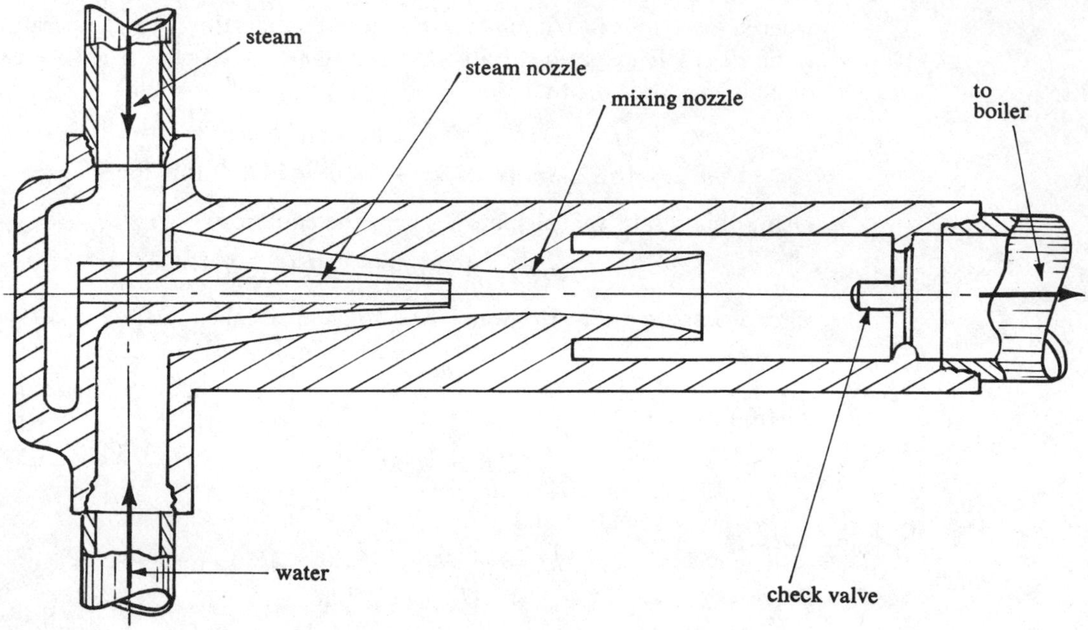

Figure 13–14 Lifting injector.

The nonlifting injector, which we have discussed here, operates with a slight vacuum pressure in the mixing chamber *a*. (See figure 13–13.) With a slightly different physical configuration, the steam and water mixture may have a higher velocity in this chamber with a corresponding decrease in pressure. This arrangement is depicted in figure 13–14 and is called a *lifting injector* since the decreased pressure in the mixing chamber allows water to be drawn up from a well or reservoir below the injector. Of course, this "lifting" of water can only occur if the pressure in the mixing chamber is less than the atmospheric pressure and consequently there is no conceptual difference between lifting and nonlifting injectors.

13.4 The Typical Steam Engine

Now that we have investigated the parts of the steam engine and the thermodynamic cycle which suitably describes the engine, let us look at two examples of the steam engine: an open cycle engine (which fits the description of the museum pieces) and a closed cycle (which has an attractive operation for modern applications).

Example 13.4

A simple double-acting steam engine is required to develop 200 horsepower at 180 rpm. The boiler can generate superheated steam at 320°C and 10.0 bars and the piston-cylinder is assumed to expand the steam reversibly and adiabatically to 1.01 bars. Assuming the injector receives water at 30°C and delivers it to the boiler at 90°C and 10.0 bars pressure, determine the mass flow rates through the cylinder and the boiler, the thermodynamic efficiency, and the recommended bore and stroke for the cylinder. The engine is diagrammed in figure 13–15.

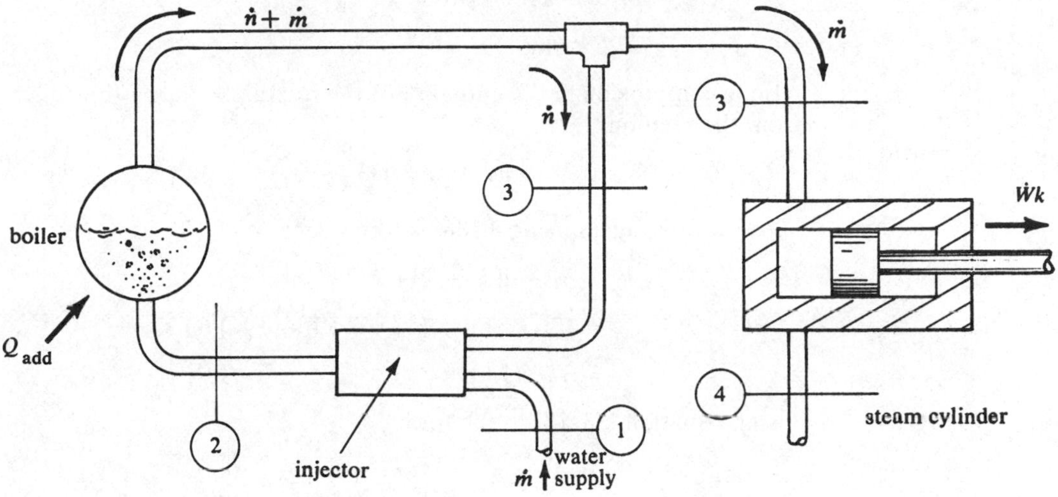

Figure 13–15 The steam engine.

Solution

We obtain the mass flow rate of the steam through the cylinder by first identifying the cylinder as an open system and writing the steady flow energy equation

$$h_4 - h_3 = -wk_{os}$$

This equation results from the assumption

$$\Delta KE = \Delta PE = 0$$

and the statement that the cylinder is adiabatic, $Q = 0$. The power generated is then found from

$$\dot{m}(h_4 - h_3) = -\dot{W}k_{os} = -200 \text{ hp}$$

Notice that we have used as the cycle work the term $h_4 - h_3$ rather than equation (13–6) because the additional enthalpy values h_1 and h_2 do not represent work. That is, the injector does not require work to operate.

We have from table B.2 that $h_3 = 3091$ kJ/kg and from table B.1 or the Mollier diagram, $h_4 = 2615$ kJ/kg since $s_3 = s_4 = 7.189$ kJ/kg · K. Then

$$\dot{m} = \frac{200 \text{ hp}}{(3091 \text{ kJ/kg} - 2615 \text{ kJ/kg})}$$

$$= 0.313 \text{ kg/s}$$

Answer

Using equation 13–15 we obtain the enthalpy of the feedwater h_1. From table B.1 we get $h_f = 419$ kJ/kg at a pressure of 1.01 bars and 100°C. Then

$$h_1 = h_f - c(T_f - T_1)$$
$$= 419 \text{ kJ/kg} - (4.1868 \text{ kJ/kg} \cdot \text{K})(373 \text{ K} - 303 \text{ K})$$
$$= 126 \text{ kJ/kg}$$

Also, we approximate the enthalpy of the mixture entering the boiler h_2 from the relation

$$h_2 - h_1 = c(T_2 - T_1)$$

This gives us, again using 4.1868 kJ/kg · K for c, for h_2

$$h_2 = h_1 + c(T_2 - T_1)$$
$$= 126 \text{ kJ/kg} + (4.1868 \text{ kJ/kg} \cdot \text{K})(60 \text{ K})$$
$$= 377 \text{ kJ/kg}$$

Using equation (13–14) in the form

$$\frac{\dot{n}}{\dot{m}} = \frac{h_1 - h_2}{h_2 - h_3}$$

we have, for the mass flow $\dot{n}$

$$\dot{n} = \dot{m}\left(\frac{h_1 - h_2}{h_2 - h_3}\right)$$
$$= (0.313 \text{ kg/s})\left(\frac{126 \text{ kJ/kg} - 377 \text{ kJ/kg}}{377 \text{ kJ/kg} - 3609 \text{ kJ/kg}}\right)$$
$$= 0.024 \text{ kg/s}$$

and the mass flow through the boiler must then be

$$\dot{n} + \dot{m} = 0.313 \text{ kg/s} + 0.024 \text{ kg/s} = 0.337 \text{ kg/s}$$

The thermodynamic efficiency is

$$\eta_T = \frac{\dot{W}k_{\text{cycle}}}{\dot{Q}_{\text{add}}} \times 100 \qquad \textbf{(13–16)}$$

and in our problem, $\dot{W}k_{\text{cycle}} = \dot{W}k_{os} = 200 \text{ hp} = 149.2 \text{ kW}$. Also

$$\dot{Q}_{\text{add}} = (\dot{n} + \dot{m})(h_3 - h_2)$$
$$= (0.337 \text{ kg/s})(3091 \text{ kJ/kg} - 377 \text{ kJ/kg})$$
$$= 914 \text{ kW}$$

From equation (13–16) we then get

$$\eta_T = \frac{(149.2\ \text{kW})}{(914\ \text{kW})}(100)$$

$$= 16.3\% \qquad \textit{Answer}$$

We may now recommend cylinder dimensions by using equation (13–9)

$$\dot{W}k_{\text{cycle}} = (\text{mep})(LAN)$$

and suggest a mean effective pressure of 6.0 bars. This mep value was reached by assuming it to be near 60% of the boiler pressure. Then, since we have a double-acting cylinder, we may assign a value of 360 cycles/min for N. Then, from equation (13–9)

$$LA = \frac{(149.2\ \text{kW})(60\ \text{s/min})}{(360\ \text{cycles/min})(6 \times 10^5\ \text{N/m}^2)}$$

$$= 0.0414\ \text{m}^3$$

If we assume the bore and stroke to be equal in value, then we get

$$\pi\left(\frac{L^3}{4}\right) = 0.0414\ \text{m}^3$$

and

$$L = \sqrt[3]{\frac{(4)(0.0414\ \text{m}^3)}{\pi}} = 0.375\ \text{m}$$

Example 13.5

Consider the engine of example 13.4 and assume that the cycle is closed. The steam exhausts from the engine to a condenser where it becomes a saturated liquid and is returned to the boiler by a feedwater pump. If the pump is reversible, determine the mass flow rate through the boiler and the thermodynamic efficiency. Assume the engine produces 200 horsepower as the net power.

Solution

In figure 13–16 is shown the configuration of the cycle. We have replaced the injector by a pump, but this was done primarily to show the diversity of arrangements possible. Of course, the injector would require an additional flow path for the steam. We now can readily identify the work and heat terms from equations (13–2), (13–3), and (13–6):

$$q_{\text{rej}} = h_1 - h_4$$

$$q_{\text{add}} = h_3 - h_2$$

and

$$wk_{\text{cycle}} = h_3 - h_2 + h_1 - h_4$$

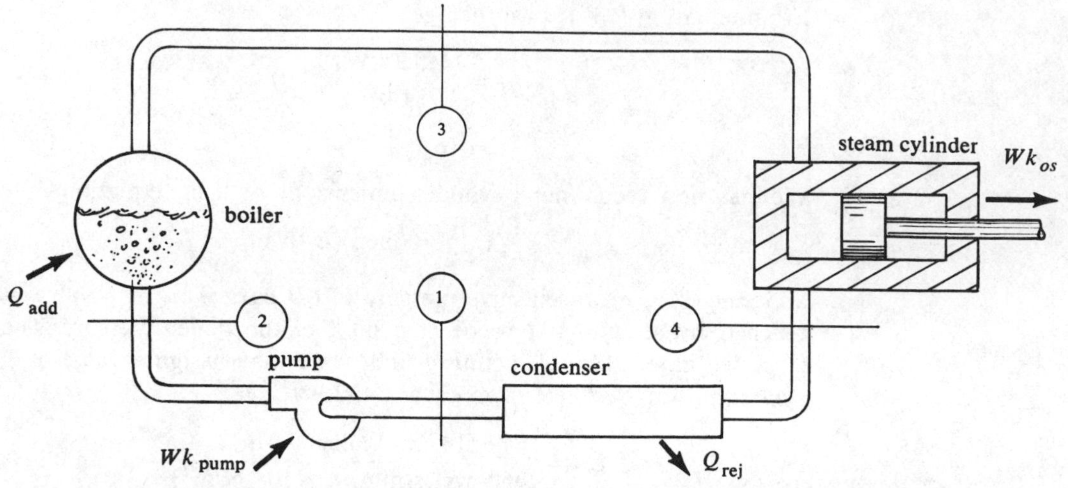

Figure 13–16 Closed cycle steam engine.

From example 13.3 we have

$$h_3 = 3091 \text{ kJ/kg} \quad \text{and} \quad h_4 = 2615 \text{ kJ/kg}$$

From table B.1 we obtain $h_1 = 419$ kJ/kg since the steam is a saturated liquid at state (1). Since the pump is reversible, we write

$$h_2 - h_1 = -wk_{\text{pump}} = \int_{p2}^{p1} vdp$$

and, assuming the steam is incompressible with a specific volume v_1 of 0.001 m^3/kg (table B.1), we get

$$h_2 = h_1 + v\int_{p_1}^{p_2} dp = h_1 + v(p_2 - p_1) = 419 \text{ kJ/kg} + (0.001 \text{ m}^3\text{/kg})$$

$$\times (8.99 \times 10^5 \text{ N/m}^2)$$

$$= 419 \text{ kJ/kg} + 0.9 \text{ kJ/kg}$$

$$= 419.9 \text{ kJ/kg}$$

Then the work of the cycle is

$$wk_{\text{cycle}} = 3091 \text{ kJ/kg} - 419.9 \text{ kJ/kg}$$

$$+ 419 \text{ kJ/kg} - 2615 \text{ kJ/kg}$$

$$= 475.1 \text{ kJ/kg}$$

The power is 200 hp or 149.2 kW, so

$$149.2 \text{ kW} = \dot{m}\ (475.1 \text{ kJ/kg}) = 149.2 \text{ kJ/s}$$

and the mass flow rate is

$$\dot{m} = \frac{149.2 \text{ kJ/s}}{475.1 \text{ kJ/kg}}$$

$$= 0.314 \text{ kg/s} \qquad \textit{Answer}$$

This is the mass flow rate through each of the cycle components.

The efficiency of the cycle can be computed from equation (13–7):

$$\eta_T = \left(1 - \frac{h_4 - h_1}{h_3 - h_2}\right) \times 100$$

$$= \left(1 - \frac{2615 \text{ kJ/kg} - 419.9 \text{ kJ/kg}}{3091 \text{ kJ/kg} - 419.9 \text{ kJ/kg}}\right) \times 100$$

$$= 17.8\% \qquad \textit{Answer}$$

Notice that we have increased the efficiency of the engine by using the closed instead of the open cycle. This increase can be attributed mainly to the energy differences between the condenser water (in the closed cycle) and the feedwater (in the open cycle).

13.5 Valving

As we have just seen, we may improve the efficiency of a steam engine by using a closed cycle rather than an open cycle. Other obvious methods for improving the efficiency of the engine are (1) increasing the superheated steam temperature and pressure, (2) increasing the vacuum (or decreasing the absolute pressure) of the region to which the steam is exhausted from the cylinder, and (3) reducing the irreversibilities in operation. Of these three methods, the first two are open-ended and limited only by the practical ability to handle high pressures and temperatures as well as high vacuums. The third method is one which concerned engineers almost from the time the steam engine was conceived. While we found efficiencies of 16.3% to 17.8% in examples 13.4 and 13.5, the efficiencies realized by actual steam engines have been closer to 4% to 15%. This pronounced discrepancy between theoretical and actual efficiencies can generally be attributed to the fact that a piston-cylinder does not expand steam reversibly and adiabatically (although higher superheats would make this more nearly realized) and to the fact that the valves are inefficient. We will here consider the common types of valves used in steam engines to emphasize that valve design improvements constituted major developments in the steam engine and would be one of the prime concerns in future designs.

The cylinder shown in figure 13–3 utilizes a valve which oscillates back and forth, shuttling steam through either of two paths. This valve is called a *slide valve* and represents one of the earliest methods of controlling or switching steam flow. It is reliable, simple to operate, and inexpensive, but it generally is limited to low speeds (due to inertia problems) and low steam temperatures. At superheated conditions, the steam will create a considerable lubrication problem in the slide valve.

In order to provide for higher steam temperatures and pressures, the *piston valve*, shown in figure 13–17, was developed. This device accomplished its purpose, but the friction between the valve piston and body was still high.

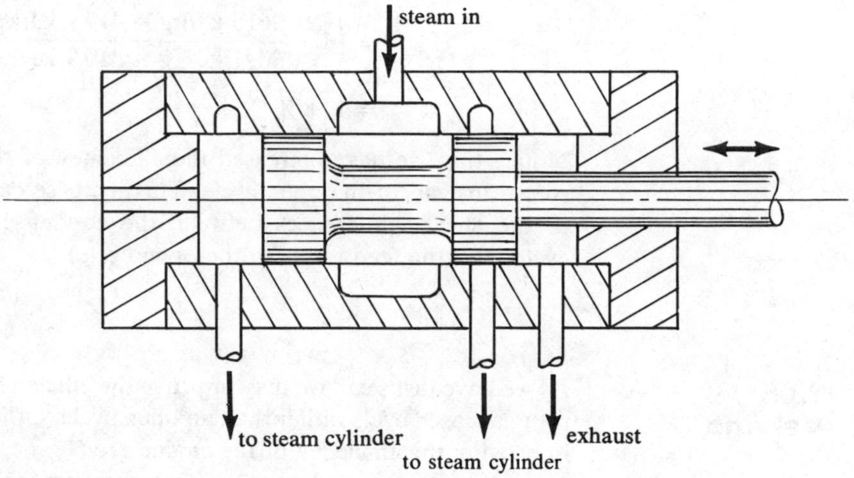

Figure 13–17 Piston valve.

The *Corliss valve* arrangement, shown in figure 13–18, represents an attempt to further reduce inefficiencies in valves. The valves *A*1, *A*2, *B*1, and *B*2 are synchronized, and with a slight rotation they can each be opened or closed. In the position shown, valves *A*1 and *B*2 are open, while the remaining two are closed, and the piston will move to the right. The devious path taken by the steam, however, in reaching the cylinder still allows for thermal inefficiencies.

The most advanced valve to be used on steam engines has been the *poppet valve* or mushroom valve, shown in figure 13–19. This valve, when used with cams and springs, as in internal combustion engines, can be very efficient and the flow path of the steam may be simplified to reduce thermal losses.

In the future, other valve designs or concepts may very well provide better control and conveyance of steam for more efficient engine operations.

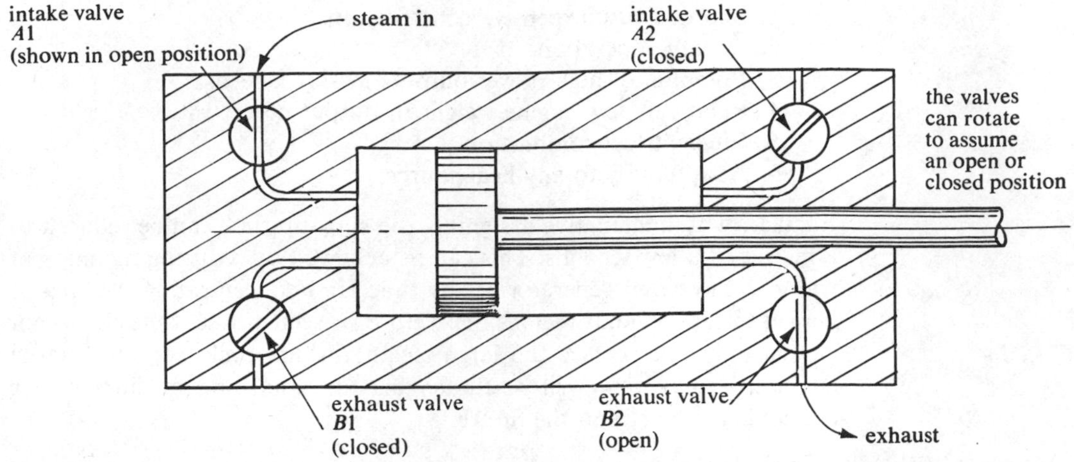

Figure 13–18 Corliss valve.

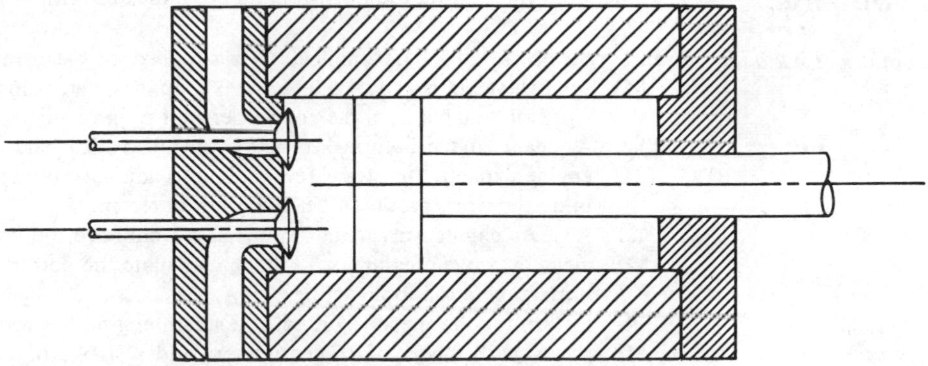

Figure 13–19 Poppet valves.

13.6 Applications

The steam engine enjoyed a long period of extensive use. It challenged and replaced the animals in power production and was the single most important heat engine for social needs for almost one hundred years. It has been used in powering ground and water transportation vehicles and various stationary power devices and in producing electric power. The reasons it was replaced by other heat engines can be listed as follows:

1. Reciprocating motion which tends to create vibrations, nonsmooth motion, and maintenance problems.
2. Low thermodynamic efficiency.
3. Inconvenience due to a slow start-up time.

The inherent advantages of the steam engine which made it one of the first developments of the industrial revolution and which could return it to the prominence are as follows:

1. Simple and inexpensive construction.
2. Reliable operation.
3. Potential of high power outputs at slow speeds.
4. Relatively low engine weight to output power ratio.
5. Efficient fuel combustion.
6. Adaptability to any heat source.

Further, with imaginative designing, the start-up time can be reduced to a matter of a few seconds. This can be achieved by reducing the mass of water in the steam generator at any time, and concentrating the area of heat transfer to small regions. Therefore, it seems that while the steam engine may not replace the large steam turbine cycles or other large engines, it may very well be the engine for small and medium power demands in vehicles of the future.

Practice Problems

Problems using SI units are indicated with the notation (M) under the problem number, and those using the English units are indicated with (E).

Sections 13.1 and 13.2

13.1. **(M)** Steam at 20.0 bars and 640°C is supplied to a steam engine. If the exhaust steam is at 1.01 bars and 90% quality, determine the maximum work obtainable from the engine per unit of mass of steam.

13.2. **(M)** A steam engine is furnished 480°C steam at 40.0 bars pressure. If the engine expands the steam reversibly and adiabatically to 0.8 bar, determine the work produced per kilogram of steam.

13.3. **(M)** A steam engine runs at 3000 rpm, has a 5-cm bore, a 6-cm stroke, and a mean effective pressure of 7.0 bars. Calculate the power produced. Assume engine is double-acting.

13.4. **(M)** Determine the speed required of a steam engine to produce 20 horsepower with a bore of 15 centimeters and a stroke of 12 centimeters. Assume the average cylinder pressure cannot exceed 8.0 bars. Assume engine is double-acting.

13.5. **(E)** Water at 60°F and 200 psia is supplied to a boiler. Determine the heat added per pound-mass water to produce steam at 200 psia and 440°F if the specific heat of water is assumed to be 1 Btu/lbm · °F between 60°F and 381°F at 200 psia.

13.6. **(E)** A boiler is supplied 2 lbm/s of saturated liquid water at 300 psia. Determine the rate of heat added to produce steam at 300 psia and 500°F.

13.7. **(M)** Determine the work per pound-mass steam produced by a steam engine in expanding steam from 150 psia and 500°F to saturated vapor at 20 psia.

13.8. **(E)** An engine diagram encloses an area of 6500 ft-lbf for a steam engine running at 1000 rpm. Determine the power produced, the mep, and the dimensions of the bore and stroke (assuming they are equal) if the volume swept by the piston is 4 ft^3.

Section 13.3

13.9. List the major components of a steam engine cycle and their functions.

13.10. Define *cutoff*.

13.11. Sketch the *p-V* and *T-s* diagrams of a steam engine.

13.12. **(E)** A nonlifting injector supplies 30 lbm of 190°F water at 100 psia to a boiler. If the water supplied to the injector is at 14.7 psia and 170°F, determine the rate of steam required if the steam is at 190 psia and 450°F.

Section 13.4

13.13. **(M)** A closed cycle, ideal steam engine with a bore of 6 centimeters and a stroke of 7 centimeters is required to deliver 95 horsepower at 1500 rpm. If steam is furnished at 20.0 bars and 800°C and is expanded reversibly and adiabatically to 1.5 bars, and the steam leaving the condenser is saturated liquid, determine the following if the engine is double-acting:

(a) Mass flow rate of steam.
(b) Thermodynamic efficiency.
(c) mep of the device.
(d) Sketch of the *p*-*V* and *T*-*s* diagrams.

13.14. **(E)** A simple, double-acting steam engine operates at 150 rpm and uses 1 lbm/s of steam. The water furnished to the boiler by an injector is at 180°F and 150 psia. The steam delivered to the engine is at 500°F and the steam is expanded reversibly and adiabatically to 14.7 psia pressure. Assume the feedwater is at 65°F and 14.7 psia. Determine

(a) Power produced.
(b) Rate of heat addition.
(c) Thermodynamic efficiency.

14

Refrigeration and Heat Pumps

In this chapter are considered the reversed heat engines and their operating cycles. The various devices that utilize the reversed heat engine cycles provide the mechanisms for today's air conditioners, refrigerators, freezers, and some heaters. Although all these devices can be classified as *heat pumps*, it is the heating devices which today are generally referred to as heat pumps.

We shall first consider the reversed Carnot cycle in order to introduce the basic concepts of refrigeration and some of the terminology associated with them and the heat pumps. Next we shall look at the reversed Brayton cycle to illustrate the mechanics of how a practical cycle might be designed. The vapor compression cycles will then be introduced and analyzed with treatments of both the dry and wet compression cycles given. Since the vapor compression is the most common of the actual heat pump cycles, example problems will be presented to demonstrate the operations of actual present-day devices. Also, various types of cycle-working media or refrigerants will be listed and briefly discussed. Another cycle which is used in actual refrigerating devices, the ammonia absorption cycle, is described. A brief description is given for the manner of gas liquifaction — one of the more recent applications of the reversed heat engines. The heat pump as applied to the problem of heating is considered and some of the operating characteristics are presented.

14.1 The Reversed Carnot Cycle

We have discussed the Carnot cycle as a heat engine operating between two temperature regions and producing work. In reverse, the Carnot cycle transfers energy from a lower temperature region to a higher temperature one; we have already been introduced to this device, called the *heat pump*. With this arrangement, the Carnot cycle is defined by the following four reversible processes:

(1–2) Isothermal heat addition.
(2–3) Adiabatic compression.
(3–4) Isothermal heat rejection.
(4–1) Adiabatic expansion to initial state (1).

The property diagrams are given in figure 14–1 and the system diagram in figure 14–2. Notice that the enclosed areas in the property diagrams must be equal for the reversible cycle so that

$$Wk_{\text{cycle}} = Q_{\text{net}} = \Sigma Q = Q_{\text{add}} + Q_{\text{rej}} \tag{14–1}$$

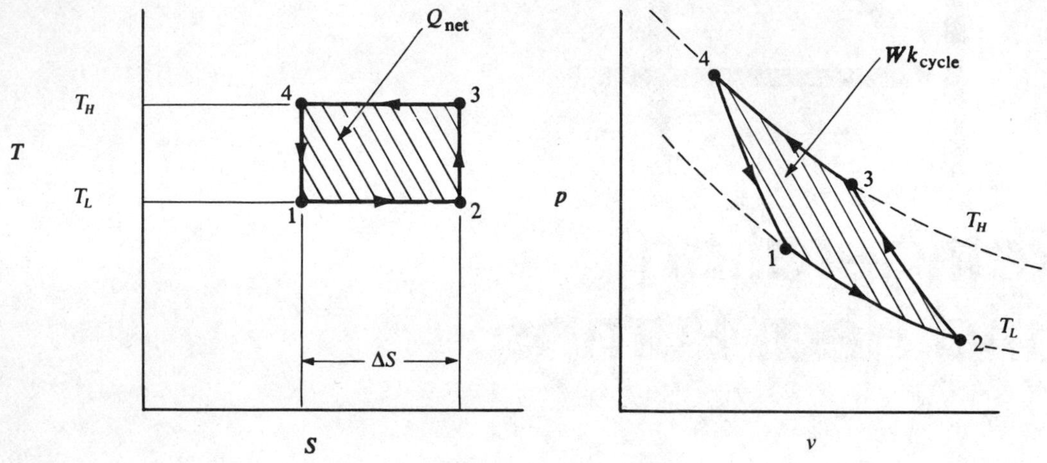

Figure 14–1 Property diagrams for Carnot heat pump.

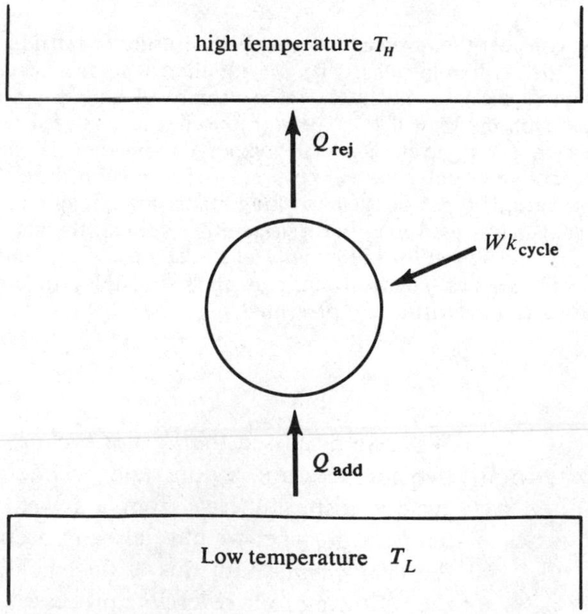

Figure 14–2 System diagram of Carnot heat pump.

For the Carnot engine we have that

$$Q_{\text{add}} = T_L \Delta S$$
$$Q_{\text{rej}} = -T_H \Delta S \qquad \textbf{(14–2)}$$

where ΔS is the absolute value of the entropy change during the isothermal heat rejection and isothermal heat addition. Then

$$Q_{\text{net}} = (T_L - T_H)\Delta S = Wk_{\text{cycle}} \qquad \textbf{(14–3)}$$

and the coefficient of performance COP defined by equation (8–22)

$$\text{COP} = \frac{\text{heat rejected}}{\text{net cycle work}} = \frac{Q_{\text{rej}}}{Wk_{\text{cycle}}}$$

reduces to

$$\text{COP} = \frac{T_H}{T_H - T_L} \qquad \textbf{(14–4)}$$

We earlier defined the coefficient of refrigeration COR by equation (8–23)

$$\text{COR} = \frac{-Q_{\text{add}}}{Wk_{\text{cycle}}}$$

which is, for the Carnot heat pump, given by

$$\text{COR} = \frac{T_L}{T_H - T_L} \qquad \textbf{(14–5)}$$

One of the common units in the English system for describing the refrigerating capacity of a heat pump device (such as an air conditioner or freezer) is the *ton*. The definition is:

1 Ton of Refrigeration: *Amount of energy removed from 1 ton of water at 32°F and 14.7 psia, in converting from a pure liquid to a solid (ice) over a period of 24 hours.*

It can be seen then that the unit ton is a description of the rate of heat transfer and we can give it further meaning by noting that 144 Btu are required to convert 1 lbm of saturated liquid H_2O at 32°F and 14.7 psia into ice at 32°F. Then

$$\begin{aligned} \text{1 ton refrigeration} &= 144 \text{ Btu/lbm} \times 2000 \text{ lbm/ton} \times 1 \text{ day/24 hr} \\ &= 12{,}000 \text{ Btu/hr} \\ &= 200 \text{ Btu/min} \\ &= 3.33 \text{ Btu/s} \end{aligned}$$

Example 14.1

A room needs to be kept at 25°C when the atmospheric temperature is 50°C. If the room requires a 4-kW air conditioner, determine the minimum amount of power required and the minimum amount of heat put in the surroundings due to this operation.

Solution

We note that the best device conceivable for air conditioning would be a reversible Carnot heat engine. For this device the coefficient of refrigeration is, from equation (14–5)

$$\text{COR} = \frac{25°\text{C} + 273 \text{ K}}{(50°\text{C} + 273 \text{ K}) - (25°\text{C} + 273 \text{ K})}$$
$$= 11.9$$

Then, since the heat added $\dot{Q}_{\text{add}}$ is given as 4 kilowatts we can write, from equation (8–23)

$$11.9 = \frac{-\dot{Q}_{\text{add}}}{\dot{W}k_{\text{cycle}}}$$

and we then get

$$\dot{W}k_{\text{cycle}} = \frac{-4 \text{ kW}}{11.9} = -0.336 \text{ kW}$$

We recall that 0.746 kW = 1.0 horsepower, so we can convert this answer to

$$\dot{W}k_{\text{cycle}} = -0.45 \text{ hp}$$ *Answer*

The amount of heat put into the surroundings is simply the heat rejected or, from equation (14–1)

$$\dot{Q}_{\text{rej}} = \dot{W}k_{\text{cycle}} - \dot{Q}_{\text{add}}$$

and substituting values into this equation gives

$$\dot{Q}_{\text{rej}} = -0.336 \text{ kW} - 4.0 \text{ kW}$$
$$= -4.336 \text{ kW}$$ *Answer*

Interestingly, although the room is cooled for a time, the net effect of an air conditioner or any other heat pump device is to increase the temperature of the surroundings or the universe, which includes the room itself. Further, there is an increasing interest in the application of the heat pump as a heating device. The ability of this mechanism to transfer heat from a cold to a hot or warmer environment represents a significant characteristic for heating.

Example 14.2

A Carnot heat pump is proposed to heat a dwelling in cold climates. If the expected minimum climatic temperature is −30°C and the dwelling is to be kept at 27°C, determine the minimum power required. It is anticipated that 10 kilowatts of heat are required.

Solution

The heat pump is used in heating when the rejected heat $\dot{Q}_{\text{rej}}$ is the desired output and is utilized to heat a dwelling or other facility. Here we use 10 kilowatts of heat for heating and by using the COP, we have

$$\frac{\dot{Q}_{\text{rej}}}{\dot{W}k_{\text{cycle}}} = \frac{T_H}{T_H - T_L}$$

Then we obtain, for the work

$$\dot{W}k_{\text{cycle}} = \frac{\dot{Q}_{\text{rej}}(T_H - T_L)}{T_H}$$

$$= \frac{(10\ \text{kW})(300\ \text{K} - 243\ \text{K})}{(27°\text{C} + 273\ \text{K})}$$

$$= 1.9\ \text{kW}$$ *Answer*

Notice that the heat pump provides more energy as heat than is supplied as work. Thus, compared to, say, electric resistance heating or an oil or gas furnace, the heat pump is more efficient. However, there are some design problems associated with the heat pump which detract significantly from its value as a heating devicc. We shall discuss this further in section 14.7.

14.2 The Reversed Brayton Cycle

While the Carnot heat pump theoretically represents a most attractive device, a practical Carnot cycle is generally not convenient. As a result, a reversed Brayton cycle (see chapter 11 for the Brayton power cycle) has often been proposed as a practical heat pump. The Brayton cycle normally functions with air or some perfect gas as a working medium so that the reversed cycle is sometimes called a *gas refrigerating* cycle. The property diagrams and the schematic of the physical components are shown in figures 14–3 and 14–4 respectively. As in the case for a Carnot heat pump, the enclosed areas on the *T-s* and *p-V* diagrams are equal if the processes are all reversible and we then have

$$Wk_{\text{cycle}} = Q_{\text{net}} = Q_{\text{rej}} + Q_{\text{add}} \tag{14-6}$$

For the ideal gas refrigeration cycle, the processes (2–3) and (4–1) are reversible adiabatic, but in actual applications they are polytropic pro-

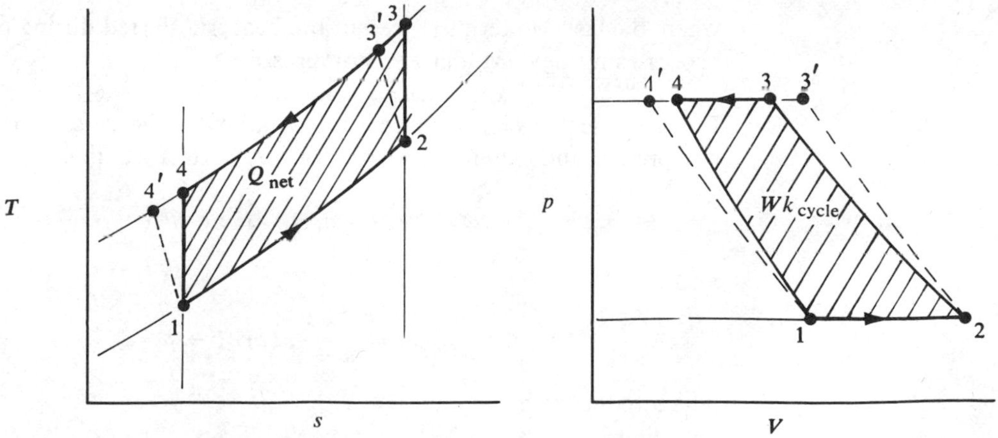

Figure 14–3 Property diagrams of ideal gas refrigeration cycles.

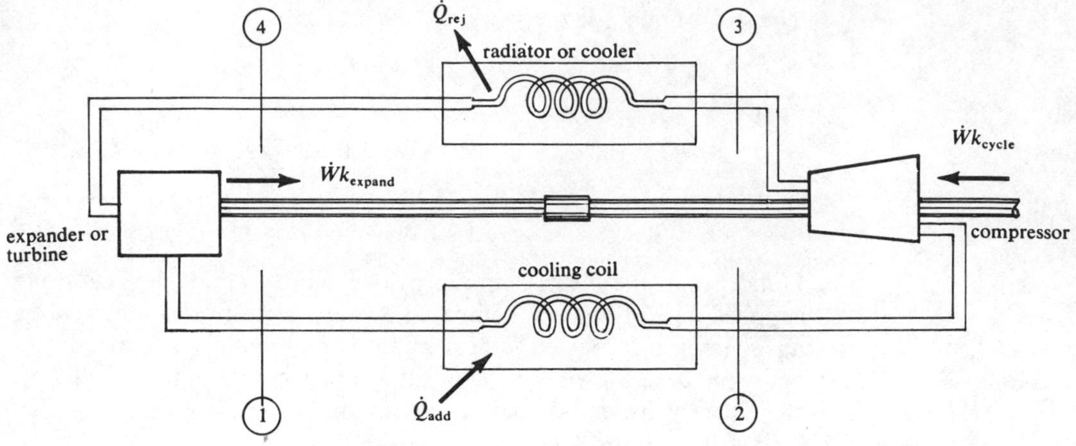

Figure 14–4 Schematic of gas refrigeration cycle.

cesses, slightly different from the adiabatic case. For this deviation we may visualize the processes to be (1–2), (2–3′), (3′–4′), and (4′–1) as shown in figure 14–3. Quite generally,

$$Wk_{cycle} = Wk_{expand} + Wk_{comp} \tag{14–7}$$

and if the gas circulating through the device is a perfect gas we may use the polytropic relations from chapter 6

$$Wk_{comp} = \frac{-n}{n-1}(p_3V_3 - p_2V_2) \tag{14–8}$$

$$Wk_{expand} = \frac{-n}{n-1}(p_1V_1 - p_4V_4) \tag{14–9}$$

to calculate the cyclic work. Also, the net heat of the cycle will be given by

$$Q_{net} = Q_{add} + Q_{rej} + Q_{expand} + Q_{comp} \tag{14–10}$$

where the last two terms represent the heat transferred during polytropic processes in the expander and compressor.

For the ideal cycle, with reversible adiabatic processes through the compressor and expander, we get for any perfect gas that

$$Wk_{comp} = \frac{-k}{k-1}(p_3V_3 - p_2V_2) \tag{14–11}$$

and

$$Wk_{expand} = \frac{-k}{k-1}(p_1V_1 - p_4V_4) \tag{14–12}$$

which reduce to

$$Wk_{comp} = mc_p(T_3 - T_2) \tag{14–13}$$

and

$$Wk_{\text{expand}} = mc_p(T_1 - T_4) \quad \textbf{(14–14)}$$

for perfect gases with constant specific heats.

Example 14.3 An ideal Brayton cycle describes the operation of a 10-kW air conditioner which keeps a room at 27°C while the surrounding atmospheric temperature is 42°C. Assume the radiator and cooling coil operate with perfect heat transfer ($t_2 = 27°C$ and $t_4 = 42°C$) and the pressure ratio across the compressor is 15 to 1. Determine the rate of heat addition from the room, the rate of heat rejection to the atmosphere, the power required, the mass flow rate of air, the COP, and the COR.

Solution The rate of heat addition from the room, $\dot{Q}_{\text{add}}$, is identified as the capacity of the air conditioner:

$$\dot{Q}_{\text{add}} = 10 \text{ kW}$$ *Answer*

From this result and the equation

$$\dot{Q}_{\text{add}} = \dot{m}q_{\text{add}} = \dot{m}(h_2 - h_1) = \dot{m}c_p(T_2 - T_1) \quad \textbf{(14–15)}$$

we can obtain the mass flow rate. First we note that, for the reversible adiabatic expansion

$$\frac{T_4}{T_3} = \left(\frac{p_4}{p_3}\right)^{(k-1)/k}$$

and then

$$T_1 = T_4\left(\frac{p_1}{p_4}\right)^{(k-1)/k} = (42°\text{C} + 273 \text{ K})(1/15)^{0.4/1.4}$$

$$= 145 \text{ K}$$

In this computation k was assumed to have a value of 1.4. From this result and assuming a value of 1.007 kJ/kg · K for c_p, we obtain from equation (14–15) that

$$10 \text{ kW} = 10 \text{ kJ/s} = \dot{m}(1.007 \text{ kJ/kg K})(300 \text{ K} - 145 \text{ K})$$

$$\dot{m} = \frac{(10 \text{ kJ/s})}{(1.007 \text{ kJ/kg} \cdot \text{K})(155 \text{ K})}$$

$$= 0.064 \text{ kg/s}$$ *Answer*

The rate of heat rejection to the surroundings $\dot{Q}_{\text{rej}}$, is obtained from

$$\dot{Q}_{\text{rej}} = \dot{m}q_{\text{rej}} = \dot{m}c_p(T_4 - T_3) \quad \textbf{(14–16)}$$

First we calculate the temperature T_3 from

$$\frac{T_3}{T_2} = \left(\frac{P_3}{P_2}\right)^{(k-1)/k} \quad \text{or} \quad T_3 = T_2\left(\frac{P_3}{P_2}\right)^{(k-1)/k}$$

so

$$T_3 = (300\ \text{K})(15)^{0.4/1.4} = 650\ \text{K}$$

Then

$$\dot{Q}_{rej} = (0.064\ \text{kg/s})(1.007\ \text{kJ/kg}\cdot\text{K})(315\ \text{K} - 650\ \text{K})$$
$$= -21.6\ \text{kJ/s} = -21.6\ \text{kW} \qquad \textit{Answer}$$

From equation (14–6) we compute the net power of the cycle:

$$\dot{W}k_{cycle} = \dot{Q}_{rej} + \dot{Q}_{add}$$
$$= -21.6\ \text{kW} + 10\ \text{kW} = -11.6\ \text{kW} \qquad \textit{Answer}$$

The coefficient of performance, COP, is

$$\text{COP} = \frac{-21.6\ \text{kW}}{-11.6\ \text{kW}} = 1.86 \qquad \textit{Answer}$$

and the coefficient of refrigeration, COR, is

$$\text{COR} = \frac{-10\ \text{kW}}{-11.6\ \text{kW}} = 0.86 \qquad \textit{Answer}$$

Notice that the Carnot heat pump operating between 27°C and 42°C would have COP and COR values of

$$\text{COP} = \frac{315\ \text{K}}{315\ \text{K} - 300\ \text{K}} = 21.0$$

and

$$\text{COR} = \frac{300\ \text{K}}{315\ \text{K} - 300\ \text{K}} = 20.0$$

A quick comparison of the gas refrigeration cycle and the Carnot heat pump reveals that the COP and COR values are much lower for the gas refrigeration cycle. This is the primary reason that the reversed Brayton or gas refrigeration cycle is not more attractive. It can be used to refrigerate at very low temperatures and it is flexible enough to be used with various working media, but the low efficiency, as reflected in the COP and COR values, detracts from its popularity.

14.3 The Vapor Compression Cycle

The Carnot heat pump represents the ultimate in refrigeration cycles. A practical application of the ideal cycle is the vapor compression cycle, defined by the following processes:

(1–2) Constant pressure heat addition during a phase change of the working medium, or refrigerant.

(2–3) Adiabatic compression.

(3–4) Constant pressure heat rejection.
(4–1) Throttling expansion at constant enthalpy.

If the cycle is considered reversible, then all these processes must also be reversible. The typical pressure-volume diagram for the ideal reversible vapor compression cycle is depicted in figure 14–5. The corresponding temperature-entropy diagram is shown in figure 14–6 (a) and (b). Notice that a vapor compression can be performed with dry vapor (superheated) or a mixture of saturated vapor and liquid. As the descriptions of the cycles imply, the dry compression cycle involves a compression process

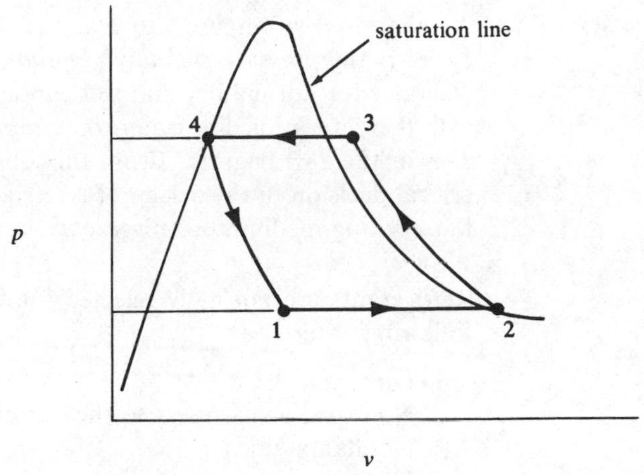

Figure 14–5 *p-v* diagram for ideal vapor compression refrigeration cycle.

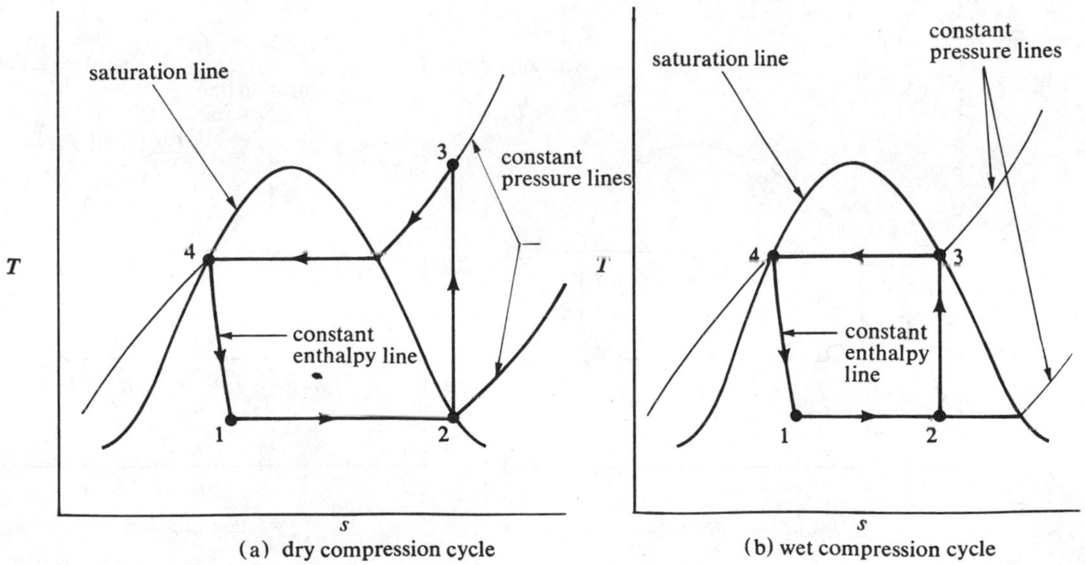

Figure 14–6 *T-s* diagram for ideal refrigeration cycles.

(2–3) with a dry vapor, while the wet vapor compression cycle involves a mixture of vapor and liquid through the compression.

The dry compression cycle seems to be more popular in application to actual refrigeration devices even though the wet compression more closely approximates the reversed Carnot cycle; that is, the COP of the wet compression cycle would be expected to exceed the COP of the dry compression, both operating between the same pressures. The reason for the success of the dry compression cycle is that compressors typically perform better with a pure vapor than with a mixture of vapor and liquid.

Another tool often used to evaluate and analyze vapor compression cycles is the *pressure-enthalpy diagram*. In figure 14–7 are shown these diagrams for normal dry and wet vapor compression refrigeration cycles with the saturation line again determining the limits of the cycle as it does in the *T-s* diagram. From this observation, we may expect that a critical decision in the design of a refrigeration unit is in the selection of the working medium or *refrigerant*.

Refrigerants are normally selected for vapor compression cycles by the following criteria:

1. Economical.
2. Nontoxic or harmless to the surroundings.
3. Nonflammable.
4. High latent heat of condensation (h_{fg}) at the refrigerating temperature.
5. Low saturation pressure at operating temperature.

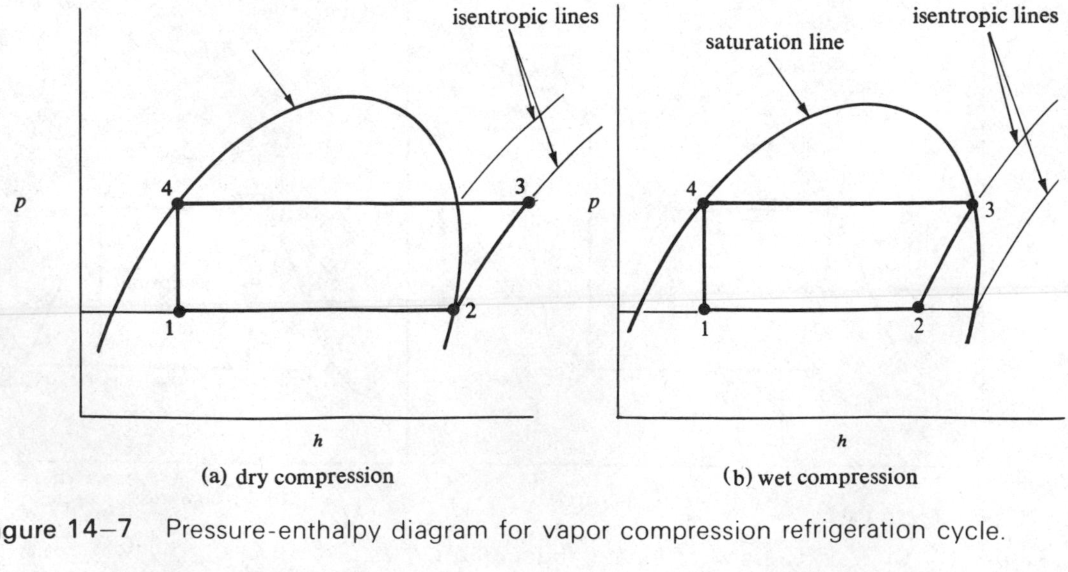

Figure 14–7 Pressure-enthalpy diagram for vapor compression refrigeration cycle.

Thermodynamically, criterion (4) represents the most significa of working media. A high latent heat of condensation reflects a cap for a high amount of heat addition to the cycle refrigerant per uni refrigerant mass. In table 14–1 are listed the heats of condensation h_{fg} fo some common refrigerants at typical saturation pressures, in order that a quick comparison can be made. Remember, however, that the heat of vaporization alone should not determine the most desirable refrigerants. Even some properties other than those listed above, such as viscosity, solubility, or thermal conductivity, could enter into the requirements for selecting a working medium for refrigeration cycles.

Table 14–1

Heat of Vaporization of Some Common Refrigerants

		Saturation Pressure		*Saturation Temperature*		h_{fg}	
Refrigerant	*Formula*	bars	psia	°C	°F	kJ/kg	Btu/lbm
Ammonia	NH_3	3.32	48.21	−6	20	1240	533.1
Sulfur dioxide	SO_2	1.18	17.18	−6	20	384	165.3
Carbon dioxide	CO_2	10.05	145.8	−40	−40	317	136.5
Methyl chloride	CH_3Cl	2.02	29.3	−6	20	412	178.4
R-12	CCl_2F_2	2.46	35.7	−6	20	158	67.9
R-22	$CHClF_2$	2.40	34.7	−20	−5	221	94.9
R-22	$CHClF_2$	3.98	57.7	−6	20	211	90.5

The most common substances used for refrigeration processes are as follows:

ammonia
sulfur dioxide
carbon dioxide
methyl chloride
dichlorodifluoromethane (*refrigerant-12, R-12*)
chlorodifluoromethane (*refrigerant-22, R-22*)
propane
butane
dichloromethane (*carrene*)

In appendix B are tables of thermodynamics properties of ammonia and Refrigerant-22 for quantitative analysis of the vapor compression cycles.

The typical physical components that constitute a vapor compression refrigeration cycle are shown in figure 14–8. Here we see that an evaporator or cooling coil absorbs energy or accepts heat addition from an external region by utilizing a working medium during its phase

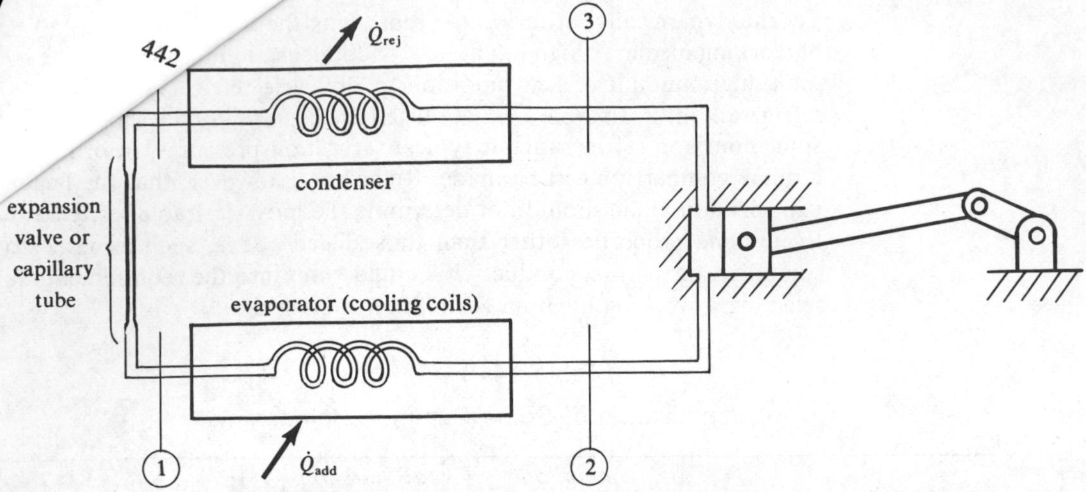

Figure 14–8 Sketch of typical vapor compression refrigerator.

change. (See figure 14–6 or 14–7.) The refrigerant collects energy and is a saturated vapor (or near to that state) as it enters the compressor. The compressor increases the pressure and, more importantly, the temperature of the refrigerant as it leaves. Here the working medium is probably a superheated vapor or, in the case of a wet vapor compression, a saturated vapor. The condenser allows the refrigerant to release much of its energy in a heat rejection. Then the refrigerant returns through an expander of restriction to the evaporator.

By applying the steady flow energy equation to the various components, we get (making appropriate assumptions on the individual items)

1. for the evaporator:

$$h_2 - h_1 = q_{add} \tag{14–17}$$

2. for the compressor:

$$h_3 - h_2 = -wk_{comp} \tag{14–18}$$

3. for the condenser:

$$h_4 - h_3 = q_{rej} \tag{14–19}$$

4. for the expander:

$$h_1 - h_4 = 0 \tag{14–20}$$

Then, for the complete cycle we apply equation (8–22)

$$\text{COP} = \frac{\dot{Q}_{rej}}{\dot{W}k_{cycle}} = \frac{q_{rej}}{wk_{cycle}}$$

which can be written as

$$\text{COP} = \frac{h_4 - h_3}{h_2 - h_3} \qquad (\ldots$$

for the vapor compression cycle. Also, the coefficient of refrigeration i

$$\text{COR} = -\frac{h_2 - h_1}{h_2 - h_3} \qquad \textbf{(14–22)}$$

Notice the difference between these last two equations and those applicable to the Carnot heat pump (equations 14–4 and 14–5).

14.4 Applications of the Vapor Compression Cycle

Attention is given here to the details of the application of the thermodynamic concepts to actual devices. First we will look at a refrigeration device which can be described by a wet vapor compression cycle, followed by a similar arrangement which is best described by a dry compression cycle.

Example 14.4

A refrigerator operates with ammonia as the working medium and the system removes 1000 Btu/hr from a freezer compartment at 26°F. If the heat is rejected to a room which is at 85°F and the refrigerator can be described by an ideal wet vapor compression cycle with saturated vapor ammonia leaving the compressor, determine the following: (1) COP and COR, (2) the flow rate of the ammonia through the cycle, and (3) the power required.

Solution

This refrigerator can be described by the sketch given in figure 14–8 and the property diagrams for the cycle as those given in figures 14–5, 14–6(b), and 14–7(b).

1. From the ammonia table, table B.11 in the appendix, we obtain directly, or through interpolation, the values

$$h_3 = h_g \text{ @ } 85°\text{F} = 631.4 \text{ Btu/lbm}$$
$$h_4 = h_f \text{ @ } 85°\text{F} = 137.8 \text{ Btu/lbm}$$
$$h_1 = h_4 \text{ and } s_3 = 1.1919 \text{ Btu/lbm} \cdot °\text{R}$$

Since $s_2 = s_3$, we can write

$$s_2 = \chi(s_{g2} - s_{f2}) + s_{f2} = 1.1919$$

Also, from table B.11 we have, since $t_2 = 26°\text{F}$

$$s_{g2} = 1.2861$$
$$s_{f2} = 0.1573$$

...nd then we can calculate the quality χ

$$\chi = \frac{1.1919 - 0.1573}{1.2761 - 0.1573} = 0.925$$

Hence

$$h_2 = \chi(h_{fg2}) + h_{f2}$$
$$= (0.925)(548.1) + 71.3 = 578.2$$

then, from equation (14–17)

$$q_{\text{add}} = h_2 - h_1 = 578.2 - 137.8 = 440.4 \text{ Btu/lbm}$$

and from equation (14–19)

$$q_{\text{rej}} = h_4 - h_3 = 137.8 - 631.4 = -493.6 \text{ Btu/lbm}$$

From equations (14–21) and (14–22) we get

$$\text{COP} = \frac{-493.6}{578.2 - 631.4} = 8.64$$ *Answer*

and

$$\text{COR} = -\frac{440.4}{578.2 - 631.4} = 7.71$$ *Answer*

2. From the given condition that

$$\dot{Q}_{\text{add}} = 1000 \text{ Btu/lbm}$$

we can then write

$$\dot{Q}_{\text{add}} = \dot{m}q_{\text{add}} = \dot{m}(440.4 \text{ Btu/lbm})$$

Then

$$\dot{m} = \frac{1000 \text{ Btu/hr}}{440.4 \text{ Btu/lbm}} = 2.27 \text{ lbm/hr}$$ *Answer*

3. The power required can be determined from the relationship

$$\dot{W}k_{\text{cycle}} = \dot{m}wk_{\text{cycle}}$$
$$= (2.27 \text{ lbm/hr})(578.2 - 631.4 \text{ Btu/lbm})$$
$$= 120.8 \text{ Btu/hr}$$

or

$$\dot{W}k_{\text{cycle}} = (120.8 \text{ Btu/lbm})(1/2545 \text{ Btu/hp} \cdot \text{hr})$$
$$= 0.047 \text{ hp}$$ *Answer*

Example 14.5 An air conditioner operates in 45°C weather to keep a room at 20°C by withdrawing 12,000 kJ/hr of heat. Assuming that the evaporator and condenser have perfect heat conduction, that the cycle is an ideal dry compression cycle, and that the working media is Refrigerant-22, deter-

mine the rate of heat rejected to the atmosphere, the po
and COP and COR.

Solution The air conditioner is physically describable by the sketch in
14–8 and the operating cycle is depicted by the property diagrams
figures 14–7(a), 14–6(a), and 14–5. To determine the heat rejection, the power, and the coefficient of performance, we must determine the enthalpy values. From table B.9 in the appendix we obtain the following values:

$$h_2 = h_g \text{ @ } 20°\text{C} = 411.9 \text{ kJ/kg}$$

$$s_2 = 1.72462 \text{ kJ/kg}$$

and the pressure at state (3) must be the saturation pressure at 45°C or 17.29 bars. Then we may obtain h_3 from a double interpolation. First we notice in table B.10 that 17.29 bars pressure lies between listed values of 15.71 and 18.98 bars. The entropy of 1.72462 kJ/kg · K lies between 55°C and 70°C at 18.98 bars. Interpolating at 55°C and 70°C we find the following enthalpies and entropies:

p, bars	55°C	70°C
17.29	$h = 426.9$	440.7
	$s = 1.724$	1.764

We see that the entropy at 55°C corresponds closely to the value of 1.72462. We then write $h_3 = 426.9 \approx 427$ kJ/kg. Also,

$$h_4 = h_f \text{ @ } 45°\text{C} = 256.4 \text{ kJ/kg}$$

and

$$h_1 = h_4$$

We can now proceed to compute the required terms. The heat rejected can be obtained from equation (14–19):

$$q_{\text{rej}} = h_4 - h_3 = 256.4 \text{ kJ/kg} - 427 \text{ kJ/kg}$$
$$= -170.6 \text{ kJ/kg}$$

and the rate of heat rejection is then

$$\dot{Q}_{\text{rej}} = \dot{m}q_{\text{rej}}$$

The mass flow rate is calculated from

$$\dot{Q}_{\text{add}} = \dot{m}q_{\text{add}} = \dot{m}(h_2 - h_1) = 12{,}000 \text{ kJ/h}$$

or

$$\dot{m} = \frac{12{,}000 \text{ kJ/hr}}{411.9 \text{ kJ/kg} - 256.4 \text{ kJ/kg}}$$
$$= 77.2 \text{ kg/hr}$$

$$\dot{Q}_{rej} = (77.2 \text{ kg/hr})(-170.6 \text{ kJ/kg})$$
$$= -13{,}170 \text{ kJ/hr} \qquad \textit{Answer}$$

The power is computed from

$$\dot{W}k_{cycle} = \dot{m}wk_{cycle} = \dot{m}(h_2 - h_3)$$
$$= (77.2 \text{ kg/hr})(411.9 \text{ kJ/kg} - 427 \text{ kJ/kg})$$
$$= -1170 \text{ kJ/hr} \qquad \textit{Answer}$$

We may write this answer in more familiar units:

$$\dot{W}k_{cycle} = -0.325 \text{ kW} \qquad \textit{Answer}$$

From equations (14–21) and (14–22) we calculate

$$\text{COP} = \frac{-13{,}170 \text{ kJ/hr}}{-1170 \text{ kJ/hr}}$$
$$= 11.26 \qquad \textit{Answer}$$

and

$$\text{COR} = \frac{-12{,}000 \text{ kJ/hr}}{-1170 \text{ kJ/hr}}$$
$$= 10.26 \qquad \textit{Answer}$$

14.5 Ammonia Absorption Refrigeration

The ammonia absorption cycle represents an attempt to reduce the mechanical work required to compress refrigerants and replace this energy demand by a heat transfer process. The typical cycle, shown in figure 14–9, indicates that the condenser, the expansion value, and the evaporator are components having identical functions in the ammonia absorption and vapor compression cycles. That is, the condenser serves to remove heat from the system; the expansion valve allows isenthalpic (constant enthalpy) expansion of the ammonia to a low pressure; and the evaporator allows for heat addition to the low pressure ammonia from a region being refrigerated or cooled. In the ammonia absorption cycle, the compressor is replaced by an auxiliary water system. The absorber receives a low pressure mixture predominately of water with some ammonia (weak liquor) and low pressure ammonia vapor. These two fluids are mixed together in the absorber so that the ammonia is "absorbed" or dissolved in the weak liquor. The fluid flowing out of the absorber is then a strong solution of water and ammonia (strong liquor) and is pumped to a generator under a high pressure. Heat is supplied to the generator by steam, electric coils, or other appropriate means, which increases the temperature of the water-ammonia liquor. A con-

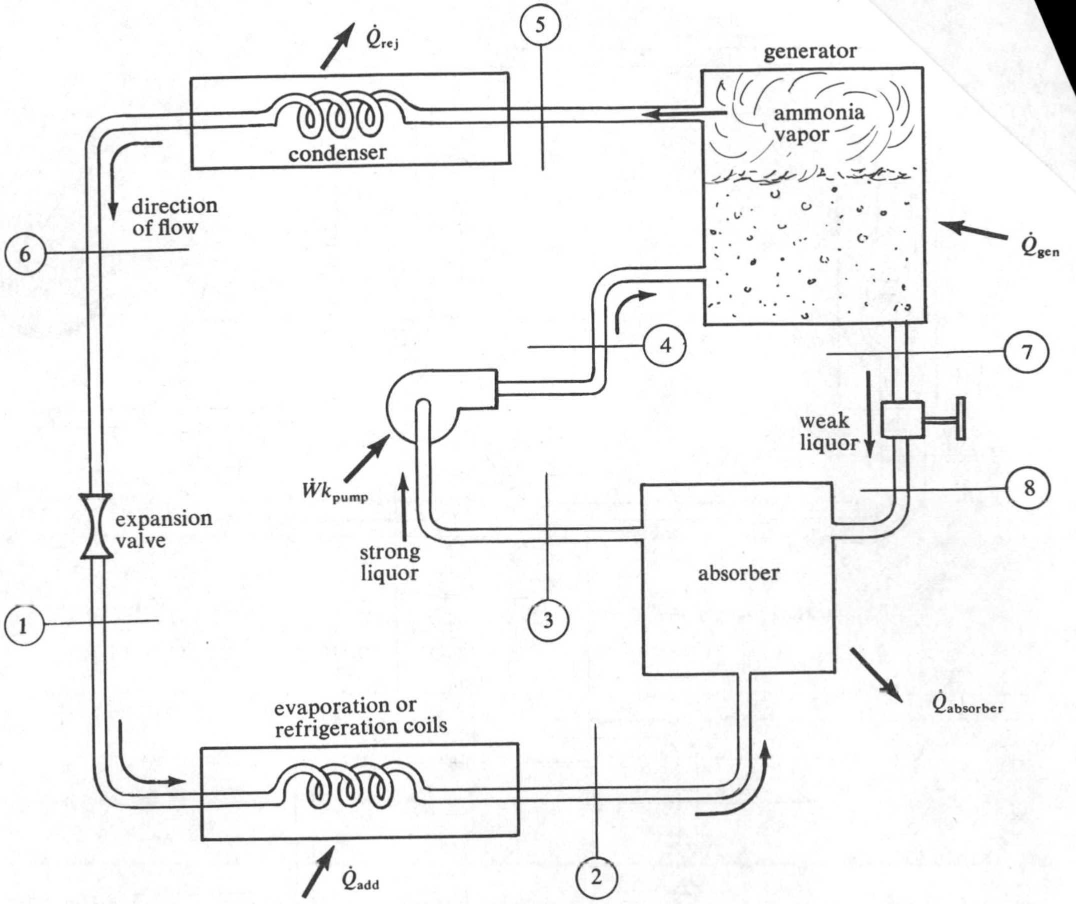

Figure 14–9 Ammonia absorption refrigeration cycle.

sequence of this is that ammonia vapor boils off and recirculates to the condenser, while the remaining liquor returns to the absorber. All of this operates because ammonia is more soluble in cold water than in hot water, and the absorber and the generator provide the means of taking advantage of this phenomenon.

For a complete analysis of the ammonia absorption refrigeration cycle, the reader is referred to publications specifically concerned with refrigeration.

14.6 Gas Liquefaction

One of the most recent applications of the heat pump is to the *liquefaction* or condensation of gases. This requires that the gases be brought to very low temperatures so that the phase change from gas to liquid may

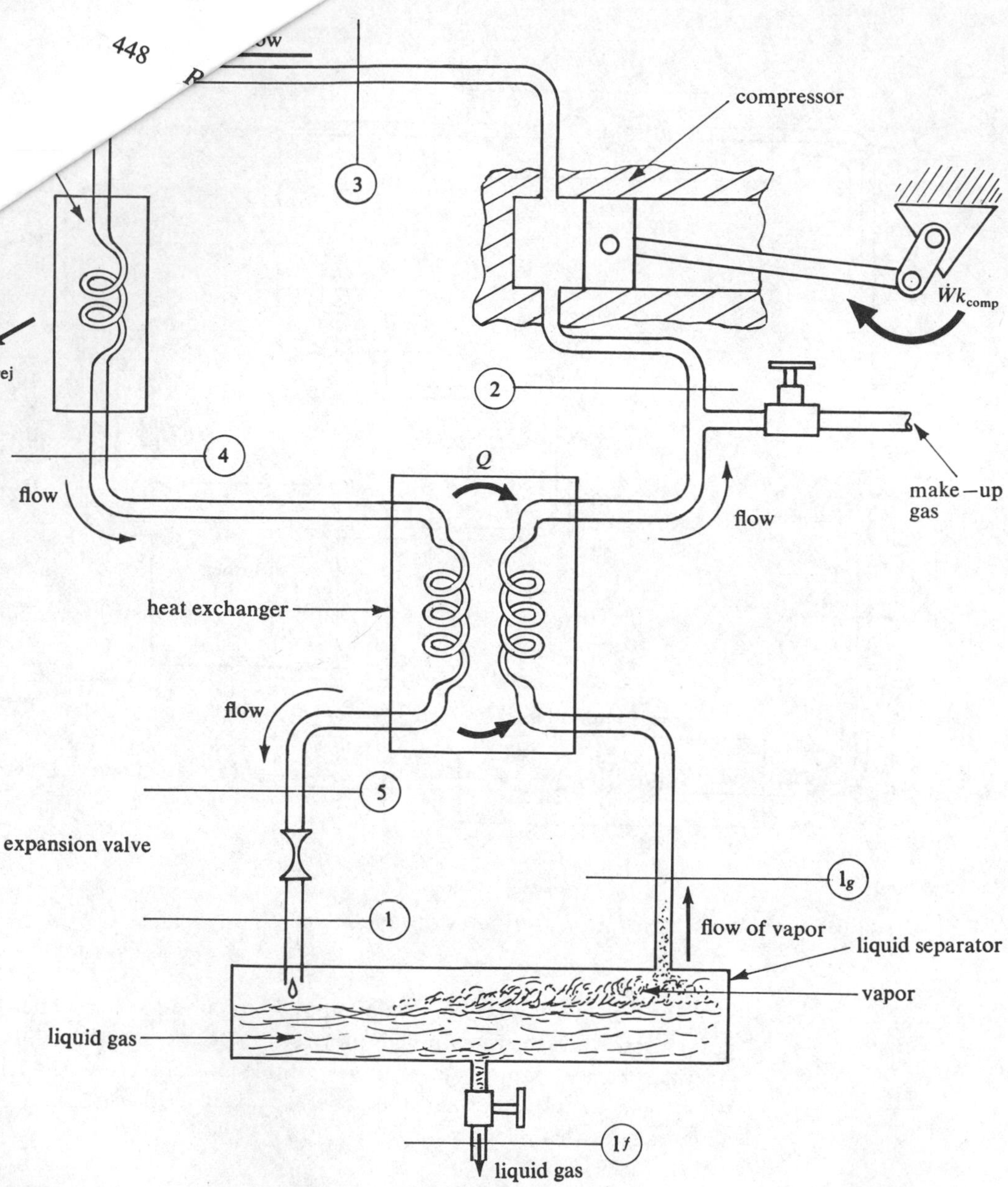

Figure 14–10 Schematic of gas-liquefaction process.

begin. Table 14–2 lists the boiling temperature t_B or temperature at which the listed gas exhibits the liquid-gaseous phase change at 1 atmosphere of pressure (1.01 bars) and it can easily be seen that very low temperatures are required.

The most common and typical method of producing liquid gases (such as oxygen, hydrogen, and air) is depicted in figure 14–10. Here the gas

Table 14–2

Saturation Temperature of Gases at 1.01 bars (14.7 psia)

Substance	*Saturation Temperature* °C	°F
Air	−194.2	−317.6
Helium	−213.3	−352.0
Hydrogen	−252.7	−422.9
Nitrogen	−195.8	−320.4
Oxygen	−183.0	−297.4
Chlorine	−34.6	−30.3
H_2O (water)	100.0	212.0

Note: Condensed from the International Critical Tables with permission of the National Academy of Sciences. Revised to SI units by author.

itself is a working medium (refrigerant) and is converted to a saturated liquid through an expansion valve after losing much of its energy through heat transfer at a high pressure and a relatively high temperature. The heat exchanger has the sole function of transferring energy from the high pressure gas to the low pressure gas through heat transfer, but frequently more than one heat exchanger is used in the actual cycle.

In figure 14–11 is presented the temperature-entropy diagram of the gas liquefaction process, assuming reversible and ideal situations for all the components. In this diagram the temperature at state (4) (entering the high pressure side of the heat exchanger) is indicated as being equal to

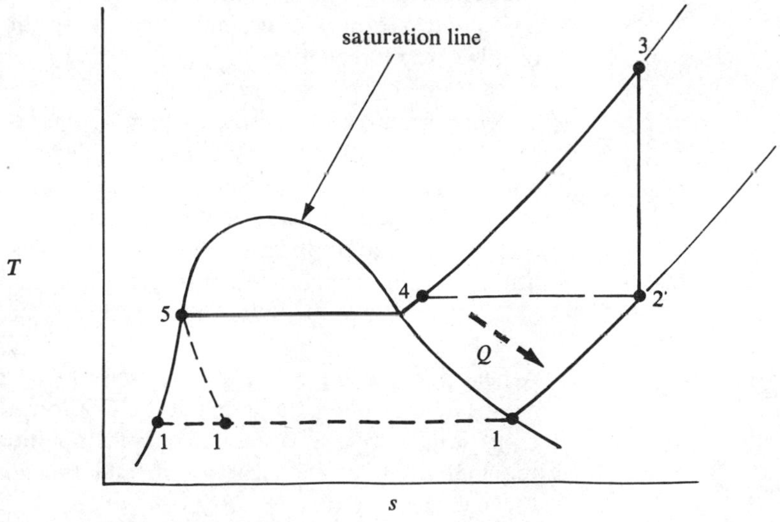

Figure 14–11 *T-s* diagram of gas liquefaction.

that at state (2), but in actual cases T_4 must be greater than T_2 if heat transfer is to be effected over a finite period of time. Also, at state (5) the gas may not be a saturated liquid but may be a mixture of liquid and vapor, a saturated vapor, or even somewhat superheated. However, if the whole cycle is to have meaning, the expansion (5–4) through the expansion valve must result in a mixture of liquid and vapor. This allows for the separation of the mixture and appearance of liquid gas at low pressure.

14.7 The Heat Pump

The use of thermodynamic devices to transfer heat from one environment to another has been a significant contribution to the modern life style. The previous sections in this chapter have been concerned mainly with the application of transferring heat out of an area so that the temperature there may be lowered. In this section we will discuss the application of heating or transferring heat into an area, such as a dwelling or chamber, from a cooler external environment. Today these devices are referred to as *heat pumps*, although we have used this term for all reversed heat engines.

The heat pump can be designed to operate as an air conditioner in warm seasons and a heating unit during the colder seasons. The device has the capability of providing more energy in the form of heat than is put into the system as work. Therefore, it represents a very attractive alternative to the common electric resistance heating. The primary difficulty associated with the heat pump is its inability to function well over a wide range of temperatures. Thus, if it is designed to operate in moderate weather, it is generally inadequate for very cold climates and conversely, a design for cold weather use would be uneconomical and even unacceptable in warmer weather. Thus, as a total heating system the heat pump requires some major technological improvements which, as of now, are unavailable.

The heat pump can operate on any of the refrigeration cycles and use most of the common refrigerants. The devices are commonly driven by an electric motor or similar power mechanism. The rejected heat $\dot{Q}_{rej}$ from the cycle represents the heating effect produced by the heat pump and in a typical application, heat is transferred from a low temperature atmosphere to a high temperature dwelling.

Example 14.6

A heat pump using a dry vapor compression cycle with Refrigerant-22 as the working media is rated at 6 kilowatts and has an optimum condition of 20°C at 0°C outside air temperature. Determine the power required to drive the heat pump and the flow rate of the refrigerant through the system.

Solution

This device has a dry vapor compression cycle, so we may obtain the enthalpy values at the four corners of the cycle as indicated in figure 14–7. From table B.8 we find $h_4 = h_1 = 224$ kJ/kg, which is the enthalpy of saturated liquid at 20°C. The pressure here, at state (4), and at state (3) is 9.1 bars. We assume ideal heat transfer in the heat pump and, further, find $h_2 = h_g$ at 0°C = 405.4 kJ/kg. At 0°C the pressure is 4.98 bars and the entropy at state (2) is s_g (0°C) or 1.75179 kJ/kg · K. At state (3) the entropy must also have this value so, from table B.9, and using $p_3 = 9.1$ bars pressure and $s_3 = 1.75179$ kJ/kg · K, we interpolate to find $h_3 = 420$ kJ/kg. We may then use equation (14–18) to determine the required work. To determine the power, we use equation (14–19) to first obtain the mass flow rate:

$$\dot{m}(h_4 - h_3) = \dot{Q}_{\text{rej}} = -6.0 \text{ kW}$$

since the rejected heat represents the heat rating. Then

$$\dot{m} = \frac{-6.0 \text{ kW}}{224 \text{ kJ/kg} - 420 \text{ kJ/kg}}$$

$$= 0.0306 \text{ kg/s}$$ *Answer*

The power is then

$$\dot{W}k_{\text{cycle}} = \dot{m}(h_2 - h_3)$$
$$= (0.0306 \text{ kg/s})(405.2 \text{ kJ/kg} - 420 \text{ kJ/kg})$$
$$= -0.45 \text{ kJ/s} = -0.45 \text{ kW}$$ *Answer*

Example 14.7

A heat pump is designed for 10°F climate and to deliver 50,000 Btu/hr into a 70°F dwelling. If this device operates on a wet compression cycle with Refrigerant-22 as the working media, determine the flow rate of the refrigerant through the system and the flow rate required if the outside temperature drops to −20°F and the 50,000 Btu/hr is still required. Assume the temperature in the dwelling is 60°F at the lower outside temperature.

Solution

For the wet compression heat pump cycle we refer to figure 14–7, and the enthalpies from table B.8 are

$$h_1 = h_4 = 30.116 \text{ Btu/lbm} = h_f \text{ @ } 70°\text{F}$$

$$h_3 = 110.414 \text{ Btu/lbm} = h_g \text{ @ } 70°\text{F}$$

The flow rate may then be computed from the relationship

$$\dot{m}(h_4 - h_3) = \dot{Q}_{\text{rej}}$$
$$= 50{,}000 \text{ Btu/hr}$$

so

$$\dot{m} = \frac{(50{,}000\ \text{Btu/lbm})}{(30.116\ \text{Btu/lbm} - 110.414\ \text{Btu/lbm})}$$

$$= -623\ \text{lbm/hr}$$ *Answer*

At the colder condition, we find

$$h_1 = h_4 = 27.172\ \text{Btu/lbm} = h_f\ @\ 60°\text{F}$$

$$h_3 = 109.7\ \text{Btu/lbm} = h_g\ @\ 60°\text{F}$$

and

$$\dot{m} = \frac{(50{,}000\ \text{Btu/hr})}{(27.172\ \text{Btu/lbm} - 109.7\ \text{Btu/lbm})}$$

$$= -606\ \text{lbm/hr}$$ *Answer*

Practice Problems

Problems that use SI units are indicated with the notation (M) under the problem number, and those using the English units are indicated with (E). Those using mixed units are marked with (C) for combined under the problem number.

Section 14.1

14.1. **(M)** A Carnot heat pump operates between a high temperature region of 80°C and a low temperature region of −10°C. Determine the COP and COR.

14.2. **(M)** A reversible Carnot heat engine receives heat at −5°C and requires 0.1 kilowatt for each kilowatt of heat received. Determine the COP.

14.3. **(M)** For the heat pump of problem 14.2 determine
(a) Temperature of sink for rejected heat.
(b) COR.

14.4. **(M)** A 200-kW Carnot freezer operates between −10°C and 30°C. Determine the rate of heat rejected to the atmosphere and the power required.

14.5. **(E)** A reversed Carnot cycle operates between 100°F and 10°F. Determine the COR and COP.

14.6. **(E)** A Carnot air conditioner has a COP of 4 and a capacity of 5 tons. Determine the power requirement of this device.

14.7. **(E)** A Carnot freezer is used to freeze water at 14.7 psia. What is the maximum COR achieved by this device if the minimum atmospheric temperature is 80°F?

14.8. Sketch the *p-V*, *T-s*, and *h-s* diagrams of the ideal Carnot cycle.

Section 14.2

14.9. A reversed ideal Brayton cycle refrigerator operates with a pressure ratio of 18 to 1. Determine the COP and COR.

14.10. **(M)** An ideal gas refrigerator keeps a freezer at −10°C when operating and rejects heat to a surrounding at 35°C. The freezer unit requires a $\frac{1}{2}$-hp electric motor to run a compressor and the pressure ratio of the compressor is 16 to 1 assuming air to be the working medium. Determine
(a) Capacity rating of the freezer in kilowatts.
(b) COP.

14.11. For the unit of problem 14.10, determine
(M) (a) COR.
(b) Rate of heat rejected to surrounding.
(c) Mass flow rate of air.

14.12. A reversed ideal Brayton cycle refrigerator uses 5000 lbm/hr of air with
(E) a pressure ratio of 10 to 1. If the refrigerator is rated at 2 tons, determine the power required by the unit.

14.13. An air refrigerating machine uses 100 lbm/min of air to cool a storage
(E) room. The air is 75°F as it leaves the radiator and 30°F as it enters the compressor. Assume the air is reversibly and adiabatically compressed from 15 psia to 55 psia in the compressor and expanded reversibly and adiabatically through the turbine. Determine
(a) Temperature of air entering the radiator.
(b) Temperature of air entering cooling coils.
(c) $\dot{Q}_{net}$
(d) COP and COR.
(e) Net work rate to cycle.

14.14. Use the same cycle as in problem 14.13 except that the compression of
(E) the air is reversible and polytropically done with $n = 1.34$ and the expansion is reversible and polytropic with $n = 1.45$. Determine
(a) Temperature of air entering the radiator.
(b) Temperature of air entering the cooling coil.
(c) $\dot{Q}_{net}$
(d) COP.
(e) Refrigeration rating of device.

Section 14.3

14.15. Saturated liquid ammonia is throttled through an expansion valve from
(M) 45°C to 1.90 bars. Determine the enthalpy and the quality of the ammonia leaving the expansion valve.

14.16. At a rate of 6.0 kg/s, Refrigerant-22 enters an evaporator as 12% vapor
(M) and at 1.316 bars. If the refrigerant leaves with a quality of 88%, determine the rate of heat addition.

14.17. An ideal dry compression refrigerator operates on the cycle diagrammed
(M) in figure 14–6(a). If the working media is Refrigerant-22 and the pressures are 11.9 bars and 2.96 bars, determine
(a) wk_{comp}.
(b) q_{add}.
(c) q_{rej}.

14.18. Determine the heat added per pound-mass of ammonia if the ammonia
(E) has 20% quality and is at −20°F entering the evaporator and saturated vapor upon leaving.

14.19. An ideal wet compression air conditioner using Refrigerant-22 uses a
(E) cycle as diagrammed in figure 14–6(b). If the refrigerant is at 30°F in the evaporator and 110°F in the condenser, determine
(a) wk_{comp}.
(b) q_{add}.
(c) q_{rej}.

14.20. Calculate the COP for
(C) (a) Cycle of problem 14.17.
(b) Cycle of problem 14.19.

Section 14.4

14.21. **(M)** Saturated liquid Refrigerant-22 at 25°C is expanded to 1.05 bars through an expansion valve. It leaves the evaporator at 100% quality and 1.05 bars pressure. If 12 kg/min of the refrigerant flow through the device, what is its rating? Is this a wet or dry compression cycle?

14.22. **(M)** A freezer using Refrigerant-22 removes 1500 kJ/hr from a compartment at −2°C. The refrigerant is at −5°C in the evaporator and the quality is 85% leaving the evaporator. If the freezer operates on an ideal wet compression cycle, determine

(a) $\dot{m}$ of refrigerant.
(b) $\dot{Q}_{rej}$.
(c) $\dot{W}k_{cycle}$.
(d) COP and COR.

14.23. **(E)** A wet compression cycle refrigerator using ammonia as a working medium operates such that the ammonia is 15°F in the evaporator and 90°F in the condenser. Determine the power required per ton of refrigeration.

14.24. **(E)** A freezer using Refrigerant-22 removes 1500 Btu/hr from a compartment at 18°F. The refrigerant is at 23°F in the evaporator and the quality is 85% leaving the evaporator. If the freezer operates on an ideal wet compression cycle determine

(a) $\dot{m}$ of refrigerant.
(b) $\dot{Q}_{rej}$.
(c) $\dot{W}k_{cycle}$.
(d) COP and COR.

14.25. **(M)** A heat pump rated at 15 kilowatts operates with a maximum flow rate of 5.2 kg/min of Refrigerant-22. The cycle is of the wet vapor compression type and can operate in climatic temperatures to −10°C. Determine the maximum inside temperature expected from this cycle.

14.26. **(E)** A heat pump provides 65,000 Btu/hr of heat to a dwelling at 80°F when the outside temperature is 0°F. The device operates on a dry compression cycle and uses Refrigerant-22 as the working media. Determine the maximum flow rate of the refrigerant, the COP, and the power required.

15

Mixtures

Various concepts and descriptions for the combining of separate systems will be presented here. We shall first consider a mixture of perfect gases and define *partial pressure*, *mass fractions*, and other terms necessary for an accurate, useful description of the system. Then the common mixture of air and water vapor will be considered in terms of the thermodynamics of meteorology. Chemical potential will be defined in an intuitive manner and related to the water-air interaction. To give the reader the barest concept of the mechanism of mixing, *diffusion* will be briefly introduced through *Fick's law* and *mass balance*.

The concluding topic to be considered will be a mixture of chemically active systems. Since the reader is not expected to be well versed in chemical systems, only an outline of the approach in handling chemical reactions can be given. This topic is introduced mainly to show how to determine heats of combustion in the fast chemical reaction of fuels and air or oxygen.

15.1 The Perfect Gas Mixture

Suppose we take 1 m^3 of oxygen at 1.01 bars and 27°C and mix it with 1 m^3 of nitrogen at 1.01 bars and 27°C in a 2-m^3 container. After these two gases have been sufficiently mixed, we notice that each gas is occupying the full volume of 2 m^3 and that the temperature is 27°C. For this final condition then, if we assume both gases are perfect, we can write

$$p_O V = m_O R T$$

for the oxygen, and

$$p_N V = m_N R T$$

for the nitrogen, where V is the total volume of the mixture. Then if we assume an "average" gas constant, R_{av}, for the mixture we have

$$p_O V = m_O R_{av} T \tag{15–1}$$

and

$$p_N V = m_N R_{av} T \tag{15–2}$$

If we add these two results we get

$$(p_O + p_N)V = (m_N + m_O)R_{av}T$$

but $m_N + m_O$ = mass of the mixture = m. As a consequence we must have that

$$p_O + p_N = \text{pressure of mixture} = p$$

which is known as *Dalton's law of partial pressures.* Using this result

$$pV = mR_{av}T \tag{15–3}$$

and equations (15–1) and (15–2) we get that

$$\frac{p_O}{p} = \frac{m_O}{m} \quad \text{and} \quad \frac{p_N}{p} = \frac{m_N}{m} \tag{15–4}$$

so that we call the pressures p_O and p_N the partial pressures of the oxygen and nitrogen, respectively. If we have a multi-component mixture (three or more substances), it is convenient to number each substance or gas for bookkeeping purposes. For example, in air we have oxygen (call it gas 1), argon (call it gas 2), and nitrogen (call it gas 3), so the partial pressure of oxygen would be denoted as p_1. Similarly for argon, the partial pressure would be p_2 and for nitrogen, p_3. Thus we could abbreviate the notation as p_i where i could be 1, 2, or 3, depending on which gas is being considered. Of course, a mixture of four or more substances then requires that i be able to be assigned any of these other numbers. If we consider an "i-th" component or gas, we mean a typical substance in the mixture. The partial pressure of a gas 1 in a certain mixture is given by

$$p_1 = \frac{m_1}{m} p$$

For a gas 2

$$p_2 = \frac{m_2}{m} p$$

and in general, the partial pressure of the i-th gas would be

$$p_i = \frac{m_i}{m} p \tag{15–5}$$

Also, Dalton's law is then written for the general case as

$$p = \sum_{i=1}^{n} p_i = \sum_{i=1}^{n} \frac{m_i}{m} p = p \sum_{i=1}^{n} \frac{m_i}{m} \tag{15–6}$$

Example 15.1

A sample of air is found to be composed of 74.9% nitrogen, 23.9% oxygen, and 1.2% argon. The sample was taken at 1.0 bar pressure and 28°C. Calculate the partial pressures of the components.

Solution

Using equation (15–5) we obtain, for nitrogen

$$p_N = (0.749)(1.0 \text{ bar}) = 0.749 \text{ bar}$$ *Answer*

Similarly, for oxygen we find

$$p_O = (0.239)(1.0 \text{ bar}) = 0.239 \text{ bar}$$ *Answer*

and for argon

$$p_{Ar} = (0.012)(1.0 \text{ bar}) = 0.012 \text{ bar}$$ *Answer*

Example 15.2

There are 2 lbm of air at 80°F and 15 psia found to be composed of 75.5% (by mass) nitrogen, 23.2% oxygen, and 1.3% argon. Calculate the partial pressures of each of the components.

Solution

Using equation (15–5)

$$p_i = \frac{m_i}{m} p$$

we have that p is the air pressure, 15 psia. The partial pressure of nitrogen p_N we then get from

$$p_N = (0.755)(15 \text{ psia}) = 11.325 \text{ psia}$$ *Answer*

Similarly, for oxygen

$$p_O = (0.232)(15) = 3.480 \text{ psia}$$ *Answer*

and for argon

$$p_{Ar} = (0.013)(15) = 0.195 \text{ psia}$$ *Answer*

Mixing is an irreversible process and accordingly the total entropy of components increases after they have been mixed. To see this we write the change in entropy of a mixture during the mixing process as

$$\Delta S = \sum_{i=1}^{n} m_i \Delta s_i \tag{15–7}$$

where we can recall, from chapter 7

$$\Delta s_i = \int \frac{c_V dT}{T} + R_i \ln \frac{V}{V_i}$$

or, if c_V is constant

$$\Delta s_i = c_V \ln \frac{T}{T_i} + R \ln \frac{V}{V_i} \tag{15–8}$$

where T_i, R_i, and V_i are the temperature, gas constant, and volume of the i-th component before mixing. From equation (15–8) it can be deduced that the sum of the various entropy changes must be positive since generally V is greater than V_i and since the sum of the ratios T/T_i will be near zero. An example should serve to show the details.

Example 15.3

At 27°C and 1.01 bars pressure 1.0 m³ of oxygen is mixed with 1.0 m³ of nitrogen. The mixing process occurs in a 2.0-m³ container also at 27°C. Determine the entropy change due to this process.

Solution

From equation (15–7) we have

$$\Delta S = m_O \Delta s_O + m_N \Delta s_N$$

where O and N denote the properties of oxygen and nitrogen respectively. From the perfect gas laws we have

$$m_O = \frac{p_O V_O}{R_O T_O} = \frac{(1.01 \times 10^5 \text{ N/m}^2)(1.0 \text{ m}^3)}{(260 \text{ N} \cdot \text{m/kg} \cdot \text{K})(300 \text{ K})} = 1.295 \text{ kg}$$

and

$$m_N = \frac{p_N V_N}{R_N T_N} = \frac{(1.01 \times 10^5 \text{ N/m}^2)(1.0 \text{ m}^3)}{(297 \text{ N} \cdot \text{m/kg} \cdot \text{K})(300 \text{ K})}$$
$$= 1.134 \text{ kg}$$

Using equation (15–8) we have

$$\Delta s_O = c_{VO} \ln \frac{T}{T_O} + R_O \ln \frac{V}{V_O}$$

and

$$\Delta s_N = c_{VN} \ln \frac{T}{T_N} + R_N \ln \frac{V}{V_N}$$

But $T = T_N = T_O$ and $V_O = V_N = V/2$ in this problem so that we obtain from equation (15–7)

$$\Delta S = m_O R_O \ln \frac{V}{V/2} + m_N R_N \ln \frac{V}{V/2}$$

$$= (1.295 \text{ kg})(260 \text{ N} \cdot \text{m/kg} \cdot \text{K})(\ln 2.0)$$
$$+ (1.134 \text{ kg})(297 \text{ N} \cdot \text{m/kg} \cdot \text{K})(\ln 2.0)$$
$$= 467 \text{ J/K}$$ *Answer*

This problem and its solution tells us that the entropy is indeed increased without reversible heat transfer — in fact, the above entropy increase is entirely due to irreversibilities in mixing.

15.2 Water and Air Mixtures

Although air is a mixture of nearly perfect gases, it is commonly considered to be a homogeneous, single component having distinct properties associated with it. These properties, such as the gas constant R, would be the "average" property, as mentioned in the previous section, or the "total" property in the case of pressure. In the cases where we consider air a uniform, homogenous substance, we are assuming it to be dry; that is, no water vapor is mixed with the air. Dry air is generally assumed to be composed of 78% nitrogen, 21% oxygen, and 1% argon by volume. In reality, however, air seeks an equilibrium with its surroundings which are commonly liquid water. (The sea and sky are common neighbors.) When equilibrium is achieved, the rate of evaporation of the liquid water equals the rate of condensation of water vapor (see figure 15–1)—the water vapor being a gas mixed with the air. To describe properly the air-water mixture, we may define, for some precise volume of this mixture, the mass ratio of water vapor to air. This ratio ω is written

$$\omega = \frac{m_V}{m_a} \tag{15–9}$$

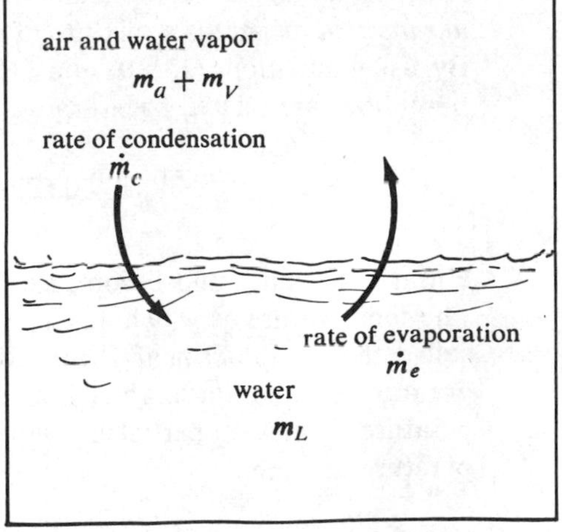

Figure 15–1 Equilibrium of liquid water and air in a closed container.

and is called *specific humidity*, where m_V is the mass of water vapor and m_a is the mass of air. If we assume the water vapor acts like a perfect gas then the specific humidity can be written

$$\omega = \frac{p_V V/R_V T}{p_a V/R_a T} = \frac{R_a}{R_V}\frac{p_V}{p_a} = 0.622\,\frac{p_V}{p_a} \tag{15–10}$$

Now, when water and its vapor are in equilibrium we say the water is in a phase change (the liquid-vapor phase change, in fact). This liquid and vapor mixture is called *saturated* and depending on the quality (see section 12.6 for a definition of *quality*), it is somewhere between a saturated liquid and a saturated vapor. If the water vapor is in a mixture with air and is in equilibrium with liquid water, then the water vapor is a saturated vapor, and we call the mixture *saturated air*.

If the air-water vapor mixture is saturated air, then the partial pressure of the water vapor p_V is equal to the saturation pressure p_g corresponding to the temperature of the air. Use of the saturated steam tables can be made (table B.1) to obtain the corresponding saturation pressure at a given temperature. Air which is not saturated is either dry air or some intermediate condition. To describe the relative condition we define relative humidity β as

$$\beta = \left(\frac{p_V}{p_g}\right) \times 100\% \tag{15–11}$$

which gives us the conditions

$$\beta = 100\% \text{ for saturated air } (p_V = p_g)$$

and

$$\beta = 0\% \text{ for dry air } (p_V = 0)$$

Air which has a relative humidity between 100% and 0% is called *unsaturated air* and is a mixture of dry air and superheated water vapor. By using equations (15–10) and (15–11) we see that relative and specific humidities are related by

$$\omega = 0.622\,\beta\left(\frac{p_g}{p_a}\right)\frac{1}{100} \tag{15–12}$$

Unsaturated air could become saturated through a temperature change. The temperature at which an air-water vapor mixture is saturated air is called the *dew point temperature.* In figure 15–2 is shown on a *T-s* diagram an example state *a* which is an unsaturated air. By reducing the temperature at constant partial vapor pressure p_V, we reach the dew point temperature at state *b*.

Another manner of reaching a saturation condition from an unsaturated state is to increase the partial vapor pressure p_V at constant temperature until the pressure p_g is reached. This process is indicated by the curve *a-c*

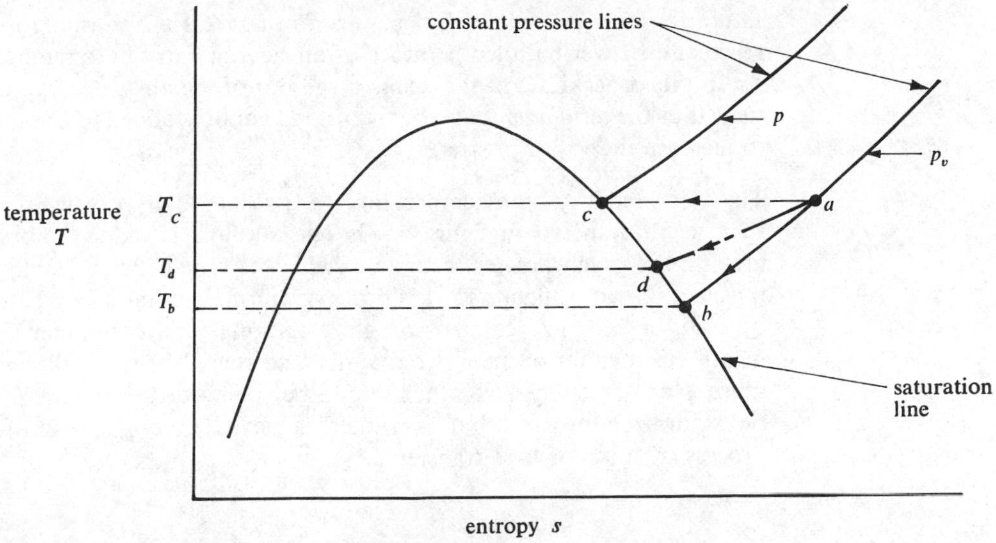

Figure 15–2 Temperature-entropy diagram for water vapor.

in figure 15–2. Notice that an increase in vapor pressure p_V is normally associated with a corresponding reduction in partial air pressure p_a since by Dalton's law we must have $p = p_a + p_V$ where p is the atmospheric pressure.

For further descriptions of the air-water mixture, the dry-bulb and wet-bulb temperatures are commonly given. The dry-bulb temperature is that which is recorded in normal atmospheric conditions. The wet-bulb temperature, however, is measured in an atmosphere of saturated air and ideally is measured in a saturated condition. This process, the adiabatic

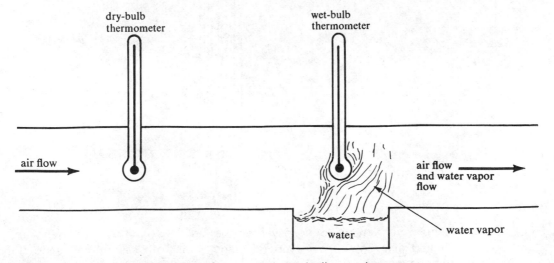

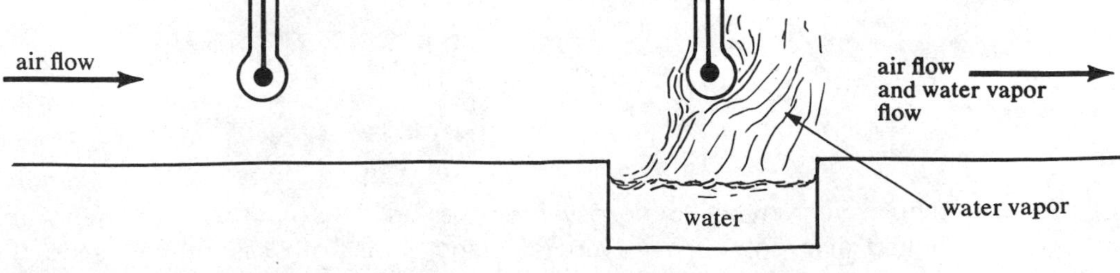

Figure 15–3 General features of wet- and dry-bulb psychrometer.

saturation one, is depicted by curve *a-d* in figure 15–2 so that T_d would represent the wet-bulb temperature of air having a dry-bulb temperature T_a. In this process, no heat is transferred out of the air-water vapor mixture (thus the term *adiabatic*), but saturated vapor is added to the mixture to increase the vapor pressure.

The wet- and dry-bulb temperatures are recorded by a psychrometer such as that shown in figure 15–3. The air flow is induced either by moving the psychrometer in the air (which is then called a *sling psychrometer*, shown in figure 15–4, since it is normally rotated by hand) or by using a fan to force air through a stationary psychrometer. When measured in either of these two manners, however, the wet-bulb temperature is not precisely that which would be achieved through an adiabatic saturation process, but is rather achieved through a saturation process with minor heat transfer.

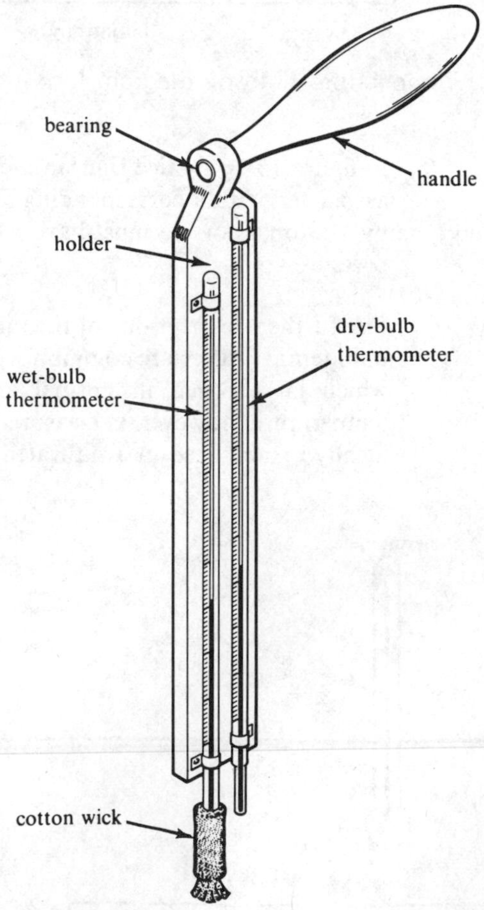

Figure 15–4 Sling psychrometer. Sling psychrometer held by handle in one hand and rotated at even speed in vertical plane until both thermometers do not change readings. Cotton wick must be water-soaked at all times during use.

The various properties of air-water vapor mixtures, which we must consider in describing the normal atmosphere, are correlated on the psychrometric chart B.2 given in appendix B. In figure 15–5 the basic elements of this chart are shown in order to assist in understanding the manner of data extraction. The chart given here and in appendix B is based on an atmospheric pressure of 14.7 psia or 1.01 bars. If the atmospheric pressure deviates greatly from this value, the chart must be corrected. (See the following references in appendix C: Lee and Sears, *Thermodynamics*, and Mooney, *Mechanical Engineering Thermodynamics*.)

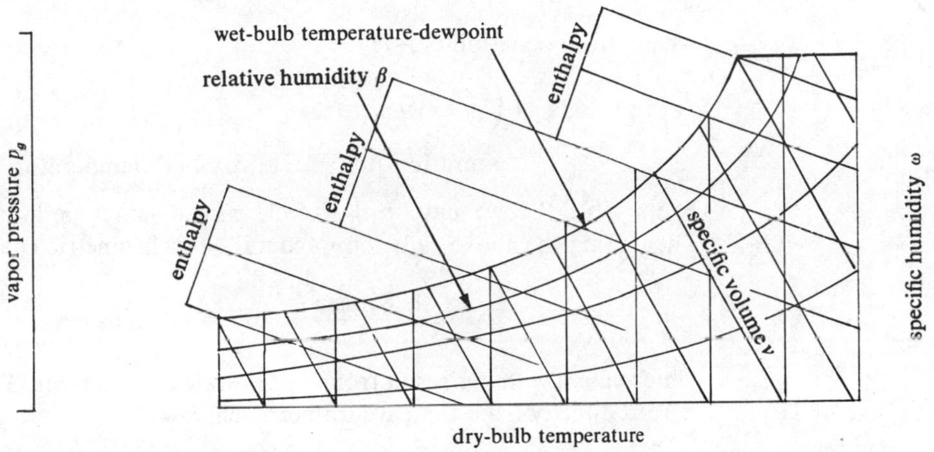

Figure 15–5 Sketch of the basic parameters constituting a psychrometric chart.

Example 15.4 Unsaturated air is at 20°C. Find the partial pressure required of the water vapor to convert this mixture to the saturated state at 20°C.

Solution Using the steam table B.1 or the psychrometric chart B.2, we obtain $p_g = 0.023$ bar at 20°C, so the pressure of the vapor, p_v, must be 0.023 bar for the saturation state of air at 20°C.

Example 15.5 Atmospheric air at 14.7 psia has a temperature of 85°F and a wet-bulb temperature of 70°F. Determine the vapor pressure, the relative humidity, the specific humidity, and the enthalpy of the atmospheric air.

Solution Since the atmospheric pressure is standard, we may directly use the data of the psychrometric chart. The dry-bulb temperature can be assumed to be 85°F. Then, using chart B.2 we obtain the vapor pressure

$$p_V = 0.286 \text{ psia} \qquad \textit{Answer}$$

The relative humidity can also be directly read from the chart,

$$\beta = 48\% \qquad \textit{Answer}$$

and the specific humidity is

$$\omega = 86 \text{ grains/lbm}$$

Since there are 7000 grains in 1 lbm we get

$$\omega = \frac{86}{7000} \text{ lbm vapor/lbm air} = 0.0123 \qquad \textit{Answer}$$

Also, we should be able to get the above answers to correlate in equation (15–12)

$$\omega = 0.622\beta \frac{p_g}{p_a} \frac{1}{100}$$

where, from equation (15–11)

$$p_g = \left(\frac{p_V}{\beta}\right) 100$$

$$= \text{saturation pressure at dry-bulb temperature}$$

From table B.1 we have $p_g = 0.5959$ psia at a temperature of 85°F. Then, using the above value of p_v from the psychrometric chart, we have

$$p_g = \frac{0.286 \text{ psia}}{48} \times 100 = 0.596 \text{ psia}$$

which agrees with our result from the saturated steam table. The enthalpy is read directly from the psychrometric chart as

$$h = 34.1 \text{ Btu/lbm air} \qquad \textit{Answer}$$

The enthalpy is based on the unit of mass of dry air and so the total enthalpy must be determined by using only the mass of dry air, not the mixture mass.

There are numerous reasons for determining the thermodynamic properties of the atmosphere, among which are *humidification* and *dehumidification processes.* The following two example problems will indicate the method of approach to these two processes.

Example 15.6

Humidification is a technique of increasing the specific humidity of air to provide human comfort. In figure 15–6 is shown the schematic of a typical humidifier which is also operating as a heater. Assuming that air enters the humidifier at 1.01 bars, 20°C, and 40% relative humidity and should be 25°C and 50% relative humidity upon leaving the unit, determine the amount of heat required in the heating unit and the amount of water required per kilogram of dry air, assuming the entering liquid water is at 15°C.

Solution

The steady flow energy equation for the humidifier can be written as

$$m_a h_2 - (m_a h_1 + m_w h_w) = Q \qquad \textbf{(15–13)}$$

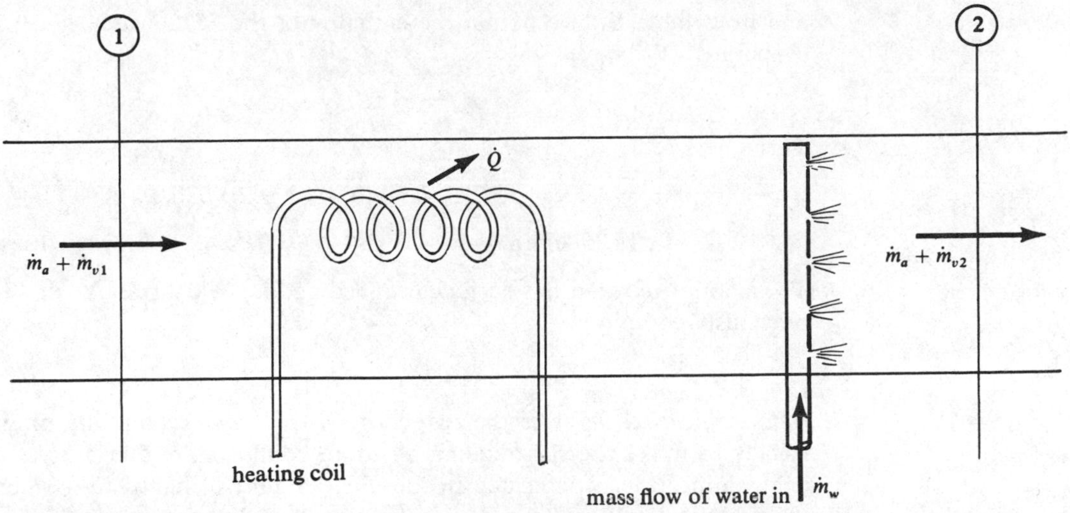

Figure 15–6 Humidifying process.

and the mass flow equation as

$$\dot{m}_a + \dot{m}_{V1} + \dot{m}_w = \dot{m}_a + \dot{m}_{V2} \tag{15–14}$$

From the psychrometric chart we read

$$h_2 = 69 \text{ kJ/kg}$$
$$h_1 = 53 \text{ kJ/kg}$$
$$\omega_2 = 10 \text{ g/kg dry air}$$
$$= 0.010 \text{ kg/kg dry air}$$

and

$$\omega_1 = 6 \text{ g/kg} = 0.006 \text{ kg/kg dry air}$$

Using equation (15–13), we find the heat required per kilogram of dry air Q/m_a:

$$\frac{Q}{m_a} = q = h_2 - h_1 - \frac{m_w}{m_a} h_w$$
$$= 69 \text{ kJ/kg} - 53 \text{ kJ/kg} - \frac{m_w}{m_a} h_a$$

From equation (15–14) we note that

$$\dot{m}_w = \dot{m}_{v2} - \dot{m}_{v1} = \dot{m}_a\omega_2 - \dot{m}_a\omega_1$$

Then

$$\frac{m_w}{m_a} = \omega_2 - \omega_1 = 0.010 \text{ kg/kg dry air} - 0.006 \text{ kg/kg dry air}$$
$$= 0.004 \text{ kg/kg dry air}$$

and from table B.1 we obtain the enthalpy of the 15°C water which is evaporated. We have

$$h_w = 63 \text{ kJ/kg}$$

Then

$$q = 69 \text{ kJ/kg} - 53 \text{ kJ/kg} - (0.004 \text{ kg/kg})(63 \text{ kJ/kg})$$
$$= 15.75 \text{ kJ/kg}$$ *Answer*

The amount of water has been determined and is the 0.004 kg/kg dry air previously computed.

Example 15.7

Dehumidification involves the reduction of the specific humidity of air, usually to make the surrounding air more comfortable. The process of dehumidification is indicated in the *T-s* diagram of figure 15–7 where the air at state (1) is reduced in its specific humidity to state (2) by first being cooled to the dew point (4), being condensed and losing moisture to state (3), and being subsequently heated to state (2). The physical operation of this is shown in figure 15–8.

Assume that air is taken in at 14.7 psia, 80°F, and 80% relative humidity. If the air is to be cooled to 72°F and reduced to 60% relative humidity, determine the heat rejected in the cooling unit, the heat added in the heating unit, and the amount of water condensed per pound-mass of dry air.

Solution

Applying the steady flow energy equation to the cooling and heating units separately, we obtain

$$m_a h_3 + m_w h_5 - m_a h_1 = Q_{\text{rej}} \tag{15–15}$$

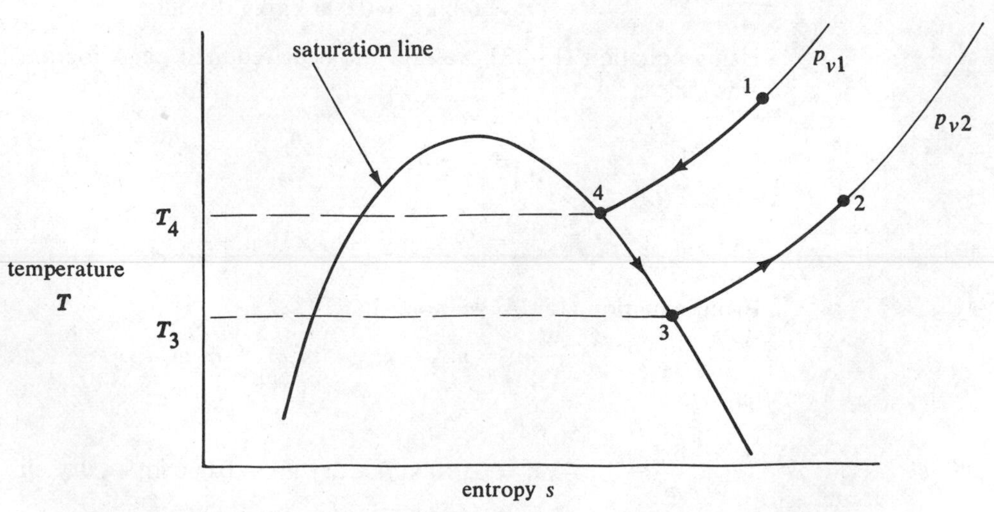

Figure 15–7 Dehumidifying process on a temperature-entropy plane.

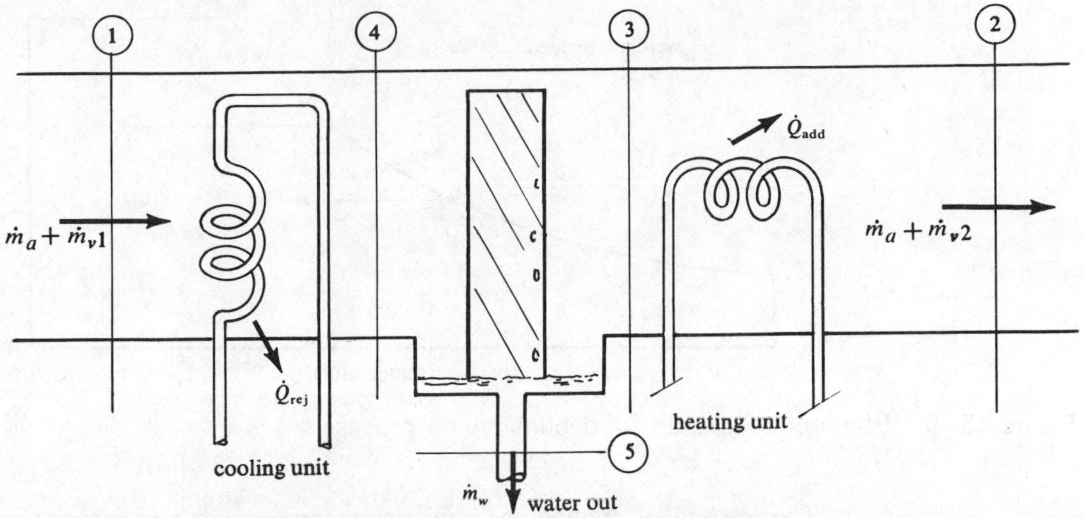

Figure 15–8 Dehumidifying process.

and

$$m_a h_2 - m_a h_3 = Q_{add} \tag{15–16}$$

Per unit mass of dry air we have

$$h_3 + \frac{m_w}{m_A} h_5 - h_1 = q_{rej} \tag{15–17}$$

and

$$h_2 - h_3 = q_{add} \tag{15–18}$$

From the psychrometric chart we read

$$h_1 = 38.8 \text{ Btu/lbm dry air}$$
$$h_2 = 28.4 \text{ Btu/lbm dry air}$$

and

$$\omega_1 = \frac{125}{7000} = 0.0178 \text{ lbm/lbm dry air}$$

$$\omega_2 = \frac{70}{7000} = 0.010 \text{ lbm/lbm dry air}$$

The amount of water condensed per pound-mass of dry air is the decrease in specific humidity

$$\frac{m_w}{m_a} = \omega_1 - \omega_2 = 0.0178 - 0.010 = 0.0078 \text{ lbm condensate/lbm dry air}$$

Answer

The enthalpy at state (5), the condensed water, is given by $h_5 = h_f$ at the dew point of the incoming air mixture. The dew point is found from the psychrometric chart to be 73.4°F and then $h_f = 41.45 = h_5$ from table B.1.

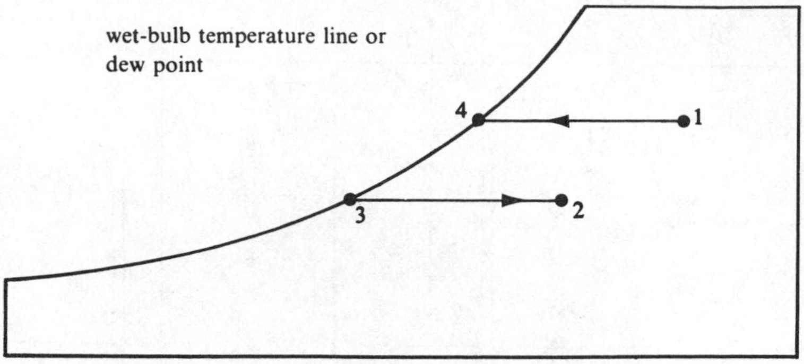

Figure 15–9 Psychrometric chart for dehumidifying process.

The method of reading the dew point corresponding to state (1) is indicated in figure 15–9 and in the same manner we obtain the dew point of the dehumidified air at state (2); $t_3 = 57.2°F$ and the enthalpy $h_3 = 24.6$. We can now calculate the heat transfer from equation (15–17);

$$q_{\text{rej}} = 24.6 + (0.0078)(41.45) - 38.8 = -13.9 \text{ Btu/lbm dry air}$$

Answer

and, from equation (15–18)

$$q_{\text{add}} = 28.4 - 24.6 = 3.8 \text{ Btu/lbm dry air}$$ *Answer*

15.3 Chemical Potential

Equilibrium is a relative term which requires precise definition to become meaningful. We have earlier seen that two bodies or systems are in *thermal equilibrium* if their temperatures are equal. *Mechanical equilibrium* is achieved between two systems when their pressures are equal at the interface. (Mechanical equilibrium is static or quasi-static.) In figure 15–10 we see two systems, A and B, which we can consider as liquids, solids, or gases. They are in thermal equilibrium when $T_A = T_B$, and in static mechanical equilibrium when $p_A = p_B$. There is also one other form of equilibrium, called *chemical equilibrium*, which we must consider for a complete definition of equilibrium. These two systems, which we assume are each mixtures of the same components, 1 and 2, are said to be in chemical equilibrium when the chemical potential of component 1 in system A (μ_{A1}) is equal to the chemical potential of component 1 in system $B(\mu_{B1})$ and in a similar manner $\mu_{B2} = \mu_{B1}$. That is, two systems which interface one another attain chemical equilibrium when both have the same components and these components have the same chemical potential in both systems. It can be shown that the chemical potential of a component 1 in a state A is

$$\mu_{A1} = h_{A1} - T_A s_A$$

$$\mu_{A1} = h_{A1} - T_A s_{A1}$$

$$\mu_{A2} = h_{A2} - T_A s_{A2}$$

$$\mu_{B1} = h_{B1} - T_B s_{B1}$$

$$\mu_{B2} = h_{B2} - T_B s_{B2}$$

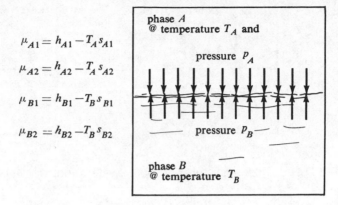

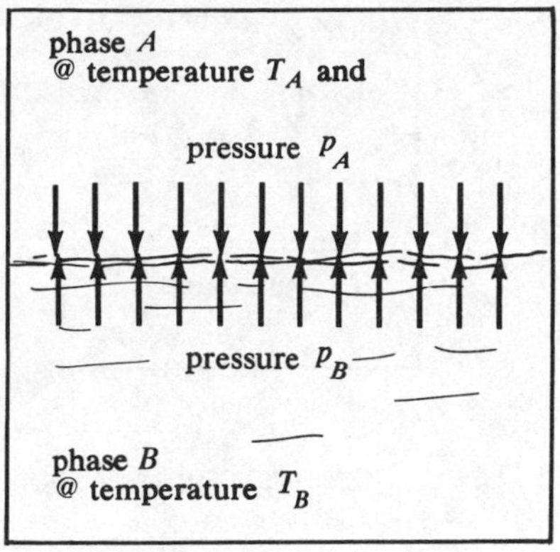

Figure 15–10 Equilibrium between two systems.

and for any general component i

$$\mu_{Ai} = h_{Ai} - T_A s_{Ai} \qquad \textbf{(15–19)}$$

Chemical equilibrium can be visualized as a balance of the mass flow of a component, say i, from a system A to another system B with the mass flow of the same component from system B to A. Chemical equilibrium between A and B is then achieved when all the components of the two systems balance the mass flow rates. It is this balance which is reflected in the criterion for chemical equilibrium

$$\mu_{Ai} = \mu_{Bi} \qquad \textbf{(15–20)}$$

Example 15.8 Determine the chemical potential of saturated liquid water at 160°C.

Solution The system (water) is a one-component mixture and its chemical potential can be computed from equation (15–19). From table B.1 we get

$$h_{A1} = 676 \text{ kJ/kg}$$

$$s_{A1} = 1.942 \text{ kJ/kg} \cdot \text{K}$$

so

$$\mu_{A1} = 676 \text{ kJ/kg} - (433 \text{ K})(1.942 \text{ kJ/kg} \cdot \text{K})$$

$$= -165 \text{ kJ/kg}$$

Answer

15.4 Diffusion

Suppose two systems are brought into contact with each other and are in thermal and mechanical equilibrium. Then, if we consider the systems A and B as shown in figure 15–11(a), the conditions follow that

$$T_A = T_B \quad \text{and} \quad p_A \text{ (at } x = 0) = p_B \text{ (at } x = 0)$$

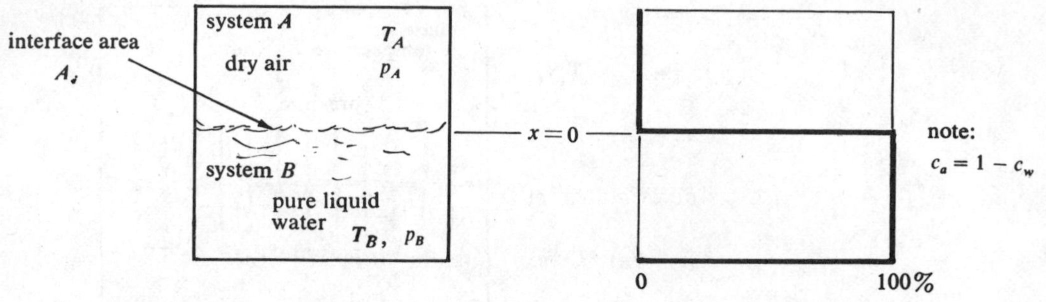

Figure 15–11(a) Closed system of pure water and dry air, initial condition (time $\tau = 0$).

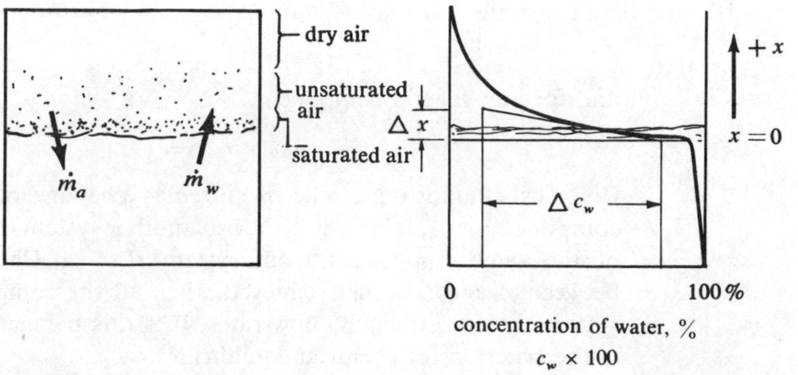

Figure 15–11(b) Closed system of pure water and dry air after a finite period of time has elapsed.

Let us then assume that system A is dry air and system B is pure water so that the mixture A can be described as 100% air and B as 100% water, as depicted in the graph of figure 15–11(a). As we have seen in section 15.3, however, for complete equilibrium to be achieved, the chemical potential of the air in A must equal the chemical potential of the air in the water ($\mu_{A(\text{air})} = \mu_{B(\text{water})}$). Likewise, $\mu_{A(\text{water})} = \mu_{B(\text{water})}$ must hold and these two statements imply that water must be present in A and air present in B. As a consequence of this natural action, water will migrate to A and air will migrate to B. The description of the mass flow rates of this migration are given by Fick's law of diffusion:

$$\text{Mass flow rate of water to } A = \dot{m}_w = -\rho_w A_i \mathfrak{D}_{wA} \frac{dc_w}{dx} \qquad \textbf{(15–21)}$$

and

$$\text{Mass flow rate of air to } B = \dot{m}_a = -\rho_a A_i \mathfrak{D}_{aB} \frac{dc_a}{dx} \qquad \textbf{(15–22)}$$

Here A_i is the cross-sectional area of the interface between the two systems; $\mathfrak{D}_{wA}$ is the diffusivity or coefficient of diffusion of water in system A;

$\mathfrak{D}_{aB}$ is the diffusivity of air in system B; and c is the concentration or mass ratio such that c_w = mass of w in volume V/total mass in volume V and c_a = mass of a in V/total mass in V.

We can approximate the equation (15–21) by taking finite increments of x so that

$$\dot{m}_w \approx -\rho_w A_i \mathfrak{D}_{wA} \frac{(\Delta c_w)}{\Delta x} \qquad \textbf{(15–23)}$$

and

$$\dot{m}_a \approx -\rho_a A_i \mathfrak{D}_{aB} \frac{(\Delta c_a)}{\Delta x} \qquad \textbf{(15–24)}$$

where Δc is the change in concentration in a distance Δx. Initially we will see an infinite mass flow rate of water to the system A and air to the system B since

$$\Delta c_a = c_{aB} - c_{aA} = 0 - 1 = -1$$
$$\Delta c_{Bw} = c_{wA} - c_{wB} = 0 - 1 = -1$$

but $\Delta x = 0$ for both, so from equations (15–23) and (15–24) we get

$$\dot{m}_w \approx -\rho_w A_i \mathfrak{D}_{wA} \left(\frac{1}{0}\right) = \infty$$

and

$$\dot{m}_a \approx -\rho_a A_i \mathfrak{D}_{aB} \left(\frac{1}{0}\right) = \infty$$

This mass flow rate will only occur for an infinitesimal time as mixing of air and water will occur in both systems A and B. After some period of time, system A will be a nonuniform mixture of dry air and water. Near the interface the mixture will be saturated air, but further up, the concentration of water will progressively decrease so that the mixture could be described as almost dry air. Similarly, air will diffuse into the water, system B, but the concentration will decrease the further and further down into the system. This state is depicted by figure 15–11(b) where the concentration is plotted as a function of the distance from the interface. Notice that $\Delta c/\Delta x$ represents the slope of the curve of concentration at a particular point. Steady state conditions will be reached when the curve of concentration remains unchanged with time, but until this condition is achieved the curve can be expected to change so that the magnitude of $\Delta c/\Delta x$ decreases. That is, diffusion will continue to seek an equality of concentration between systems.

Diffusion can occur between a liquid and a gas (as we have just indicated), between a gas and a gas, a liquid and a liquid, a liquid and a solid, a gas and a solid, or a solid and a solid. Of course, the rates of mass flow vary extremely between these combinations, but the mass flow can be predicted through the coefficient of diffusion. In table 15–1 is shown the

Table 15–1

The Orders of Magnitude for Diffusivity at 27°C, 1.01 bars

Diffusing Phase (Transported Material) i	*Diffused Phase, A (System Invaded by Material)*	$\mathfrak{D}_{ia}$ cm²/s
gas	gas	0.1
gas	liquid	1×10^{-5}
gas	solid	1×10^{-10}
liquid	gas	1×10^{-5}
liquid	liquid	1×10^{-5}
liquid	solid	1×10^{-15}
solid	gas	1×10^{-10}
solid	liquid	1×10^{-15}
solid	solid	1×10^{-20}

order of magnitudes of $\mathfrak{D}$ for the various combinations of systems. Remember that the numbers are merely approximate values to describe the variation in diffusion. It can easily be seen from the table that diffusion of a solid to a solid is much less than a gas to a gas or any other combination.

In table 15–2 are given some nominal values of diffusivity $\mathfrak{D}$ for common materials.

Table 15–2

Some Common Materials at 20°C and 1 Atmosphere Pressure*

A	*B*	$\mathfrak{D}_{ab}$	
Diffusing Material	*Diffused Material*	cm²/s	ft²/hr
Ammonia	air	0.236	0.914
Carbon dioxide	air	0.164	0.636
Water vapor	air	0.256	0.992
Ethyl ether	air	0.093	0.360
Helium	SiO_2	3×10^{-10}	1.16×10^{-9}
Helium	pyrex	4.5×10^{-11}	1.74×10^{-10}
Bismuth	Pb	1.1×10^{-16}	4.26×10^{-16}
Mercury	Pb	2.5×10^{-15}	9.67×10^{-15}
Ethanol	water	1.13×10^{-5}	4.37×10^{-5}
Water (liquid)	n-Butanol	1.25×10^{-5}	4.80×10^{-5}

Notes: Data abstracted from M. Jakob and G. Hawkins, *Elements of Heat Transfer*, 3rd. ed. (New York, 1957), p. 276; and from R. B. Bird, W. Stewart, and E. Lightfoot, *Transport Phenomena* (New York, 1960), pp. 504–5; with permission of John Wiley & Sons, Inc.

Example 15.9 A 70-cm-diameter closed steel drum contains water and air at 10°C and 1 atmosphere pressure. At a given time the concentration curve is found to be that shown in figure 15–12. Determine the net mass flow rate across the liquid-air interface.

Solution From figure 15–12 we see that the slope of the concentration curves at the interface are measured to be

$$\frac{\Delta c_w}{\Delta x} = \frac{-80}{10 \times 100} = -0.08/\text{cm}$$

$$\frac{\Delta c_a}{\Delta x} = \frac{80}{-10 \times 100} = -0.08/\text{cm}$$

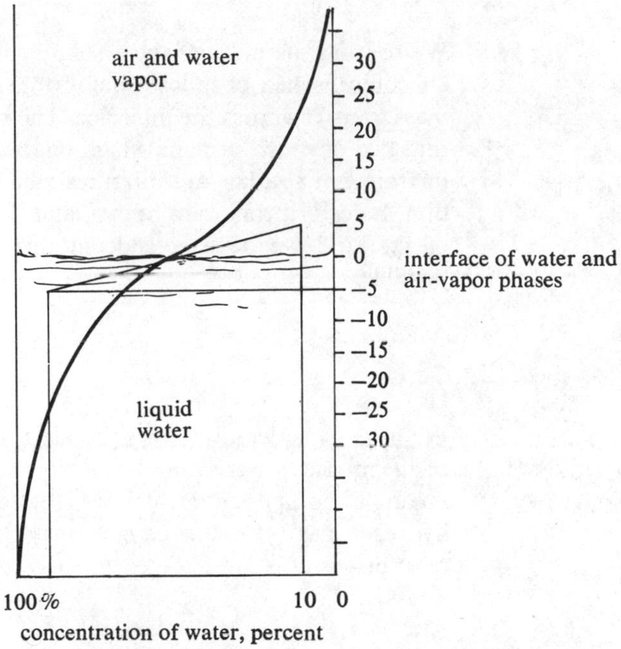

Figure 15–12 Relation of the concentration of water in a closed steel drum.

The density of the water at the interface we assume to be 1 g/cm³. We shall assume the air to have a density of 0.00125 g/cm³, found from the perfect gas relation. The area of the interface A_i is obtained from

$$A_i = \pi \times \frac{\text{dia}}{4} = \pi \times \frac{70^2}{4}\ \text{cm}^2 = 3847\ \text{cm}^2$$

and we assume the diffusivity is the same for air and water, giving us

$$\mathfrak{D}_{wA} = \mathfrak{D}_{aB} = 0.256\ \text{cm}^2/\text{s}$$

from table 15–2. Then from equations (15–23) and (15–24) we get

$$\dot{m}_w = -(1\ g/cm^3)(3847\ cm^2)(0.256\ cm^2/s)(-0.08/cm)$$
$$= 78.787\ g/s$$

and

$$\dot{m}_a = -(0.00125\ g/cm^2)(3847\ cm^2)(0.256\ cm^2/s)(-0.08/cm)$$
$$= 0.098\ g/s$$

yielding as a net mass flow rate

$$\dot{m} = \dot{m}_w - \dot{m}_a$$
$$= 78.787 - 0.098$$
$$= 78.689\ g/s \qquad \textit{Answer}$$

We see that the mass of the system is moving from the liquid to the gas. Of course, when chemical equilibrium is achieved, there will be no net mass transfer across the interface. The reader should recognize, however, that two or more systems which interact are not homogeneous, uniform mixtures, but rather are mixtures which contain gradients of concentration. Indeed, if this were not so, and if diffusion did not occur, mixtures of the kind we have considered throughout this book would not be possible.

15.5 Chemical Reactions

We have considered mixtures of inert materials; that is, the components of the mixtures were not chemically reacting. Here we will consider the mixing of components which do react chemically. Suppose that we have hydrogen gas (H_2) and oxygen gas (O_2) mixed together. The chemical reaction

$$1\ H_2 + 1\ O_2 \rightarrow 2\ H_2O \qquad \textbf{(15–25)}$$

will follow, which means water will be produced. Consider the two elements carbon (C) and oxygen (O_2). The chemical reaction resulting will be

$$1\ C + 1\ O_2 \rightarrow 1\ CO_2 \qquad \textbf{(15–26)}$$

Finally, if we had octane fuel (C_8H_{18}), called a *hydrocarbon* or *fossil fuel*, mixed with oxygen the reaction

$$1\ C_8H_{18} + 12.5\ O_2 \rightarrow 8\ CO_2 + 9\ H_2O \qquad \textbf{(15–27)}$$

follows. There are many, many more examples which we could write for chemical reactions, but let us further investigate the above three examples. In equation (15–25) we see that 1 mole of hydrogen and 1 mole of oxygen produces 2 moles of water. This is equivalent to saying that 2 lbm of hydrogen and 32 lbm of oxygen produce 34 lbm of water;

that is, the molecular weight MW converts the chemical reaction equation to a mass balance equation. From equation (15–26) we get

$$(1 \text{ mole})(12 \text{ lbm C/mole}) + (1 \text{ mole})(32 \text{ lbm } O_2\text{/mole}) = (1 \text{ mole})(44 \text{ lbm } CO_2\text{/mole})$$

or

$$12 \text{ lbm} + 32 \text{ lbm} = 44 \text{ lbm}$$

Similarly, from equation (15–27) we see that a mass of 1 C_8H_{18} = $(12 \times 8) + (1 \times 18) = 114$ grams requires a mass of $(12.5)(32) = 400$ grams of oxygen. We have earlier found that 23.5% of air is normally considered to be oxygen, so

$$400 \text{ g} \times \frac{1}{0.235} = 1702 \text{ g air}$$

is required to furnish sufficient oxygen to burn or react with octane fuel. The ratio of the mass of air m_A to the mass of fuel (octane in this case), m_F, is called the air/fuel ratio and is a parameter frequently used in analysis of combustion processes. We then can write

$$\text{Air/fuel ratio} = \frac{m_A}{m_F} \tag{15–28}$$

and for a complete reaction of octane fuel, or complete combustion, we have

$$\text{Air/fuel ratio} = \frac{1702}{114} = 14.9 \text{ g air/g fuel}$$

Normally to insure combustion, an excessive amount of air is used with a fuel mixture.

A chemical reaction can be visualized as a rearrangement of atoms and molecules through a conservation of mass and energy. We have already mentioned the balance of mass in a chemical reaction so let us now consider the balance of energy. A molecule such as H_2, O_2, N_2, C_8H_{18}, H_2O, and CO_2 can be considered as having stored energy* in the amount required to form the molecule from the separate atoms. Each type of molecule has a different amount of stored energy so that during a chemical reaction there could be an imbalance in stored energy between the reactants (atoms or molecules on the left side of the chemical equation) and the products (molecules and atoms on the right side). If the reactants have more energy than the products, there will be an excess of energy after the reaction, which will be liberated. This type of reaction is called *exothermic* to describe the release of energy during the process. The combustion of octane by reacting with oxygen as described by equation (15–27) represents an exothermic reaction. The energy released, called

*This is called "free energy" or "free energy of formation" by many authors.

the *heat of combustion* ΔQ_c or heating value HV, is accounted for and the energy balance fulfilled by rewriting equation (15–27) to read

$$C_8H_{18} + 12.5\ O_2 \rightarrow 8\ CO_2 + 9\ H_2O + \Delta Q_c$$

The energy released then represents the desired product of a combustion reaction to produce power with heat engines.

The reader should be aware that there are many exothermic reactions, some fast (called *combustion*) and some slow, and a characteristic speed can be given to each one. In designing or analyzing heat engines, we must account for these speeds of combustion if completion of reactions is to be achieved, along with a more efficient release of the energy for thermal uses.

Some chemical reactions require energy to drive them; that is, the energy of the reactants is less than the energy of the products, in which case there exists the possibility of using this reaction as a heat sink or cooling device. These types of reactions are called *endothermic* to indicate that the reaction tends to cool or lower the system's temperature.

Practice Problems

Problems that use SI units are indicated with the notation (M) under the problem number and those problems using English units are indicated with (E). Those using mixed units are marked with (C) for combined under the problem number.

Section 15.1

15.1. **(M)** Determine the partial pressures of the constituents in a mixture of 0.05 kilogram oxygen, 0.10 kilogram CO_2, 0.10 kilogram argon, and 0.062 kilogram helium at 2.0 bars pressure.

15.2. **(M)** Calculate the entropy increase as 8 grams of water vapor mix with 70 grams of air at 1 atmosphere pressure and 20°C.

15.3. **(E)** Determine the partial pressure of water vapor in a mixture of 8% (by mass) H_2O, 80% air, and 12% CO_2 if the mixture is at 14.7 psia and 80°F.

15.4. **(E)** Determine the entropy increase due to mixing of 2 ft^3 of helium at 10 psia and 60°F and 3 ft^3 of argon at 20 psia and 60°F in a 3-ft^3 container at 60°F.

Section 15.2

15.5. **(C)** Obtain the saturation temperature of steam (water vapor) at
(a) 20 psia.
(b) 2.0 bars.

15.6. **(C)** Obtain the saturation pressure of water vapor at
(a) 60°F.
(b) 80°C.

15.7. **(C)** Determine the enthalpy and specific humidity for air having a relative humidity of 60% and a dry-bulb temperature of
(a) 90°F.
(b) 12°C.

15.8. (M) Air at 1.01 bars pressure and 20°C has a wet-bulb temperature of 15°C. Determine
(a) Vapor pressure.
(b) ω.
(c) β.

15.9. (M) A humidifier operates with air entering at 1.01 bars and 25°C. If the wet-bulb temperature is 10°C and the leaving air must be at 20°C and 50% relative humidity, determine the amount of heat transfer required and the water required per unit mass of air. Assume supplied water is 5°C.

15.10. (M) A dehumidifier having a maximum cooling capacity of 100 kJ/min provides 6 kilograms ot air per minute at 25°C and 55% relative humidity. If the incoming air is at 35°C, determine the maximum relative humidity it may have and still allow the humidifier to furnish the 6 kg/min of air at 25°C, 55% relative humidity.

15.11. (E) Atmospheric air is at a pressure of 29.4 millimeters Hg and 80°F. If the wet-bulb temperature is 68°F, determine
(a) ω.
(b) β.
(c) Dew point temperature.

15.12. (E) Air at 95°F, 14.7 psia, and 90% relative humidity must be brought to 75°F and 60% relative humidity. Determine the amount of heat rejected, the amount of heat added, and the amount of water condensed per pound-mass of dry air for an air conditioner-dehumidifier serving this purpose.

15.13. (E) There are 1500 ft³/min of air at 92°F, 14.7 psia, and 10% relative humidity to be conditioned to 70°F and 60% relative humidity. Determine the amount of heat transfer required and the rate of water flow necessary to humidify this air.

Section 15.3

15.14. (C) Determine the chemical potential of saturated water vapor at
(a) 500°F.
(b) 240°C.

15.15. (C) Determine the chemical potential of saturated ammonia vapor at
(a) 6.0 bars.
(b) 23.74 psia.

15.16. (C) Calculate the chemical potential of saturated Refrigerant-22 liquid at
(a) 60°C.
(b) 140°F.

15.17. (E) Determine the chemical potential of water at 70°F which is in chemical equilibrium with 70°F air and water mixture at 60% relative humidity.

Section 15.4

15.18. (M) Predict the concentration of ammonia in air 1 centimeter from an interface of liquid ammonia and air if the interface area is 15 cm² and the evaporation rate of ammonia is 2 g/min. Assume the air and ammonia system is at 20°C.

15.19. (M) Liquid mercury (Hg) is contained in a lead beaker as shown in figure 15–13. Calculate the amount of mercury lost through diffusion to the beaker after 48 hours, if the concentration of mercury is 2% at a distance 0.01 centimeter into the lead beaker wall from the inside surface. Assume the system is at a temperature of 20°C.

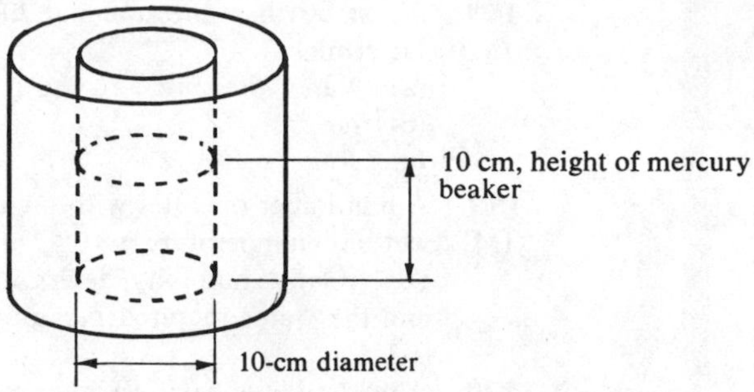

Figure 15–13

15.20. **(E)** A 3-ft-radius sphere contains helium gas at 20 psia and 95°F. Determine the amount of helium lost through diffusion in 24 hours if the sphere is made of Pyrex and if it is assumed that at $\frac{1}{8}$ inch from the inside wall, helium is totally absent.

15.21. **(E)** Water is evaporating at 3 lbm/hr in a 3-inch square container. Estimate the specific humidity 1 inch from the water-air surface in the air if the dry-bulb temperature is 70°F.

Section 15.5

15.22. For the fuel C_8H_{17}, having the reaction

$$2\,C_8H_{17} + 24\tfrac{1}{2}\,O_2 \rightarrow 16\,CO_2 + 17\,H_2O$$

determine the ideal air/fuel ratio.

15.23. For propane gas C_3H_8, having the reaction

$$C_3H_8 + 5\,O_2 \rightarrow 3\,CO_2 + 4\,H_2O$$

determine the ideal air/fuel ratio.

15.24. For butane gas having the reaction

$$C_4H_{10} + 6\tfrac{1}{2}\,O_2 \rightarrow 4\,CO_2 + 5\,H_2O$$

calculate the air/fuel ratio.

16

Other Power Devices

In this chapter we are concerned with some diversified examples of systems to which the concepts of thermodynamics can be applied. Of course, any volume in space can be identified as a system and thus subject to the laws of thermodynamics, but the examples presented here are typical devices (or systems) which might be encountered by engineers and technologists and which thus should be considered from a thermodynamic approach. *Electric generators*, *motors*, and *batteries* are described, and a brief discussion is given to present the electrical analogies of work, heat, and energy. We then consider the operation of the *hydrogen-oxygen fuel cell* and some simple *thermoelectric devices*. Finally, *magnetohydrodynamic* and *electrohydrodynamic systems* are described in very brief terms.

Engineers and technologists must apply their knowledge and expertise to all phases of social needs, and to emphasize this trend, some *biological systems* are identified for thermodynamic analysis.

16.1 Electric Generators, Motors, and Batteries

Elementary to a consideration of electrical devices is an understanding of the concept of *current*. We visualize atoms as being comprised of a nucleus surrounded by electrons, *e*. Each electron, when stripped from its nuclear influence, possesses a charge, called a *coulomb;* the amount of charge possessed by each electron has been found to be 1.6×10^{-19} coulomb. If a metal or other material contains an abundance of free electrons, it is said to have a negative charge and if there is a shortage of electrons it is positive. Free electrons will migrate from an area in which they are dense to an area in which they are scarce; that is, electrons flow from negative to positive, as shown in figure 16–1, where we also see that

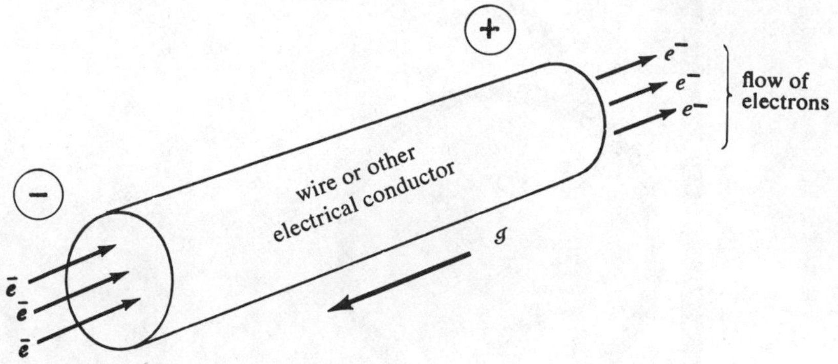

Figure 16–1 Electric current flow.

current $\mathcal{I}$ is depicted as flowing from positive to negative. Current is defined by the equation

$$\mathcal{I} = \frac{d\mathcal{Q}}{d\tau} \tag{16–1}$$

where $\mathcal{Q}$ is amount of charge and where the unit of current, coulomb/s, is called the *ampere*. Since electrons possess negative charges, we then say the positive current of electricity must travel from positive to negative.

Another concept we need to consider briefly in electrical phenomena is that of the electrical potential $\mathcal{E}$. We write

$$\mathcal{E} = \frac{dE_e}{d\mathcal{Q}} \tag{16–2}$$

where E_e is the electrical energy. The unit for the electrical potential can easily be seen to be Btu/coulomb or more commonly, joule/coulomb. The volt is defined as the joule/coulomb.

The essential components of a common electric generator are a *rotor* and a *stator*. The stator is a stationary housing consisting of magnets or electromagnets, as shown in figure 16–2. The rotor is situated between the stator magnets so that as it rotates, it interrupts the magnetic lines of force and thus induces an electric current through a conducting wire. Frequently the stator is composed of the conducting wires, and the rotor is then the source of magnetic fields, but here we will consider the arrangement as shown in figure 16–2. We first write the first law for the generator

$$\Delta E_e = Q_f - Wk_{\text{gen}} \tag{16–3}$$

and then, if we assume the process to be reversible, $Q_f = 0$. This gives us

$$\Delta E_e = -Wk_{\text{gen}}$$

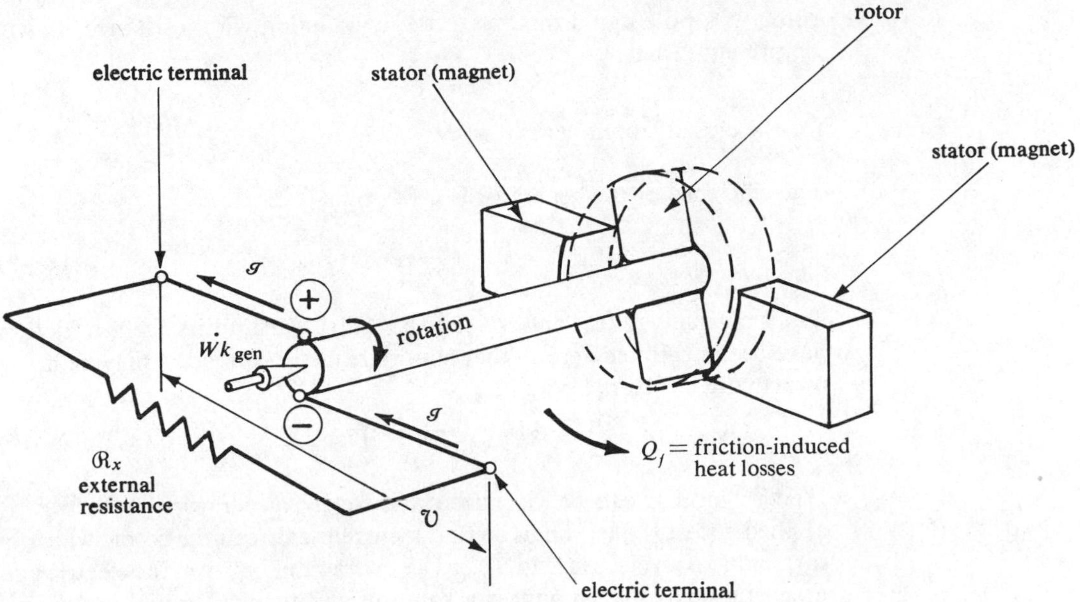

Figure 16–2 The electric generator.

and, from equation (16–2), we obtain

$$\Delta E_e = \int dE_e = \int \mathcal{E}\, d\mathcal{Q} = -Wk_{gen} \quad \textbf{(16–4)}$$

Since

$$\dot{E}_e = \frac{dE_e}{d\tau} = \mathcal{E}\frac{d\mathcal{Q}}{d\tau}$$

we can use the definition of current, equation (16–1), to obtain

$$\mathcal{E}\mathcal{I} = -\dot{W}k_{gen} \quad \textbf{(16–5)}$$

If we measure the potential or voltage across the terminals, its value will be $\mathcal{E}$, provided there exists no external connection between the poles. If, on the other hand, we insert a load or resistance $\mathcal{R}_x$ between the poles, as depicted in figure 16–2, the voltage will be less than $\mathcal{E}$. Ohm's law states that

$$\text{Potential drop} = \text{resistance} \times \text{current} \quad \textbf{(16–6)}$$

so that, calling the voltage $\mathcal{V}$, we have across the resistance that

$$\mathcal{V} = \mathcal{R}_x\mathcal{I} \quad \textbf{(16–7)}$$

where $\mathcal{R}_x$ is the resistance measured in ohms, $\mathcal{I}$ is the current in amperes, and $\mathcal{V}$ is the voltage. The current is also flowing through the generator rotor so a potential drop, due to the conducting wire resistance $\mathcal{R}_I$, will be present. Then

$$\mathcal{V} = \mathcal{E} - \mathcal{I}\mathcal{R}_I \tag{16–8}$$

for the closed circuit generator.

The efficiency of the generator is given by

$$\eta = \frac{\mathcal{V}\mathcal{I}}{\dot{W}k_{gen}} \times 100 \tag{16–9}$$

and is normally in the range of 85 to 95%. Irreversibilities such as friction have been neglected here so that a more realistic correction may be made to actual generators by writing

$$\dot{W}k_{gen} = \mathcal{E}\mathcal{I} + \dot{W}k_{friction} \tag{16–10}$$

Electric motors can be visualized as reversed generators. A voltage is applied across the terminals so that a current exists in the rotor, which in turn induces an electric field, $\mathcal{E}_{ind}$. The interaction between the electric and magnetic fields then produces a rotation and torque through the rotor. For the motor we have

$$\mathcal{E}_{ind} = \mathcal{V} - \mathcal{I}\mathcal{R}_I \tag{16–11}$$

and the mechanical efficiency is

$$\eta_{mech} = \frac{\dot{W}k_{motor}}{\mathcal{V}\mathcal{I}} \times 100 \tag{16–12}$$

where the electric power used by the motor is $\mathcal{V}\mathcal{I}$. The motor output power $\dot{W}k_{motor}$ is given by

$$\dot{W}k_{motor} = \mathcal{E}_{ind}\mathcal{I} - \dot{W}k_{friction} \tag{16–13}$$

While the electric generator and the motor involve energy transfers between electrical and mechanical systems, the common electrical battery utilizes a transfer of energy between chemical and electrical systems.

The Daniel cell is a device which produces electrical energy from chemical reactions. It is not exactly like the common batteries, but its workings involve the essential concepts of the typical battery. The Daniel cell, shown in figure 16–3, is composed of zinc sulfate ($ZnSO_4$) and copper sulfate ($CuSO_4$) solutions separated by a barrier which prevents mixing of these two solutions, yet which allows ions (such as Cu^{++}, Zn^{++}, and SO_4^{--}) to pass. A solid bar of zinc (anode) is placed in the $ZnSO_4$ solution and a bar of copper (cathode) in the $CuSO_4$ solution. The chemical reaction at the anode is

$$Zn \rightarrow Zn^{++} + 2e^- \tag{16–14}$$

and at the cathode is

$$Cu^{++} + 2e^- \rightarrow Cu \tag{16–15}$$

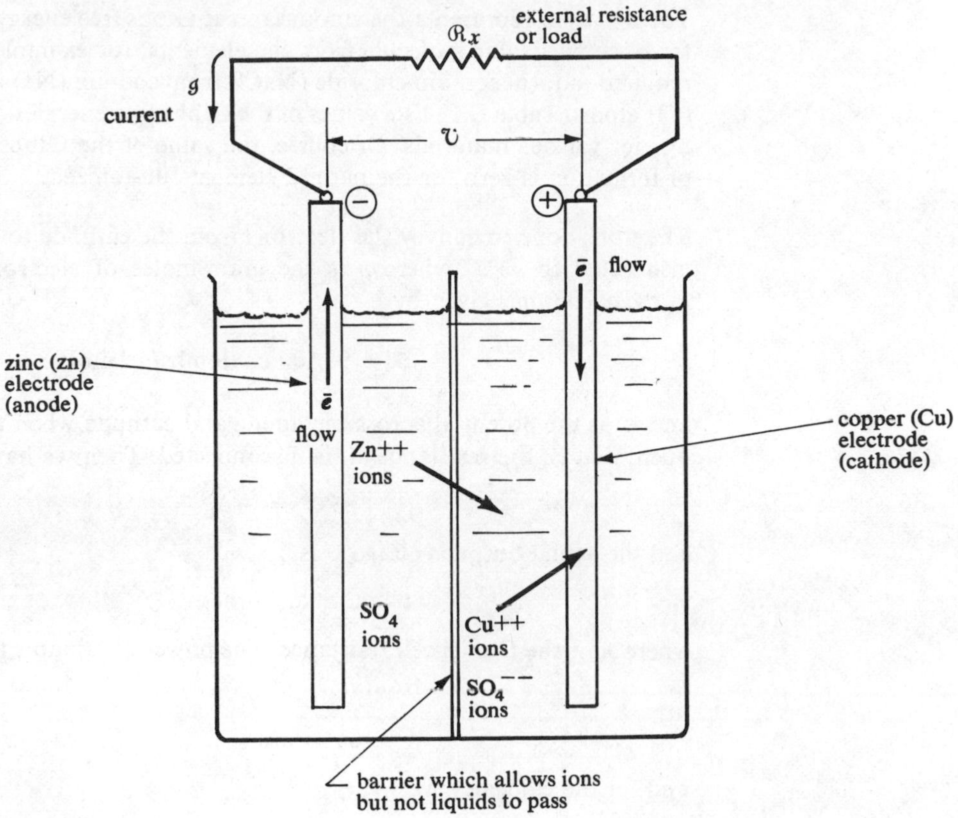

Figure 16–3 Daniel cell battery.

The total chemical reaction for the Daniel cell is found by adding equations (16–14) and (16–15)

$$\underbrace{Zn + Cu^{++}}_{\text{reactants}} \rightarrow \underbrace{Cu + Zn^{++}}_{\text{products}} \tag{16–16}$$

with the result that the cathode becomes plated with copper, the anode loses zinc, and the electrons are allowed to flow through an external load. The maximum work or electrical energy obtainable from the chemical reaction (16–16) is given by equation (9–28)

$$Wk_{\text{use}} = -\Delta\mathcal{G}$$

and since the useful work is also the maximum work obtainable for useful purposes, we write

$$Wk_{\max} = -\Delta\mathcal{G}$$

where the Gibbs free energy can be shown to be calculated from the summation

$$\Delta\mathcal{G} = \underbrace{\Sigma\Delta\mathcal{G}_f^\circ}_{\text{products}} - \underbrace{\Sigma\Delta\mathcal{G}_f^\circ}_{\text{reactants}} \tag{16–17}$$

The $\Delta\mathcal{G}_f^\circ$ term represents the amount of the Gibbs free energy required to form the particular molecule from the elements, for example, the energy required to form sodium chloride (NaCl) from sodium (Na) and chlorine (Cl) atoms. Table B.19 lists values of the Gibbs free energies of formation $\Delta\mathcal{G}_f^\circ$ for various materials. Of course, the value of the Gibbs free energy of formation is zero for the natural elements un-ionized.

The work done to convey the electrons from the cathode to the anode is also equal to $n\mathfrak{F}\mathcal{V}^\circ$ where n is the gram-moles of electrons, $\mathfrak{F}$ is the *Faraday constant* given by

$$\mathfrak{F} = 96{,}500 \text{ coulombs/g} \cdot \text{mole}$$

and $\mathcal{V}^\circ$ is the potential across the anode and cathode when the circuit is open; that is, the resistance $\mathcal{R}_x$ is disconnected. Then we have

$$n\mathfrak{F}\mathcal{V}^\circ = -\Delta\mathcal{G} \tag{16–18}$$

and the actual output voltage $\mathcal{V}$ is

$$\mathcal{V} = \mathcal{V}^\circ - \mathcal{I}\mathcal{R}_\text{I} \tag{16–19}$$

where $\mathcal{R}_\text{I}$ is the Daniel cell resistance. The power obtainable from the cell is

$$\mathcal{V}\mathcal{I} = \mathcal{I}^2\mathcal{R}_x \tag{16–20}$$

and for the efficiency we write

$$\eta_\text{battery} = \frac{\mathcal{V}\mathcal{I}}{\Delta\mathcal{G}} \tag{16–21}$$

Example 16.1

Determine the maximum work obtainable from a Daniel cell, the open circuit voltage, and the operating voltage if 0.1 ampere of current is drawn and the internal resistance is 0.05 ohm.

Solution

We calculate the maximum work from the sum of the Gibbs free energy values. Using equation (16–17) and the values from table B.19, we have

$$\Delta\mathcal{G} = \text{Cu} + \text{Zn}^{++} - \text{Zn} + \text{Cu}^{++}$$

or

$$\Delta\mathcal{G} = \underbrace{0 + (-35.18)}_{\text{products}} - \underbrace{0 - 15.53}_{\text{reactants}} = -50.71$$

Then

$$Wk_\text{max} = -\Delta\mathcal{G} = 50.71 \text{ kcal/g} \cdot \text{mole}$$ *Answer*

The open circuit voltage can be calculated from equation (16–18):

$$\mathcal{V}^\circ = \frac{-\Delta\mathcal{G}}{n\mathfrak{F}} = \frac{50.71}{(96{,}500)}$$

and $n = 2$ g · moles of electrons per gram-mole of reaction, so

$$\mathcal{V}^\circ = \frac{50.71 \text{ kcal/g} \cdot \text{mole} \times 4184 \text{ J/kcal}}{2 \text{ g} \cdot \text{mole/g} \cdot \text{mole} \times 96{,}500 \text{ coulombs/g} \cdot \text{mole}}$$

$$= 1.098 \text{ J/coulomb} = 1.098 \text{ volts} \qquad \textit{Answer}$$

If we now use equation (16–19), we can calculate the operating voltage when 0.1 amp of current is drawn:

$$\mathcal{V} = 1.098 \text{ volts} - (0.1 \text{ amp})(0.05 \text{ ohm})$$
$$= 1.093 \text{ volts} \qquad \textit{Answer}$$

This example problem and the preceding discussion are items and areas which engineers and engineering technologists in particular may encounter in practice — they were presented to illustrate the application of the thermodynamic concepts to electrical devices. In no way, however, should the reader infer that this brief presentation is a sufficient treatment of electrical concepts nor should he assume that other thermodynamic ideas such as entropy cannot be applied to electrical phenomena.

16.2 Fuel Cells

The fuel cell has gathered increased attention due to its attractiveness as a source of electric power. The power is developed by means of a "controlled" chemical reaction during which free electrons are forced to travel an external path to complete the reaction. Many fuels have been investigated for use in fuel cells, including biological or organic fuels, but one of the most successful combinations has been hydrogen and oxygen. The device using these particular fuels is called the *hydrogen-oxygen fuel cell* and is shown in figure 16–4. As can be seen, pure hydrogen and oxygen are added to the cell, and water and electrical power are produced. Pumps recirculate the portions of the oxygen and hydrogen which do not react in the cell. The chemical reaction at the anode is

$$H_2 - 2\,OH^- \rightarrow 2\,H_2O + 2e^- \qquad \textbf{(16–22)}$$

and at the cathode

$$H_2O + \frac{1}{2}O_2 + 2e^- \rightarrow 2\,OH \qquad \textbf{(16–23)}$$

The full chemical reaction of the fuel cell is then given by the addition of equations (16–22) and (16–23)

$$H_2O + H_2 + \frac{1}{2}O_2 + 2\,OH^- + 2e^- \rightarrow 2\,OH^- + 2\,H_2O + 2e^-$$

or

$$\underbrace{H_2 + \frac{1}{2}O_2}_{\text{reactants}} \rightarrow \underbrace{H_2O}_{\text{product}} \qquad \textbf{(16–24)}$$

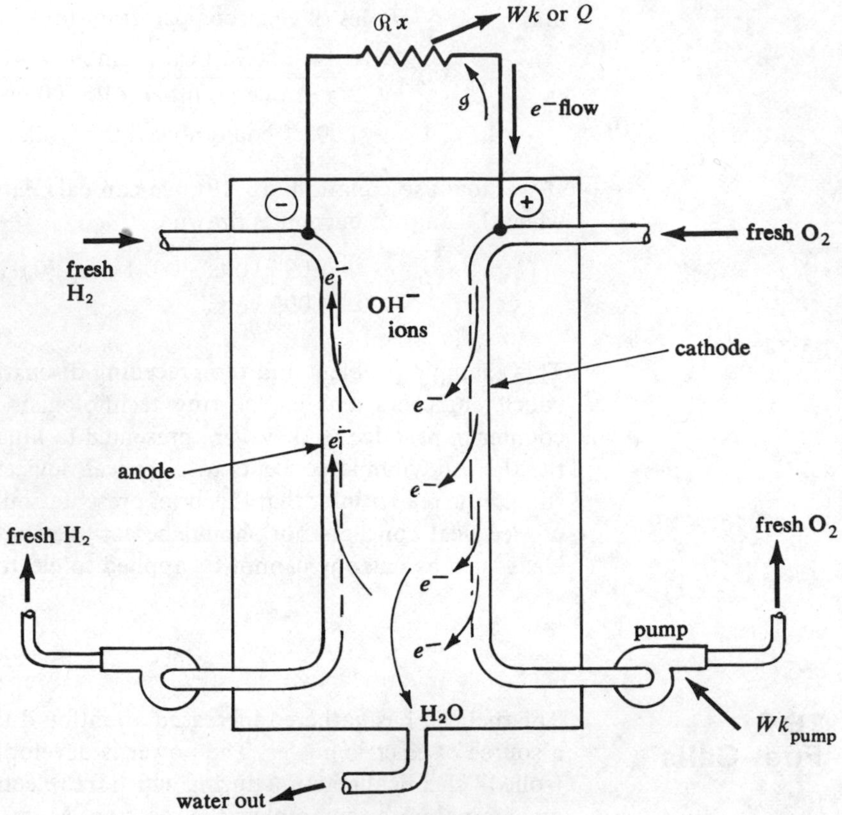

Figure 16–4 The hydrogen-oxygen fuel cell.

The maximum work is obtained from equation (9–28)

$$Wk_{max} = -\Delta\mathcal{G}$$

and the open circuit voltage from equation (16–18)

$$\mathcal{V}^\circ = \frac{-\Delta\mathcal{G}}{n\mathcal{F}}$$

Example 16.2 Calculate the maximum work obtainable from the hydrogen-oxygen fuel cell and calculate the open circuit voltage.

Solution From table B.19 we have

$$\Delta\mathcal{G}_f^\circ = -56.69 \text{ for } H_2O \text{ (the products)}$$
$$= 0 \text{ for } H_2 \text{ and } O_2 \text{ (the reactants)}$$

so, using equation (16–17) we get

$$\Delta\mathcal{G} = -56.69 \text{ kcal/g} \cdot \text{mole}$$

Since the molecular weight (MW) of water is 18 g/g · mole we have

$$\Delta\mathcal{G} = -\frac{56.69}{18}\ \text{kcal/g } H_2O = -3.15\ \text{kcal/g } H_2O$$

Then

$$Wk_{\text{fuel cell}} = 3.15\ \text{kcal/g} \qquad \textit{Answer}$$

We may express this answer in more familiar units by using the energy conversions listed in table B.15. We then get

$$Wk_{\text{fuel cell}} = 3.15\ \text{kcal/g} \times 4.19\ \text{kJ/kcal}$$
$$= 13.2\ \text{kJ/g} \qquad \textit{Answer}$$

The open circuit voltage is

$$\mathcal{V}^\circ = \frac{-\Delta\mathcal{G}}{n\mathcal{F}} = \frac{56.69}{2 \times 96{,}500}\ \text{kcal/coulomb}$$
$$= 0.000294\ \text{kcal/coulomb} \times 4184\ \text{J/kcal}$$
$$= 1.23\ \text{J/coulomb}$$
$$= 1.23\ \text{volts} \qquad \textit{Answer}$$

Fuel cells are attractive devices for providing power. A major technological problem associated with them, however, is in the fabricating of the electrodes (anode and cathode) with reasonable effort, and until this difficulty is alleviated, the fuel cell cannot be expected to replace the battery for common uses.

16.3 Thermoelectric Devices

We have seen that resistance in a conducting wire produces a voltage drop given by Ohm's law, equation (16–7)

$$\mathcal{V} = \mathcal{I}\mathcal{R}$$

The power represented by this is given by

$$\mathcal{V}\mathcal{I} = \mathcal{I}^2\mathcal{R} \qquad \textbf{(16–25)}$$

and this power is dissipated in a resistor $\mathcal{R}$ in the form of an increase in the temperature of the resistor. If the resistor is in equilibrium and steady state conditions with the surroundings, the power $\mathcal{I}^2\mathcal{R}$ is reflected in heat transfer to the surrounding as shown in figure 16–5. We call this effect *Joule heating* $\dot{Q}_J$, which represents the most common means of electrical energy dissipation. Joule heating is used to advantage in any electric heater and is present in any conductor of electricity having resistance.

If two wires or conductors, A and B, composed of different materials, are joined as shown in figure 16–6 and a current is impressed through

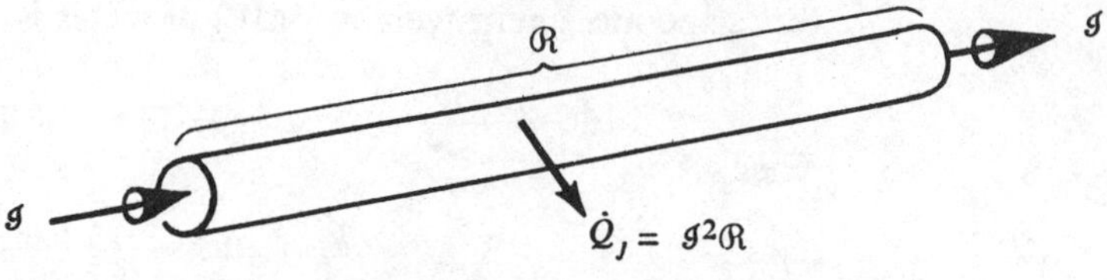

Figure 16–5 Joule heating.

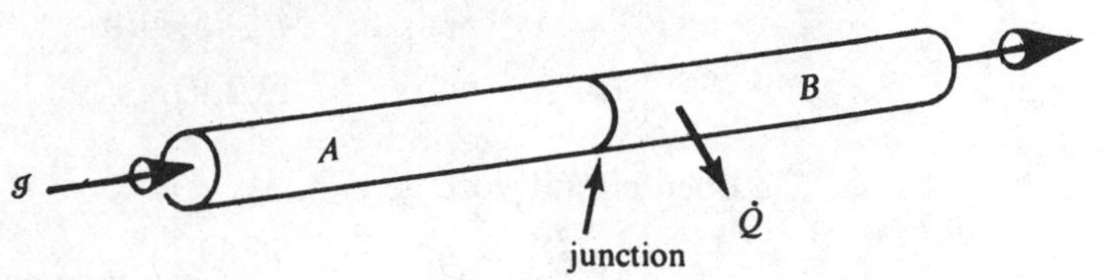

Figure 16–6 Peltier effect.

them, another heat transfer can be induced. This heat, called *Peltier heating*, is due to thermoelectric phenomena called the *Peltier* and *Seebeck* effects and is the basis for the thermocouple. The heat emanating from the joint $\dot{Q}$ is given by the sum of the Joule and Peltier heats

$$\dot{Q} = \mathcal{I}^2\mathcal{R} + \mathcal{I}(\alpha_A - \alpha_B) \tag{16–26}$$

where α_A and α_B are the Seebeck coefficients of wires A and B. The Seebeck coefficients are the mathematical values allowing for the Peltier heat and are generally found to be functions of the material temperature. A further discussion of the Seebeck coefficient can be found in publications devoted to direct energy conversion devices. It may be noted that the Peltier effect can be reversed by impressing heat transfer into the joint and thus inducing a current in the wires. This is the essential operation of the temperature measuring device called the *thermocouple*. In the thermocouple, the current is measured, allowing for a prediction of the joint temperature.

Another thermoelectric phenomenon found in all actual conductors is the *Thomson effect*. A wire or other electric conductor subject to a temperature change in its volume, as shown in figure 16–7, will have a

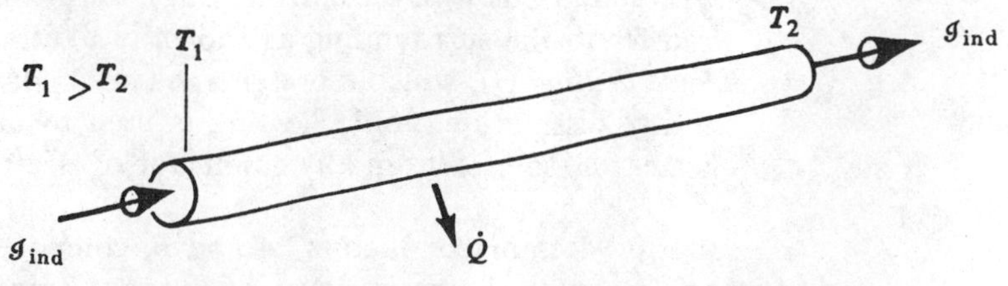

Figure 16–7 Thomson effect.

current induced in it. The heat transfer due to this effect is called the *Thomson heat* and is generally written as $\mathcal{I}_{\text{ind}}\mathfrak{I}\Delta T$ where $\mathfrak{I}$ is the Thomson coefficient (determined by experimental means) and ΔT is $T_1 - T_2$. The heat transfer on the sides of the conductor is then

$$\dot{Q} = \mathcal{I}_{\text{ind}}^2\mathcal{R} + \mathcal{I}_{\text{ind}}\mathfrak{I}\Delta T \tag{16–27}$$

Semiconductor materials are utilized in some useful thermoelectric devices, called *thermoelectric generators.* These materials (semiconductors) exhibit a selectivity in allowing current flow, generally by prohibiting current to pass in one of the two directions and they also have wide varieties of values for Seebeck coefficients. In figure 16–8 is shown a thermoelectric device composed of two distinct types of semiconductor materials, N-type and P-type. The bar labeled A is a conductor of electricity and a receiver of external heat transfer. From the previously discussed concept of Peltier heating, we see that in this case the Peltier heat is added to the junctions between A and the N-type material and the P-type material. At these two joints we then have, from equation (16–26),

$$\dot{Q}_{\text{add}} = \mathcal{I}_{\text{ind}}^2\mathcal{R} + \mathcal{I}_{\text{ind}}(\alpha_n - \alpha_A) + \mathcal{I}_{\text{ind}}(\alpha_A - \alpha_p) \tag{16–28}$$

where $\mathcal{I}_{\text{ind}}$ is the current induced in the circuit due to the *addition* of Peltier heat. The semiconductor materials, N- and P-types, have properties such that α_n is negative and α_p is positive. This means that the Seebeck effect at the two junctions, instead of canceling, will add and give us, from equation (16–28)

$$\dot{Q}_{\text{add}} = \mathcal{I}_{\text{ind}}^2\mathcal{R} - \mathcal{I}_{\text{ind}}(|\alpha_n| + |\alpha_p|) \tag{16–29}$$

Thermoelectric generators are attractive as sources of electric power since they are not inherently dependent on the source of thermal energy. Heat may be transferred from a nuclear reactor, steam, hot exhaust gas, the sun, or numerous other sources.

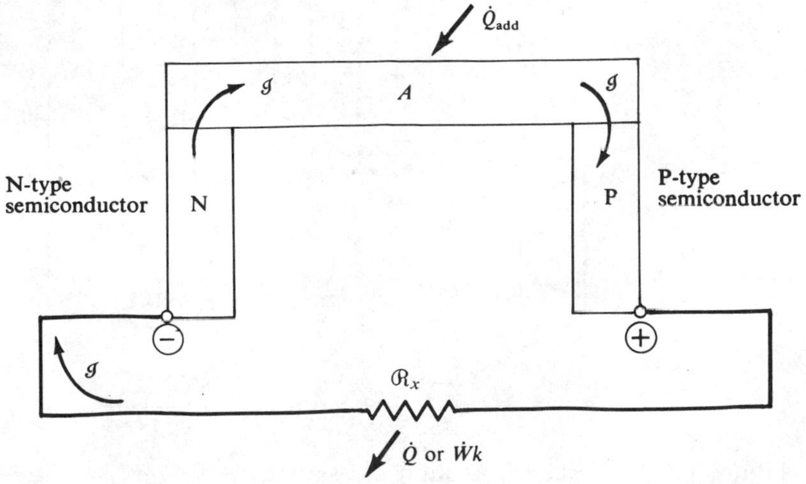

Figure 16–8 Elements of a simple thermoelectric generator.

16.4 Magnetohydrodynamics (MHD)

The electric generator produces electric power by passing a conducting wire through a magnetic field. A magnetohydrodynamic (MHD) generator, such as shown in figure 16–9, produces electric power in much the same way except the conductor passing through the magnetic field is a hot gas or fluid. The basic operation of this device is indicated in figure 16–10 where hot gases (ionized gases if possible) flow through a chamber having two sides which conduct electricity (sides *a-b-c-d* and *f-e-g-h*) and two sides which are insulated (sides *b-e-g-c* and *a-f-h-d*). An electric field $\mathcal{E}_e$ is applied across the two conducting sides and this field induces a magnetic field having lines of force perpendicular to the gas flow. The flow of hot gases through these fields induces an electric potential $\mathcal{E}_{ind}$

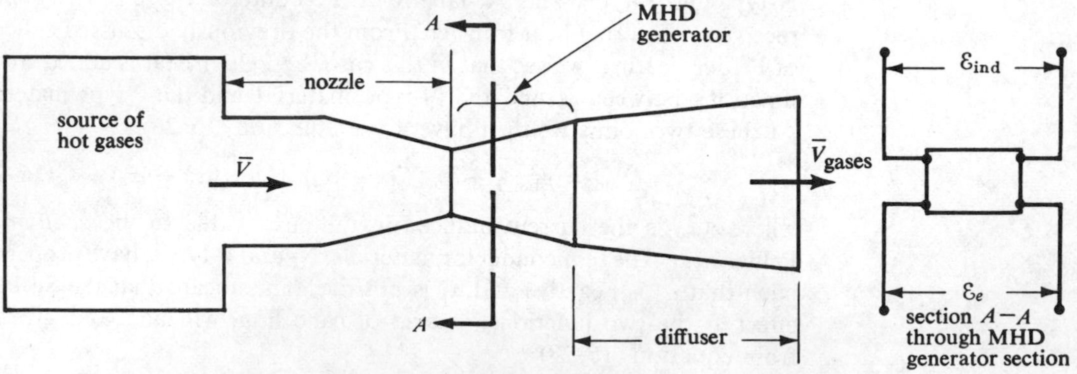

Figure 16–9 Sketch of magnetohydrodynamic power generator.

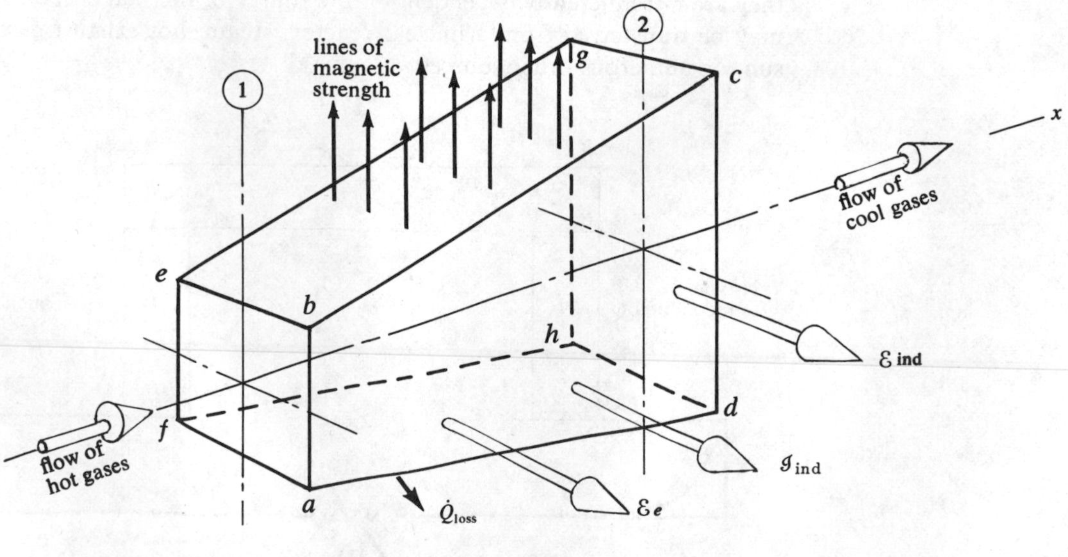

Figure 16–10 Basic operation of MHD generator.

of significant strength. By applying the first law of thermodynamics to this system we have

$$\dot{m}\left(\frac{V_2^2 - V_1^2}{2}\right) + \dot{m}(h_2 - h_1) = \dot{Q}_{loss} - \mathcal{I}_{ind}\mathcal{E}_{ind} + \mathcal{I}_{ind}\mathcal{E}_e \tag{16-30}$$

where $\mathcal{I}_{ind}$ is the induced current due to the MHD effect and $\dot{m}$ is the mass flow rate of the gases.

The MHD generator represents a device that can utilize very high temperature gases with present materials and produce significant amounts of power. In addition, the exhaust gases from the device can be used to provide a heat source for a conventional heat engine, such as a gas turbine or a steam turbine. When used tandem in this manner with other devices, the MHD generator can represent a significant improvement in thermodynamic efficiencies.

Another interesting aspect of the phenomena being considered here is the reversed procedure; that is, if magnetic and electric fields are applied across a conduit containing a fluid which can conduct electricity, a velocity or flow is induced in the fluid. This technique is responsible for the concept of pumps having no moving parts.

The primary disadvantages of the MHD device are twofold:

1. The gas or fluid must have a sufficiently low resistance and must be a good electrical conductor. Most gases do not have these properties.
2. The source of hot gases must be exceedingly large if significant amounts of power are to be generated.

Significant treatises exist which the reader may refer to for a more complete treatment of magnetohydrodynamic devices.

16.5 Biological Systems

Engineers and technologists have been applying engineering and scientific concepts to inorganic, inanimate objects or systems without too much hesitation. On the other hand, organic, biological systems have been avoided with few exceptions. The following two systems, the muscle and the heart, are biological devices which have been investigated from engineering viewpoints and are presented here to set examples for application of thermodynamic concepts to other biological systems, heretofore avoided.

The muscle represents an organic system which converts chemical energy to mechanical work and this is achieved at essentially constant temperature. Many have attempted to explain scientifically the details of this conversion with no great success, and indeed the muscle represents the only known device which directly converts chemical energy to work. It appears that the source of chemical energy resides in a macromolecular substance called adenosine triphosphate (ATP) and other simpler phosphates. This reaction releases energy to be used in work and heat. The muscle system is shown in figure 16–11, along with a weight W representing an external force applied to the muscle. It is viewed in a process of contracting or shortening while exerting a force; the actual process has been simplified by replacing the muscle force by a weight W. We may apply the first law to the muscle and get

$$H_2 - H_1 = Q - Wk \qquad \textbf{(16–31)}$$

where H_2 is the enthalpy of the ATP and phosphates and H_1 is the enthalpy of the resulting chemical, adenosine diphosphate (ADP). The work gotten from a muscle is Wk and the maximum work is predicted from equation (9–28)

$$Wk_{max} = -\Delta G'$$

We may consider the mechanical efficiency of the muscle system to be

$$\eta_{mech} = \frac{Wk}{-\Delta G'} \times 100 \qquad \textbf{(16–32)}$$

and this seems to be around 30% to 40% for most healthy muscles. The actual work obtained from a muscle is, of course, dependent on the

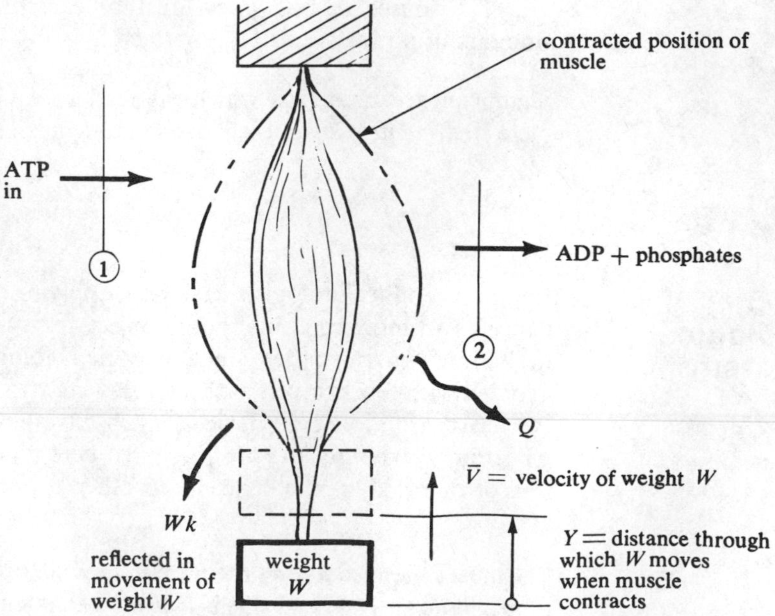

Figure 16–11 The muscle as a system.

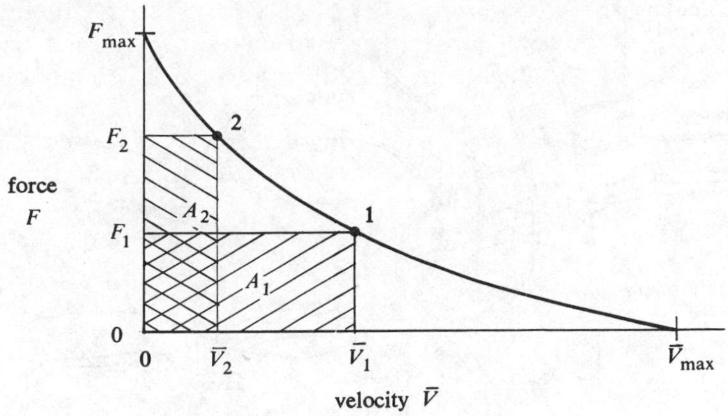

Figure 16–12 Force-velocity relationship for typical muscle contraction.

amount of contraction or distance through which force is applied. If the force applied to the muscle is more than the muscle can move, there will be no contraction of the muscle and no work done. (The muscle will use up $\Delta \mathcal{G}$, however.) We can also say that the weight W will have no velocity in this case. Now, if the force applied to the muscle is reduced by using a smaller weight W, the muscle will then pull the weight through a distance and with some velocity. Continuing, if the force is reduced further the weight will be pulled faster by the muscle until it happens that when no weight or force is applied to the muscle, contraction will proceed at a rapid rate. This relationship, the force applied to the muscle versus the velocity at which the weight moves, is shown in figure 16–12. The velocity $\bar{V}_{max}$ represents the velocity of muscle contraction when no force ($F = 0$) is applied and the force F_{max} represents the greatest amount of force a muscle can pull. The power gotten from a muscle pulling a constant force or weight is

$$\dot{W}k = F\bar{V} \tag{16–33}$$

and is represented by the rectangular area under the curve in figure 16–12. As an example, the power gotten from a muscle working under a pull of F_1 and with velocity $\bar{V}_1$ is the area A_1. Similarly, for a force F_2 and velocity $\bar{V}_2$, the power is the area A_2.

The heart is an organ which is composed of various muscles acting together. While the actual configuration of the heart is quite complex and blood flowing through it is equally complicated, as indicated in figure 16–13, for our purposes we can consider it to be a form of pump. The heart functions as essentially two separate blood pumps. The solid arrows in the diagram indicate directions of blood flow. Oxygen-poor blood comes from the body to the *right atrium*. When the heart expands this blood is drawn into the *right ventricle*. Then the blood is expelled into the pulmonary artery when the heart contracts. The pulmonary

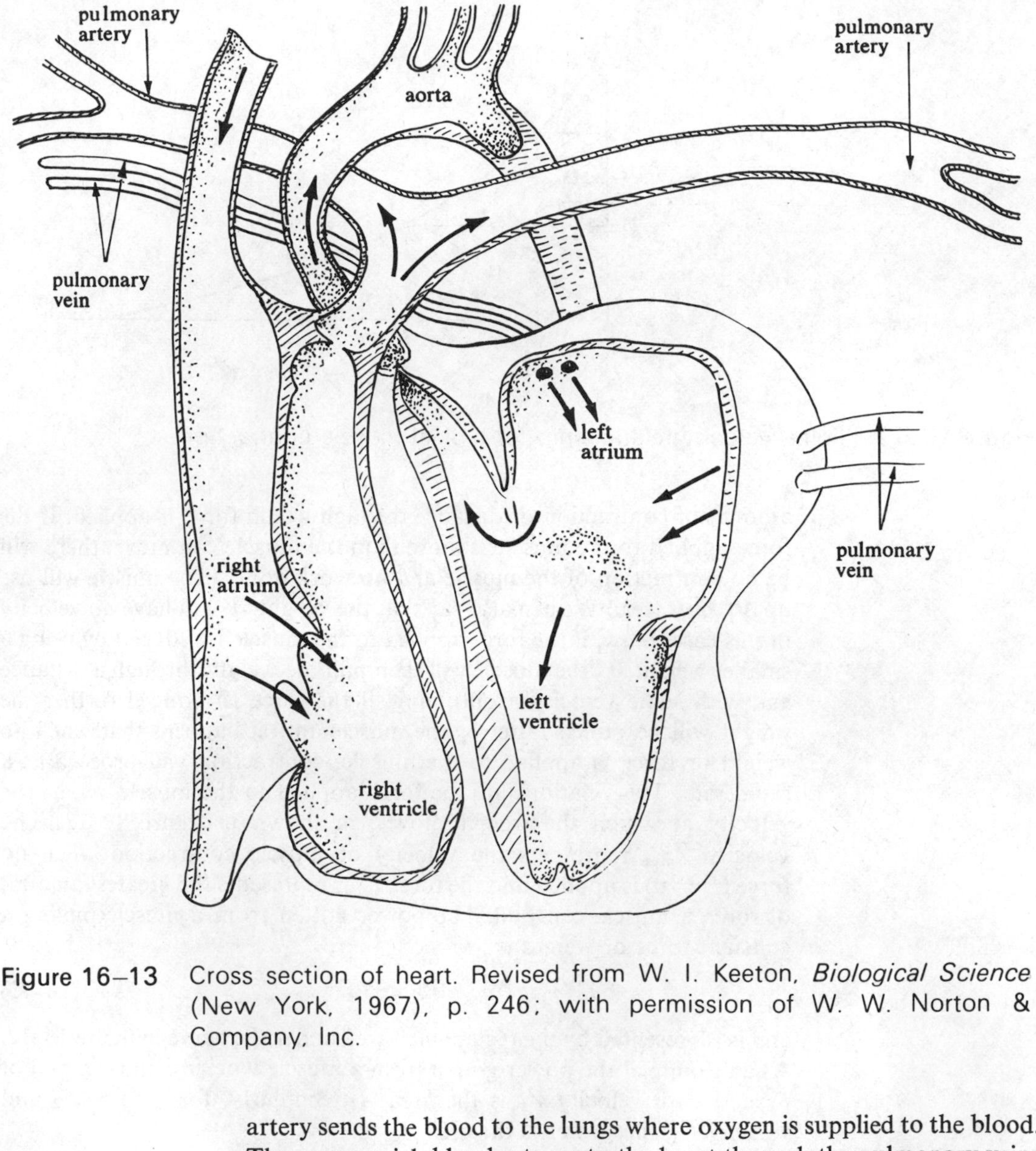

Figure 16–13 Cross section of heart. Revised from W. I. Keeton, *Biological Science* (New York, 1967), p. 246; with permission of W. W. Norton & Company, Inc.

artery sends the blood to the lungs where oxygen is supplied to the blood. The oxygen-rich blood returns to the heart through the pulmonary veins and into the *left atrium*. Upon heart expansion the blood flows into the *left ventricle* and then, when the heart contracts, the *left ventricle* reduces in volume and sends the oxygen-rich blood into the body via the *aorta*.

We can simplify the heart to the system as sketched in figure 16–14 and the energy equation gives us, approximately

$$\dot{m}\left(\frac{\bar{V}_2^2 - \bar{V}_1^2}{2}\right) + (h_1 - h_2)\dot{m} = \dot{Q} - \dot{W}k \qquad \textbf{(16–34)}$$

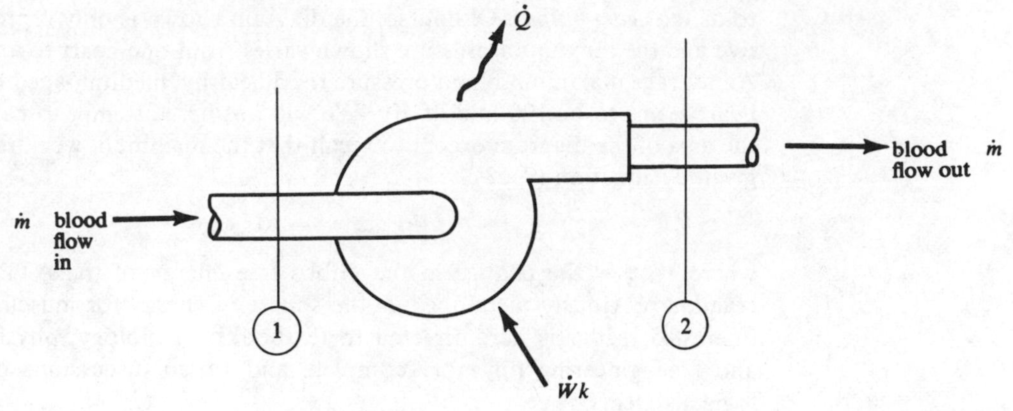

Figure 16–14 Simplistic representation of the heart by a blood pump.

where $\dot{m}$ is the mass flow of the blood. It should be mentioned that blood flow through the heart is not steady, although over a long period of time it may be approximated as such; nor is the flow of blood as simple an analogy as water flowing through a rigid pipe. The mechanism of the heart is so complex that man will never fully understand its workings, but our simplistic approach is better than none in attempting to analyze the heart. The pressure-volume diagram of the blood passing through the heart is shown in figure 16–15. The enclosed area is frequently referred

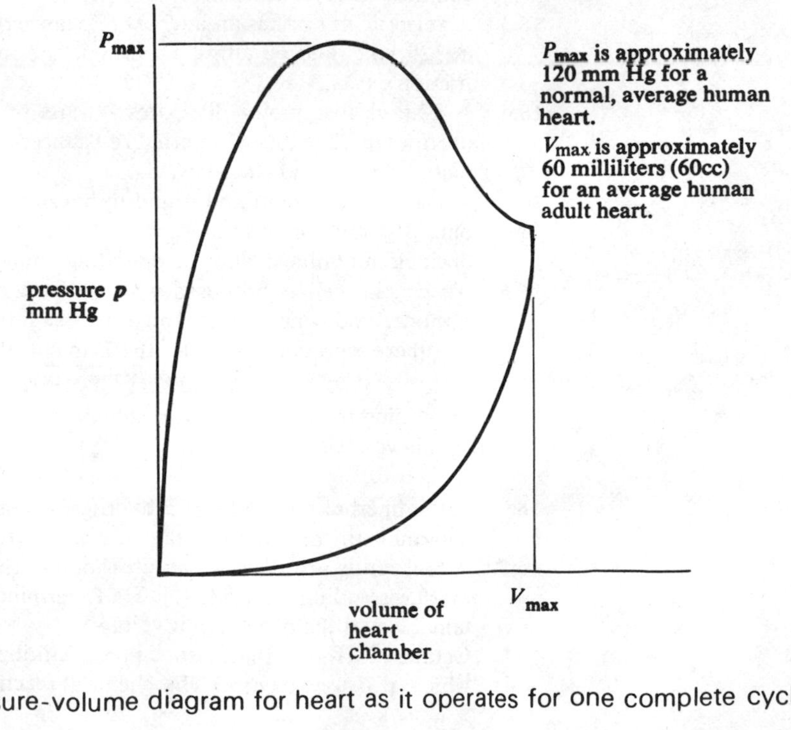

Figure 16–15 Pressure-volume diagram for heart as it operates for one complete cycle.

to as the *heart output*. Of course, the diagram shown is only representative and the maximum pressure shown varies from one heart to another. An average maximum blood pressure for a healthy, medium-aged human heart seems to be 120 mm of Hg. We will not here attempt to calculate the area of the diagram except to recall that the maximum work must be given by equation (9–28)

$$Wk_{max} = -\Delta\mathcal{G}$$

where $-\Delta\mathcal{G}$ is the change in the Gibbs free energy of the ATP/ADP reaction previously mentioned as the source of energy for muscles. The interested reader is here directed to textbooks on biology, physiology, and bioengineering for more complete and varied discussions of biological systems.

Practice Problems

Problems designated with an asterisk * should only be attempted by those having a background which includes the knowledge of calculus.

Section 16.1

16.1. An electric generator is found to have an efficiency of 93% when delivering 60 amps of electric current. If 15 horsepower are required to drive the generator, determine the generator's voltage output.

16.2. An electric generator has an internal resistance of 0.03 ohm and delivers 100 amps at 120 volts. Determine the induced electric potential and the generator efficiency.

16.3. An electric motor has an internal resistance of 0.06 ohm and operates at 120 volts and 60 amps. Determine the motor output power if no friction exists.

16.4. A 3-hp electric motor dissipates 2 watts of frictional resistance, and operates at 220 volts. If internal resistance is 0.09 ohm, determine the motor current and efficiency.

16.5. A Daniel cell is built and found to have an internal resistance of 0.06 ohm. If 1 amp of current is drawn, calculate the maximum work, the open circuit voltage, and the operating voltage.

16.6. An electric cell is proposed which uses a zinc anode, a silver (Ag) cathode, and zinc sulfate and silver sulfate (Ag_2SO_4) solutions. In all other respects it resembles the Daniel cell. The cathode reaction is

$$2Ag^+ + 2e^- \rightarrow Ag$$

Determine the maximum work obtainable from this cell and its open circuit voltage.

Section 16.2

16.7. If the internal resistance of a hydrogen-oxygen fuel cell is 0.05 ohm, determine the operating voltage if 2 amps are drawn.

16.8. A fuel cell is proposed which uses sodium (Na) and chlorine (Cl) and produces sodium chloride (NaCl). Determine the maximum work obtainable and the open circuit voltage.

16.9. Octane (C_8H_{18}) in the gaseous phase is utilized as a fuel in a fuel cell which also uses oxygen. The chemical reaction is $C_8H_{18} - 12\frac{1}{2}\ O_2 \rightarrow$

$8\ CO_2 + 9\ H_2O$. Determine the maximum work obtainable, the open circuit voltage, and the operating voltage when 16 amps are drawn with no internal resistance.

Section 16.3

16.10. A 10-ohm resistor carries 8 amps. Determine the steady state Joule heating.

16.11. A resistor radiates 50 Btu/hr when a current of 0.7 amp passes through it. Determine the resistance in ohms.

16.12. An electric heater is required to conduct 200 Btu/min. Determine the resistance and current if 120 volts are applied across the heater coil.

16.13. A P-N type thermoelectric generator receives 10 cal/s of heat from an external supply. If the resistance of the generator is 0.05 ohm and the Seebeck coefficients are $\alpha_n = -0.13$ volt and $\alpha_p = 0.20$ volt, determine the induced current and the voltage of the device.

Section 16.4

16.14. (E) Hot gases having properties of air enter an MHD generator at 2000°F and 700 ft/s. If they leave the generator at 1050°F and 400 ft/s, estimate the work generated per pound-mass of gases if heat losses are neglected.

16.15. (M) An MHD power generator receives 1200 g/s of 1600°C gases having a velocity of 35 m/s and discharges them at 10 m/s and 650°C. A potential of 300 volts is applied across the generator section where the gases have a resistance of 1.5 ohms. If heat losses are 500 cal/s and the gases have properties of air, determine the net power produced by the MHD generator.

Section 16.5

16.16. (M) A muscle lifts 50 grams through 2 centimeters. If the efficiency is found to be 32%, what is the change in Gibbs free energy of the ATP-ADP reaction?

16.17. (M) A heart is pumping 5000 g/min of blood which has a density of 1 g/cc. If the thermal effects are neglected (no temperature change and $Q = 0$) and if the velocity change is negligible between entering and leaving blood, estimate the power produced by the heart to pump this blood if the pressure is 110 mm Hg leaving and 0 mm Hg entering.

***16.18. (M)** A muscle is found to have the following force-velocity relationship:

$$F \text{ (grams)} = 60 - 6\bar{V}$$

where $\bar{V}$ is in centimeters per second. Plot the F-$\bar{V}$ curve and determine the maximum power obtainable from the muscle. Then sketch the graph of power versus velocity of muscle.

16.19. (M) A heart which has a volume of 65 cc is using 7.5 kJ/min of power to pump blood to a pressure of 170 mm of Hg. Predict the average or mean effective blood pressure in the heart if the heart is pumping at 120 beats per minute.

APPENDIXES

A

Formulas and Equations

A.1 Logarithmic Relations

The following algebraic relations and definitions are useful in the manipulation of logarithmic quantities, including the algebra of powers:

$$\log X = N \qquad 10^N = X$$

$$\log Y = M \qquad 10^M = Y$$

$$\log X + \log Y = \log XY$$

$$(X)(Y) = 10^{N+M}$$

$$\log X - \log Y = \log \frac{X}{Y}$$

$$\frac{X}{Y} = 10^{N-M}$$

$$\ln X = L \qquad e^L = X$$

$$\ln Y = K \qquad e^K = Y$$

$$e = 2.303\ldots$$

$$\ln X = (2.303\ldots)\log X$$

Using x and y as variables, independent of the exponents n and m, we have,

$$\frac{x^n}{y^m} = x^n y^{-m}$$

$$\left(\frac{x}{y}\right)^n\left(\frac{x}{y}\right) = \left(\frac{x}{y}\right)^{n+1}$$

$$\left(\frac{x}{y}\right)^n\left(\frac{y}{x}\right) = \left(\frac{x}{y}\right)^n\left(\frac{x}{y}\right)^{-1} = \left(\frac{x}{y}\right)^{n-1}$$

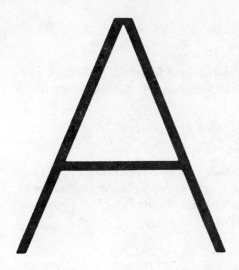

A.2 Raising Numbers to Powers

Frequently in thermodynamic calculations the following problem arises: Given a value for x and n, determine the result of x^n. Using logarithms we have

$$(\log x^n) = n(\log x) = y$$

$$\text{antilog } y = x^n$$

By using a slide rule or log table, we can obtain the log of x. Then multiplying this value by n yields a new value, y. Using log tables or appropriate slide rule scales, we can obtain the antilog of y. This value corresponds to the desired quantity, x^n.

A.3 Integral Formulas

For equations of reversible work, the integrals $\int p\,dv$ or $\int v\,dp$ were evaluated. In order to complete the integration, however, the variable p must be written in terms of V or V in terms of p. To generalize to any variable, x, the following indefinite integrals (or antiderivatives) result from the given quantity:

$$\int dx = x + c_1$$

$$\int x\,dx = \frac{1}{2}x^2 + c_2$$

$$\int x^n\,dx = \frac{1}{n+1}x^{n+1} + c_n$$

$$\int \frac{dx}{x} = \ln x + c_3$$

If the integral is to be evaluated between two given values (say x_1 and x_2), the following results are obtained:

$$\int_{x_1}^{x_2} dx = x_2 - x_1$$

$$\int_{x_1}^{x_2} x\,dx = \frac{1}{2}x_2^2 - \frac{1}{2}x_1^2 = \frac{1}{2}(x_2^2 - x_1^2)$$

$$\int_{x_1}^{x_2} x^n\,dx = \frac{1}{n+1}x_2^{n+1} - \frac{1}{n+1}x_1^{n+1} = \frac{1}{n+1}(x_2^{n+1} - x_1^{n+1})$$

$$\int_{x_1}^{x_2} \frac{dx}{x} = \ln x_2 - \ln x_1 = \ln \frac{x_2}{x_1}$$

B

Tables and Charts

Table B.1

Saturated Steam Table
SI UNITS

Temp °C t	Pres-sure bars* p	Specific Volume m^3/kg v_f	v_g	h_f	Enthalpy kJ/kg h_{fg}	h_g	s_f	Entropy $kJ/kg \cdot K$ s_{fg}	s_g
0.01	0.006	0.000102	206.3	0.00	2501	2501	0.000	9.154	9.154
5	0.009	0.0010001	147.2	21.1	2489	2510	0.076	8.948	9.024
10	0.012	0.0010004	106.4	42.0	2477	2519	0.151	8.748	8.899
20	0.023	0.001002	57.8	83.9	2454	2537	0.296	8.371	8.667
30	0.042	0.001004	32.9	125.7	2430	2556	0.437	8.015	8.452
40	0.074	0.001008	19.6	167.5	2406	2574	0.572	7.684	8.256
50	0.123	0.001012	12.0	209	2383	2592	0.704	7.371	8.075
60	0.199	0.001017	7.68	251	2358	2609	0.831	7.077	7.908
70	0.312	0.00102	5.05	293	2333	2626	0.955	6.799	7.754
80	0.474	0.00103	3.41	335	2308	2643	1.075	6.537	7.612
90	0.701	0.00104	2.36	377	2282	2659	1.193	6.286	7.479
100	1.013	0.00104	1.67	419	2257	2676	1.307	6.048	7.355
110	1.433	0.00105	1.210	461	2230	2691	1.418	5.821	7.239
120	1.985	0.00106	0.892	504	2202	2706	1.528	5.602	7.130
130	2.701	0.00107	0.668	546	2175	2721	1.635	5.397	7.027
140	3.614	0.00108	0.509	589	2145	2734	1.739	5.191	6.930
150	4.76	0.00109	0.393	632	2114	2746	1.842	4.996	6.838
160	6.18	0.00110	0.307	676	2082	2758	1.943	4.808	6.751
170	7.92	0.00111	0.243	719	2050	2769	2.042	4.625	6.667
180	10.03	0.00113	0.194	763	2015	2778	2.140	4.446	6.586
190	12.55	0.00114	0.156	808	1978	2786	2.236	4.271	6.507
200	15.55	0.00115	0.127	852	1941	2793	2.331	4.101	6.432
210	19.08	0.00117	0.104	898	1900	2798	2.425	3.933	6.358
220	23.20	0.00119	0.0860	944	1858	2802	2.518	3.767	6.285
230	27.98	0.00121	0.0715	990	1813	2803	2.610	3.603	6.213
240	33.48	0.00123	0.0597	1038	1765	2803	2.702	3.441	6.143
250	39.78	0.00125	0.0501	1086	1715	2801	2.793	3.279	6.072
260	46.94	0.00128	0.0422	1135	1661	2796	2.885	3.116	6.001
270	55.1	0.00130	0.0356	1185	1605	2790	2.976	2.954	5.930
280	64.2	0.00133	0.0301	1237	1543	2780	3.068	2.789	5.857
290	74.5	0.00137	0.0255	1290	1476	2766	3.161	2.622	5.783
300	85.9	0.00140	0.0216	1345	1404	2749	3.255	2.450	5.705
310	98.7	0.00145	0.0183	1402	1325	2727	3.351	2.272	5.623
320	112.9	0.0015	0.0154	1462	1238	2700	3.450	2.085	5.535
330	128.7	0.0016	0.0130	1526	1140	2666	3.552	1.889	5.441
340	146.1	0.0016	0.0108	1595	1027	2622	3.661	1.675	5.336
350	165.4	0.0017	0.0088	1671	894	2565	3.779	1.433	5.212
360	186.7	0.0019	0.0069	1762	719	2481	3.916	1.137	5.053
370	210.5	0.0022	0.0049	1893	438	2331	4.114	0.681	4.795
374.15	221.3	0.00326	0.00326	2100	0	2100	4.430	0.000	4.430

*1 bar = 10^5 Pa = 10^2 kPa.

Table B.1 (continued).

Pressure bars*	Temp °C	Specific Volume m³/kg		Enthalpy kJ/kg			Entropy kJ/kg·K		
p	T	v_f	v_g	h_f	h_{fg}	h_g	s_f	s_{fg}	s_g
0.01	6.9	0.0010001	129.9	29	2484	2513	0.105	8.875	8.98
0.02	17.5	0.0010007	87.9	73	2460	2533	0.261	8.459	8.72
0.04	28.9	0.001004	39.5	121	2433	2554	0.423	8.047	8.49
0.06	36.2	0.001006	23.7	152	2415	2567	0.521	7.809	8.33
0.08	41.5	0.001009	18.1	174	2402	2576	0.593	7.637	8.23
0.10	45.8	0.001010	13.4	192	2392	2584	0.649	7.500	8.149
0.15	54.0	0.001014	10.0	226	2373	2599	0.755	7.252	8.007
0.20	60.1	0.001017	7.65	251	2358	2609	0.832	7.075	7.907
0.40	75.9	0.001026	3.99	318	2318	2636	1.026	6.644	7.670
0.60	86.0	0.001033	2.73	360	2293	2653	1.145	6.386	7.531
0.80	93.5	0.001039	2.09	392	2273	2665	1.233	6.201	7.434
1.0	99.6	0.00104	1.69	417	2258	2675	1.303	6.057	7.360
1.013	100.0	0.00104	1.67	419	2257	2676	1.307	6.048	7.355
1.5	111.4	0.00105	1.16	467	2226	2693	1.434	5.789	7.223
2.0	120.2	0.00106	0.89	505	2202	2707	1.530	5.597	7.127
4.0	143.6	0.00108	0.462	605	2133	2738	1.777	5.120	6.897
6.0	158.8	0.00110	0.316	657	2086	2757	1.931	4.830	6.761
8.0	170.4	0.00111	0.240	721	2048	2769	2.046	4.617	6.663
10.0	179.9	0.00113	0.195	763	2015	2778	2.138	4.449	6.587
20	212.4	0.00118	0.0996	909	1890	2799	2.447	3.893	6.340
30	233.8	0.00122	0.0667	1008	1796	2804	2.646	3.540	6.186
40	250.3	0.00125	0.0498	1088	1713	2801	2.796	3.274	6.070
50	263.9	0.00129	0.0394	1154	1640	2794	2.921	3.052	5.973
60	275.6	0.00132	0.0324	1214	1571	2785	3.027	2.863	5.890
70	285.8	0.00135	0.0274	1267	1505	2772	3.122	2.692	5.814
80	295.0	0.00138	0.0235	1317	1441	2758	3.208	2.537	5.745
90	304.1	0.00142	0.0205	1364	1379	2743	3.287	2.391	5.678
100	311.0	0.00145	0.0180	1408	1317	2725	3.360	2.155	5.515
110	318.0	0.00149	0.0160	1450	1255	2705	3.430	2.123	5.553
120	324.6	0.00153	0.0143	1491	1194	2685	3.496	1.996	5.492
130	330.8	0.00157	0.0128	1532	1130	2662	3.560	1.871	5.431
140	336.6	0.00161	0.0115	1571	1067	2638	3.624	1.748	5.372
150	342.1	0.00166	0.0104	1610	1001	2611	3.685	1.625	5.310
160	347.3	0.00171	0.0093	1650	932	2582	3.746	1.501	5.247
170	352.3	0.00177	0.0084	1690	858	2548	3.807	1.370	5.177
180	357.0	0.00184	0.0075	1732	778	2510	3.871	1.236	5.107
190	361.4	0.0019	0.0067	1776	690	2460	3.938	1.089	5.027
200	365.7	0.0020	0.0059	1827	583	2410	4.015	0.913	4.928
210	369.8	0.0022	0.0050	1888	449	2336	4.108	0.695	4.803
220	373.7	0.0027	0.0037	2016	152	2168	4.303	0.288	4.591
221.3	374.15	0.00326	0.00326	2100	0	2100	4.430	0.000	4.430

Table B.1 (continued).

ENGLISH UNITS

Temp °F t	Pressure psia p	Specific Volume ft³/lbm v_f	v_g	Enthalpy Btu/lbm h_f	h_{fg}	h_g	Entropy Btu/lbm·°F s_f	s_{fg}	s_g
32.00	0.089	0.0160	3304.7	−0.018	1075.5	1075.5	0.000	2.1873	2.1873
36.00	0.104	0.0160	2839.0	4.008	1073.2	1077.2	0.0081	2.1651	2.1732
40.00	0.122	0.0160	2445.8	8.027	1071.0	1079.0	0.0162	2.1432	2.1594
50.00	0.178	0.0160	1704.8	18.05	1065.3	1083.4	0.0361	2.0901	2.1262
60.00	0.256	0.0160	1207.6	28.06	1059.7	1087.7	0.0555	2.0391	2.0946
80.00	0.506	0.0161	633.3	48.04	1048.4	1096.4	0.0932	1.9426	2.0359
100.00	0.949	0.0161	350.4	68.0	1037.1	1105.1	0.1295	1.8530	1.9825
101.74	1.000	0.0161	333.60	69.7	1036.1	1105.8	0.1326	1.8455	1.9781
120.00	1.693	0.0162	203.26	89.0	1025.6	1113.6	0.1646	1.7693	1.9339
140.00	2.889	0.0163	123.00	108.0	1014.0	1122.0	0.1985	1.6910	1.8895
160.00	4.741	0.0164	77.29	128.0	1002.2	1130.2	0.2313	1.6174	1.8487
162.24	5.000	0.0164	73.532	130.2	1000.9	1131.1	0.2349	1.6094	1.8443
180.00	7.51	0.0165	50.225	148.0	990.2	1138.2	0.2631	1.5480	1.8111
193.21	10.000	0.0166	38.420	161.2	982.1	1143.3	0.2836	1.5043	1.7879
200.00	11.53	0.0166	33.639	168.1	977.9	1146.0	0.2940	1.4824	1.7764
212.00	14.696	0.0167	26.799	180.2	970.3	1150.5	0.3121	1.4447	1.7568
213.03	15.000	0.0167	26.290	181.2	969.7	1150.9	0.3137	1.4415	1.7552
220.00	17.19	0.0168	23.148	188.2	965.2	1153.4	0.3241	1.4201	1.7442
227.96	20.00	0.0168	20.087	196.2	960.1	1156.3	0.3358	1.3962	1.7320
240.00	24.97	0.0169	16.321	208.5	952.1	1160.6	0.3533	1.3609	1.7142
250.34	30.00	0.0170	13.744	218.9	945.2	1164.1	0.3682	1.3313	1.6995
260.00	35.43	0.0171	11.762	228.8	938.6	1167.4	0.3819	1.3043	1.6862
267.25	40.00	0.0172	10.497	236.1	933.6	1169.8	0.3921	1.2844	1.6765
280.00	49.20	0.0173	8.644	249.2	924.6	1173.8	0.4098	1.2501	1.6599
292.71	60.00	0.0174	7.174	262.2	915.4	1177.6	0.4273	1.2167	1.6440
300.00	67.01	0.0175	6.466	269.7	910.0	1179.7	0.4372	1.1979	1.6351
312.04	80.00	0.0176	5.471	282.1	901.0	1183.1	0.4534	1.1675	1.6208
327.82	100.00	0.0177	4.431	298.5	888.6	1187.2	0.4743	1.1284	1.6027
340.00	118.0	0.0179	3.788	311.3	878.8	1190.1	0.4902	1.0990	1.5892
358.43	150.0	0.0181	3.014	330.6	863.4	1194.1	0.5141	1.0554	1.5695
380.00	195.7	0.0184	2.335	353.6	844.5	1198.0	0.5416	1.0057	1.5473
400.97	250.0	0.0187	1.843	376.1	825.0	1201.1	0.5679	0.9585	1.5264
420.00	308.8	0.0189	1.500	396.9	806.2	1203.1	0.5915	0.9165	1.5080
444.60	400.0	0.0193	1.161	424.2	780.4	1204.6	0.6217	0.8630	1.4847
500.00	680.9	0.0204	0.675	487.9	714.3	1202.2	0.6890	0.7443	1.4333
518.21	800.0	0.0209	0.569	509.8	689.6	1199.4	0.7111	0.7051	1.4163
550.00	1045.4	0.0218	0.424	549.4	641.8	1191.2	0.7501	0.6356	1.3856
567.19	1200.0	0.0223	0.362	571.9	613.0	1184.8	0.7714	0.5969	1.3683
600.00	1543.2	0.0236	0.267	617.1	550.6	1167.7	0.8134	0.5196	1.3330
650.00	2208.4	0.0268	0.162	696.4	425.0	1121.4	0.8837	0.3830	1.2667
700.00	3094.3	0.0366	0.0752	822.5	172.7	995.2	0.9901	0.1490	1.1390
705.47	3208.2	0.0508	0.0508	906.0	000.0	906.0	1.0612	0.0000	1.0612

Note: Data abstracted from ASME Steam Tables, copyright 1967, and reproduced with permission of the American Society of Mechanical Engineers.

Table B.2

Steam Properties of the Superheated States
SI UNITS
v (m³/kg); h (kJ/kg); s (kJ/kg · K)

Pressure, bars,* p (Sat. Temp., °C)		Sat. Vapor	80	160	240	320	400	480	560	640	720	800	880	960
			Temperature, °C, t											
0.01	v	129.9	164.0	201.1	238.3	275.4	312.6	349.8	386.9	424.1				
(8.9)	h	2513	2651	2803	2958	3117	3280	3448	3619	3796				
	s	8.98	9.406	9.793	10.121	10.408	10.665	10.902	11.122	11.325				
0.1	v	13.4	16.3	20.0	23.7	27.4	30.2	34.8	38.5	42.2				
(45.8)	h	2584	2649	2802	2957	3117	3280	3448	3619	3796				
	s	8.149	8.337	8.727	9.056	9.343	9.601	9.838	10.058	10.262				
0.4	v	3.99	4.09	4.98	5.91	6.84	7.77	8.69	9.61	10.54				
(75.9)	h	2636	2645	2800	2956	3116	3279	3447	3619	3795				
	s	7.670	7.690	8.086	8.415	8.703	8.962	9.199	9.419	9.622				
1.0	v	1.69		1.98	2.36	2.73	3.10	3.472	3.84	4.21	4.58	4.95	5.32	5.69
(99.6)	h	2675		2796	2954	3114	3278	3446	3618	3795	3974	4175	4348	4539
	s	7.360		7.654	7.988	8.281	8.541	8.777	8.995	9.195	9.384	9.563	9.733	9.895
1.5	v	1.16		1.32	1.57	1.82	2.07	2.31	2.56	2.81	3.06	3.30	3.55	3.79
(111.4)	h	2693		2793	2952	3112	3277	3445	3617	3794	3973	4157	4347	4539
	s	7.223		7.462	7.799	8.092	8.353	8.589	8.807	9.007	9.197	9.376	9.546	9.708
2.0	v	0.89		0.984	1.18	1.36	1.55	1.73	1.91	2.10	2.30	2.48	2.66	2.85
(120.2)	h	2707		2790	2950	3111	3276	3445	3617	3794	3973	4157	4347	4539
	s	7.127		7.324	7.663	7.957	8.219	8.456	8.673	8.873	9.064	9.242	9.414	9.575
3.0	v	0.61		0.651	0.780	0.906	1.03	1.16	1.28	1.40	1.53	1.65	1.77	1.89
(133.5)	h	2725		2783	2946	3109	3274	3444	3616	3793	3972	4157	4346	4538
	s	6.992		7.126	7.470	7.766	8.030	8.268	8.486	8.686	8.876	9.056	9.226	9.388
4.0	v	0.46		0.484	0.583	0.678	0.772	0.866	0.959	1.052	1.145	1.237	1.330	1.422
(143.6)	h	2738		2776	2941	3106	3273	3443	3615	3792	3971	4156	4346	4538
	s	6.897		6.980	7.332	7.631	7.895	8.134	8.352	8.553	8.744	8.923	9.093	9.255
5.0	v	0.37		0.384	0.464	0.541	0.617	0.692	0.767	0.841	0.916	0.990	1.062	1.138
(151.8)	h	2749		2767	2937	3104	3272	3441	3614	3791	3971	4156	4346	4537
	s	6.822		6.864	7.224	7.525	7.791	8.030	8.249	8.450	8.640	8.819	8.990	9.152

Table B.2 (continued).

Pressure, bars,* p (Sat. Temp., °C)		Sat. Vapor	80	160	240	320	400	480	560	640	720	800	880	960
			Temperature, °C, t											
10	v	0.195			0.227	0.268	0.307	0.345	0.382	0.420	0.457	0.494	0.531	0.569
(179.9)	h	2778			2918	3091	3263	3435	3609	3787	3968	4154	4344	4536
	s	6.587			6.877	7.189	7.461	7.703	7.924	8.127	8.318	8.498	8.669	8.831
20	v	0.0996			0.108	0.131	0.151	0.171	0.190	0.209	0.228	0.247	0.265	0.284
(212.4)	h	2799			2875	3065	3246	3423	3600	3780	3963	4150	4340	4533
	s	6.340			6.491	6.837	7.122	7.371	7.596	7.802	7.994	8.174	8.346	8.509
30	v	0.0667			0.0683	0.085	0.099	0.113	0.126	0.139	0.152	0.164	0.177	0.188
(233.8)	h	2804			2823	3038	3229	3411	3592	3773	3957	4145	4337	4531
	s	6.186			6.225	6.615	6.916	7.172	7.400	7.608	7.803	7.984	8.157	8.320
40	v	0.0498				0.062	0.073	0.084	0.094	0.104	0.113	0.123	0.132	0.142
(250.3)	h	2801				3010	3210	3399	3583	3766	3952	4141	4334	4528
	s	6.070				6.446	6.762	7.028	7.259	7.470	7.666	7.848	8.023	8.187
50	v	0.0394				0.048	0.058	0.067	0.075	0.083	0.090	0.098	0.106	0.113
(263.9)	h	2794				2980	3193	3386	3574	3759	3946	4136	4329	4524
	s	5.973				6.304	6.640	6.912	7.148	7.362	7.558	7.742	7.916	8.080
60	v	0.0324				0.039	0.047	0.055	0.062	0.069	0.075	0.082	0.088	0.094
(275.6)	h	2785				2948	3174	3373	3564	3751	3940	4132	4326	4521
	s	5.890				6.177	6.535	6.815	7.056	7.271	7.471	7.655	7.830	7.994
70	v	0.0274				0.032	0.040	0.047	0.053	0.059	0.064	0.070	0.075	0.081
(285.8)	h	2772				2913	3155	3360	3554	3744	3935	4127	4322	4518
	s	5.814				6.058	6.442	6.731	6.976	7.194	7.395	7.580	7.755	7.919
80	v	0.0235				0.027	0.034	0.040	0.046	0.051	0.056	0.061	0.066	0.071
(295.0)	h	2758				2874	3135	3347	3544	3736	3929	4122	4318	4515
	s	5.745				5.943	6.358	6.657	6.905	7.126	7.329	7.515	7.690	7.856
90	v	0.0205				0.023	0.030	0.036	0.040	0.046	0.050	0.054	0.058	0.063
(304.1)	h	2743				2829	3114	3334	3534	3728	3922	4117	4314	4512
	s	5.678				5.827	6.280	6.589	6.843	7.092	7.270	7.459	7.635	7.801

Table B.2 (continued).

Pressure, bars,* p (Sat. Temp., °C)		Sat. Vapor	Temperature, °C, t: 80	160	240	320	400	480	560	640	720	800	880	960
100 (311.0)	v	0.0180				0.0193	0.0265	0.0316	0.0362	0.0404	0.0446	0.0486	0.0525	0.0564
	h	2725				2778	3093	3320	3524	3719	3915	4111	4309	4507
	s	5.515				5.705	6.207	6.527	6.786	7.011	7.217	7.406	7.585	7.752
110 (318.0)	v	0.0160				0.0163	0.0236	0.0284	0.0327	0.0366	0.0404	0.0441	0.0477	0.0512
	h	2705				2719	3071	3305	3513	3711	3909	4103	4305	4504
	s	5.553				5.579	6.138	6.469	6.733	6.961	7.168	7.357	7.536	7.704
120 (324.6)	v	0.0143					0.0211	0.0258	0.0298	0.0334	0.0369	0.0403	0.0436	0.0469
	h	2685					3049	3291	3503	3703	3903	4102	4301	4500
	s	5.492					6.071	6.415	6.684	6.915	7.123	7.314	7.493	7.661
130 (330.8)	v	0.0128					0.0191	0.0235	0.0273	0.0307	0.0340	0.0371	0.0402	0.0432
	h	2662					3026	3277	3493	3696	3897	4097	4297	4497
	s	5.431					6.006	6.364	6.638	6.872	7.082	7.274	7.454	7.623
140 (336.6)	v	0.0115					0.0173	0.0216	0.0252	0.0284	0.0307	0.0344	0.0373	0.0401
	h	2638					3000	3262	3482	3688	3841	4092	4293	4493
	s	5.372					5.942	6.314	6.594	6.832	6.992	7.238	7.418	7.587
150 (342.1)	v	0.0104					0.0157	0.0199	0.0233	0.0264	0.0286	0.0320	0.0348	0.0374
	h	2611					2973	3248	3472	3680	3855	4087	4289	4490
	s	5.310					5.878	6.268	6.554	6.794	6.956	7.204	7.385	7.554
160 (347.3)	v	0.0093					0.0143	0.0184	0.0217	0.0247	0.0267	0.0300	0.0326	0.0351
	h	2582					2945	3233	3461	3672	3829	4082	4284	4486
	s	5.247					5.816	6.223	6.515	6.758	6.922	7.171	7.353	7.523
170 (352.3)	v	0.0084					0.0131	0.0171	0.0203	0.0231	0.0251	0.0282	0.0306	0.0330
	h	2548					2915	3218	3450	3664	3873	4077	4280	4482
	s	5.177					5.753	6.179	6.477	6.725	6.942	7.140	7.323	7.493
180 (357.0)	v	0.0075					0.0119	0.0160	0.0190	0.0217	0.0242	0.0266	0.0289	0.0311
	h	2510					2884	3206	3440	3656	3867	4072	4276	4479
	s	5.107					5.688	6.137	6.441	6.691	6.911	7.110	7.294	7.465

Table B.2 (continued).

Pressure, bars,* p (Sat. Temp., °C)		Sat. Vapor	80	160	240	320	400	480	560	640	720	800	880	960
			Temperature, °C, t											
190	v	0.0067					0.0109	0.0149	0.0179	0.0205	0.0229	0.0251	0.0273	0.0295
(361.4)	h	2466					2851	3188	3429	3648	3861	4068	4272	4476
	s	5.027					5.622	6.095	6.407	6.660	6.881	7.082	7.266	7.437
200	v	0.0059					0.0100	0.0140	0.0169	0.0194	0.0217	0.0238	0.0259	0.0280
(365.7)	h	2410					2816	3170	3418	3640	3855	4063	4268	4473
	s	4.928					5.553	6.055	6.374	6.631	6.853	7.056	7.240	7.412
210	v	0.0050					0.0091	0.0132	0.0160	0.0184	0.0206	0.0216	0.0246	0.0266
(369.8)	h	2336					2778	3152	3407	3632	3849	4058	4264	4470
	s	4.803					5.481	6.015	6.342	6.603	6.826	7.031	7.215	7.388
220	v	0.0037					0.0083	0.0124	0.0151	0.0175	0.0196	0.0216	0.0235	0.0254
(373.7)	h	2168					2736	3135	3396	3624	3843	4053	4260	4466
	s	4.591					5.406	5.975	6.312	6.576	6.801	7.007	7.192	7.365
221.3	v	0.00326					0.0082	0.0123	0.0150	0.0174	0.0195	0.0215	0.0234	0.0253
(374.15)	h	2100					2730	3133	3395	3623	3842	4052	4259	4466
	s	4.430					5.395	5.970	6.308	6.572	6.798	7.004	7.189	7.362
230	v	n.a.					0.0075	0.0117	0.0143	0.0166	0.0187	0.0206	0.0225	0.0243
	h						2690	3117	3385	3616	3837	4048	4256	4463
	s						5.324	5.936	6.283	6.549	6.776	6.983	7.170	7.343
250	v	n.a.					0.0060	0.0104	0.0130	0.0152	0.0171	0.0189	0.0206	0.0223
	h						2579	3080	3362	3600	3824	4038	4248	4457
	s						5.137	5.860	6.225	6.498	6.729	6.938	7.126	7.300

* 1 bar = 10^5 Pa = 10^2 kPa.

Table B.2 (continued).

ENGLISH UNITS

v (ft³/lbm); h (Btu/lbm); s (Btu/lbm °R)

Pressure psia, p (Sat. Temp.)		Sat. Vapor	Temperature, °F, t 200	300	400	500	600	700	800	1000	1200	1400	1500
1	v	333.60	392.50	452.3	511.90	571.5	631.1	690.7					
(101.74)	h	1105.8	1150.2	1195.7	1241.8	1288.6	1336.1	1384.5					
	s	1.9781	2.0209	2.1152	2.1722	2.2237	2.2708	2.3144					
4	v	90.63	97.79	112.86	127.85	142.79	157.71	172.62	187.53	217.33	247.13	276.92	291.82
(152.97)	h	1127.3	1149.0	1195.0	1241.4	1288.3	1335.9	1384.3	1433.6	1534.8	1639.6	1748.0	1803.5
	s	1.8625	1.8967	1.9617	2.0190	2.0707	2.1178	2.1615	2.2022	2.2767	2.3440	2.4057	2.4347
10	v	38.42	38.84	44.98	51.03	57.04	63.03	69.00	74.98	86.91	98.84	110.76	116.72
(193.21)	h	1143.3	1146.6	1193.7	1240.6	1287.8	1335.5	1384.0	1433.4	1534.6	1639.5	1747.9	1803.4
	s	1.7879	1.7928	1.8593	1.9173	1.9692	2.0166	2.0603	2.1011	2.1757	2.2430	2.3046	2.3337
15	v	26.290		29.899	33.963	37.985	41.986	45.978	49.964	57.926	65.882	73.833	77.807
(213.03)	h	1150.9		1192.5	1239.9	1287.3	1335.2	1383.8	1433.2	1534.5	1639.4	1747.8	1803.4
	s	1.7552		1.8134	1.8720	1.9242	1.9717	2.0155	2.0563	2.1309	2.1982	2.2599	2.2890
20	v	20.087		22.356	25.428	28.457	31.466	34.465	37.458	43.435	49.405	55.370	58.352
(227.96)	h	1156.3		1191.4	1239.2	1286.9	1334.9	1383.5	1432.9	1534.3	1639.3	1747.8	1803.3
	s	1.7320		1.7805	1.8397	1.8921	1.9397	1.9836	2.0244	2.0991	2.1665	2.2282	2.2572
30	v	13.744		14.810	16.892	18.929	20.945	22.951	24.952	28.943	32.927	36.907	38.896
(250.34)	h	1164.1		1189.0	1237.8	1286.0	1334.2	1383.0	1432.5	1534.0	1639.0	1747.6	1803.2
	s	1.6995		1.7334	1.7937	1.8467	1.8946	1.9386	1.9795	2.0543	2.1217	2.1834	2.2125
50	v	8.515		8.769	10.062	11.306	12.529	13.741	14.947	17.350	19.746	22.137	23.332
(281.01)	h	1174.1		1184.1	1234.9	1284.1	1332.9	1382.0	1431.7	1533.4	1638.6	1747.3	1802.9
	s	1.6585		1.6720	1.7349	1.7890	1.8374	1.8816	1.9227	1.9977	2.0652	2.1270	2.1561
75	v	5.816			6.645	7.494	8.320	9.135	9.945	11.553	13.155	14.752	15.550
(307.60)	h	1181.9			1231.2	1281.7	1331.3	1380.7	1430.7	1532.7	1638.1	1746.9	1802.6
	s	1.6259			1.6868	1.7424	1.7915	1.8361	1.8774	1.9526	2.0202	2.0821	2.1113

Table B.2 (continued).

Pressure psia, p (Sat. Temp.)		Sat. Vapor	200	300	400	500	600	700	800	1000	1200	1400	1500
								Temperature, °F, t					
100	v	4.431			4.935	5.588	6.216	6.833	7.443	8.655	9.860	11.060	11.659
(327.82)	h	1187.2			1227.4	1279.3	1329.6	1379.5	1429.7	1532.0	1637.6	1746.5	1802.2
	s	1.6027			1.6516	1.7088	1.7586	1.8036	1.8451	1.9205	1.9883	2.0502	2.0794
150	v	3.014			3.2208	3.6799	4.1112	4.5298	4.9421	5.7568	6.5642	7.3671	7.7674
(358.43)	h	1194.1			1219.1	1274.3	1326.1	1376.9	1427.6	1530.5	1636.5	1745.7	1801.7
	s	1.5695			1.5993	1.6602	1.7115	1.7573	1.7992	1.8751	1.9431	2.0052	2.0344
200	v	2.287			2.3598	2.7247	3.0583	3.3783	3.6915	4.3077	4.9165	5.5209	5.8219
(381.80)	h	1198.3			1210.1	1269.0	1322.6	1374.3	1425.5	1529.1	1635.4	1745.0	1800.9
	s	1.5454			1.5593	1.6242	1.6773	1.7239	1.7663	1.8426	1.9109	1.9732	2.0025
250	v	1.843				2.1504	2.4262	2.6872	2.9410	3.4382	3.9278	4.4131	4.6546
(400.97)	h	1201.1				1263.5	1319.0	1371.6	1423.4	1527.6	1634.4	1744.2	1800.2
	s	1.5264				1.5951	1.6502	1.6976	1.7405	1.8173	1.8858	1.9482	1.9776
300	v	1.543				1.7665	2.0044	2.2263	2.4407	2.8585	3.2688	3.6746	3.8764
(417.35)	h	1202.9				1257.7	1315.2	1368.9	1421.3	1526.2	1633.3	1743.4	1799.6
	s	1.5105				1.5703	1.6274	1.6758	1.7192	1.7964	1.8652	1.9278	1.9572
350	v	1.326				1.4913	1.7028	1.8970	2.0832	2.4445	2.7980	3.1471	3.3205
(431.73)	h	1204.0				1251.5	1311.4	1366.2	1419.2	1524.7	1632.3	1742.6	1798.9
	s	1.4968				1.5483	1.6077	1.6571	1.7009	1.7787	1.8477	1.9105	1.9400
400	v	1.161				1.2841	1.4763	1.6499	1.8151	2.1339	2.4450	2.7515	2.9037
(444.60)	h	1204.6				1245.1	1307.4	1363.4	1417.0	1523.3	1631.2	1741.9	1798.2
	s	1.4847				1.5282	1.5901	1.6406	1.6850	1.7632	1.8325	1.8955	1.9250
500	v	0.928				0.9919	1.1584	1.3037	1.4397	1.6992	1.9507	2.1977	2.3200
(461.01)	h	1204.7				1231.2	1299.1	1357.7	1412.7	1520.3	1629.1	1740.3	1796.9
	s	1.4639				1.4921	1.5652	1.6176	1.6578	1.7371	1.8069	1.8702	1.8998
600	v	0.770				0.7944	0.9456	1.0726	1.1892	1.4093	1.6211	1.8284	1.9309
(486.20)	h	1203.7				1215.9	1290.3	1351.8	1408.3	1517.4	1627.0	1738.8	1795.6
	s	1.4461				1.4590	1.5329	1.5884	1.6351	1.7155	1.7859	1.8494	1.8792

Table B.2 (continued).

Pressure psia, p (Sat. Temp.)		Sat. Vapor	200	300	400	500	600	700	800	1000	1200	1400	1500
							Temperature, °F, t						
700	v	0.656					0.7928	0.9072	1.0102	1.2023	1.3858	1.5647	1.6530
(503.08)	h	1201.8					1281.0	1345.6	1403.7	1514.4	1624.8	1737.2	1794.3
	s	1.4304					1.5090	1.5673	1.6154	1.6970	1.7679	1.8318	1.8617
800	v	0.569					0.6774	0.7828	0.8759	1.0470	1.2093	1.3669	1.4446
(518.21)	h	1199.4					1271.1	1339.3	1399.1	1511.4	1622.7	1735.7	1792.9
	s	1.4163					1.4869	1.5484	1.5980	1.6807	1.7522	1.8164	1.8464
900	v	0.501					0.5869	0.6858	0.7713	0.9262	1.0720	1.2131	1.2825
(531.95)	h	1196.4					1260.6	1332.7	1394.4	1508.5	1620.6	1734.1	1791.6
	s	1.4032					1.4659	1.5311	1.5822	1.6662	1.7382	1.8028	1.8329
1000	v	0.446					0.5137	0.6080	0.6875	0.8295	0.9622	1.0901	1.1529
(544.58)	h	1192.9					1249.3	1325.9	1389.6	1505.4	1618.4	1732.5	1790.3
	s	1.3910					1.4457	1.5149	1.5677	1.6530	1.7256	1.7905	1.8207
1200	v	0.362					0.4016	0.4905	0.5615	0.6845	0.7974	0.9055	0.9584
(567.19)	h	1184.8					1224.2	1311.5	1379.7	1499.4	1614.2	1729.4	1787.6
	s	1.3683					1.4061	1.4851	1.5415	1.6298	1.7035	1.7691	1.7996
1400	v	0.302					0.3176	0.4059	0.4712	0.5809	0.6798	0.7737	0.8195
(587.07)	h	1175.3					1194.1	1296.1	1369.3	1493.2	1609.9	1726.3	1785.0
	s	1.3474					1.3652	1.4575	1.5182	1.6096	1.6845	1.7508	1.7815
1600	v	0.255						0.3415	0.4032	0.5031	0.5915	0.6748	0.7153
(604.87)	h	1164.5						1279.4	1358.5	1486.9	1605.6	1723.2	1782.3
	s	1.3274						1.4312	1.4968	1.5916	1.6678	1.7347	1.7657
1800	v	0.219						0.2906	0.3500	0.4426	0.5229	0.5980	0.6343
(621.02)	h	1152.3						1261.1	1347.2	1480.6	1601.2	1720.1	1779.7
	s	1.3079						1.4054	1.4768	1.5753	1.6528	1.7204	1.7516

Table B.2 (continued).

Pressure psia, p (Sat. Temp.)		Sat. Vapor	200	300	400	500	600	700	800	1000	1200	1400	1500
2000	*v*	0.188						0.2488	0.3072	0.3942	0.4680	0.5365	0.5695
(635.80)	*h*	1138.3						1240.9	1335.4	1474.1	1596.9	1717.0	1777.1
	s	1.2881						1.3794	1.4578	1.5603	1.6391	1.7075	1.7389
2500	*v*	0.131						0.1681	0.2293	0.3068	0.3692	0.4259	0.4529
(668.11)	*h*	1093.3						1176.7	1303.4	1457.5	1585.9	1709.2	1770.4
	s	1.2345						1.3076	1.4129	1.5269	1.6094	1.6796	1.7116
3000	*v*	0.085						0.0982	0.1759	0.2484	0.3033	0.3522	0.3753
(695.33)	*h*	1020.3						1060.5	1267.0	1440.2	1574.8	1701.4	1763.8
	s	1.1619						1.1966	1.3692	1.4976	1.5841	1.6561	1.6888
3200	*v*	0.0566							0.1588	0.2301	0.2827	0.3291	0.3510
(705.08)	*h*	931.6							1250.9	1433.1	1570.3	1698.3	1761.2
	s	1.0832							1.3515	1.4866	1.5749	1.6477	1.6806

Note: Data abstracted from ASME Steam Tables, copyright 1967, and reproduced with permission of the American Society of Mechanical Engineers.

Table B.3

Isentropic Work of Compression (Ideal Pump Work) for Water at Saturated Liquid ($h - h_f$, kJ/kg)
SI UNITS

Pressure, p		*Saturation Temperature, °C*						
bars	MPa	0	30	90	150	210	270	330
20	2.0	1.9	1.9	2.1	1.9	0.2		
30	3.0	2.6	2.8	2.8	2.8	1.9		
40	4.0	3.9	4.2	4.4	4.2	2.3		
80	8.0	7.4	7.7	7.7	7.9	6.0	2.8	
100	10.0	9.8	10.2	10.2	10.5	9.3	6.5	
120	12.0	11.9	12.1	12.3	12.1	11.6	9.1	
140	14.0	14.2	14.0	14.2	14.4	14.2	11.6	3.0
160	16.0	15.8	15.8	16.1	16.8	16.5	13.7	5.4
180	18.0	17.7	17.7	18.1	18.8	18.6	16.3	8.4
200	20.0	20.0	20.0	20.5	21.2	20.9	19.5	11.9
220	22.0	22.1	22.1	22.6	24.2	24.2	22.1	14.2
250	25.0	25.1	25.1	25.4	26.3	27.0	25.6	18.6

Note: Data abstracted from ASME Steam Tables, copyright 1967, and reproduced with permission of the American Society of Mechanical Engineers.

Table B.3 (continued).

Steam Table
Compressed Liquid
ENGLISH UNITS

Pressure psia p		Temperature, °F, t 32	100	200	300	400	500	600
200	$(v - v_f) \times 10^5$ ft³/lbm	−1.2	−1.0	−0.7	−1.0			
	$(h - h_f)$ Btu/lbm	+0.61	+0.52	+0.42	+0.26			
	$(s - s_f) \times 10^3$ Btu/lbm-°F	+0.0	−0.1	−0.2	−0.3			
400	$(v - v_f) \times 10^5$ ft³/lbm	−2.2	−1.9	−1.7	−2.0	−2.0		
	$(h - h_f)$ Btu/lbm	+1.21	+1.05	+0.88	+0.63	+0.17		
	$(s - s_f) \times 10^3$ Btu/lbm-°F	+0.0	−0.2	−0.5	−0.6	−0.4		
600	$(v - v_f) \times 10^5$ ft³/lbm	−3.2	−3.0	−3.7	−4.0	−4.0		
	$(h - h_f)$ Btu/lbm	+1.82	+0.58	+1.33	+1.00	+0.39		
	$(s - s_f) \times 10^3$ Btu/lbm-°F	+0.0	−0.3	−0.7	−1.0	−1.0		
1000	$(v - v_f) \times 10^5$ ft³/lbm	−5.2	−5.0	−5.7	−7.0	−9.0	−7.0	
	$(h - h_f)$ Btu/lbm	+3.02	+1.37	+2.24	+1.74	+0.86	−0.11	
	$(s - s_f) \times 10^3$ Btu/lbm-°F	+0.1	−0.6	−1.2	−1.7	−2.0	−1.4	
2000	$(v - v_f) \times 10^5$ ft³/lbm	−11.2	−10.0	−10.7	−14.0	−20.0	−29.0	−32.0
	$(h - h_f)$ Btu/lbm	+6.01	+5.26	+4.51	+3.62	+2.09	−0.37	−2.62
	$(s - s_f) \times 10^3$ Btu/lbm-°F	+0.2	−1.2	−2.4	−3.5	−4.6	−5.6	−4.3
3000	$(v - v_f) \times 10^5$ ft³/lbm	−16.2	−14.0	−15.7	−21.0	−31.0	−48.0	−88.0
	$(h - h_f)$ Btu/lbm	+8.97	+7.88	+6.79	+5.52	+3.37	−0.38	−7.02
	$(s - s_f) \times 10^3$ Btu/lbm-°F	+0.2	−1.8	−3.6	−5.2	−7.0	−9.4	−12.5

Note: Data abstracted from ASME Steam Tables, copyright 1967, and reproduced with permission of the American Society of Mechanical Engineers.

Table B.4

Gas Constants
SI UNITS

Substance	*Symbol*	*M*	*R* J/kg · K	c_p kJ/kg · K at 25°C	c_v kJ/kg · K at 25°C	*k* c_p/c_v
Acetylene	C_2H_2	26.038	320	1.687	1.368	1.234
Air		28.967	287	1.007	0.719	1.399
Ammonia	NH_3	17.032	488	2.096	1.607	1.304
Argon	Ar	39.944	208	0.521	0.312	1.668
Benzene	C_6H_6	78.114	106	1.045	0.939	1.113
n-Butane	C_4H_{10}	58.124	143	1.676	1.533	1.093
Isobutane	C_4H_{10}	58.124	143	1.666	1.523	1.094
1-Butene	C_4H_3	56.108	148	1.527	1.374	1.111
Carbon dioxide	CO_2	44.011	189	0.844	0.655	1.288
Carbon monoxide	CO	28.011	297	1.040	0.744	1.399
Carbon tetrachloride	CCl_4	153.839				
n-Deuterium	D_2	4.029				
Dodecane	$C_{12}H_{26}$	170.340	49	1.646	1.597	1.031
Ethane	C_2H_6	30.070	277	1.751	1.475	1.188
Ethyl ether	$C_4H_{10}O$	74.124				
Ethylene	C_2H_4	28.054	297	1.552	1.256	1.236
Freon, F-12	CCl_2F_2	120.925	69	0.573	0.504	1.136
Helium	He	4.003	2079	5.196	3.117	1.667
n-Heptane	C_7H_{16}	100.205	83	1.656	1.573	1.053
n-Hexane	C_6H_{14}	86.178	96	1.660	1.564	1.062
Hydrogen	H_2	2.016	4124	14.302	10.178	1.405
Hydrogen sulfide	H_2S	34.082				
Mercury	Hg	200.610				
Methane	CH_4	16.043	519	2.227	1.708	1.304
Methyl fluoride	CH_3F	34.035				
Neon	Ne	20.183	412	1.030	0.618	1.667
Nitric Oxide	NO	30.008	277	0.995	0.718	1.386
Nitrogen	N_2	28.016	297	1.040	0.743	1.400
Octane	C_8H_{18}	114.232	73	1.653	1.581	1.046
Oxygen	O_2	32.000	260	0.917	0.657	1.396
n-Pentane	C_5H_{12}	72.151	115	1.666	1.551	1.074
Isopentane	C_5H_{12}	72.151	115	1.663	1.548	1.074
Propane	C_3H_8	44.097	189	1.667	1.478	1.128
Propylene	C_3H_6	42.081	198	1.519	1.279	1.187
Sulfur dioxide	SO_2	64.066	130	0.621	0.491	1.264
Water vapor	H_2O	18.016	462	1.864	1.402	1.329
Xenon	Xe	131.300	63	0.158	0.095	1.667

Note: 1 J/kg · K = 2.388×10^{-4} Btu/lbm °R.

Table B.4 (continued).

ENGLISH UNITS

Substance	Symbol	M	R ft-lb$_f$ / lb$_m$·°R	c_p Btu / lb$_m$·°R at 77°F	c_v Btu / lb$_m$·°R at 77°F	k $\frac{c_p}{c_v}$
Acetylene	C_2H_2	26.038	59.39	0.4030	0.3267	1.234
Air	....	28.967	53.36	0.2404	0.1718	1.399
Ammonia	NH_3	17.032	90.77	0.5006	0.3840	1.304
Argon	A	39.944	38.73	0.1244	0.0746	1.668
Benzene	C_6H_6	78.114	19.78	0.2497	0.2243	1.113
n-Butane	C_4H_{10}	58.124	26.61	0.4004	0.3662	1.093
Isobutane	C_4H_{10}	58.124	26.59	0.3979	0.3637	1.094
1-Butene	C_4H_8	56.108	27.545	0.3646	0.3282	1.111
Carbon dioxide	CO_2	44.011	35.12	0.2015	0.1564	1.288
Carbon monoxide	CO	28.011	55.19	0.2485	0.1776	1.399
Carbon tetrachloride	CCl_4	153.839				
n-Deuterium	D_2	4.029				
Dodecane	$C_{12}H_{26}$	170.340	9.074	0.3931	0.3814	1.031
Ethane	C_2H_6	30.070	51.43	0.4183	0.3522	1.188
Ethyl ether	$C_4H_{10}O$	74.124				
Ethylene	C_2H_4	28.054	55.13	0.3708	0.3000	1.236
Freon, F-12	CCl_2F_2	120.925	12.78	0.1369	0.1204	1.136
Helium	He	4.003	386.33	1.241	0.7446	1.667
n-Heptane	C_7H_{16}	100.205	15.42	0.3956	0.3758	1.053
n-Hexane	C_6H_{14}	86.178	17.93	0.3966	0.3736	1.062
Hydrogen	H_2	2.016	766.53	3.416	2.431	1.405
Hydrogen sulfide	H_2S	34.082				
Mercury	Hg	200.610				
Methane	CH_4	16.043	96.40	0.5318	0.4079	1.304
Methyl fluoride	CH_3F	34.035				
Neon	Ne	20.183	76.58	0.2460	0.1476	1.667
Nitric oxide	NO	30.008	51.49	0.2377	0.1715	1.386
Nitrogen	N_2	28.016	55.15	0.2483	0.1774	1.400
Octane	C_8H_{18}	114.232	13.54	0.3949	0.3775	1.046
Oxygen	O_2	32.000	48.29	0.2191	0.1570	1.396
n-Pentane	C_5H_{12}	72.151	21.42	0.3980	0.3705	1.074
Isopentane	C_5H_{12}	72.151	21.42	0.3972	0.3697	1.074
Propane	C_3H_8	44.097	35.07	0.3982	0.3531	1.128
Propylene	C_3H_6	42.081	36.72	0.3627	0.3055	1.187
Sulfur dioxide	SO_2	64.066	24.12	0.1483	0.1173	1.264
Water vapor	H_2O	18.016	85.80	0.4452	0.3349	1.329
Xenon	Xe	131.300	11.78	0.03781	0.02269	1.667

Source: E. F. Obert, *Concepts of Thermodynamics*, copyright 1960, McGraw-Hill. Used with permission of McGraw-Hill Book Company.

Note: 1 Btu/lbm·°R = 1 calorie/g·K

Note: Data selected from J.F. Masi, Trans. ASME, 76:1067 (October, 1954); National Bureau of Standards (U.S.). Circ. 500, February 1952; "Selected Values of Properties of Hydrocarbons and Related Compounds," American Petroleum Institute Research Project 44, Thermodynamics Research Center, Texas A & M University; College Station, Texas (Loose Leaf Data Sheets, extant 1972).

Table B.5

Critical Properties

Substance	Ref. date	Sym- bol	M	T_c K	p_c atm	v_c cm³/g mole	v_c ft³/ mole	z_c
Acetylene	1928	C_2H_2	26.038	309.5	61.6	113	......	0.274
Air	1917	....	28.967	132.41	37.25	93.25		
Ammonia	1920	NH_3	17.032	405.4	111.3	72.5	1.16	0.243
Argon	1910	A	39.944	150.72	47.996	75	1.20	0.291
Benzene	1948	C_6H_6	78.114	562.6	48.6	260	4.17	0.274
n-Butane	1939	C_4H_{10}	58.124	425.17	37.47	255	4.08	0.274
Isobutane	1910	C_4H_{10}	58.124	408.14	36.00	263	4.21	0.283
1-Butene	1950	C_4H_8	56.108	419.6	39.7	240	3.84	0.277
Carbon dioxide	1950	CO_2	44.011	304.20	72.90	94	1.51	0.275
Carbon monoxide	1936	CO	28.011	132.91	34.529	93	1.49	0.294
Carbon tetrachloride	1931	CCl_4	153.839	556.4	45.0	276	......	0.272
n-Deuterium	1951	D_2	4.029	38.43	16.421			
Dodecane	1953	$C_{12}H_{26}$	170.340	659	17.9	......	11.5	0.237
Ethane	1939	C_2H_6	30.070	305.43	48.20	148	2.37	0.285
Ethyl ether	1929	$C_4H_{10}O$	74.124	467.8	35.6	282.9		
Ethylene	1939	C_2H_4	28.054	283.06	50.50	124	1.99	0.270
Freon, F-12	1957	CCl_2F_2	120.925	385.16	40.0	217	3.47	0.279
Helium	1936	He	4.003	5.19	2.26	58	0.929	0.308
n-Heptane	1937	C_7H_{16}	100.205	540.17	27.00	426	6.82	0.260
n-Hexane	1946	C_6H_{14}	86.178	507.9	29.94	368	5.89	0.264
Hydrogen	1951	H_2	2.016	33.24	12.797	65	1.04	0.304
Hydrogen sulfide	1948	H_2S	34.082	373.7	88.8	98	1.57	0.284
Mercury	1953	Hg	200.610					
Methane	1953	CH_4	16.043	190.7	45.8	99	1.59	0.290
Methyl fluoride	1932	CH_3F	34.035	317.71	58.0			
Neon	1936	Ne	20.183	44.39	26.86	41.7	0.668	0.308
Nitric oxide	1951	NO	30.008	179.2	65.0	58	0.929	0.256
Nitrogen	1951	N_2	28.016	126.2	33.54	90	0.144	0.291
Octane	1931	C_8H_{18}	114.232	569.4	24.64	486	7.77	0.256
Oxygen	1948	O_2	32.000	154.78	50.14	74	1.19	0.292
n-Pentane	1899	C_5H_{12}	72.151	469.78	33.31	311	4.98	0.269
Isopentane	1910	C_5H_{12}	72.151	461.0	32.92	308	4.93	0.268
Propane	1940	C_3H_8	44.097	370.01	42.1	200	3.20	0.277
Propylene	1953	C_3H_6	42.081	365.1	45.40	181	2.90	0.274
Sulfur dioxide	1945	SO_2	64.066	430.7	77.8	122	......	0.269
Water	1934	H_2O	18.016	647.27	218.167	56	0.897	0.230
Xenon	1951	Xe	131.300	289.81	58.0	118.8	1.90	0.290

Table B.6

SI UNITS

T K	h kJ/kg	p_r	u kJ/kg	v_r	ϕ kJ/kg · K
60	59.7	0.00503	42.4	7961	0.903
80	79.7	0.01372	56.8	3887	1.191
100	99.8	0.02990	71.1	2230	1.414
120	119.8	0.0565	85.3	1416	1.597
140	139.8	0.0968	99.7	964	1.752
160	159.9	0.1543	114.0	691	1.885
180	179.9	0.2328	128.3	516	2.003
200	200.0	0.336	142.6	397	2.109
220	220.0	0.469	156.8	313	2.204
240	240.0	0.636	171.1	252	2.292
260	260.1	0.841	185.5	206.3	2.372
280	280.1	1.089	199.8	171.5	2.446
300	300.2	1.386	214.1	144.3	2.512
320	320.3	1.738	228.1	122.8	2.578
340	340.4	2.15	242.9	105.5	2.641
360	360.6	2.63	257.2	91.4	2.699
380	380.8	3.18	271.7	79.8	2.753
400	401.0	3.81	286.2	70.1	2.805
420	421.3	4.52	300.7	61.9	2.855
440	444.6	5.33	315.3	55.0	2.902
460	462.0	6.25	330.0	49.1	2.947
480	482.5	7.27	344.8	44.0	2.991
500	503.0	8.41	359.5	39.6	3.033
550	554.8	11.86	396.9	30.9	3.131
600	607.0	16.28	434.8	24.6	3.222
650	659.8	21.9	473.3	19.83	3.307
700	713.3	28.8	512.4	16.21	3.386
750	767.3	37.4	552.1	13.39	3.461
800	821.9	47.8	592.3	11.17	3.531
850	877.2	60.3	633.2	9.40	3.598
900	932.9	75.3	674.6	7.97	3.662
950	989.2	93.1	716.6	6.81	3.723
1000	1046.0	114.0	759.0	5.85	3.781
1100	1161.1	167.1	845.3	4.39	3.891
1200	1277.8	238	933.4	3.36	3.992
1300	1395.0	331	1022.9	2.62	4.087
1400	1515.4	451	1113.6	2.07	4.175
1500	1636.0	602	1205.5	1.662	4.258
1600	1757.5	791	1298.3	1.349	4.337
1700	1880.1	1025	1392.2	1.106	4.411
1800	2003.3	1310	1486.8	0.916	4.482
1900	2127.4	1655	1582.1	0.766	4.549
2000	2252.1	2068	1678.1	0.645	4.613
2200	2503.2	3138	1871.8	0.468	4.732
2400	2756.6	4607	2067.8	0.347	4.843
2600	3011.7	6575	2265.5	0.264	4.947
2800	3268.4	9159	2464.8	0.2039	5.040
3000	3526.5	12490	2665.5	0.1602	5.129
3200	3786.0	16720	2867.5	0.1276	5.212
3600	4308.2	28569	3275.0	0.0840	5.366

Table B.6 (continued).

Air Table
Properties of Air at Low Pressure*
ENGLISH UNITS

T °R	h Btu/lbm	p_r	u Btu/lbm	v_r	ϕ Btu/lbm·°R
200	47.67	.04320	33.96	1714.9	0.36303
250	59.64	.09415	42.50	983.6	0.41643
300	71.61	.17795	51.04	624.5	0.46007
350	83.57	.3048	59.58	425.4	0.49695
400	95.53	.4858	68.11	305.0	0.52890
450	107.50	.7329	76.65	227.45	0.55710
500	119.48	1.0590	85.20	174.90	0.58233
525	125.47	1.2560	89.48	154.84	0.59403
550	131.46	1.4779	93.76	137.85	0.60518
575	137.47	1.7269	98.05	123.34	0.61586
600	143.47	2.005	102.34	110.88	0.62607
625	149.49	2.313	106.64	100.08	0.63589
650	155.50	2.655	110.94	90.69	0.64533
675	161.54	3.032	115.26	82.47	0.65443
700	167.56	3.446	119.58	75.25	0.66321
725	173.60	3.900	123.91	68.86	0.67169
750	179.66	4.396	128.25	63.20	0.67991
800	191.81	5.526	136.97	53.63	0.69558
850	204.01	6.856	145.74	45.92	0.71037
900	216.26	8.411	154.57	39.64	0.72438
950	228.58	10.216	163.46	34.45	0.73771
1000	240.98	12.298	172.43	30.12	0.75042
1050	253.45	14.686	181.47	26.48	0.76259
1100	265.99	17.413	190.58	23.40	0.77426
1150	278.61	20.51	199.78	20.771	0.78548
1200	291.30	24.01	209.05	18.514	0.79628
1250	304.08	27.96	218.40	16.563	0.80672
1300	316.94	32.39	227.83	14.868	0.81680
1350	329.88	37.35	237.34	13.391	0.82658
1400	342.90	42.88	246.93	12.095	0.83604
1450	356.00	49.03	256.60	10.954	0.84523
1500	369.17	55.86	266.34	9.948	0.85416
1550	382.42	63.40	276.17	9.056	0.86285
1600	395.74	71.73	286.06	8.263	0.87130
1650	409.13	80.89	296.03	7.556	0.87954
1700	422.59	90.95	306.06	6.924	0.88758
1750	436.12	101.98	316.16	6.357	0.89542
1800	449.71	114.03	326.32	5.847	0.90308
1900	477.09	141.51	346.85	4.974	0.91788

Source: J. H. Keenan and J. Kaye, Gas Tables (New York, 1948), John Wiley & Sons, Inc., with permission of authors and publisher. Conversion to SI by author.
*Per pound-mass.

Table B.6 (continued).

T °R	h Btu/lbm	p_r	u Btu/lbm	v_r	ϕ Btu/lbm·°R
2000	504.71	174.00	367.61	4.258	0.93205
2200	560.59	256.6	409.78	3.176	0.95868
2400	617.22	367.6	452.70	2.419	0.98331
2600	674.49	513.5	496.26	1.8756	1.00623
3000	790.68	941.4	585.04	1.1803	1.04779
3500	938.40	1829.3	698.48	0.7087	1.09332
4000	1088.26	3280	814.06	0.4518	1.13334
4500	1239.86	5521	931.39	0.3019	1.16905
5000	1392.87	8837	1050.12	0.20959	1.20129
5500	1547.07	13568	1170.04	0.15016	1.23068
6000	1702.29	20120	1291.00	0.11047	1.25769
6500	1858.44	28974	1412.87	0.08310	1.28268

Table B.7

Heat of Combustion ($-\Delta H^\circ$ at 77°F or 25°C)

Substance	Symbol	h (h_{fg}) of vaporization, Btu/lbm	HHV $H_2O(l)$ and $CO_2(g)$		LHV $H_2O(g)$ and $CO_2(g)$	
			kJ/kg	Btu/lbm	kJ/kg	Btu/lbm
Acetylene	$C_2H_2(g)$	...	49,916	21,460	48,227	20,734
Benzene	$C_6H_6(g)$	186	42,268	18,172	40,579	17,446
n-Butane	$C_4H_{10}(g)$	156	49,504	21,283	45,718	19,655
Isobutane	$C_4H_{10}(g)$	141	49,360	21,221	45,573	19,593
1-Butene	$C_4H_{10}(g)$	156	48,436	20,824	45,299	19,475
Carbon	C(graphite)	...	32,764	14,086		
Carbon monoxide	$CO(g)$	...	10,103	4,343.6		
n-Decane	$C_{10}H_{22}(g)$	155	48,004	20,638	44,601	19,175
n-Dodecane	$C_{12}H_{26}(g)$	155	47,832	20,564	44,473	19,120
Ethane	$C_2H_6(g)$	...	51,879	22,304	47,488	20,416
Ethylene	$C_2H_4(g)$	...	50,300	21,625	47,162	20,276
n-Heptane	$C_7H_{16}(g)$	157	48,439	20,825	44,924	19,314
n-Hexane	$C_6H_{14}(g)$	157	48,679	20,928	45,103	19,391
Hydrogen	$H_2(g)$	...	141,786	60,957	119,954	51,571
Methane	$CH_4(g)$	...	55,500	23,861	50,014	21,502
n-Nonane	$C_9H_{20}(g)$	156	48,118	20,687	44,685	19,211
n-Octane	$C_8H_{18}(g)$	156	48,258	20,747	44,789	19,256
n-Pentane	$C_5H_{12}(g)$	157	49,013	21,072	45,355	19,499
Isopentane	$C_5H_{12}(g)$	147	48,904	21,025	45,243	19,451
Propane	$C_3H_8(g)$	147	50,349	21,646	46,355	19,929
Propylene	$C_3H_6(g)$	...	48,920	21,032	45,783	19,683

Source: E. F. Obert, Concepts of Thermodynamics, McGraw-Hill, copyright 1960; by permission of McGraw-Hill Book Company.
Note: Data from "Selected Values of Properties of Hydrocarbons and Related Compounds," America Petroleum Institute Research Project 44, Thermodynamics Research Center, Texas A & M University, College Station, Texas (Loose Leaf Data Sheets, extant 1972).

Table B.8

Monochlorodifluoromethane, $CHClF_2$ (Refrigerant-22)
Properties of Saturation States

SI UNITS

Temp °C	Pressure bars	Specific Volume v_f	v_g	Enthalpy h_f	h_g	Entropy s_f	s_g
−50	0.6439	0.69526	324.557	144.959	383.921	0.77919	1.85000
−45	0.8271	0.70219	256.990	150.153	386.282	0.80216	1.83708
−40	1.0495	0.70936	205.745	155.414	388.609	0.82490	1.82504
−35	1.3168	0.71680	166.400	160.747	390.896	0.84743	1.81380
−30	1.6348	0.72452	135.844	166.140	393.138	0.86976	1.80329
−25	2.0098	0.73255	111.859	171.606	395.330	0.89190	1.79342
−20	2.4483	0.74091	92.8432	177.142	397.467	0.91386	1.78415
−15	2.9570	0.74964	77.6254	182.749	399.544	0.93564	1.77540
−10	3.5430	0.75876	65.3399	188.426	401.555	0.95725	1.76713
− 5	4.2135	0.76831	55.3394	194.176	403.496	0.97870	1.75928
0	4.9757	0.77834	47.1354	200.000	405.361	1.00000	1.75179
5	5.8378	0.78889	40.3556	205.899	407.143	1.02116	1.74463
10	6.8070	0.80002	34.7136	211.877	408.835	1.04218	1.73775
15	7.8915	0.81180	29.9874	217.937	410.430	1.06309	1.73109
20	9.0993	0.82431	26.0032	224.084	411.918	1.08390	1.72462
25	10.439	0.83765	22.6242	230.324	413.289	1.10462	1.71827
30	11.919	0.85193	19.7417	236.664	414.530	1.12530	1.71200
35	13.548	0.86729	17.2686	243.114	415.627	1.14594	1.70576
40	15.335	0.88392	15.1351	249.686	416.561	1.16659	1.69946
45	17.290	0.90203	13.2841	256.396	417.308	1.18730	1.69305
50	19.423	0.92193	11.6693	263.264	417.839	1.20811	1.68643
55	21.744	0.94400	10.2521	270.318	418.116	1.22910	1.67208
60	24.266	0.96878	9.00062	277.594	418.089	1.25038	1.67208
65	26.999	0.99702	7.88749	285.142	417.687	1.27206	1.66402
70	29.959	1.02987	6.88899	293.038	416.809	1.29436	1.65504
75	33.161	1.06916	5.98334	301.399	415.299		

Table B.8 (continued).

ENGLISH UNITS

TEMP.	PRESSURE	VOLUME cu ft/lb		ENTHALPY Btu/lb			ENTROPY Btu/(lb)(°R)	
°F	PSIA	LIQUID v_f	VAPOR v_g	LIQUID h_f	LATENT h_{fg}	VAPOR h_g	LIQUID s_f	VAPOR s_g
-100	2.3983	0.010664	18.433	-14.564	107.935	93.371	-0.03734	0.26274
- 90	3.4229	0.010771	13.235	-12.216	106.759	94.544	-0.03091	0.25787
- 80	4.7822	0.010881	9.6949	- 9.838	105.548	95.710	-0.02457	0.25342
- 70	6.5522	0.010995	7.2318	- 7.429	104.297	96.868	-0.01832	0.24932
- 60	8.818	0.011113	5.4844	- 4.987	103.001	98.014	-0.01214	0.24556
- 50	11.674	0.011235	4.2224	- 2.511	101.656	99.144	-0.00604	0.24209
- 40	15.222	0.011363	3.2957	0.000	100.257	100.257	0.00000	0.23888
- 30	19.573	0.011495	2.6049	2.547	98.801	101.348	0.00598	0.23591
- 25	22.086	0.011564	2.3260	3.834	98.051	101.885	0.00894	0.23451
- 20	24.845	0.011634	2.0926	5.131	97.285	102.415	0.01189	0.23315
- 15	27.865	0.011705	1.8695	6.436	96.502	102.939	0.01483	0.23184
- 10	31.162	0.011778	1.6825	7.751	95.704	103.455	0.01776	0.23058
- 5	34.754	0.011853	1.5177	9.075	94.889	103.964	0.02067	0.22936
0	38.657	0.011930	1.3723	10.409	94.056	104.465	0.02357	0.22817
5	42.888	01012008	1.2434	11.752	93.206	104.958	0.02645	0.22703
10	47.464	0.012088	1.1290	13.104	92.338	105.442	0.02932	0.22592
15	52.405	0.012171	1.0272	14.466	91.451	105.917	0.03218	0.22484
20	57.727	0.012255	0.93631	15.837	90.545	106.383	0.03503	0.22379
25	63.450	0.012342	0.85500	17.219	89.620	106.839	0.03787	0.22277
30	69.591	0.012431	0.78208	18.609	88.674	107.284	0.04070	0.22178
40	83.206	0.012618	0.65753	21.422	86.720	108.142	0.04632	0.21986
50	98.727	0.012815	0.55606	24.275	84.678	108.953	0.05190	0.21803
60	116.31	0.013025	0.47272	27.172	82.540	109.712	0.05745	0.21627
70	136.12	0.013251	0.40373	30.116	80.298	110.414	0.06296	0.21456
80	158.33	0.013492	0.34621	33.109	77.943	111.052	0.06846	0.21288
90	183.09	0.013754	0.20789	36.158	75.461	111.619	0.07394	0.21122
100	210.60	0.014038	0.25702	39.267	72.838	112.105	0.07942	0.20956
120	274.60	0.014694	0.19238	45.705	67.077	112.782	0.09042	0.20613
140	351.94	0.015518	0.14418	52.528	60.403	112.931	0.10163	0.20235
160	444.53	0.016627	0.10701	59.948	52.316	112.263	0.11334	0.19776
180	554.78	0.018332	0.07679	68.498	41.570	110.068	0.12635	0.19133
200	686.35	0.022436	0.047438	80.862	21.990	102.853	0.14460	0.17794
204.81	721.91	0.030525	0.30525	91.329	0.000	91.329	0.16016	0.16016

Note: Data abstracted from "Thermodynamic properties of Freon-22," copyright 1964; with permission of E. I. duPont de Nemours & Co.

Table B.9

Monochlorodifluoromethane, $CHClF_2$ (Refrigerant-22)
Properties of Superheated States
SI UNITS

v (10^3 m³/kg); h (kJ/kg); s (kJ/kg · K)

Pressure bars, p (Sat. Temp., °C)		Sat. Properties	Temperature, °C, t: −80	−65	−50	−35	−20	−5	10	25	40	55	70
0.049	v	3580	3778	4075	4371	4667	4962	5257	5552				
(−90)	h	364.4	369.5	377.5	385.7	394.2	403.0	412.0	421.3				
	s	1.998	2.026	2.065	2.103	2.140	2.176	2.211	2.244				
0.091	v	2018	2039	2201	2361	2522	2682	2842	3002	3162	3322		
(−82)	h	368.3	369.4	377.3	385.6	394.1	402.9	411.9	421.3	430.8	440.7		
	s	1.961	1.966	2.006	2.044	2.081	2.117	2.151	2.185	2.218	2.250		
0.159	v	1199		1255	1348	1440	1532	1624	1716	1807	1900		
(−74)	h	372.3		377.1	385.4	393.9	402.7	411.8	421.2	430.7	440.6		
	s	1.927		1.951	1.990	2.027	2.063	2.097	2.131	2.164	2.197		
0.205	v	940.9		965.0	1036.9	1108.4	1179.7	1250.8	1321.8	1392.6	1463.3		
(−70)	h	374.2		376.9	385.2	393.8	402.6	411.7	421.1	430.7	440.5		
	s	1.912		1.926	1.964	2.001	2.037	2.072	2.106	2.139	2.171		
0.263	v	746.3		750.1	806.5	862.6	918.4	974.0	1029.5	1084.9	1140.1	1195.3	
(−66)	h	376.2		376.7	385.1	393.7	402.5	411.6	421.0	430.6	440.5	450.6	
	s	1.898		1.901	1.940	1.977	2.013	2.048	2.082	2.115	2.147	2.179	
0.334	v	598.0			633.8	678.3	722.5	766.5	810.3	854.1	897.8	941.3	
(−62)	h	378.1			384.9	393.5	402.4	411.5	420.9	430.5	440.4	450.5	
	s	1.885			1.916	1.954	1.990	2.025	2.059	2.092	2.124	2.156	
0.420	v	483.6			502.8	538.4	573.9	609.1	644.1	679.1	713.9	748.7	
(−58)	h	380.1			384.6	393.3	402.2	411.3	420.7	430.4	440.3	450.4	
	s	1.873			1.893	1.931	1.967	2.002	2.036	2.070	2.102	2.134	
0.522	v	394.6			402.3	431.3	460.0	488.4	516.7	544.9	573.0	601.1	
(−53)	h	382.0			384.3	393.0	401.9	411.3	420.6	430.2	440.1	450.3	
	s	1.861			1.871	1.909	1.950	1.981	2.015	2.048	2.081	2.112	
0.644	v	324.6			324.6	348.3	371.7	395.0	418.1	441.1	463.9	486.7	509.5
(−50)	h	383.9			383.9	392.7	401.7	410.9	420.4	430.0	440.0	450.1	460.5
	s	1.850			1.850	1.888	1.925	1.960	1.994	2.028	2.060	2.092	2.123

Table B.9 (continued).

Pressure bars, p (Sat. Temp., °C)		Sat. Properties	Temperature, °C, t −50	−35	−20	−5	10	25	40	55	70	85	100
0.788	v	269.0		283.4	302.8	322.0	341.0	359.9	378.7	397.4	416.0		
(−46)	h	385.8		392.3	401.4	410.6	420.1	429.8	439.8	450.0	460.4		
	s	1.840		1.868	1.904	1.940	1.974	2.008	2.040	2.072	2.103		
0.960	v	224.6		232.2	248.4	264.4	280.2	295.9	311.4	326.9	342.4		
(−39)	h	387.7		391.9	401.0	410.3	419.8	429.6	439.6	449.8	460.2		
	s	1.830		1.848	1.885	1.920	1.955	1.989	2.021	2.053	2.084		
1.10	v	197.0		200.9	215.1	229.1	242.9	256.6	270.2	283.8	297.2		
(−39)	h	389.0		391.5	400.7	410.0	419.6	429.4	439.4	449.6	460.1		
	s	1.823		1.833	1.870	1.906	1.941	1.975	2.007	2.039	2.070		
1.32	v	166.4		166.4	178.5	190.3	202.0	213.5	225.0	236.3	247.6	258.9	
(−35)	h	390.9		390.9	400.2	409.6	419.2	429.1	439.1	449.4	459.8	470.5	
	s	1.814		1.814	1.852	1.888	1.923	1.957	1.989	2.021	2.053	2.083	
1.57	v	141.4			148.9	159.0	169.0	178.8	188.4	198.1	207.6	217.1	
(−31)	h	392.7			399.6	409.1	418.8	428.7	438.8	449.1	459.6	470.3	
	s	1.805			1.833	1.870	1.905	1.939	1.972	2.004	2.035	2.066	
1.85	v	120.8			124.9	133.6	142.1	127.4	158.8	167.0	175.1	183.2	
(−27)	h	394.5			398.9	408.5	418.3	427.8	438.4	448.7	459.3	470.1	
	s	1.797			1.815	1.852	1.888	1.905	1.955	1.987	2.019	2.049	
2.18	v	103.7			105.3	112.8	120.2	108.41	134.6	141.6	148.6	155.5	
(−23)	h	396.2			398.1	407.9	417.7	427.2	438.0	448.4	459.0	469.8	
	s	1.790			1.797	1.835	1.871	1.889	1.938	1.971	2.002	2.033	
2.54	v	89.5				95.70	102.13	92.66	114.59	120.68	126.70	132.67	138.60
(−19)	h	397.9				407.1	417.1	426.6	437.5	447.9	458.6	469.4	480.4
	s	1.782				1.818	1.854	1.873	1.922	1.955	1.987	2.018	2.048
2.96	v	77.6				81.50	87.16		98.05	103.35	108.59	113.8	118.9
(−15)	h	399.5				406.2	416.4		436.9	447.5	458.1	469.0	480.1
	s	1.775				1.801	1.838		1.907	1.939	1.971	2.002	2.033

Table B.9 (continued).

Pressure bars, *p* (Sat. Temp., °C)		Sat. Properties	Temperature, °C, *t* −5	10	25	40	55	70	85	100	115	130	145
3.93	*v*	59.1	59.71	64.20	68.51	71.71	76.81	80.84	84.82	88.74	92.64		
(−7)	*h*	402.7	404.1	414.6	425.0	435.6	446.3	457.1	468.1	479.3	490.6		
	s	1.762	1.768	1.805	1.842	1.876	1.909	1.942	1.973	2.004	2.033		
4.51	*v*	51.9		55.36	59.23	62.97	66.61	70.18	73.69	77.16	80.59		
(−3)	*h*	404.3		413.5	424.1	434.8	445.6	456.5	467.6	478.8	490.2		
	s	1.756		1.790	1.826	1.861	1.895	1.927	1.959	1.990	2.019		
5.14	*v*	45.7		47.86	51.35	54.71	57.97	61.15	64.27	67.35	70.38		
(1)	*h*	405.7		412.2	423.1	433.9	444.8	455.8	466.9	478.2	489.7		
	s	1.750		1.774	1.811	1.846	1.880	1.913	1.945	1.976	2.006		
5.84	*v*	40.4		41.46	44.64	47.68	50.61	53.46	56.25	59.00	61.70		
(5)	*h*	407.1		410.9	421.9	432.9	443.9	455.0	466.3	477.6	489.1		
	s	1.745		1.758	1.796	1.832	1.866	1.899	1.931	1.963	1.993		
6.60	*v*	35.8		35.96	38.89	41.66	44.31	46.89	49.40	51.85	54.27	56.66	
(9)	*h*	408.5		409.3	420.6	431.8	443.0	454.2	465.5	477.0	488.5	500.2	
	s	1.739		1.742	1.781	1.817	1.852	1.886	1.918	1.949	1.980	2.009	
7.44	*v*	31.8			33.93	36.48	38.90	41.23	43.50	45.72	47.90	50.04	
(13)	*h*	409.8			419.1	430.5	441.9	453.2	464.7	476.2	487.9	499.6	
	s	1.734			1.766	1.803	1.838	1.872	1.905	1.936	1.967	1.997	
8.36	*v*	28.3			29.64	32.00	34.22	36.35	38.42	40.43	42.00	44.33	
(17)	*h*	411.0			417.4	429.1	440.7	452.2	463.8	475.4	487.1	499.0	
	s	1.728			1.750	1.788	1.825	1.859	1.892	1.924	1.954	1.984	
9.36	*v*	25.3			25.90	28.10	30.16	32.12	34.01	35.84	37.63	39.38	
(21)	*h*	412.2			415.5	427.5	439.3	451.0	462.7	474.5	486.3	498.2	
	s	1.723			1.734	1.774	1.811	1.846	1.879	1.911	1.942	1.972	
10.44	*v*	22.6			22.6	24.7	26.6	28.4	30.2	31.8	33.48	35.08	36.64
(25)	*h*	413.3			413.3	425.7	437.8	449.7	461.6	473.5	485.4	497.4	509.6
	s	1.718			1.718	1.759	1.797	1.832	1.866	1.899	1.930	1.960	1.990

Table B.9 (continued).

Pressure bars, p (Sat. Temp., °C)		Sat. Properties	Temperature, °C, t: 55	70	85	100	115	130	145	160	175	190	205
12.88 (33)	v	18.2	20.80	22.38	23.88	25.31	26.69	27.8	29.34				
	h	415.2	434.3	446.7	459.0	471.2	483.4	495.6	507.9				
	s	1.708	1.768	1.806	1.841	1.874	1.906	1.937	1.967				
15.71 (41)	v	14.7	16.24	17.67	18.99	20.24	21.43	22.58	23.69	24.78			
	h	416.7	429.7	443.0	455.8	468.4	480.9	493.4	506.0	518.5			
	s	1.698	1.739	1.778	1.815	1.849	1.882	1.914	1.944	1.974			
18.98 (49)	v	11.97	12.58	13.93	15.14	16.26	17.30	18.31	19.27	20.21			
	h	417.8	423.9	438.3	451.9	465.1	478.0	490.8	503.6	516.4			
	s	1.688	1.707	1.750	1.788	1.824	1.858	1.891	1.922	1.952			
22.73 (57)	v	9.733		10.902	12.050	13.075	14.021	14.911	15.761	16.679	17.372		
	h	418.1		432.2	447.0	461.0	474.5	487.8	500.9	514.0	527.1		
	s	1.677		1.719	1.761	1.799	1.834	1.868	1.900	1.931	1.960		
27.00 (65)	v	7.887		8.351	9.515	10.494	11.370	12.179	12.940	13.667	14.366		
	h	417.7		424.0	440.8	455.9	470.2	484.1	497.7	511.1	524.5		
	s	1.664		1.683	1.731	1.772	1.810	1.845	1.878	1.909	1.939		
31.85 (73)	v	6.336			7.367	8.357	9.197	9.950	10.646	11.302	11.927	12.529	
	h	416.0			432.5	449.5	465.0	479.6	493.8	507.7	521.4	535.0	
	s	1.649			1.696	1.743	1.783	1.820	1.855	1.887	1.919	1.948	
37.35 (81)	v	4.988			5.413	6.538	7.383	8.106	8.756	9.358	9.926	10.47	
	h	412.3			419.8	441.1	458.5	474.3	489.2	503.7	517.8	531.8	
	s	1.630			1.651	1.709	1.755	1.795	1.831	1.865	1.897	1.928	
40.37 (85)	v	4.358			4.358	5.707	6.580	7.298	7.933	8.514	9.058	9.573	10.068
	h	409.1			409.1	435.7	454.5	471.1	486.6	501.4	515.8	530.0	544.1
	s	1.617			1.617	1.690	1.739	1.781	1.819	1.854	1.886	1.918	1.947

Table B.9 (continued).

ENGLISH UNITS

v (ft³/lbm); *h* (Btu/lbm); *s* (Btu/lbm · °R) (saturation properties in parentheses)

ABSOLUTE PRESSURE, psi

TEMP. °F	**5.0**		
	19.7411*		
	(−78.62 °F)		
	v	*h*	*s*
	(9.3011)	(95.871)	(0.25283)
−80	—	—	—
−70	9.5237	97.018	0.25581
−60	9.7810	98.362	0.25921
−50	10.038	99.721	0.26257
−40	10.293	101.094	0.26588
−30	10.549	102.482	0.26915
−20	10.803	103.885	0.27238
−10	11.058	105.302	0.27557
0	11.311	106.735	0.27872
10	11.565	108.182	0.28183
20	11.818	109.643	0.28491
30	12.070	111.120	0.28796
40	12.323	112.611	0.29097
50	12.575	114.117	0.29396
60	12.826	115.638	0.29691
70	13.078	117.174	0.29984
80	13.329	118.724	0.30274
90	13.580	120.288	0.30561
100	13.831	121.867	0.30845
110	14.082	123.461	0.31128
120	14.333	125.069	0.31407
130	14.583	126.691	0.31685
140	14.833	128.327	0.31960
150	15.084	129.977	0.32233
160	15.334	131.642	0.32504
170	15.584	133.320	0.32772
180	15.834	135.012	0.33039
190	16.083	136.718	0.33304
200	16.333	138.437	0.33566
210	16.583	140.170	0.33827
220	16.832	141.916	0.34086
230	17.082	143.676	0.34343
240	—	—	—

TEMP. °F	**10.0**		
	9.561*		
	(−55.59 °F)		
	v	*h*	*s*
	(4.8778)	(98.515)	(0.24339)
−50	4.9518	99.291	0.24590
−40	5.0838	100.690	0.24927
−30	5.2152	102.101	0.25260
−20	5.3460	103.526	0.25588
−10	5.4762	104.963	0.25911
0	5.6060	106.414	0.26230
10	5.7353	107.878	0.26545
20	5.8643	109.356	0.26856
30	5.9929	110.847	0.27164
40	6.1212	112.353	0.27468
50	6.2492	113.872	0.27769
60	6.3769	115.404	0.28067
70	6.5044	116.951	0.28362
80	6.6316	118.512	0.28654
90	6.7586	120.086	0.28943
100	6.8855	121.674	0.29229
110	7.0122	123.276	0.29513
120	7.1387	124.892	0.29794
130	7.2651	126.522	0.30073
140	7.3913	128.165	0.30349
150	7.5174	129.822	0.30623
160	7.6434	131.493	0.30895
170	7.7693	133.177	0.31165
180	7.8951	134.875	0.31432
190	8.0208	136.586	0.31697
200	8.1464	138.310	0.31961
210	8.2719	140.048	0.32222
220	8.3974	141.799	0.32482
230	8.5228	143.562	0.32739
240	8.6481	145.339	0.32995
250	8.7734	147.129	0.33249

TEMP. °F	**15**		
	0.304		
	(−40.57 °F)		
	v	*h*	*s*
	(3.3412)	(100.194)	(0.23906)
−40	3.3463	100.276	0.23925
−30	3.4365	101.712	0.24263
−20	3.5261	103.159	0.24596
−10	3.6152	104.618	0.24924
0	3.7037	106.088	0.25248
10	3.7918	107.570	0.25567
20	3.8794	109.065	0.25882
30	3.9667	110.571	0.26192
40	4.0537	112.091	0.26500
50	4.1404	113.623	0.26803
60	4.2268	115.168	0.27103
70	4.3129	116.727	0.27400
80	4.3989	118.298	0.27694
90	4.4846	119.882	0.27985
100	4.5701	121.480	0.28273
110	4.6554	123.091	0.28559
120	4.7406	124.715	0.28841
130	4.8256	126.352	0.29121
140	4.9105	128.003	0.29399
150	4.9952	129.667	0.29674
160	5.0799	131.344	0.29947
170	5.1644	133.034	0.30217
180	5.2488	134.737	0.30486
190	5.3332	136.454	0.30752
200	5.4174	138.183	0.31016
210	5.5016	139.925	0.31278
220	5.5857	141.680	0.31538
230	5.6697	143.448	0.31797
240	5.7537	145.229	0.32053
250	5.8376	147.022	0.32307
260	5.9215	148.828	0.32560
270	—	—	—

TEMP. °F	**20**		
	5.304		
	(−29.12 °F)		
	v	*h*	*s*
	(2.5527)	(101.444)	(0.23566)
−30	—	—	—
−20	2.6156	102.785	0.23874
−10	2.6841	104.266	0.24207
0	2.7521	105.756	0.24535
10	2.8196	107.257	0.24858
20	2.8867	108.769	0.25177
30	2.9534	110.292	0.25491
40	3.0198	111.826	0.25801
50	3.0858	113.372	0.26107
60	3.1516	114.930	0.26410
70	3.2171	116.500	0.26709
80	3.2823	118.082	0.27005
90	3.3474	119.677	0.27298
100	3.4122	121.284	0.27588
110	3.4769	122.904	0.27874
120	3.5414	124.536	0.28159
130	3.6058	126.182	0.28440
140	3.6700	127.840	0.28719
150	3.7341	129.510	0.28995
160	3.7981	131.194	0.29269
170	3.8619	132.890	0.29540
180	3.9257	134.599	0.29810
190	3.9894	136.321	0.30077
200	4.0529	138.055	0.30342
210	4.1164	139.802	0.30604
220	4.1799	141.562	0.30865
230	4.2432	143.334	0.31124
240	4.3065	145.119	0.31381
250	4.3697	146.916	0.31636
260	4.4329	148.725	0.31889
270	4.4960	150.547	0.32141
280	4.5591	152.381	0.32390

TEMP. °F	**25**		
	10.304		
	(−19.73 °F)		
	v	*h*	*s*
	(2.0704)	(102.444)	(0.23308)
−20	—	—	—
−10	2.1251	103.907	0.23637
0	2.1808	105.419	0.23970
10	2.2360	106.939	0.24297
20	2.2908	108.469	0.24619
30	2.3452	110.008	0.24937
40	2.3992	111.558	0.25250
50	2.4529	113.118	0.25559
60	2.5063	114.689	0.25864
70	2.5594	116.271	0.26166
80	2.6123	117.865	0.26464
90	2.6650	119.470	0.26758
100	2.7175	121.087	0.27050
110	2.7698	122.716	0.27338
120	2.8219	124.357	0.27624
130	2.8738	126.010	0.27907
140	2.9257	127.675	0.28187
150	2.9774	129.353	0.28464
160	3.0289	131.043	0.28739
170	3.0804	132.745	0.29012
180	3.1318	134.460	0.29282
190	3.1830	136.187	0.29550
200	3.2342	137.926	0.29815
210	3.2853	139.678	0.30079
220	3.3363	141.443	0.30340
230	3.3873	143.219	0.30600
240	3.4382	145.008	0.30857
250	3.4890	146.809	0.31113
260	3.5397	148.622	0.31367
270	3.5905	150.447	0.31619
280	3.6411	152.284	0.31869
290	3.6917	154.134	0.32117

TEMP. °F	**30**		
	15.304		
	(−11.71 °F)		
	v	*h*	*s*
	(1.7439)	(103.279)	(0.23101)
−10	1.7521	103.541	0.23159
0	1.7997	105.076	0.23497
10	1.8467	106.616	0.23828
20	1.8933	108.165	0.24154
30	1.9395	109.721	0.24475
40	1.9853	111.286	0.24792
50	2.0308	112.861	0.25104
60	2.0760	114.445	0.25412
70	2.1209	116.040	0.25716
80	2.1655	117.645	0.26016
90	2.2100	119.261	0.26312
100	2.2542	120.888	0.26606
110	2.2982	122.526	0.26896
120	2.3421	124.176	0.27183
130	2.3858	125.837	0.27467
140	2.4294	127.510	0.27748
150	2.4728	129.194	0.28027
160	2.5162	130.891	0.28303
170	2.5594	132.600	0.28576
180	2.6024	134.320	0.28848
190	2.6455	136.053	0.29116
200	2.6884	137.797	0.29383
210	2.7312	139.554	0.29647
220	2.7739	141.323	0.29909
230	2.8166	143.104	0.30169
240	2.8592	144.897	0.30427
250	2.9018	146.701	0.30684
260	2.9443	148.518	0.30938
270	2.9867	150.347	0.31190
280	3.0291	152.187	0.31441
290	3.0715	154.040	0.31689

Table B.9 (continued).

ABSOLUTE PRESSURE, psi

40 — 25.304 (1.63 °F)

TEMP. °F	v	h	s
	(1.3285)	(104.627)	(0.22780)
10	1.3594	105.953	0.23064
20	1.3959	107.541	0.23399
30	1.4319	109.134	0.23728
40	1.4675	110.732	0.24051
50	1.5028	112.337	0.24369
60	1.5378	113.950	0.24682
70	1.5724	115.570	0.24991
80	1.6068	117.199	0.25296
90	1.6410	118.837	0.25596
100	1.6749	120.485	0.25893
110	1.7087	122.142	0.26187
120	1.7423	123.810	0.26477
130	1.7757	125.487	0.26764
140	1.8090	127.176	0.27048
150	1.8421	128.875	0.27329
160	1.8751	130.585	0.27607
170	1.9080	132.306	0.27883
180	1.9407	134.039	0.28156
190	1.9734	135.783	0.28426
200	2.0060	137.538	0.28694
210	2.0385	139.304	0.28960
220	2.0709	141.082	0.29223
230	2.1033	142.872	0.29485
240	2.1356	144.673	0.29744
250	2.1678	146.486	0.30001
260	2.2000	148.310	0.30257
270	2.2321	150.145	0.30510
280	2.2641	151.992	0.30761
290	2.2961	153.851	0.31011
300	2.3281	155.721	0.31259
310	2.3600	157.602	0.31505

50 — 35.304 (12.61 °F)

TEMP. °F	v	h	s
	(1.0744)	(105.692)	(0.22535)
10	—	—	—
20	1.0968	106.897	0.22788
30	1.1269	108.529	0.23125
40	1.1564	110.163	0.23455
50	1.1857	111.800	0.23780
60	1.2145	113.443	0.24099
70	1.2431	115.091	0.24413
80	1.2714	116.745	0.24722
90	1.2994	118.406	0.25027
100	1.3272	120.075	0.25328
110	1.3548	121.752	0.25625
120	1.3822	123.438	0.25918
130	1.4095	125.133	0.26208
140	1.4366	126.838	0.26495
150	1.4635	128.552	0.26778
160	1.4903	130.276	0.27059
170	1.5170	132.010	0.27337
180	1.5436	133.755	0.27611
190	1.5701	135.510	0.27884
200	1.5965	137.276	0.28153
210	1.6228	139.053	0.28421
220	1.6491	140.840	0.28686
230	1.6752	142.638	0.28948
240	1.7013	144.448	0.29209
250	1.7274	146.268	0.29467
260	1.7533	148.100	0.29723
270	1.7792	149.943	0.29978
280	1.8051	151.797	0.30230
290	1.8309	153.661	0.30480
300	1.8567	155.537	0.30729
310	1.8824	157.424	0.30976
320	1.9081	159.322	0.31221

60 — 45.304 (22.03 °F)

TEMP. °F	v	h	s
	(0.90222)	(106.569)	(0.22337)
30	0.92300	107.904	0.22612
40	0.94863	109.577	0.22950
50	0.97385	111.249	0.23282
60	0.99871	112.923	0.23607
70	1.0232	114.600	0.23927
80	1.0475	116.281	0.24241
90	1 0715	117.967	0.24551
100	1.0952	119.658	0.24855
110	1.1187	121.356	0.25156
120	1.1420	123.061	0.25453
130	1.1652	124.774	0.25746
140	1.1882	126.495	0.26035
150	1.2111	128.225	0.26321
160	1.2338	129.963	0.26604
170	1.2564	131.711	0.26884
180	1.2788	133.468	0.27161
190	1.3012	135.235	0.27435
200	1.3235	137.012	0.27706
210	1.3457	138.799	0.27975
220	1.3678	140.596	0.28241
230	1.3898	142.403	0.28505
240	1.4118	144.221	0.28767
250	1.4337	146.050	0.29027
260	1.4556	147.889	0.29284
270	1.4773	149.739	0.29539
280	1.4991	151.600	0.29792
290	1.5208	153.471	0.30044
300	1.5424	155.353	0.30293
310	1.5640	157.246	0.30541
320	1.5856	159.149	0.30786
330	1.6071	161.063	0.31030

70 — 55.304 (30.32 °F)

TEMP. °F	v	h	s
	(0.77766)	(107.312)	(0.22172)
30	—	—	—
40	0.79981	108.972	0.22507
50	0.82224	110.682	0.22846
60	0.84428	112.391	0.23178
70	0.86598	114.098	0.23503
80	0.88736	115.807	0.23823
90	0.90846	117.519	0.24137
100	0.92932	119.234	0.24446
110	0.94995	120.953	0.24751
120	0.97038	122.679	0.25051
130	0.99063	124.410	0.25347
140	1.0107	126.148	0.25639
150	1.0306	127.893	0.25928
160	1.0504	129.647	0.26213
170	1.0701	131.408	0.26495
180	1.0896	133.178	0.26774
190	1.1091	134.957	0.27050
200	1.1284	136.745	0.27323
210	1.1477	138.543	0.27594
220	1.1669	140.350	0.27862
230	1.1860	142.166	0.28127
240	1.2050	143.993	0.28390
250	1.2239	145.830	0.28650
260	1.2428	147.677	0.28909
270	1.2617	149.534	0.29165
280	1.2805	151.401	0.29419
290	1.2992	153.279	0.29672
300	1.3179	155.167	0.29922
310	1.3365	157.066	0.30170
320	1.3551	158.975	0.30416
330	1.3737	160.894	0.30661
340	1.3923	162.824	0.30904

80 — 65.304 (37.76 °F)

TEMP. °F	v	h	s
	(0.68318)	(107.954)	(0.22029)
40	0.68782	108.347	0.22107
50	0.70822	110.098	0.22454
60	0.72820	111.843	0.22793
70	0.74780	113.584	0.23125
80	0.76708	115.323	0.23450
90	0.78605	117.061	0.23770
100	0.80477	118.801	0.24083
110	0.82325	120.544	0.24392
120	0.84152	122.290	0.24696
130	0.85960	124.040	0.24995
140	0.87751	125.796	0.25290
150	0.89626	127.558	0.25582
160	0.91286	129.326	0.25869
170	0.93034	131.102	0.26154
180	0.94770	132.885	0.26435
190	0.96495	134.677	0.26712
200	0.98209	136.476	0.26987
210	0.99915	138.284	0.27259
220	1.0161	140.101	0.27529
230	1.0330	141.928	0.27795
240	1.0498	143.763	0.28060
250	1.0666	145.608	0.28322
260	1.0833	147.463	0.28581
270	1.0999	149.328	0.28838
280	1.1165	151.202	0.29094
290	1.1330	153.087	0.29347
300	1.1495	154.981	0.29598
310	1.1659	156.885	0.29847
320	1.1823	158.800	0.30094
330	1.1987	160.725	0.30339
340	1.2150	162.660	0.30583

90 — 75.304 (44.53 °F)

TEMP. °F	v	h	s
	(0.60897)	(108.516)	(0.21903)
50	0.61924	109.496	0.22096
60	0.63766	111.280	0.22443
70	0.65568	113.056	0.22781
80	0.67334	114.827	0.23112
90	0.69069	116.594	0.23437
100	0.70777	118.360	0.23755
110	0.72459	120.127	0.24068
120	0.74120	121.894	0.24376
130	0.75760	123.665	0.24678
140	0.77383	125.439	0.24977
150	0.78989	127.218	0.25271
160	0.80581	129.002	0.25561
170	0.82159	130.793	0.25848
180	0.83725	132.589	0.26131
190	0.85279	134.393	0.26411
200	0.86824	136.205	0.26687
210	0.88359	138.024	0.26961
220	0.89885	139.851	0.27232
230	0.91403	141.687	0.27500
240	0.92914	143.532	0.27766
250	0.94418	145.385	0.28029
260	0.95916	147.248	0.28289
270	0.97408	149.120	0.28548
280	0.98894	151.002	0.28804
290	1.0038	152.893	0.29058
300	1.0185	154.794	0.29309
310	1.0332	156.704	0.29559
320	1.0479	158.624	0.29807
330	1.0626	160.554	0.30053
340	1.0772	162.494	0.30297
350	1.0917	164.444	0.30540

Table B.9 (continued).

ENGLISH UNITS

v (ft³/lbm); h (Btu/lbm); s (Btu/lbm · °R) (saturation properties in parentheses)

ABSOLUTE PRESSURE, psi

TEMP. °F	100 85.304 (50.77 °F)		
	v	h	s
	(0.54908)	(109.013)	(0.21790)
60	0.56498	110.700	0.22117
70	0.58177	112.514	0.22463
80	0.59818	114.319	0.22801
90	0.61425	116.117	0.23131
100	0.63003	117.911	0.23454
110	0.64555	119.702	0.23771
120	0.66084	121.492	0.24083
130	0.67592	123.284	0.24389
140	0.69081	125.077	0.24691
150	0.70554	126.874	0.24988
160	0.72011	128.674	0.25281
170	0.73454	130.480	0.25570
180	0.74885	132.290	0.25855
190	0.76304	134.107	0.26137
200	0.77712	135.931	0.26415
210	0.79111	137.761	0.26691
220	0.80510	139.599	0.26963
230	0.81883	141.445	0.27233
240	0.83257	143.299	0.27500
250	0.84624	145.161	0.27764
260	0.85985	147.032	0.28026
270	0.87340	148.912	0.28285
280	0.88689	150.800	0.28542
290	0.90033	152.698	0.28797
300	0.91372	154.605	0.29050
310	0.92707	156.522	0.29300
320	0.94038	158.448	0.29549
330	0.95365	160.383	0.29796
340	0.96688	162.328	0.30040
350	0.98008	164.283	0.30283
360	0.99324	166.248	0.30525

TEMP. °F	110 95.304 (56.55 °F)		
	v	h	s
	(0.49969)	(109.456)	(0.21687)
60	0.50526	110.101	0.21812
70	0.52109	111.956	0.22165
80	0.53651	113.798	0.22510
90	0.55156	115.628	0.22846
100	0.56631	117.451	0.23174
110	0.58078	119.269	0.23496
120	0.59501	121.083	0.23812
130	0.60901	122.897	0.24122
140	0.62282	124.710	0.24427
150	0.63646	126.525	0.24727
160	0.64994	128.342	0.25023
170	0.66327	130.163	0.25314
180	0.67648	131.988	0.25602
190	0.68956	133.818	0.25886
200	0.70254	135.654	0.26166
210	0.71542	137.496	0.26443
220	0.72821	139.345	0.26717
230	0.74091	141.201	0.26989
240	0.75354	143.064	0.27257
250	0.76609	144.935	0.27522
260	0.77858	146.814	0.27785
270	0.79101	148.702	0.28046
280	0.80338	150.598	0.28304
290	0.81570	152.502	0.28560
300	0.82798	154.416	0.28813
310	0.84020	156.339	0.29065
320	0.85239	158.271	0.29314
330	0.86453	160.212	0.29561
340	0.87664	162.162	0.29807
350	0.88871	164.121	0.30050
360	0.90075	166.091	0.30292

TEMP. °F	120 105.304 (61.95 °F)		
	v	h	s
	(0.45822)	(109.853)	(0.21593)
70	0.47032	111.381	0.21884
80	0.48494	113.262	0.22236
90	0.49918	115.128	0.22578
100	0.51309	116.982	0.22912
110	0.52670	118.827	0.23239
120	0.54005	120.667	0.23559
130	0.55318	122.503	0.23873
140	0.56610	124.337	0.24182
150	0.57884	126.171	0.24485
160	0.59142	128.005	0.24784
170	0.60384	129.842	0.25078
180	0.61613	131.682	0.25368
190	0.62830	133.526	0.25654
200	0.64036	135.375	0.25936
210	0.65232	137.229	0.26215
220	0.66418	139.088	0.26490
230	0.67596	140.954	0.26763
240	0.68766	142.827	0.27033
250	0.69929	144.707	0.27299
260	0.71085	146.595	0.27564
270	0.72235	148.490	0.27825
280	0.73379	150.394	0.28084
290	0.74517	152.306	0.28341
300	0.75651	154.226	0.28595
310	0.76780	156.155	0.28848
320	0.77905	158.092	0.29098
330	0.79026	160.039	0.29346
340	0.80144	161.995	0.29592
350	0.81257	163.959	0.29836
360	0.82368	165.933	0.30078
370	0.83475	167.916	0.30319

TEMP. °F	140 125.304 (71.83 °F)		
	v	h	s
	(0.39243)	(110.535)	(0.21425)
70	—	—	—
80	0.40342	112.143	0.21725
90	0.41646	114.086	0.22082
100	0.42911	116.009	0.22428
110	0.44143	117.915	0.22766
120	0.45346	119.809	0.23096
130	0.46524	121.694	0.23418
140	0.47679	123.573	0.23734
150	0.48814	125.447	0.24044
160	0.49932	127.319	0.24348
170	0.51033	129.189	0.24648
180	0.52121	131.060	0.24943
190	0.53195	132.933	0.25233
200	0.54258	134.808	0.25520
210	0.55309	136.686	0.25802
220	0.56351	138.569	0.26081
230	0.57385	140.456	0.26357
240	0.58409	142.349	0.26629
250	0.59427	144.247	0.26899
260	0.60437	146.152	0.27165
270	0.61441	148.064	0.27429
280	0.62439	149.983	0.27690
290	0.63432	151.909	0.27949
300	0.64419	153.843	0.28205
310	0.65402	155.784	0.28459
320	0.66380	157.734	0.28711
330	0.67354	159.692	0.28960
340	0.68325	161.658	0.29208
350	0.69292	163.633	0.29453
360	0.70255	165.616	0.29696
370	0.71216	167.608	0.29938
380	0.72173	169.609	0.30178

TEMP. °F	160 145.304 (80.71 °F)		
	v	h	s
	(0.34249)	(111.095)	(0.21276)
80	—	—	—
90	0.35387	112.984	0.21623
100	0.36568	114.986	0.21984
110	0.37710	116.961	0.22334
120	0.38820	118.917	0.22674
130	0.39901	120.856	0.23006
140	0.40958	122.783	0.23330
150	0.41992	124.701	0.23647
160	0.43008	126.612	0.23958
170	0.44006	128.519	0.24263
180	0.44989	130.423	0.24563
190	0.45958	132.326	0.24858
200	0.46914	134.229	0.25149
210	0.47859	136.133	0.25435
220	0.48794	138.040	0.25718
230	0.49720	139.949	0.25997
240	0.50637	141.863	0.26272
250	0.51546	143.781	0.26544
260	0.52447	145.704	0.26814
270	0.53342	147.632	0.27080
280	0.54231	149.567	0.27343
290	0.55115	151.507	0.27604
300	0.55993	153.455	0.27862
310	0.56866	155.410	0.28117
320	0.57735	157.372	0.28371
330	0.58599	159.341	0.28622
340	0.59460	161.319	0.28870
350	0.60317	163.304	0.29117
360	0.61170	165.297	0.29362
370	0.62020	167.298	0.29604
380	0.62868	169.308	0.29845
390	0.63712	171.326	0.30084

TEMP. °F	180 165.304 (88.81 °F)		
	v	h	s
	(0.30323)	(111.555)	(0.21142)
90	0.30461	111.808	0.21188
100	0.31587	113.904	0.21566
110	0.32669	115.959	0.21930
120	0.33713	117.983	0.22282
130	0.34724	119.983	0.22624
140	0.35708	121.964	0.22957
150	0.36668	123.930	0.23282
160	0.37607	125.885	0.23600
170	0.38527	127.831	0.23912
180	0.39430	129.770	0.24217
190	0.40319	131.706	0.24518
200	0.41194	133.638	0.24813
210	0.42057	135.570	0.25103
220	0.42910	137.501	0.25390
230	0.43752	139.434	0.25672
240	0.44586	141.369	0.25951
250	0.45411	143.307	0.26226
260	0.46229	145.249	0.26497
270	0.47040	147.195	0.26766
280	0.47845	149.145	0.27031
290	0.48644	151.102	0.27294
300	0.49437	153.063	0.27554
310	0.50225	155.032	0.27811
320	0.51009	157.006	0.28066
330	0.51788	158.988	0.28319
340	0.52564	160.976	0.28569
350	0.53335	162.972	0.28817
360	0.54103	164.976	0.29063
370	0.54868	166.986	0.29307
380	0.55630	169.005	0.29549
390	0.56389	171.032	0.29789

Table B.9 (continued).

ABSOLUTE PRESSURE, psi

TEMP. °F	200 — 185.304 (96.27 °F) v	h	s
	(0.27150)	(111.934)	(0.21018)
100	0.27553	112.750	0.21165
110	0.28596	114.900	0.21545
120	0.29595	117.004	0.21911
130	0.30556	119.073	0.22265
140	0.31487	121.114	0.22608
150	0.32390	123.133	0.22942
160	0.33270	125.134	0.23268
170	0.34130	127.122	0.23586
180	0.34972	129.100	0.23898
190	0.35798	131.070	0.24203
200	0.36609	133.034	0.24503
210	0.37408	134.995	0.24798
220	0.38196	136.953	0.25089
230	0.38973	138.910	0.25374
240	0.39741	140.867	0.25656
250	0.40500	142.826	0.25934
260	0.41251	144.787	0.26209
270	0.41995	146.751	0.26480
280	0.42733	148.719	0.26747
290	0.43465	150.691	0.27012
300	0.44190	152.668	0.27274
310	0.44911	154.650	0.27533
320	0.45627	156.637	0.27790
330	0.46339	158.631	0.28044
340	0.47046	160.631	0.28296
350	0.47749	162.638	0.28545
360	0.48449	164.652	0.28792
370	0.49146	166.672	0.29037
380	0.49839	168.700	0.29280
390	0.50530	170.736	0.29521
400	0.51218	172.779	0.29760

TEMP. °F	240 — 225.304 (109.67 °F) v	h	s
110	(0.22327)	(112.487)	(0.20793)
120	0.22360	112.564	0.20806
130	0.23318	114.873	0.21208
140	0.24225	117.113	0.21591
150	0.25089	119.299	0.21959
160	0.25920	121.443	0.22314
170	0.26721	123.554	0.22657
180	0.27498	125.639	0.22991
190	0.28253	127.702	0.23316
200	0.28990	129.749	0.23633
210	0.29710	131.783	0.23944
220	0.30416	133.807	0.24248
230	0.31109	135.822	0.24547
240	0.31790	137.832	0.24841
250	0.32461	139.838	0.25130
260	0.33122	141.842	0.25414
270	0.33775	143.844	0.25694
280	0.34420	145.847	0.25970
290	0.35059	147.850	0.26243
300	0.35690	149.855	0.26512
310	0.36316	151.863	0.26778
320	0.36936	153.874	0.27041
330	0.37551	155.889	0.27301
340	0.38161	157.908	0.27559
350	0.38767	159.932	0.27814
360	0.39369	161.962	0.28066
370	0.39967	163.997	0.28315
380	0.40562	166.037	0.28563
390	0.41153	168.084	0.28808
400	0.41741	170.138	0.29051
410	0.42327	172.198	0.29292
420	0.42910	174.265	0.29531
	—	—	—

TEMP. °F	280 — 265.304 (121.52 °F) v	h	s
	(0.18821)	(112.814)	(0.20586)
130	0.19583	114.912	0.20944
140	0.20427	117.294	0.21345
150	0.21224	119.600	0.21726
160	0.21983	121.848	0.22092
170	0.22711	124.051	0.22445
180	0.23413	126.217	0.22786
190	0.24092	128.354	0.23118
200	0.24753	130.468	0.23441
210	0.25396	132.564	0.23756
220	0.26025	134.644	0.24064
230	0.26641	136.713	0.24366
240	0.27246	138.773	0.24663
250	0.27840	140.825	0.24954
260	0.28424	142.873	0.25241
270	0.29000	144.917	0.25523
280	0.29569	146.958	0.25801
290	0.30130	148.999	0.26075
300	0.30685	151.040	0.26345
310	0.31234	153.082	0.26612
320	0.31778	155.125	0.26876
330	0.32317	157.172	0.27137
340	0.32851	159.221	0.27395
350	0.33381	161.274	0.27650
360	0.33907	163.332	0.27902
370	0.34429	165.394	0.28152
380	0.34948	167.460	0.28400
390	0.35463	169.533	0.28645
400	0.35976	171.611	0.28888
410	0.36486	173.695	0.29129
420	0.36994	175.784	0.29368
430	0.37499	177.881	0.29605

TEMP. °F	290 — 275.304 (124.29 °F) v	h	s
	(0.18088)	(112.865)	(0.20536)
130	0.18600	114.312	0.20783
140	0.19445	116.755	0.21194
150	0.20239	119.110	0.21583
160	0.20992	121.398	0.21955
170	0.21712	123.635	0.22313
180	0.22404	125.830	0.22659
190	0.23073	127.992	0.22995
200	0.23722	130.128	0.23321
210	0.24354	132.244	0.23639
220	0.24970	134.342	0.23950
230	0.25573	136.426	0.24255
240	0.26164	138.500	0.24553
250	0.26745	140.566	0.24846
260	0.27315	142.625	0.25135
270	0.27877	144.680	0.25418
280	0.28432	146.732	0.25697
290	0.28979	148.782	0.25973
300	0.29519	150.831	0.26244
310	0.30054	152.881	0.26512
320	0.30583	154.932	0.26777
330	0.31107	156.986	0.27039
340	0.31626	159.041	0.27297
350	0.32141	161.101	0.27553
360	0.32652	163.164	0.27807
370	0.33160	165.231	0.28057
380	0.33664	167.303	0.28306
390	0.34165	169.380	0.28551
400	0.34662	171.463	0.28795
410	0.35157	173.551	0.29037
420	0.35650	175.645	0.29276
430	0.36139	177.745	0.29513

TEMP. °F	300 — 285.304 (126.98 °F) v	h	s
	(0.17400)	(112.904)	(0.20487)
130	0.17670	113.688	0.20620
140	0.18520	116.199	0.21042
150	0.19313	118.607	0.21441
160	0.20062	120.938	0.21820
170	0.20776	123.210	0.22184
180	0.21460	125.436	0.22534
190	0.22120	127.625	0.22874
200	0.22759	129.784	0.23204
210	0.23379	131.919	0.23525
220	0.23984	134.035	0.23839
230	0.24575	136.136	0.24145
240	0.25154	138.225	0.24446
250	0.25722	140.304	0.24741
260	0.26280	142.375	0.25031
270	0.26829	144.441	0.25316
280	0.27370	146.503	0.25597
290	0.27904	148.563	0.25873
300	0.28431	150.621	0.26146
310	0.28952	152.679	0.26415
320	0.29467	154.738	0.26681
330	0.29978	156.798	0.26944
340	0.30483	158.861	0.27203
350	0.30985	160.926	0.27460
360	0.31482	162.995	0.27714
370	0.31975	165.068	0.27965
380	0.32465	167.145	0.28214
390	0.32952	169.227	0.28460
400	0.33436	171.314	0.28705
410	0.33917	173.407	0.28947
420	0.34395	175.505	0.29186
430	0.34871	177.609	0.29424

Note: Data abstracted from "Thermodynamic properties of Freon-22," copyright 1964; with permission of E. I. duPont de Nemours & Co.

Table B.10

Mercury Vapor Properties
SI UNITS

Pressure bars p	Temp °C t	Specific Volume m^3/kg v_g	Enthalpy kJ/kg h_f	h_g	Entropy kJ/kg·K s_f	s_g
0.03	208.0	6.75	32.49	330.3	0.0884	0.708
0.06	231.6	3.48	36.17	331.9	0.0958	0.682
0.09	245.7	2.45	38.36	332.6	0.1002	0.667
0.12	257.1	1.83	40.15	333.3	0.1035	0.656
0.15	265.8	1.49	41.52	334.0	0.1061	0.649
0.2	278.0	1.13	43.43	334.7	0.1096	0.638
0.3	295.8	0.77	46.19	335.9	0.1146	0.624
0.4	309.1	0.588	48.29	336.8	0.1182	0.613
0.5	320.0	0.478	48.96	337.3	0.1211	0.606
0.6	329.1	0.403	49.75	337.5	0.1235	0.600
0.8	343.2	0.318	53.61	338.9	0.1271	0.590
1.0	355.9	0.254	55.57	339.6	0.1303	0.582
1.2	366.0	0.216	57.15	340.3	0.1327	0.576
1.4	375.3	0.184	58.62	341.0	0.1350	0.570
1.6	383.2	0.164	59.85	341.5	0.1369	0.566
2.0	397.1	0.133	62.01	342.2	0.1401	0.559
2.5	411.7	0.108	64.29	343.3	0.1435	0.551
3.0	423.9	0.091	66.20	344.0	0.1463	0.545
3.5	434.6	0.079	67.87	344.7	0.1487	0.540
4.0	444.1	0.070	69.34	345.4	0.1507	0.535
5.0	460.7	0.057	71.92	346.3	0.1543	0.530
6.0	474.8	0.048	74.13	347.3	0.1573	0.523
7.0	487.2	0.042	76.06	348.0	0.1599	0.517
8.0	498.3	0.037	77.71	348.9	0.1621	0.514
9.0	508.3	0.0336	79.20	349.6	0.1642	0.510
10.0	517.4	0.0304	80.83	350.1	0.1660	0.506
11.0	526.3	0.0278	82.18	350.8	0.1677	0.503
12.0	534.3	0.0257	83.43	351.2	0.1693	0.500

Table B.10 (continued).
ENGLISH UNITS

Pressure psia	Temp °F	Specific Volume ft³/lbm	Enthalpy Btu/lbm			Entropy Btu/lbm · °F		
p	t	v_g	h_f	h_{fg}	h_g	s_f	s_{fg}	s_g
0.4	402.3	114.5	13.81	128.1	141.9	0.02094	0.1486	0.1696
0.6	426.1	78.23	14.70	127.6	142.3	0.02195	0.1441	0.1660
0.8	443.8	59.71	15.36	127.2	142.6	0.02269	0.1408	0.1635
1.0	458.1	48.45	15.89	126.9	142.8	0.02328	0.1382	0.1615
1.5	485.1	33.14	16.90	126.3	143.2	0.02436	0.1337	0.1580
2	505.2	25.31	17.65	125.8	143.5	0.02514	0.1304	0.1556
3	535.4	17.34	18.78	125.2	144.0	0.02629	0.1258	0.1521
4	558.0	13.26	19.62	124.7	144.3	0.02714	0.1225	0.1497
5	576.2	10.77	20.30	124.3	144.6	0.02780	0.1200	0.1478
6	591.4	9.096	20.87	123.9	144.8	0.02834	0.1179	0.1462
7	605.0	7.882	21.37	123.6	145.0	0.02882	0.1161	0.1450
8	616.8	6.963	21.81	123.4	145.2	0.02923	1.1146	0.1439
9	627.5	6.244	22.21	123.2	145.4	0.02960	0.1133	0.1429
10	637.3	5.661	22.58	122.9	145.5	0.02993	0.1121	0.1420
15	676.5	3.892	24.04	122.1	146.1	0.03124	0.1074	0.1387
20	706.2	2.983	25.15	121.4	146.6	0.03220	0.1041	0.1363
25	730.4	2.429	26.05	120.9	146.9	0.03297	0.1016	0.1345
30	750.9	2.053	26.81	120.4	147.2	0.03360	0.09953	0.1331
35	769.0	1.781	27.49	120.0	147.5	0.03416	0.09774	0.1319
40	784.8	1.576	28.08	119.7	147.8	0.03464	0.09621	0.1308
45	799.3	1.414	28.62	119.4	148.0	0.03507	0.09486	0.1299
50	812.5	1.284	29.11	119.1	148.2	0.03546	0.09364	0.1291
60	836.1	1.086	29.99	118.6	148.6	0.03614	0.09154	0.1276
70	856.6	0.9436	30.75	118.1	148.9	0.03672	0.08976	0.1264
80	874.8	0.8349	31.43	117.7	149.1	0.03725	0.08824	0.1254
90	891.6	0.7497	32.06	117.3	149.4	0.03771	0.08687	0.1245
100	906.9	0.6811	32.63	117.0	149.6	0.03813	0.08565	0.1237
120	934.4	0.5767	33.60	116.4	150.1	0.03887	0.08353	0.1224
140	958.3	0.5012	34.55	115.9	150.4	0.03951	0.08175	0.1212
160	979.9	0.4438	35.35	115.4	150.8	0.04007	0.08019	0.1202
180	999.6	0.3990	36.09	115.0	151.1	0.04058	0.07881	0.1193

Source: *Standard Handbook for Mechanical Engineers*. 7th ed. Edited by T. Baumeister and L. S. Marks, Copyright 1967; with permission of McGraw-Hill Book Company. Conversion to SI by author.
Note: h_f and s_f are measured from 32°F or 0°C.

Table B.11

Properties of Ammonia (NH_3) at Saturation
SI UNITS

Temp °C	Pressure bars*	Specific Volume m^3/kg		Enthalpy kJ/kg		Entropy kJ/kg·K	
t	p	v_f	v_g	h_f	h_g	s_f	s_g
−70	0.109	0.0014	9.21	−129.1	1335.4	−0.595	6.619
−65	0.16	0.0014	6.47	−107.9	1344.4	−0.490	6.490
−60	0.22	0.0014	4.70	−86.5	1353.3	−0.389	6.368
−55	0.30	0.0014	3.48	−65.1	1362.6	−0.289	6.255
−50	0.41	0.0014	2.63	−43.7	1371.6	−0.193	6.150
−45	0.55	0.0014	2.01	−21.9	1380.9	−0.096	6.054
−40	0.72	0.0014	1.55	0.0	1390.0	0.000	5.962
−35	0.93	0.0015	1.22	22.3	1397.9	0.096	5.874
−30	1.20	0.0015	0.963	44.7	1405.6	0.205	5.874
−25	1.52	0.0015	0.772	67.2	1413.0	0.281	5.702
−20	1.90	0.0015	0.624	89.8	1420.0	0.368	5.623
−15	2.36	0.0015	0.509	112.3	1426.5	0.456	5.548
−10	2.91	0.0015	0.418	135.4	1433.0	0.544	5.476
−5	3.55	0.0015	0.347	158.2	1438.9	0.632	5.422
0	4.29	0.0016	0.290	181.2	1444.4	0.716	5.338
5	5.16	0.0016	0.243	204.4	1449.6	0.799	5.275
10	6.15	0.0016	0.205	227.7	1454.2	0.883	5.213
15	7.28	0.0016	0.175	251.4	1458.6	0.963	5.154
20	8.57	0.0016	0.149	275.2	1462.6	1.043	5.095
25	10.0	0.0017	0.129	298.9	1465.8	1.126	5.041
30	11.7	0.0017	0.110	323.1	1468.9	1.206	4.982
35	13.5	0.0017	0.0955	347.5	1471.4	1.281	4.932
40	15.5	0.0017	0.0832	371.9	1473.3	1.361	4.878
45	17.8	0.0018	0.0726	396.8	1474.5	1.436	4.823
50	20.3	0.0018	0.0635	421.9	1474.7	1.516	4.773

*1 bar = 10^2 kPa = 10^5 Pa.

Table B.11 (continued).

ENGLISH UNITS

Temp °F	Pressure psia	Specific Volume ft³/lbm		Enthalpy Btu/lbm		Entropy Btu/lbm·°R	
t	p	v_f	v_g	h_f	h_g	s_f	s_g
−100	1.24	0.022	182.90	−61.5	571.4	−0.1579	1.6025
−90	1.86	0.022	124.28	−51.4	575.9	−0.1309	1.5667
−80	2.74	0.022	86.54	−41.3	580.1	−0.1036	1.5336
−70	3.94	0.023	61.65	−31.1	584.4	−0.0771	1.5026
−60	5.55	0.023	44.73	−20.9	588.8	−0.0514	1.4747
−50	7.67	0.023	33.08	−10.5	593.2	−0.0254	1.4487
−40	10.41	0.023	24.86	0.0	597.6	0.0000	1.4242
−30	13.90	0.023	18.97	10.7	601.4	0.0250	1.4001
−20	18.30	0.024	14.68	21.4	605.0	0.0497	1.3774
−10	23.74	0.024	11.50	32.1	608.5	0.0738	1.3558
0	30.42	0.024	9.116	42.1	611.8	0.0975	1.3352
10	38.51	0.024	7.309	53.8	614.9	0.1208	1.3157
20	48.21	0.025	5.910	64.7	617.8	0.1437	1.2969
30	59.74	0.025	4.825	75.7	620.5	0.1663	1.2790
40	73.32	0.025	3.971	86.8	623.0	0.1885	1.2618
50	89.19	0.026	3.294	97.9	625.2	0.2105	1.2453
60	107.60	0.026	2.751	109.2	627.3	0.2322	1.2294
70	128.80	0.026	2.312	120.5	629.1	0.2537	1.2140
80	153.00	0.027	1.955	132.0	630.7	0.2749	1.1991
90	180.60	0.027	1.661	143.5	632.0	0.2958	1.1846
100	211.90	0.027	1.419	155.2	633.0	0.3166	1.1705
110	247.00	0.028	1.217	167.0	633.7	0.3372	1.1566
120	286.40	0.028	1.047	179.0	634.0	0.3576	1.1427
125	307.80	0.029	0.973	185.1	634.0	0.3679	1.1358

Source: "Bulletin on Thermodynamic Properties of Refrigerants," The American Society of Heating, Refrigerating, and Air Conditioning Engineers, 1969; with permission of ASHRAE.

Table B.12

Ammonia Properties at Superheat (Datum of −40°C)
SI UNITS

v (m³/kg); h (kJ/kg); s (kJ/kg · K)

Pressure bars,* p (Sat. Temp., °C)		Sat. Properties	Temperature, °C, t −40	−20	0	20	40	60	80	100	120	140
0.4	v	2.66	2.82	3.07	3.33	3.58	3.83	4.08	4.33			
(50.3)	h	1371.2	1398.8	1437.9	1479.8	1521.9	1564.5	1607.5	1651.2			
	s	6.155	6.305	6.439	6.608	6.753	6.896	7.025	7.151			
0.7	v	1.58		1.73	1.88	2.02	2.17	2.30	2.45			
(40.4)	h	1389.3		1434.4	1477.5	1520.3	1563.1	1606.3	1650.3			
	s	5.970		6.146	6.318	6.469	6.615	6.745	6.875			
1.0	v	1.14		1.20	1.32	1.42	1.52	1.62	1.72			
(−33.6)	h	1400.0		1431.2	1474.9	1518.4	1561.7	1604.9	1649.4			
	s	5.849		5.966	6.142	6.293	6.439	6.573	6.699			
2.0	v	0.595			0.649	0.699	0.755	0.805	0.855	0.905	0.986	
(−18.9)	h	1421.4			1466.5	1511.9	1556.6	1600.9	1645.6	1690.8	1736.6	
	s	5.606			5.778	5.937	6.083	6.222	6.351	6.477	6.582	
3.0	v	0.397			0.425	0.461	0.497	0.532	0.566	0.600	0.634	0.667
(−9.2)	h	1434.0			1457.5	1505.2	1551.4	1596.8	1641.9	1687.7	1733.8	1780.1
	s	5.468			5.572	5.723	5.874	6.016	6.146	6.276	6.393	6.523
4.0	v	0.310			0.313	0.341	0.370	0.396	0.422	0.448	0.478	0.499
(−1.9)	h	1442.4			1447.5	1498.2	1545.9	1592.6	1638.7	1684.7	1731.5	1778.0
	s	5.367			5.384	5.564	5.719	5.866	6.000	6.125	6.247	6.364

Table B.12 (continued)

Pressure bars,* p (Sat. Temp., °C)		Sat. Properties	−40	−20	0	20	40	60	80	100	120	140
								Temperature, °C, t				
5.0	v	0.251				0.270	0.293	0.315	0.334	0.351	0.378	0.398
(3.8)	h	1448.6				1490.7	1540.7	1588.0	1635.2	1681.9	1728.2	1776.1
	s	5.288				5.439	5.598	5.707	5.882	6.012	6.134	6.251
6.0	v	0.210				0.222	0.242	0.260	0.278	0.291	0.313	0.330
(8.1)	h	1453.5				1483.3	1534.9	1581.9	1631.5	1678.7	1726.1	1773.6
	s	5.234				5.326	5.497	5.707	5.786	5.941	6.037	6.159
7.0	v	0.181				0.187	0.204	0.220	0.236	0.247	0.266	0.281
(13.9)	h	1457.7				1475.1	1532.8	1579.1	1627.7	1675.7	1723.3	1771.2
	s	5.167				5.225	5.401	5.602	5.698	5.836	5.958	6.075
8.0	v	0.160				0.162	0.177	0.192	0.206	0.216	0.233	0.246
(17.8)	h	1461.0				1467.2	1523.1	1574.9	1624.2	1672.6	1718.9	1769.2
	s	5.120				5.141	5.326	5.485	5.631	5.765	5.891	6.008
9.0	v	0.142					0.155	0.169	0.181	0.194	0.205	0.217
(21.7)	h	1463.8					1516.8	1570.1	1620.3	1669.6	1717.9	1766.6
	s	5.079					5.250	5.418	5.564	5.698	5.823	5.945
10.0	v	0.129					0.139	0.151	0.163	0.174	0.185	0.195
(24.9)	h	1465.8					1511.0	1565.4	1616.8	1652.9	1715.7	1764.7
	s	5.041					5.187	5.355	5.506	5.644	5.769	5.891
12.0	v	0.108					0.113	0.124	0.134	0.144	0.155	0.163
(30.9)	h	1469.3					1497.9	1555.6	1609.1	1660.3	1710.3	1760.1
	s	4.970					5.066	5.246	5.401	5.543	5.673	5.795
14.0	v	0.0924					0.0943	0.1043	0.1136	0.1217	0.1299	0.1380
(36.3)	h	1471.9					1484.2	1545.9	1601.2	1653.8	1705.2	1755.4
	s	4.919					4.957	5.146	5.309	5.455	5.585	5.711
16.0	v	0.0805						0.0899	0.0980	0.1055	0.1123	0.1199
(41.1)	h	1473.3						1535.4	1593.3	1647.5	1699.6	1751.0
	s	4.865						5.058	5.225	5.376	5.510	5.640
18.0	v	0.0718						0.0780	0.0855	0.0924	0.0993	0.1055
(45.4)	h	1474.5						1524.5	1585.2	1641.0	1694.0	1746.1
	s	4.823						4.974	5.150	5.305	5.443	5.573
20.0	v	0.0643						0.0687	0.0762	0.0824	0.0886	0.0943
(49.4)	h	1474.7						1513.1	1562.6	1634.2	1688.7	1741.7
	s	4.781						4.899	5.079	5.242	5.384	5.514

Table B.12 (continued).

ENGLISH UNITS

Pressure psia, *p*		Temperature, °F, *t* −40	−20	0	20	40	60	80	100	120	160	200	300
5.0	*v*	52.36	54.97	57.55	60.12	62.69	65.24	67.79	70.33	72.87	77.95	—	—
	h	600.3	610.4	620.4	630.4	640.4	650.5	660.6	670.7	680.9	701.6	—	—
	s	1.5149	1.5385	1.5608	1.5821	1.6026	1.6223	1.6413	1.6598	1.6778	1.7122	—	—
10.0	*v*	25.90	27.26	28.58	29.90	31.20	32.49	33.78	35.07	36.35	38.90	41.45	—
	h	597.8	608.5	618.9	629.1	639.3	649.5	659.7	670.0	680.3	701.1	722.2	—
	s	1.4293	1.4542	1.4773	1.4992	1.5200	1.5400	1.5593	1.5779	1.5960	1.6307	1.6637	—
20.0	*v*	—	—	14.09	14.78	15.45	16.12	16.78	17.43	18.08	19.37	20.66	—
	h	—	—	615.5	626.4	637.0	647.5	658.0	668.5	678.9	700.0	721.2	—
	s	—	—	1.3907	1.4138	1.4356	1.4562	1.4760	1.4950	1.5133	1.5485	1.5817	—
30.0	*v*	—	—	9.25	9.731	10.20	10.65	11.10	11.55	11.99	12.87	13.73	—
	h	—	—	611.9	623.5	634.6	645.5	656.2	666.9	677.5	698.8	720.3	—
	s	—	—	1.3371	1.3618	1.3845	1.4059	1.4261	1.4456	1.4642	1.4998	1.5334	—
40.0	*v*	—	—	—	7.203	7.568	7.922	8.268	8.609	8.945	9.609	10.27	11.88
	h	—	—	—	620.4	632.1	643.4	654.4	665.3	676.1	697.7	719.4	774.6
	s	—	—	—	1.3231	1.3470	1.3692	1.3900	1.4098	1.4288	1.4648	1.4987	1.5766
50.0	*v*	—	—	—	—	5.988	6.280	6.564	6.843	7.117	7.655	8.185	9.489
	h	—	—	—	—	629.5	641.2	652.6	663.7	674.7	696.6	718.5	774.0
	s	—	—	—	—	1.3169	1.3399	1.3613	1.3816	1.4009	1.4374	1.4716	1.5500
60.0	*v*	—	—	—	—	4.933	5.184	5.428	5.665	5.897	6.352	6.787	7.892
	h	—	—	—	—	626.8	639.0	650.7	662.1	673.3	695.5	717.5	773.7
	s	—	—	—	—	1.2913	1.3152	1.3373	1.3581	1.3778	1.4148	1.4493	1.5281
70.0	*v*	—	—	—	—	4.177	4.401	4.615	4.822	5.025	5.420	5.807	6.750
	h	—	—	—	—	623.9	636.6	648.7	660.6	671.8	694.3	716.6	772.7
	s	—	—	—	—	1.2688	1.2937	1.3166	1.3378	1.3579	1.3954	1.4302	1.5095
80.0	*v*	—	—	—	—	—	3.812	4.005	4.190	4.371	4.722	5.063	5.894
	h	—	—	—	—	—	634.3	646.7	658.7	670.4	693.2	715.6	772.1
	s	—	—	—	—	—	1.2745	1.2981	1.3199	1.3404	1.3784	1.4136	1.4933

Table B.12 (continued).

Pressure psia, p		Temperature, °F, t −40	−20	0	20	40	60	80	100	120	160	200	300
90.0	v	—	—	—	—	—	3.353	3.529	3.698	3.862	4.178	4.484	5.228
	h	—	—	—	—	—	631.8	644.7	657.0	668.9	692.0	714.7	771.5
	s	—	—	—	—	—	1.2571	1.2814	1.3038	1.3247	1.3633	1.3988	1.4789
100.0	v	—	—	—	—	—	2.985	3.149	3.304	3.454	3.743	4.021	4.695
	h	—	—	—	—	—	629.3	642.6	655.2	667.3	690.8	713.7	770.8
	s	—	—	—	—	—	1.2409	1.2661	1.2891	1.3104	1.3495	1.3854	1.4660
120.0	v	—	—	—	—	—	—	2.575	2.712	2.842	3.089	3.326	3.895
	h	—	—	—	—	—	—	638.3	651.6	664.2	688.4	711.8	769.6
	s	—	—	—	—	—	—	1.2386	1.2628	1.2850	1.3254	1.3620	1.4435
160.0	v	—	—	—	—	—	—	—	1.969	2.075	2.272	2.457	2.895
	h	—	—	—	—	—	—	—	643.9	657.8	683.5	707.9	767.1
	s	—	—	—	—	—	—	—	1.2186	1.2429	1.2859	1.3240	1.4076
200.0	v	—	—	—	—	—	—	—	1.520	1.612	1.780	1.935	2.295
	h	—	—	—	—	—	—	—	635.6	650.9	678.4	703.9	764.5
	s	—	—	—	—	—	—	—	1.1809	1.2077	1.2537	1.2935	1.3791
250.0	v	—	—	—	—	—	—	—	—	1.240	1.386	1.518	1.815
	h	—	—	—	—	—	—	—	—	641.5	671.8	698.8	761.3
	s	—	—	—	—	—	—	—	—	1.1690	1.2195	1.2617	1.3501
300.0	v	—	—	—	—	—	—	—	—	—	1.123	1.239	1.496
	h	—	—	—	—	—	—	—	—	—	664.7	693.5	758.1
	s	—	—	—	—	—	—	—	—	—	1.1894	1.2344	1.3257

Source: "Bulletin on Thermodynamic Properties of Refrigerants," The American Society of Heating, Refrigerating, and Air Conditioning Engineers, 1969, with permission of ASHRAE.

Table B.13

Internal Energy of Gases (Btu/lb·mole)(datum, 520°R)

°R	O_2	N_2	Air	CO_2	H_2O	H_2	CO	C_8H_{18}	$C_{12}H_{26}$	$pv/778.16$
520	0	0	0	0	0	0	0	0	0	1,033
536.7	83	81	81	115	101	80	81	640	980	1,066
540	100	97	97	139	122	96	97	756	1,181	1,072
560	200	196	196	280	244	193	196	1,536	2,491	1,112
580	301	295	295	424	357	291	295	2,340	3,931	1,152
600	402	395	395	570	490	390	396	3,167	5,481	1,192
700	920	896	897	1,320	1,110	887	896	7,668	13,223	1,390
800	1,449	1,399	1,403	2,120	1,734	1,386	1,402	12,768	22,044	1,589
900	1,989	1,905	1,915	2,965	2,366	1,886	1,913	18,471	31,771	1,787
1000	2,539	2,416	2,431	3,852	3,009	2,387	2,430	24,773	42,277	1,986
1100	3,101	2,934	2,957	4,778	3,666	2,889	2,954	31,677	53,468	2,185
1200	3,675	3,461	3,492	5,736	4,339	3,393	3,485	39,182	65,290	2,383
1300	4,262	3,996	4,036	6,721	5,030	3,899	4,026	47,288	77,706	2,582
1400	4,861	4,539	4,587	7,731	5,740	4,406	4,580	55,995	90,688	2,780
1500	5,472	5,091	5,149	8,764	6,468	4,916	5,145	65,303	104,209	2,979
1600	6,092	5,652	5,720	9,819	7,212	5,429	5,720	74,825	118,240	3,178
1700	6,718	6,224	6,301	10,896	7,970	5,945	6,305	84,901	132,757	3,376
1800	7,349	6,805	6,889	11,993	8,741	6,464	6,899	95,503	147,735	3,575
1900	7,985	7,393	7,485	13,105	9,526	6,988	7,501	......	....	3,773
2000	8,629	7,989	8,087	14,230	10,327	7,517	8,109	......	....	3,972
2100	9,279	8,592	8,698	15,368	11,146	8,053	8,722	......	....	4,171
2200	9,934	9,203	9,314	16,518	11,983	8,597	9,339	......	....	4,369
2300	10,592	9,817	9,934	17,680	12,835	9,147	9,961	......	....	4,568
2400	11,252	10,435	10,558	18,852	13,700	9,703	10,588	......	....	4,766
2500	11,916	11,056	11,185	20,033	14,578	10,263	11,220	......	....	4,965
2600	12,584	11,682	11,817	21,222	15,469	10,827	11,857	......	....	5,164
2700	13,257	12,313	12,453	22,419	16,372	11,396	12,499	......	....	5,362
2800	13,937	12,949	13,095	23,624	17,288	11,970	13,144	......	....	5,561
2900	14,622	13,590	13,742	24,836	18,217	12,549	13,792	......	....	5,759
3000	15,309	14,236	14,394	26,055	19,160	13,133	14,443	......	....	5,958
3100	16,001	14,888	15,051	27,281	20,117	13,723	15,097	......	....	6,157
3200	16,693	15,543	15,710	28,513	21,086	14,319	15,754	......	....	6,355
3300	17,386	16,199	16,369	29,750	22,066	14,921	16,414	......	....	6,554
3400	18,080	16,855	17,030	30,991	23,057	15,529	17,078	......	....	6,752
3500	18,776	17,512	17,692	32,237	24,057	16,143	17,744	......	....	6,951
3600	19,475	18,171	18,356	33,487	25,067	16,762	18,412	......	....	7,150
3700	20,179	18,833	19,022	34,741	26,085	17,385	19,082	......	....	7,348
3800	20,887	19,496	19,691	35,998	27,110	18,011	19,755	......	....	7,547
3900	21,598	20,162	20,363	37,258	28,141	18,641	20,430	......	....	7,745
4000	22,314	20,830	21,037	38,522	29,178	19,274	21,107	......	....	7,944
4100	23,034	21,500	21,714	39,791	30,221	19,911	21,784	......	....	8,143
4200	23,757	22,172	22,393	41,064	31,270	20,552	22,462	......	....	8,341
4300	24,482	22,845	23,073	42,341	32,326	21,197	23,143	......	....	8,540
4400	25,209	23,519	23,755	43,622	33,389	21,845	23,823	......	....	8,738
4500	25,938	24,194	24,437	44,906	34,459	22,497	24,503	......	....	8,937
4600	26,668	24,869	25,120	46,193	35,535	23,154	25,186	......	....	9,136
4700	27,401	25,546	25,805	47,483	36,616	23,816	25,868	......	....	9,334
4800	28,136	26,224	26,491	48,775	37,701	24,480	26,533	......	....	9,533
4900	28,874	26,905	27,180	50,069	38,791	25,418	27,219	......	....	9,731
5000	29,616	27,589	27,872	51,365	39,885	25,819	27,907	......	....	9,930
5100	30,361	28,275	28,566	52,663	40,983	26,492	28,597	......	....	10,129
5200	31,108	28,961	29,262	53,963	42,084	27,166	29,288	......	....	10,327
5300	31,857	29,648	29,958	55,265	43,187	27,842	29,980	......	....	10,526
5400	32,607	30,337	30,655	56,569	44,293	28,519	30,674	......	....	10,724
5500	33,386	31,026	31,353	57,875	45,402	29,298	31,369	......	....	10,923
5600	34,161	31,726	32,051	59,183	46,513	29,978	32,065	......	....	11,121
5700	34,900	32,428	32,750	60,491	47,627	30,659	32,762	......	....	11,320
5800	35,673	33,130	33,449	61,891	48,744	31,342	33,461	......	....	11,519
5900	36,412	33,833	34,150	63,293	49,863	32,026	34,161	......	....	11,717
6000	37,149	34,537	34,852	64,297	50,985	32,712	34,863	......	....	11,916
6500	......	......	38,364	......	......	......	......	......	....	12,908
7000	......	......	41,893	......	......	......	...	......	....	13,901

Source: L. C. Lichty, "Internal Combustion Engines," McGraw-Hill Book Company, Inc., New York, 1939, and based upon data of A. Hershey, J. Eberhardt, and H. Hottel, *Trans. SAE*, 39: 409 (October, 1936). Used with permission of McGraw-Hill Book Company.

Table B.14

Specific Heat Equations for Some Common Gases

Gas or vapor	Equation c_p, Btu/mole·°R	Range °R	Maximum % error
O_2	$c_p = 11.515 - \frac{172}{\sqrt{T}} + \frac{1,530}{T}$	540–5000	1.1
N_2	$c_p = 9.47 - \frac{3.47(10^3)}{T} + \frac{1.16(10^6)}{T^2}$	540–9000	1.7
CO	$c_p = 9.46 - \frac{3.29(10^3)}{T} + \frac{1.07(10^6)}{T^2}$	540–9000	1.1
H_2O	$c_p = 19.86 - \frac{597}{\sqrt{T}} + \frac{7,500}{T}$	540–5400	1.8
CO_2	$c_p = 16.2 - \frac{6.53(10^3)}{T} + \frac{1.41(10^6)}{T^2}$	540–6300	0.8

$c_p = a + b(10^{-3})T + c(10^{-6})T^2 + d(10^{-9})T^3$ (T, K)
= cal/(g mole)(K) or closely Btu/lbm · mole · °R

Gas or vapor	a	b	c	d	Range K	Maximum % error
Air	6.713	0.4697	1.147	−0.4696	273–1800	0.72
	6.557	1.477	−0.2148	0	273–3800	1.64
CO	6.726	0.4001	1.283	−0.5307	273–1800	0.89
	6.480	1.566	−0.2387	0	273–3800	1.86
CO_2	5.316	14.285	−8.362	1.784	273–1800	0.67
	$c_p = 18.036 - 0.00004474T - 158.08(T)^{-1/2}$				273–3800	2.65
H_2	6.952	−0.4576	0.9563	−0.2079	273–1800	1.01
	6.424	1.039	−0.07804	0	273–3800	2.14
H_2O	7.700	0.4594	2.521	−0.8587	273–1800	0.53
	6.970	3.464	−0.4833	0	273–3800	2.03
O_2	6.085	3.631	−1.709	0.3133	273–1800	1.19
	6.732	1.505	−0.1791	0	273–3800	3.24
N_2	6.903	−0.3753	1.930	−6.861	273–1800	0.59
	6.529	1.488	−0.2271	0	273–3800	2.05
NH_3	6.5846	6.1251	2.3663	−1.5981	273–1500	0.91
CH_4	4.750	12.00	3.030	−2.630	273–1500	1.33
C_3H_8	−0.966	72.79	−37.55	7.580	273–1500	0.40
C_4H_{10}	0.945	88.73	−43.80	8.360	273–1500	0.54
C_6H_6	−8.650	115.78	−75.40	18.54	273–1500	0.34
C_2H_2	5.21	22.008	−15.59	4.349	273–1500	1.46
CH_3OH	4.55	21.86	−2.91	−1.92	273–1000	0.18

Source: E.F. Obert, *Concepts of Thermodynamics*, copyright 1960, Mcgraw-Hill. With permission of McGraw-Hill Book Company.

Table B.15

Universal Constants and Conversion Factors

Universal Constants

mR ≈ R_u Universal gas constant = 1544 ft-lbf/lb · mole · °R
= 8.31 joule/gm · mole · K

σ Stefan-Boltzmann constant = 0.174×10^{-8} Btu/hr · ft^2 · °R^4
= 5.67×10^{-8} W/m^2 · °K^4

N Avogadro's number = 6.02×10^{23} atoms/gm · mole

Conversion Factors

Force
2000 pounds-force (lbf) = 1 ton
2.248×10^{-6} lbf = 1 dyne
8.89644×10^{8} dynes = 1 ton
0.2248 lbf = 1 newton (N)
105 dynes = 1 N

Mass
454 grams (g) = 1 pound-mass (lbm)
2.2 lbm = 1 kg = 1000 g
1 slug = 32.174 lbm

Length
3.28 feet (ft) = 1 meter (m)
39.37 inches (in) = 1 m
100 centimeters (cm) = 1 m
91.44 cm = 1 yard (yd)
5280 ft = 1 mile
3 ft = 1 yd
12 in = 1 ft
0.3048 meter = 1 ft

Pressure
0.1934 psi = 1 cm Hg (0°C) (on earth's surface)
0.014 psi = 1 cm H_2O (4°C) (on earth's surface)
0.036 psi = 1 in H_2O (4°C) (on earth's surface)
27.68 in H_2O (4°C) = 1 psi (on earth's surface)
33,864 dynes/cm^2 = 1 in Hg (4°C) (on earth's surface)
10^6 dynes/cm^2 = 1 bar
1 bar = 0.9869 atmosphere (atm)
14.696 psi = 1 atm
0.49 psi = 1 in Hg (4°C) (on earth's surface)
1×10^5 newton/meters^2 (N/m^2) = 1 bar
1 N/m^2 = 1 pascal (Pa)
0.06895 bar = 1 psi
14.50 psi = 1 bar

Table B.15 (continued).

Volume	231 in^3 =	1 gallon (gal) (U.S. liquid)
	0.1337 ft^3 =	1 gal (U.S. liquid)
	3.785 liters (l) =	1 gal (U.S. liquid)
	3785.4 cm^3 =	1 gal (U.S. liquid)
	0.001 m^3 =	1 l
	1000 cm^3 =	1 l
	61.02 in^3 =	1 l
	0.0353 ft^3 =	1 l
Energy	1 joule (J) =	1 newton-meter (N · m)
	1 J =	1 watt-second (W · s)
	1054 J =	1 British thermal unit (Btu)
	1.356 J =	1 foot-pound (ft-lbf)
	4.1868 J =	1 calorie
	1 dyne · cm =	1 erg
	1×10^7 ergs =	1 J
	0.2388 calorie =	1 J
	252 calories =	1 Btu
	3414 Btu =	1 kilowatt-hour (kW · hr)
	3.414 Btu =	1 W · hr
	2545 Btu =	1 horsepower · hour (hp · hr)
	0.9488 Btu =	1 kilojoule (kJ)
	778 ft-lbf =	1 Btu
	0.737 ft-lbf =	1 J
	0.4299 Btu/lbm =	1 kJ/kg
	334.5 ft-lbf/lbm =	1 kJ/kg
	2.326 kJ/kg =	1 Btu/lbm
Power	1 joule/second (J/s) =	1 watt (W)
	1000 W =	1 kW
	0.746 kW =	1 horsepower (hp)
	1.34 hp =	1 kW
	1.41 hp =	1 Btu/s
	550 ft-lbf/s =	1 hp
	33,000 ft-lbf/min =	1 hp
	1.054 kw =	1 Btu/s
	3414 Btu/hr =	1 kW
	2545 Btu/hr =	1 hp
	12,000 Btu/hr =	1 ton (refrigeration)
	3.515 kW =	1 ton (refrigeration)

Table B.16

U.S. Standard Atmospheric Conditions*

SI UNITS

Altitude meters	*Temperature K*	*Pressure bars*	*Density kg/m^3*
−500	291.400	1.07478	1.2849
0	288.150	1.01325	1.2250
500	284.900	0.954612	1.1673
1000	281.651	0.898762	1.1117
2000	275.154	0.795014	1.0066
4000	262.166	0.616604	0.81935
6000	249.187	0.472176	0.66011
8000	236.215	0.356516	0.52579
10,000	223.252	0.264999	0.41351
20,000	216.650	0.055293	0.088910
40,000	250.350	0.00287143	0.039957
60,000	255.772	0.000224606	0.0030592
80,000	180.65	0.000010366	0.00001999
90,000	180.65	0.0000016438	0.00000317

ENGLISH UNITS

Altitude feet	*Temperature °R*	*Pressure inches mercury*	*Density lbm/ft^3*
—1000	522.236	31.0185	0.078737
0	518.670	29.9213	0.076474
500	516.887	29.3846	0.075362
1000	515.104	28.8557	0.074261
2000	511.538	27.8212	0.072098
4000	504.408	25.8426	0.067917
6000	497.279	23.9798	0.063925
8000	490.152	22.2276	0.060116
10,000	483.025	20.5808	0.056483
20,000	447.415	13.7612	0.040773
40,000	389.970	5.5584	0.018895
60,000	389.970	2.1354	0.007259
80,000	397.693	0.8273	0.002758
100,000	408.572	0.3290	0.001068
120,000	433.578	0.1358	0.000415
140,000	463.923	0.0595	0.000170
200,000	456.999	0.0058	0.000017

Source: Abstracted from *U.S. Standard Atmosphere*, prepared under sponsorship of NASA, USAF, and USWB. Printed by U.S. Government Printing Office, 1962.

*U.S. Standard Atmosphere: Ideal air devoid of moisture or dust rotating with the earth. The data represent approximately 45° latitude annual average conditions in the U. S.

Table B.17

Gravitational Acceleration near the Earth

SI UNITS

Altitude above Sea Level meters	Location on Earth's Surface degrees latitude					
	Equator 0°	20°	40°	60°	80°	90°
	meters per second per second					
0	9.7804	9.7865	9.8018	9.8191	9.8307	9.8322
300	9.7795	9.7856	9.8009	9.8182	9.8298	9.8313
600	9.7786	9.7847	9.8000	9.8173	9.8289	9.8304
1000	9.7774	9.7835	9.7988	9.8161	9.8277	9.8292
1500	9.7759	9.7820	9.7976	9.8161	9.8277	9.8292
2000	9.7749	9.7809	9.7961	9.8144	9.8259	9.8276
3000	9.7714	9.7775	9.7928	9.8101	9.8226	9.8232
30,000	9.6879	9.6964	9.7116	9.7290	9.7415	9.7421

ENGLISH UNITS

Altitude above Sea Level feet	Location on Earth's Surface degrees latitude					
	Equator 0°	20°	40°	60°	80°	90°
	feet per second per second					
0	32.088	32.108	32.158	32.215	32.253	32.258
1000	32.085	32.105	32.155	32.212	32.250	32.255
2000	32.082	32.102	32.152	32.209	32.247	32.252
4000	32.076	32.096	32.146	32.203	32.241	32.246
6000	32.070	32.090	32.140	32.203	32.241	32.246
8000	32.064	32.084	32.134	32.191	32.229	32.234
10,000	32.058	32.078	32.128	32.185	32.226	32.228
100,000	31.780	31.808	31.858	31.915	31.953	31.958

Source: U.S. Coast and Geodetic Survey, 1912.

Table B.18

Table of the Elements

IA	IIA	IIIB	IVB	VB	VIB	VIIB	VIII(B)			IB	IIB	IIIA	IVA	VA	VIA	VIIA	O
1 H 1.008																1 H 1.008	2 He 4.003
3 Li 6.94	4 Be 9.01											5 B 10.8	6 C 12.0	7 N 14.0	8 O 16.0	9 F 19.0	10 Ne 20.2
11 Na 23.0	12 Mg 24.3											13 Al 27.0	14 Si 28.1	15 P 31.0	16 S 32.1	17 Cl 35.45	18 Ar 40.0
19 K 39.1	20 Ca 40.1	21 Sc 45.0	22 Ti 47.9	23 V 50.9	24 Cr 52.0	25 Mn 54.9	26 Fe 55.8	27 Co 58.9	28 Ni 58.7	29 Cu 63.5	30 Zn 65.4	31 Ga 69.7	32 Ge 72.6	33 As 74.9	34 Se 79.0	35 Br 79.9	36 Kr 83.8
37 Rb 85.5	38 Sr 87.6	39 Y 88.9	40 Zr 91.2	41 Nb 92.9	42 Mo 95.9	43 Tc (97)	44 Ru 101.1	45 Rh 102.9	46 Pd 106.4	47 Ag 107.9	48 Cd 112.4	49 In 114.8	50 Sn 118.7	51 Sb 121.8	52 Te 127.6	53 I 126.9	54 Xe 131.3
55 Cs 132.9	56 Ba 137.3	(57– 71)	72 Hf 178.5	73 Ta 180.9	74 W 183.9	75 Re 186.2	76 Os 190.2	77 Ir 192.2	78 Pt 195.1	79 Au 197.0	80 Hg 200.6	81 Tl 204.4	82 Pb 207.2	83 Bi 209.0	84 Po (210)	85 At (210)	86 Rn (222)
87 Fr (223)	88 Ra (226)	(89– 103)															

57 La 139	58 Ce 140	59 Pr 141	60 Nd 144	61 Pm (145)	62 Sm 150	63 Eu 152	64 Gd 157	65 Tb 159	66 Dy 163	67 Ho 165	68 Er 167	69 Tm 169	70 Yb 173	71 Lu 175
89 Ac (227)	90 Th 232	91 Pa (231)	92 U 238	93 Np (237)	94 Pu (244)	95 Am (243)	96 Cm (247)	97 Bk (247)	98 Cf (251)	99 Es (254)	100 Fm (253)	101 Md (256)	102 No (254)	103 Lw (257)

Table B.19

Gibbs Free Energies of Formation*

Compound	ΔG_f° (*kJ/g·mole*)	*Ion*	ΔG_f° (*kJ/g·mole*)
AgCl (*s*)	−109.78	Ag^+ (*aq*)	77.16
AgCl (*g*)	70.30	Br^- (*aq*)	−102.87
CO (*g*)	−137.37	Cl^- (*aq*)	−131.26
CO_2 (*g*)	−394.65	CO_3^{--} (*aq*)	−528.46
CO_2 (*aq*)	−386.48	Cu^{++} (*aq*)	65.02
CH_4 (*g*)	−50.83	H^+	0
C_8H_{18} (*g*)	17.83	Na^{++} (*aq*)	−262.05
C_8H_{18} (*l*)	7.41	OH^- (*aq*)	−157.42
H_2SO_4 (*aq*)	−742.49	SO_4^{--} (*aq*)	−742.49
H_2O (*l*)	−249.91	Zn^{++} (*aq*)	−147.29
H_2O (*g*)	−228.77		
NaCl (*aq*)	−393.31		

Source: Rossini, F. D.; Wagman, D. D.; Evan, W. H.; Levine, S.; and Jaffe, I., "Selected Values of Chemical Thermodynamic Properties," Circular of the National Bureau of Standards 500, Washington, D.C., 1952.

Note: *s* = solid phase; *l* = liquid; *g* = gaseous; *aq* = aqueous.

*At standard state of 298 K and 1 atm.

Chart B.1

Mollier Diagram
SI UNITS

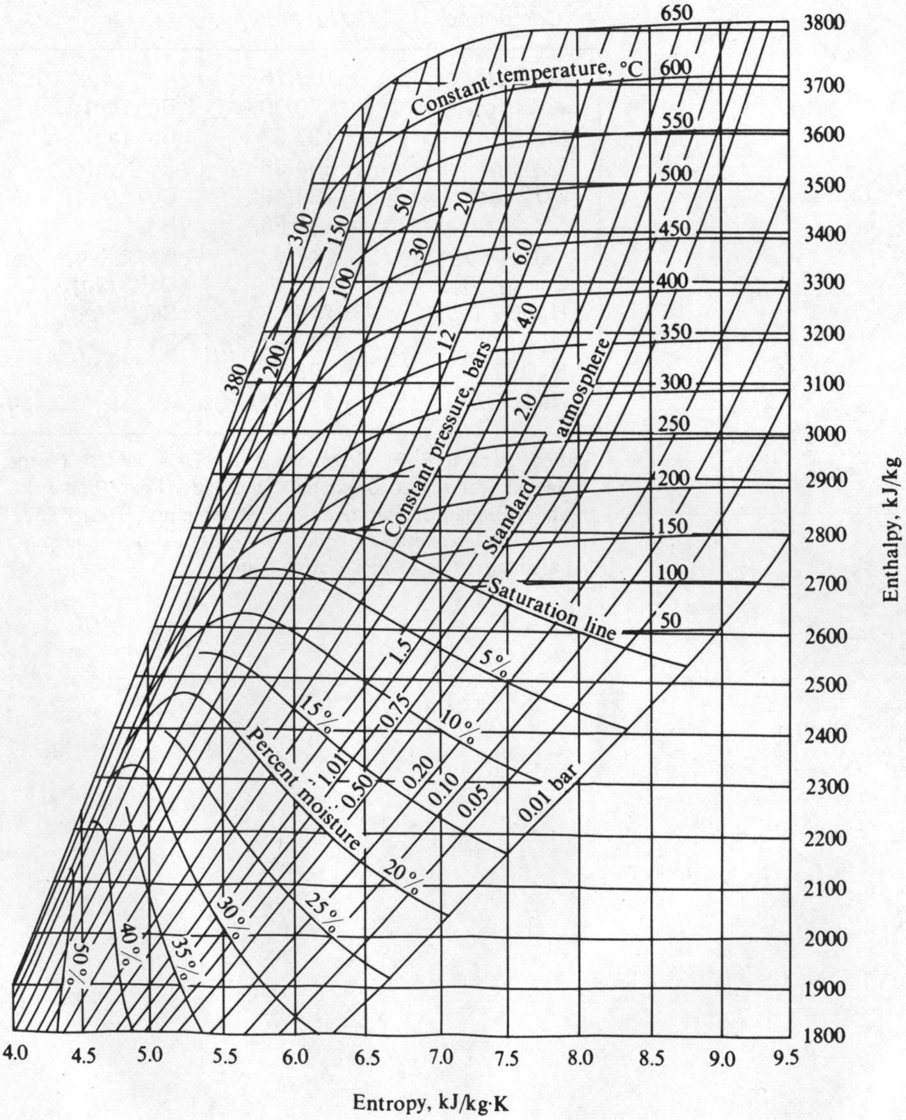

Chart B.1 (continued).

ENGLISH UNITS

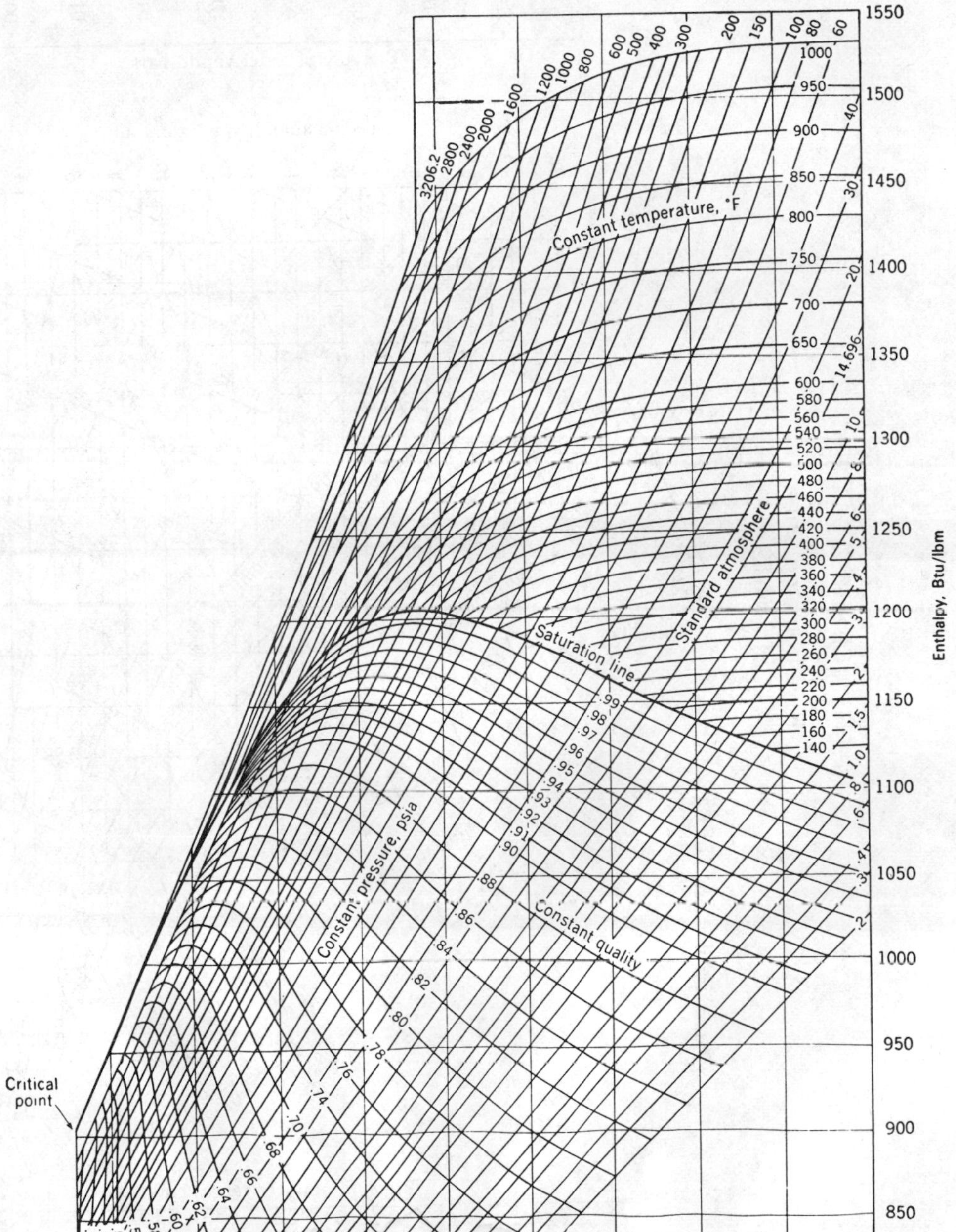

Chart B.2

Psychrometric Chart
SI UNITS

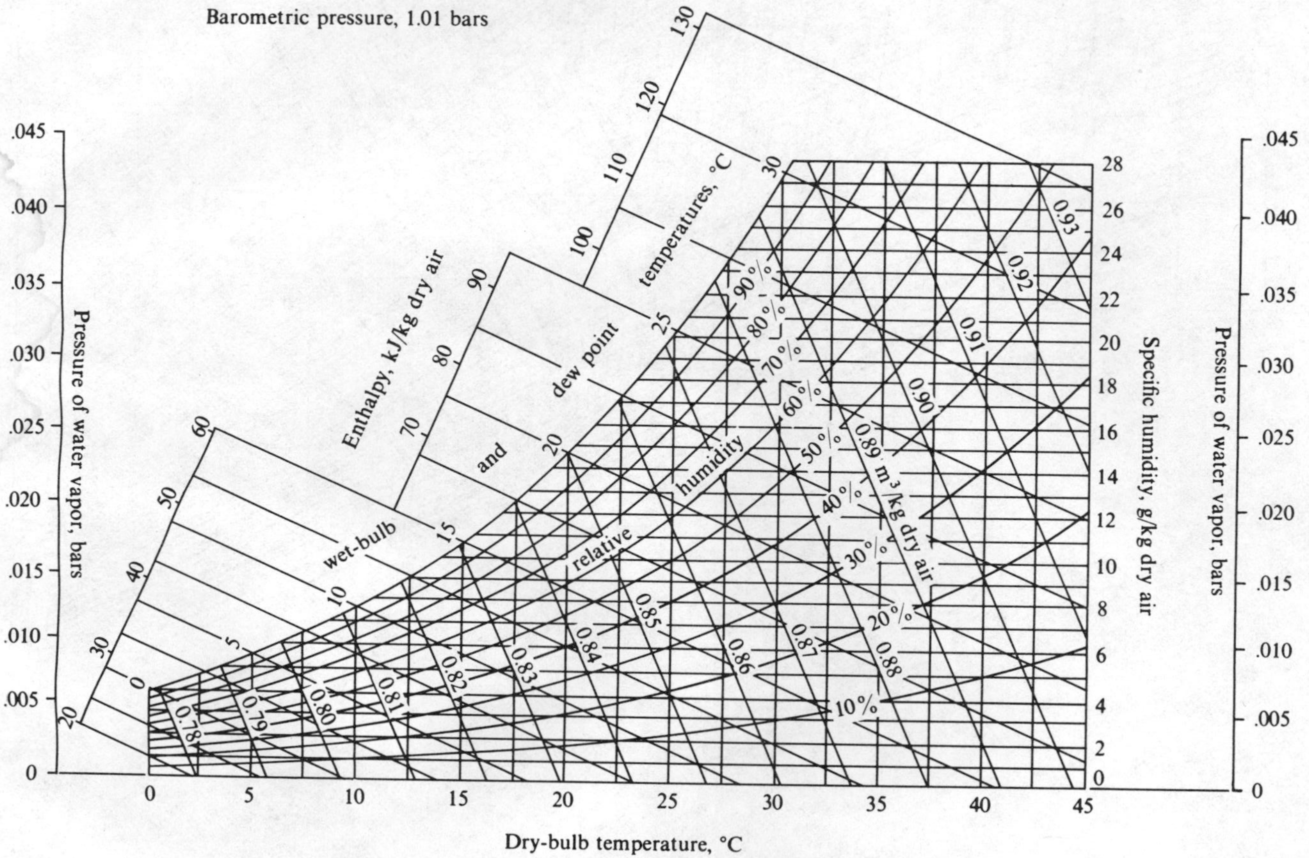

Chart B.2 (continued).

ENGLISH UNITS

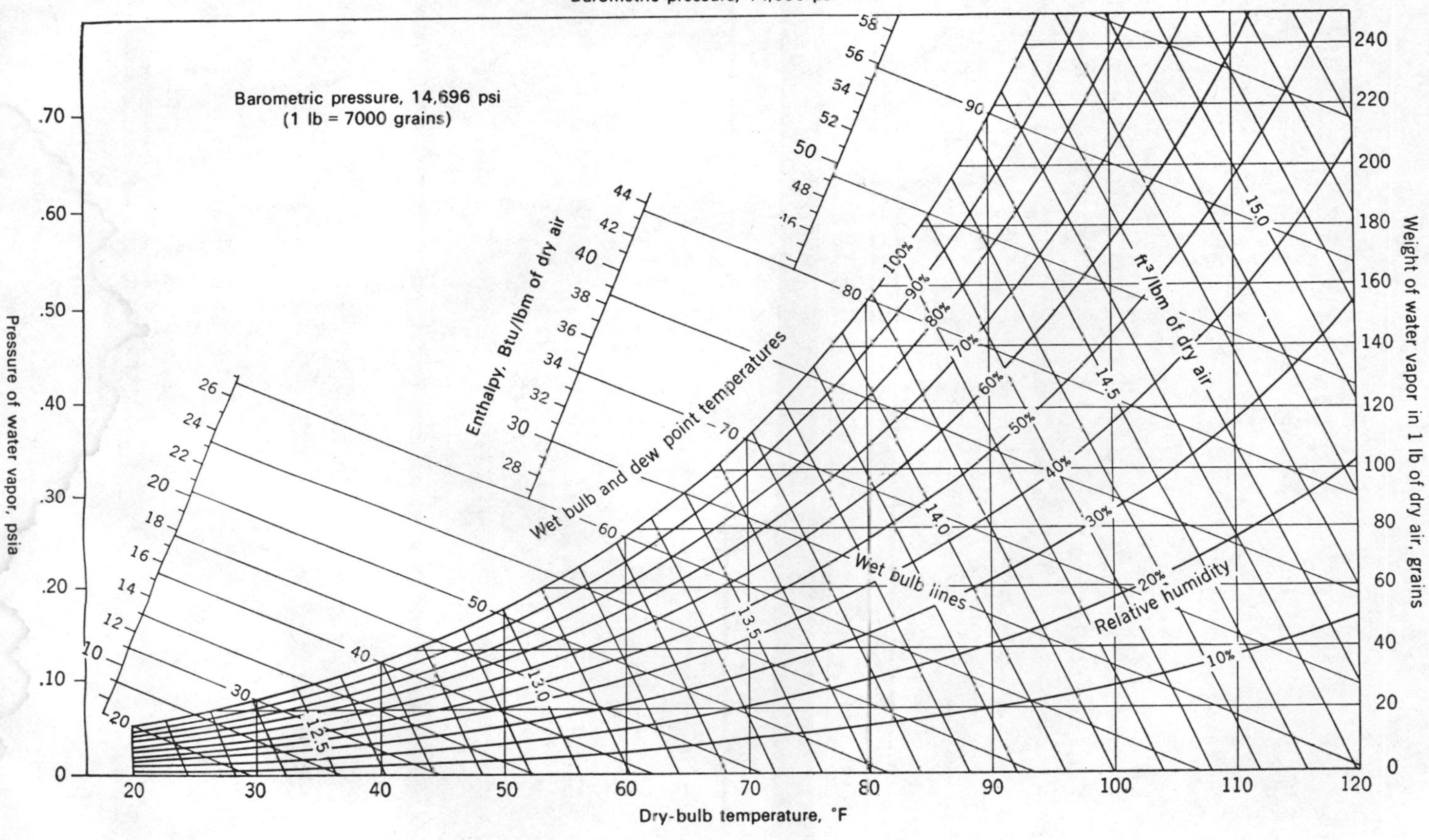

C

Selected References

The following publications were used directly in the preparation of this text. This list is not intended as a complete bibliography of the literature of thermodynamics but can be used as a source for alternate readings for the interested student.

C.1 General Thermodynamics

Bent, H. 1965. *Second law: an introduction to classical & statistical thermodynamics*. New York: Oxford U Pr.

Crawford, F. H., and Van Vorst, W. D. 1968. *Thermodynamics for engineers*. New York: Harcourt, Brace & World.

Faires, V. M. 1970. *Thermodynamics*. New York: Macmillan.

Hatsopoulos, G. N., and Keenan, J. H. *Principles of general thermodynamics*. New York: Wiley.

Hawkins, G. A. 1951. *Thermodynamics*. New York: Wiley.

Keenan, J. H. 1951. *Thermodynamics*. New York: Wiley.

Keenan, J. H., and Kaye, J. 1948. *Gas tables*. New York: Wiley.

Keenan, J. H., and Keyes, F. G. 1937. *Thermodynamic properties of steam*. New York: Wiley.

Lee, J. F., and Sears, F. W. 1963. *Thermodynamics*. Reading: Addison-Wesley.

Mooney, D. A. 1953. *Mechanical engineering thermodynamics*. Englewood Cliffs: Prentice-Hall.

Obert, E. F. 1960. *Concepts of thermodynamics*. New York: McGraw-Hill.

Reynolds, W. C., and Perkins, H. C. 1977. *Engineering thermodynamics*. New York: McGraw-Hill.

Skrotzki, B. G. A. 1963. *Basic thermodynamics: elements of energy systems*. New York: McGraw-Hill.

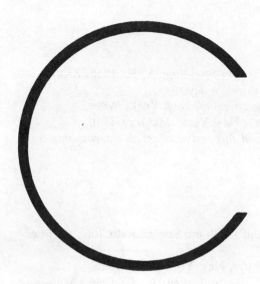

Solberg, H. L.; Cromer, O. C.; and Spalding; A. R. 1956. *Elementary heat power*. 2nd ed. New York: Wiley.

Van Wylen, G. J. 1959. *Thermodynamics*. New York: Wiley.

Wark, K. 1977. *Thermodynamics*. New York: McGraw-Hill.

Zemansky, M. W. 1957. *Heat and thermodynamics: an intermediate textbook*. New York: Wiley.

C.2 Fluid Mechanics and Heat Transfer

Binder, R. C. 1962. *Fluid mechanics*. 4th ed. Englewood Cliffs: Prentice-Hall.

Chapman, A. J. 1974. *Heat transfer*. New York: Macmillan.

Hirschfelder, J. O.; Curtis, C. F.; and Bird, R. B. 1966. *Molecular theory of gases and liquids*. New York: Wiley.

Jacob, M., and Hawkins, G. A. 1957. *Elements of heat transfer*. 3rd ed. New York: Wiley.

MacAdams, W. H. 1954. *Heat transmission*. 3rd ed. New York: McGraw-Hill.

Owczarek, J. A. 1964. *Fundamentals of gas dynamics*. Scranton: International Textbook.

Özisik, M. N. 1977. *Basic heat transfer*. New York: McGraw-Hill.

Pao, R. H. F. 1961. *Fluid mechanics*. New York: Wiley.

C.3 History and General Discussions of Matter and Energy

The way things work: an illustrated encyclopedia of modern technology. New York: Simon and Schuster. 1963.

Toulmin, S., and Goodfield, J., 1962. *The architecture of matter: the physics, chemistry & physiology of matter, both animate & inanimate, as it has evolved since the beginnings of science*. New York: Harper & Row.

Ubbelohde, A. R. 1963. *Man and energy*. Baltimore: Penguin.

C.4 Internal Combustion Engines

Lichty, L. C. 1951. *Internal combustion engines*. 6th ed. New York: McGraw-Hill.

Obert, E. F. 1959. *Internal combustion engines*. Scranton: International Textbook.

Obert, E. F. 1973. *Internal combustion engines and air pollution*. New York: Harper & Row.

Taylor, C. F. 1960. *The internal-combustion engine in the theory and practice: thermodynamics, fluid flow, performance*. Cambridge, Mass.: MIT Press.

Taylor, C. F., and Taylor, E. S. 1948. *The internal combustion engine.* Scranton: International Textbook.

C.5 Gas Turbines, Jet Propulsion

Bonney, E. A.; Zucrow, M. J.; and Besserer, C. W. 1956. *Aerodynamics, propulsion, and design practice.* New York: Van Nostrand.

Sawyer, R. T. 1945. *The modern gas turbine.* Englewood Cliffs: Prentice-Hall.

Zucrow, M. J. 1948. *Principles of jet propulsion.* New York: Wiley.

C.6 Steam Turbines and Power Generation

Babcock and Wilcox Co. 1972. *Steam, its generation and use.* 38th ed.

Combustion Engineering. 1957. *Combustion engineering.* New York.

Morse, F. T. 1953. *Power plant engineering and design.* New York: Van Nostrand.

C.7 Steam Engines

Bruce, A. W. 1952. *The steam locomotive in America.* New York: Norton.

Morrison, L. H. 1930. Valve setting. *Power.* New York: McGraw-Hill.

Peabody, C. H. 1909. *Thermodynamics of the steam engine.* New York: Wiley.

Sinclair, A. 1970. *Development of the locomotive engine.* Annotated edition prepared by John H. White. Cambridge, Mass.: MIT Press.

C.8 Refrigeration

Jordan, R. C., and Priester, G. B. 1948. *Refrigeration.* Englewood Cliffs: Prentice-Hall.

Sparks, N. R. 1938. *Theory of mechanical refrigeration.* New York: McGraw-Hill.

Threlkeld, J. L. 1970. *Thermal environmental engineering.* Englewood Cliffs: Prentice-Hall.

C.9 Direct Energy Conversion and Electrical Theory

Angrist, S. W. 1965. *Direct energy conversion.* Boston: Allyn & Bacon.

Bridgeman, P. W. 1961. *The thermodynamics of electrical phenomena in metals.* New York: Dover.

Kraus, J. D. 1953. *Electromagnetics.* New York: McGraw-Hill.

Shercliff, J. A. 1965. *A textbook of magnetohydrodynamics.* Elmsford: Pergamon.

Sutton, G. W., and Sherman, A. 1965. *Engineering magnetohydrodynamics.* New York: McGraw-Hill.

C.10 Biology, Natural Science

Keeton, W. T. 1967. *Biological science*. New York: Norton.

King, B. G., and Showers, M. J. 1963. *Human anatomy and physiology*. 5th ed. Philadelphia: Saunders.

Langley, L. L., and Cheraskin, E. 1958. *The physiology of man*, 2nd ed. New York: McGraw-Hill.

Tuttle, W. W., and Schottelius, B. A. 1969. *Textbook of physiology*. 16th ed. St. Louis: Mosby.

Answers to Selected Problems

Chapter 1

1.7 0.001
1.8 476,000
1.9 67,600
1.10 4.02
1.11 99.8
1.12 −1670

Chapter 2

2.1 19.6 N where $g = 9.8$ m/s²
2.3 0.259 slugs, 8.333 lbm
2.5 (a) 454 gm
(b) 0.908 kg
(c) 643.4 lbm
(d) 9.8×10^4 dynes
(e) 4400 lbf
2.7 0.052 N
2.9 (a) 0.13 bar
(b) 0.37 bar
(c) 1.14 bars, 1.48 bars
2.11 0.34 ft
2.13 (a) 14.8 psia
(b) 14 in Hg
2.15 Blocks *A* and *B* are not in thermal equilibrium. When *A* and *B* have same temperature, they will be in thermal equilibrium.
2.17 −1 degree D/degree R
2.19 (a) 600 °R
(b) 548 °R
(c) 110 °C
(d) 156.6 °R
2.21 (a) 17.4×10 kJ
(b) 13.2×10^5 kJ
2.23 (a) 392 J
(b) 588 J
2.25 288 J/kg
2.27 (a) 305 kJ
(b) 270 kJ
2.29 180 Btu
2.31 21,000 Btu, 2100 Btu/lbm
2.33 0.062 ft-lbf/lbm
2.35 kg/m·s, lbm/ft·s

Chapter 3

3.1 400 Joules (J)
3.3 1.414 cm
3.5 140 kJ
3.7 210 in-lbf
3.9 136 in-lbf
3.11 (a) Conduction
(b) Radiation
(c) Convection
(d) Radiation
(e) Convection

3.13 519 K (246 °C)
3.15 66.6 W/m (using k of 69.2 W/m·k for iron and 0.062 W/m·k for rock wool)
3.17 0.060 Btu/ft²·hr
3.19 267,000 Btu/ft·hr
3.21 2037 kW/m²
3.23 245 W/m²
3.25 2.08×10^7 Btu/hr·ft²
3.27 (a) 13,226 ft-lbf/lbm
(b) 4.306 Btu
(c) 13.4 Btu
(d) 27,800 N·m
(e) 3×10^9 kg·m²/s²

Chapter 4
4.1 12 m/s
4.3 (a) 11.52 kg/s
(b) 205.7 m/s
4.5 0.073 m, 0.094 m, 0.119 m
4.7 2.5 cm
4.9 0.667 kg/s loss of mass
4.11 (a) 60 lbm/s
(b) 50.1 ft/s
(c) 8.02 ft/s
4.13 0.077 lbm/min, 1.923 lbm/min
4.15 1020 ft/s
4.17 0.417 s
4.19 10 kJ
4.21 5 kJ
4.23 150 kJ/s, 1.5 kJ/kg·s
4.25 1.75 Btu/lbm
4.27 0 (zero)
4.29 351 Btu
4.31 19 kJ
4.33 100 kW
4.35 (a) no change
(b) no change
never any change
4.37 2.32×10^2 kJ/kg
4.39 0.27 ft-lbf/s
4.41 151 kJ/kg, 1556.6 kJ/kg, 3113.2 kJ
4.43 77.7 kJ/kg, 257.7 kJ/kg, 19,327.5 kJ
4.45 370 Btu/lbm, 1370 Btu/lbm, 1370 Btu
4.47 (a) 10 kJ/kg
(b) 0.008 m³/kg
4.49 1319 kJ/kg
4.51 7896 hp
4.53 + 16.8 Btu/lbm

Chapter 5
5.1 56.7 K
5.3 0.146 m³/kg
5.5 370 K
5.7 82.8 ft-lbf/lbm·°R
5.9 8.94 ft³/lbm
5.11 0.692 ft³
5.13 313 °R
5.15 −36.8 kJ/kg
5.17 618 kJ/kg
5.19 5.87 Btu/lbm
5.21 160.8 Btu/lbm
5.23 60 kJ/kg
5.25 49.6 kJ/kg·K, 35.3 kJ/kg·K
5.27 74.5 Btu/lbm
5.29 0.46 Btu/lbm·°R, 0.36 Btu/lbm·°R
5.31 344 kJ/kg, 316.7 kJ/kg
5.33 167.5 kJ/kg
5.35 2676 kJ/kg
5.37 148.2 Btu/lbm, 136.3 Btu/lbm
5.39 59.948 Btu/lbm, 58.580 Btu/lbm
5.41 1150.5 Btu/lbm

Chapter 6
6.1 600 kJ
6.3 156.5 K, 0.0112 m³, 479.7 in-lbf
6.5 0 (none)
6.7 $Wk_{rev} = 0$, $Wk_{irrev} = -420.8$ kJ
6.9 0.153 kJ
6.11 593×10^{-6} in-lbf
6.13 96.2 kJ/kg
6.15 $\dot{W}k = 33.7$ kW, $\dot{Q} = 0$
6.17 54.1 Btu
6.19 $\dot{W}k = 3.53$ hp, $\dot{Q} = 0$
6.21 $\dot{W}k = 105.7$ kW
6.23 0.568 lbm/ft³, 0.080 lbm/ft³, 0, +97.1 Btu, −97.1 Btu
6.25 (a) 664.2 °R
(b) 91.0 psia
(c) q = 0
(d) 0.37 lbm/ft³
6.29 1.20, 98 kJ/kg
6.31 67.4 Btu
6.33 417,000 Btu/min, 30,000 Btu/min, 387,000 Btu/min

Chapter 7
7.1 0.394 J/K, 5.47 J/kg·K
7.3 840 cal/gm
7.5 1.296 Btu/°R
7.7 228 °C
7.9 608 °F
7.11 0.0752 kJ/K
7.13 (a) 0.5862 kJ/kg·K
(b) 0.435 kJ/kg·K
(c) 0.340 kJ/kg·K
7.15 (a) −0.01615 kJ/kg·K
(b) +0.08685 kJ/kg·K
7.17 −0.0364 kJ/kg·K
7.19 0.0737 Btu/lbm·°R
7.21 −0.179 Btu/lbm·°R
7.23 (a) 0.00905 Btu/lbm·°R
(b) 0.00950 Btu/lbm·°R
7.25 24.8 psia
7.27 1.06×10^6 Btu

Chapter 8
8.1 1027 °C
8.3 19,000 Btu
8.5 62.8%
8.7 97.9%

8.9 19.4 Btu/s
8.13 1269 kJ/kg, 1757 kJ/kg, 488 kJ/kg, 72.2%
8.15 (a) 974.5 kJ/kg
(b) 1145.8 kJ/kg
(c) 171.3 kJ/kg
(d) 6.69, 5.69
8.17 (a) 72.2%
(b) 166,600 ft-lbf/lbm
(c) 231,000 ft-lbf/lbm, − 64,400 ft-lbf/lbm
8.19 (a) − 29.3 hp
(b) 70.6 Btu/s
(c) − 80.3 Btu/s
(d) 2.74, 2.41

Chapter 9
9.1 (a) 1134.2 kJ
(b) 1884 kJ
(c) 454.8 kJ/kg
9.3 (a) 247.2 kJ/kg
(b) 568.4 kJ/kg
(c) 213.2 kJ/kg
(d) 281.1 kJ/kg
9.5 (a) 430 Btu
(b) 70.07 Btu
9.7 (a) 134.8 Btu
(b) 139.3 Btu
(c) 1909 Btu
(d) 76.9 Btu
9.9 534 kW
9.11 94.1 Btu, zero (0)
9.15 − 199 Btu/lbm

Chapter 10
10.1 0.00098 m³, 0.00588 m³
10.3 (a) 5.56:1
(e) 0.744 kJ/cycle
(g) 59.6 kW
10.5 199 in³
10.7 (b) $T_2 = 1254\,°R$, $T_3 = 3593\,°R$, $T_4 = 1605\,°R$
(c) 1478 ft-lbf
(d) − 1167 ft-lbf
10.9 49.6%
10.11 41.3 bars
10.13 97.7 psi
10.15 79.3%, 5.37 bars, 0.2 kg/Bhp-hr
10.21 (a) $V_1 = V_4 = 1.98 \times 10^{-3}$ m³, $V_2 = 0.132X10^{-3}$ m³, $V_3 = 0.396 \times 10^{-3}$ m³
(b) 0
(c) 1.106 kJ/kg·K
(d) 13.86 kJ/cycle
(e) 55.8%
10.23 (a) 100.3 kJ
(b) 13.2 bars
(c) 250.8 kW
(d) 41.8%
10.25 (a) 70.5%
(b) 40.9%
(c) 57.9 psi
(d) 0.48 lbm/hp-hr
10.27 (a) 13.46:1
(b) 1.419:1
(c) 34,018 ft-lbf
(d) 927 hp
(e) 48%
(f) 34.4 psi

Chapter 11
11.1 57.7%
11.3 2.42:1
11.5 $p_1 = p_4 = 1.01$ bars, $p_2 = p_3 = 8.08$ bars
$T_1 = 303$ K
$T_2 = 549$ K
$T_3 = 1100$ K
$T_4 = 607$ K
11.7 $p_1 = p_4 = 14.7$ psia, $p_2 = p_3 = 147$ psia
$T_1 = 560\,°R$, $T_2 = 1066\,°R$, $T_3 = 2318\,°R$, $T_4 = 1215\,°R$
$\rho_1 = 0.0708$ lbm/ft³, $\rho_4 = 0.03265$ lbm/ft³
11.11 60 m/s
11.13 12.98 kg/s, $\overline{V}_1 = 618$ m/s, $\overline{V}_b = 299$ m/s
11.15 10,740 kJ/s
11.17 159.6 hp
11.19 2910 °R, 308,000 ft-lbf/lbm
11.21 29.1 psia, 34.4 Btu/lbm
11.23 0.0957 m², .0026 m²
11.25 177 m/s, 106 kg/s
11.27 0.292 in²
11.29 269 ft/s, 0.132 lbm/s
11.31 (a) .822 m³/s
(b) 29.8 ft³/s
11.33 (b) 1286 kJ/kg air
(c) 802.7 kJ/kg
(d) 237.1 kJ/kg
(e) 565.6 kJ/kg
(f) 44%
11.35 (a) 79.7 lbm air/lbm fuel
(b) 182 Btu/lbm
(c) 84.4 Btu/lbm
(d) 40.6%
11.37 (a) 44.8%
(b) 57298 kJ/s
(c) − 31620 kJ/s
(d) 25678 kW
11.39 (a) 74 lbm air/lbm fuel
(b) 187 Btu/lbm
(c) − 84.6 Btu/lbm
(d) 39.4%
11.41 (a) 19.8 lbm/s, 0.903 lbm/s
(b) 30.4%
11.43 (a) 7×10^7 Btu/hr
(b) 3.57×10^7 Btu/hr
(c) 1169 °R
(d) 3.16×10^7 Btu/lbm
(e) 51.1%
11.45 (a) 856 kJ/kg, − 347 kJ/kg, 775 K, 1125 K, 59.5%
(b) 360 Btu/lbm, − 129.6 Btu/lbm, 1836 °R, 2386 °R, 64.1%

11.47 459.6 ft-lbf·s/lbm

Chapter 12

12.1 47.3%
12.5 210 kg/hr
12.7 71.5 lbm/min
12.9 1187 kJ/kg, 94.7%
12.11 5058 hp
12.13 1053 ft/s
12.15 (a) 635 kJ/kg
(b) 132.04 Btu/lbm
12.19 63.6%
12.21 −1.7693 Btu/lbm °R, +1.9340 Btu/lbm °R, +0.1647 Btu/lbm °R
12.23 (a) 6.74 kJ/kg·K, 2580 kJ/kg
(b) 1.629 Btu/lbm °R, 1118 Btu/lbm
12.25 (a) 2960 kJ/kg, 7 bars
(b) 1272.5 Btu/lbm, 165 psia
12.27 (a) 1688 kW
(b) −2185 kJ/kg
(c) −3.94 kJ/kg
(d) 3286 kJ/kg
(e) 33.5% 3.26 kg/kW-hr
12.29 (a) 1.86×10^6 lbm/hr
(b) 1419.7 Btu/lbm
(c) 1029 Btu/lbm
(d) 1.444 Btu/lbm
(e) 27.1%
12.31 (a) 1.6 kg/kW-hr
(b) 44.6%
12.33 (a) 0.138 kg/kg
(b) 1889 kJ/kg
(c) 48.5%

Chapter 13

13.1 1330 kJ/kg
13.3 8.25 kW
13.5 1176.1 Btu/lbm
13.7 118 Btu/lbm
13.13 (a) 0.0722 kg/s
(b) 26.7%
(c) 72.4 bars

Chapter 14

14.1 3.92, 2.92
14.3 (a) 21.8 °C
(b) 10
14.5 5.2, 6.2
14.7 10.25
14.9 1.78, 0.78
14.11 (a) 0.824
(b) 0.682 kJ/s
(c) 0.0025 kg/s
14.13 (a) 712 °R
(b) 369 °R
(c) −1345 Btu/min
(d) 3.16, 2.16
(e) 1345 Btu/min
14.15 396.8 kJ/kg, 23%
14.17 (a) 31.4 kJ/kg
(b) 169.2 kJ/kg
(c) 200.6 kJ/kg
14.19 (a) 12.14 Btu/lbm
(b) 57.76 Btu/lbm
(c) 69.9 Btu/lbm
14.21 31.66 kW, dry compression cycle
14.23 0.82 hp/ton
14.25 27 °C

Chapter 15

15.1 0.32 bar, 0.64 bar, 0.64 bar, 0.40 bar
15.3 1.176 psia
15.5 (a) 227.96 °F
(b) 120.2 °C
15.7 (a) 42 Btu/lbm, 129 grains/lbm
(b) 44.5 kJ/kg, 5.5 g/kg
15.9 9.58 kJ/kg, 0.9 g/kg dry air
15.11 (a) 82 grains/lbm
(b) 55%
(c) 62.1 °F
15.13 1.17 Btu/lbm, 0.679 lbm/min
15.15 (a) −17.8 kJ/kg
(b) −1.1 Btu/lbm

Chapter 16

16.1 173.4 volts
16.3 6.98 kW
16.5 50.7 kcal/g-mole, 1.098 volts, 1.038 volts
16.7 1.13 volts
16.9 1250 kcal/g-mole, 2.17 volts, 2.17 volts
16.11 29.9 ohms
16.13 32.48 amps, 1.624 volts
16.15 63.6 kW
16.17 1.2 W
16.19 0.943 bar

Index